THE ORGANIC CHEMIST'S BOOK OF ORBITALS

THE ORGANIC CHEMIST'S BOOK OF ORBITALS

WILLIAM L. JORGENSEN

Department of Chemistry
Harvard University
Cambridge, Massachusetts

LIONEL SALEM

Université de Paris–Sud
Centre d'Orsay
Orsay, France

New York and London 1973

A Subsidiary of Harcourt Brace Jovanovich, Publishers

ACADEMIC PRESS, INC.
111 Fifth Avenue, New York, New York 10003

United Kingdom Edition published by
ACADEMIC PRESS, INC. (LONDON) LTD.
24/28 Oval Road, London NW1

Library of Congress Cataloging in Publication Data

Jorgensen, William L
The organic chemist's book of orbitals.

Includes bibliographical references.
1. Molecular orbitals--Charts, diagrams, etc.
2. Chemistry, Organic--Charts, diagrams, etc.
I. Salem, Lionel, joint author. II. Title.
QD461.J68 547'.1'28 72-9990
ISBN 0-12-390250-9
ISBN 0-12-390256-8 (pbk)

PRINTED IN THE U...

To Alice and Axel

To Catherine

To the people of Vietnam

December 25, 1972

Science sans conscience n'est que ruine de l'âme

RABELAIS, Pantagruel, 1532

Contents

Preface

The last decade has witnessed an unprecedented strengthening of the bond between theory and experiment in organic chemistry. Much of this success may be credited to the development of widely applicable, unifying concepts, such as the symmetry rules of Woodward and Hoffmann, and the frontier orbital theory of Fukui. Whereas the theoretical emphasis had historically been on detailed structure and spectroscopy, the new methods are designed to solve problems of special importance to organic chemists: reactivity, stereochemistry, and mechanisms.

These theories are inevitably based upon analyses of the interactions and transformations of molecular orbitals, and consequently the accurate construction and representation of molecular orbitals has become essential. Furthermore, although the forms of molecular orbitals in diatomics and of delocalized π orbitals in conjugated systems are familiar, a general, non-computational method for determining the qualitative nature of σ and π orbitals in arbitrary molecules has been lacking.

In the present work a theory for the facile construction of complex molecular orbitals from bond and group orbitals is presented and complemented by accurate drawings of the valence molecular orbitals for over one hundred molecules representing a wide range of connectivities and functional groupings. Direct applications to phenomena in organic chemistry are also discussed.

This book is addressed to all those for whom orbitals have ceased to be an abstract concept, but have instead become concrete and useful in the daily practice of chemistry. It is especially directed to the new generation of chemists eager to understand molecular structure at the electronic level.

We are greatly indebted to Professor E. J. Corey for much stimulating advice and for the unswerving encouragement which he gave us. We wish to thank Dr. Donald Barth for consultation on the graphic aspects of the project, and also many friends for their advice. Receipt of numerous unpublished results from Dr. Warren J. Hehre, Prof. J. M. Lehn, and Dr. Georges Wipff was greatly appreciated.

The greater part of the book was written while L. S. was Visiting Professor at Harvard University. He sincerely thanks the Chemistry Department for their kind hospitality.

William L. Jorgensen
Lionel Salem

THE ORGANIC CHEMIST'S BOOK OF ORBITALS

I. How Molecular Orbitals Are Built by Delocalization

A Unified Approach Based on Bond Orbitals and Group Orbitals

1. Bond Orbitals and Group Orbitals

The delocalization of molecular orbitals lies at the heart of modern chemistry. The concept that the π orbitals of benzene or naphthalene cover the entire carbon skeleton promoted the successful understanding of conjugated molecules. The work of R. Hoffmann and others has proven that in saturated molecules σ orbitals are also delocalized over several bonds, often reaching opposite ends of a sizable molecule. The purpose of this chapter is to provide some enlightenment as to the mechanism by which delocalization of orbitals occurs in organic molecules.

By far the most obvious, but also the most significant characteristic of the molecular orbitals which are pictured in Chapter III is that they are composed almost exclusively of a very small number of typical *group* and *bond* orbitals which recur endlessly from molecular orbital to molecular orbital and from molecule to molecule. Generally, in a given molecular orbital only a given type of group or bond orbital occurs, from which the label of the molecular orbital is derived (see also Chapter II). The reader will soon become proficient at recognizing the basic types of localized orbitals. The following sections are aimed at helping him in this exercise, and at allowing him to go one step further to construct, qualitatively, the delocalized molecular orbitals from the localized group orbitals. The concept which we use is an old one: localized bond orbitals were first introduced by Lennard-Jones and Hall, and the wave-mechanical properties of localized electron pairs have been widely discussed. However, for the first time accurate drawings of molecular orbitals show that these localized orbitals subsist as basic units even in large molecules with no symmetry. These drawings allow the reader to recognize and label, in almost every case, the localized building blocks for the molecular orbitals of arbitrary organic molecules. We shall now try and provide in a qualitative manner the theoretical framework for the understanding of these drawings.

Let us take a typical hydrocarbon skeleton (what will be said throughout this chapter on orbitals around carbon is of course equally valid for orbitals around nitrogen or oxygen). In hydrocarbons we can recognize basic bonds, such as

```
\     /       \
—C—C—   or   —C—H
/     \       /
```

and also basic groups such as

```
                      H                          H
                   \ /                          /
methylene groups    C      or methyl groups  —C—H
                   / \                          \
                      H                          H
```

At a given carbon atom we will recognize a CH *bond* if there is only one hydrogen atom attached to it and all other substituents are different from hydrogen; if there are two hydrogen atoms attached to it we recognize a CH_2 *group*; if there are three, a CH_3 *group*. In other words, a given carbon atom carries only one type of CH_n unit. Now we can think of the CC or CH bonds and of the CH_2 or CH_3 groups of a hydrocarbon, not only as the building blocks from which the overall molecular geometry is derived, but also as carrying basic sets of localized orbitals—CC bond orbitals (two for a single bond, four for a double bond), CH bond orbitals (two), CH_2 group orbitals (four), CH_3 group orbitals (six). Again, at a given carbon atom we recognize only one set of CH_n (CH or CH_2 or CH_3) orbitals. And while the molecule is built by linking these bonds or groups together, so are the molecular orbitals constructed by adding, subtracting or combining these bond and group orbitals. Although the basic orbitals are essentially *localized* on their own group, there is generally sufficient *overlap* between the orbitals of adjacent groups—sometimes even non-adjacent ones—to ensure that these combinations are possible. The three-dimensional "waves" on interacting adjacent groups can interfere to form "standing waves" for the entire molecule.

With respect to a local symmetry plane, the basic orbitals are either σ, *symmetric* or π, *antisymmetric*. This is true not only for the CC and CH bond orbitals but also for the CH_2 and CH_3 group orbitals. If the local symmetry elements are preserved in the full molecule, the π (or σ) local orbitals can combine to give π (or σ) molecular orbitals. The reader should, therefore, not be surprised to find, for instance, π type molecular orbitals in cyclopropane which are delocalized over the CH_2 groups.

Let us now construct in turn the various basic localized orbitals and consider their most significant features.

2. CC Single-Bond Orbitals

The valence atomic orbitals which are available to form the orbitals of a CC single bond, directed along the x axis, are the 2s and $2p_x$ atomic orbitals on each carbon atom. Their admixture—in proportions which depend on the number of neighbors at each carbon and on the subsequent hybridization—creates two (s, p_x) hybrids on each atom. One of these hybrids points away from the other atom and can be used for bonding to additional atoms. The pair of hybrids which point at each other overlap and interact in the conventional fashion [we symbolize the non-interacting orbitals by an interruption of the bond axis (Fig. 1)]. The two bond orbitals which are formed in this manner both have σ symmetry, i.e., rigor-

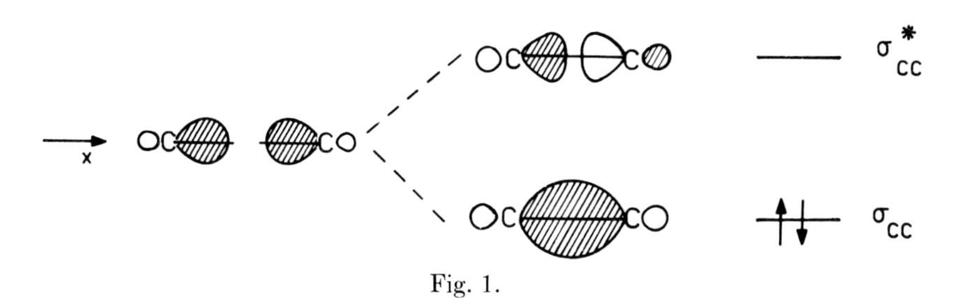

Fig. 1.

ously they are cylindrically symmetric around the bond axis. *A fortiori*, they are symmetric with respect to any plane containing the bonds. The bonding σ_{CC} orbital has two electrons, which form the "Lewis pair" in that bond, while the antibonding σ^*_{CC} orbital is empty.

3. CC Double-Bond Orbitals

If the pair of carbon atoms shown above each have only two neighbors so that they are doubly-bonded in the conventional sense, there is an extra p orbital available on each atom. These p orbitals point along the (z) direction, perpendicular to the plane of the molecular fragment. The interaction of these two atomic orbitals via overlap creates a new pair of bond orbitals with local π symmetry (Fig. 2, where again we have symbolized the non-interacting or-

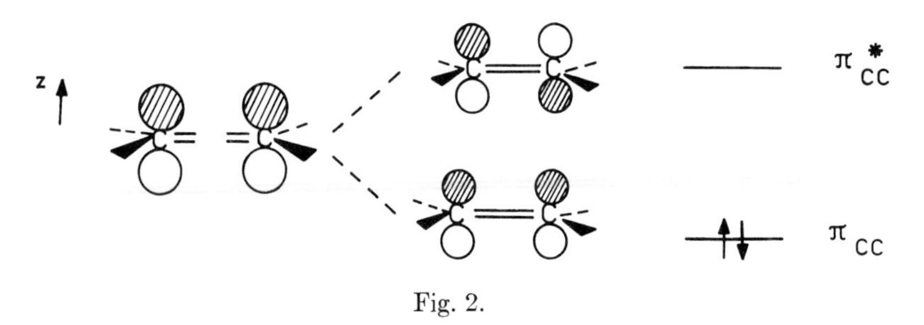

Fig. 2.

bitals by an interruption of the CC axis). Again the bonding combination is full while the other orbital is empty. The reader may wish to compare the schematic, conventional drawings with the actual three-dimensional graphics of the ethylene π orbitals (III.18).

In acetylene or triple-bonded systems, there will be a second pair of (π_{CC}, π^*_{CC}) orbitals with axes parallel to the y direction.

4. CH Single-Bond Orbitals

The construction of the pair of (σ_{CH}, σ^*_{CH}) CH bond orbitals is carried out by combining a carbon hybrid with the 1s orbital on hydrogen in a manner similar to the construction of the CC bond orbitals. The interaction diagram is shown below in Fig. 3. The bonding orbital is occupied by the two bond electrons. These two

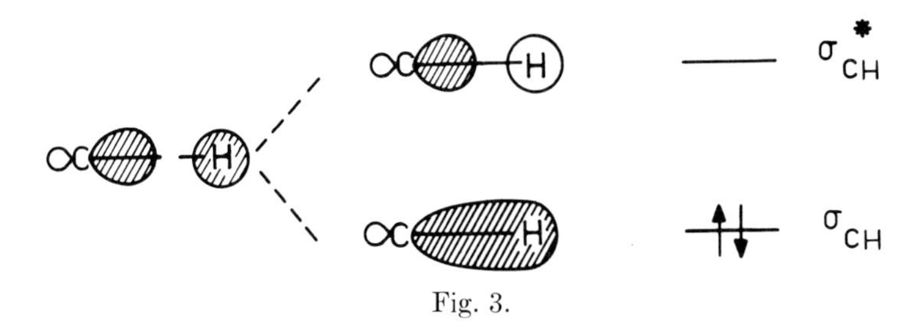

Fig. 3.

CH bond orbitals should be used for building molecular orbitals *only* if the other substituents on the carbon atom are not hydrogen atoms.

5. The Localized Orbitals of a CH_2 Group

Let us now turn to a CH_2 fragment, with two conventional CH single bonds lying in the xz plane (Fig. 4). The xy plane is a local symmetry plane for the group and we can, therefore, classify orbitals as σ-like (symmetric) or π-like (antisymmetric) with respect to this plane. In doing so, we are using the same language which presides over the σ, π classification in conjugated molecules.

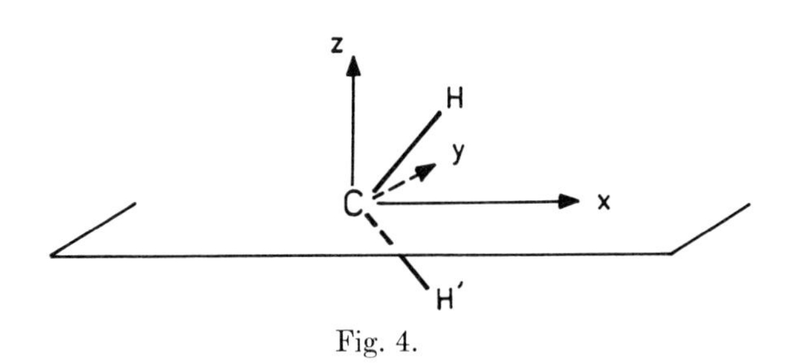

Fig. 4.

The appropriate valence atomic orbitals which must be considered are 2s, $2p_x$, $2p_z$ on carbon, and the 1s orbitals $1s_H$ and $1s'_H$. The orbital $2p_y$ is clearly nonbonding (n) relative to the carbon-hydrogen interactions and need not be considered further. The hydrogen orbitals can be combined into a $(1s_H + 1s'_H)$ combination of σ symmetry and a $(1s_H - 1s'_H)$ combination of π symmetry. (Fig. 5). Although there are three available basic σ orbitals, only two of these belong to the CH_2 group proper. We can first eliminate the ("out")

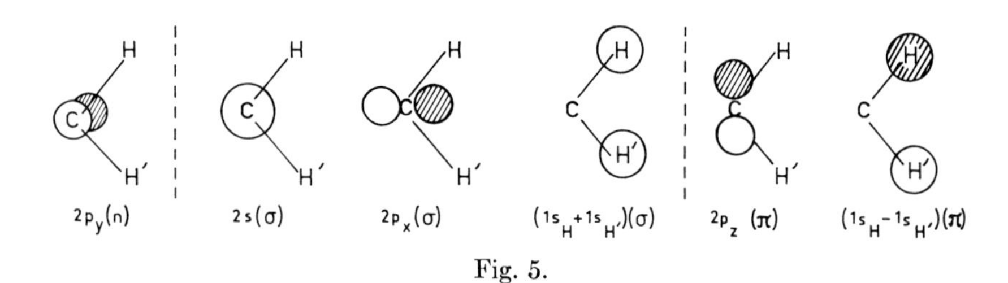

Fig. 5.

combination of 2s and $2p_x$ which points away from the hydrogen atoms, along the outer bisector of the CH_2 group, and which serves (together with the n orbital) to bind the CH_2 fragment to the remainder of the molecule. We are left with a (2s, $2p_x$) "in" hybrid pointing towards the hydrogens and with the $(1s_H + 1s_{H'})$ combination. Their interaction is shown in Fig. 6, and leads to a bonding σ_{CH_2} group orbital and an antibonding $\sigma^*_{CH_2}$ group orbital. Again these drawings are very schematic: proper σ_{CH_2} and $\sigma^*_{CH_2}$ orbitals can be found in methylene (III.2). For delocalized combinations see cyclopropane (III.56). The σ_{CH_2} bonding orbital has two electrons, while the $\sigma^*_{CH_2}$ orbital is empty.

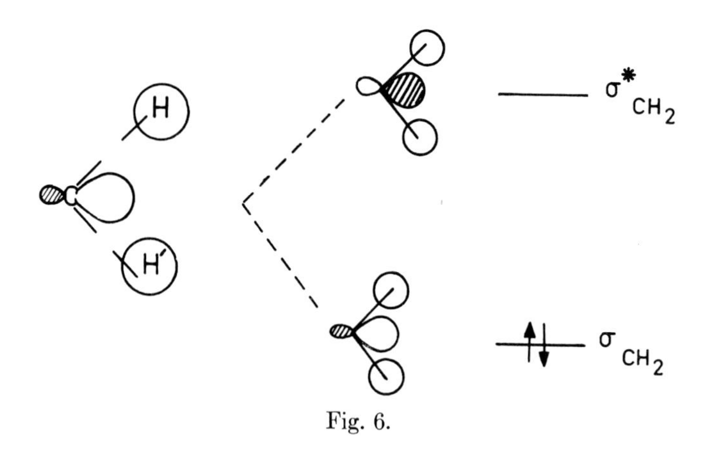

Fig. 6.

Two π-type CH_2 group orbitals can be constructed in a similar manner (Fig. 7). The two electrons in the bonding π_{CH_2} combination bring the total number of CH_2 electrons to four, corresponding to the two electron pairs in the CH bonds. The reader will find π_{CH_2}

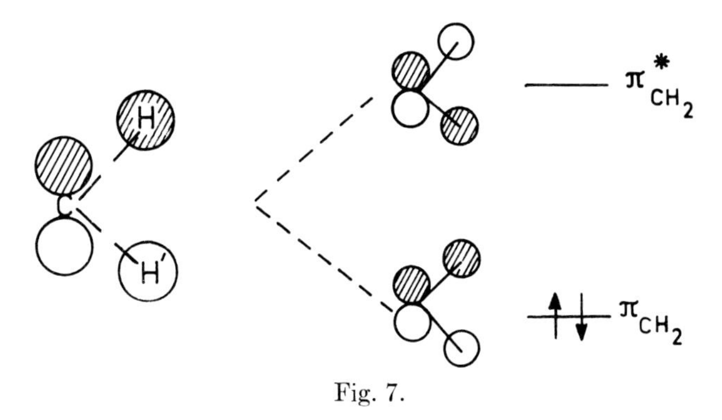

Fig. 7.

and $\pi^*_{CH_2}$ group orbitals in methylene (III.2) and delocalized combinations for planar cyclobutane (III.82).

Having derived the general forms of the four CH_2 localized orbitals, we can guess at their relative energies. Clearly the bonding pair lie below the antibonding pair. In both pairs the σ orbital should lie under the π orbital, since the π orbitals have an additional nodal plane. For instance, σ_{CH_2} is bonding between the carbon atom and each hydrogen atom, and *also* bonding between the two CH bonds—whereas π_{CH_2}, although bonding between carbon and each hydrogen atom, is antibonding between the two CH bonds. The proper ordering is shown in Fig. 8. This ordering agrees with that observed in methylene itself (III.2 and III.3).

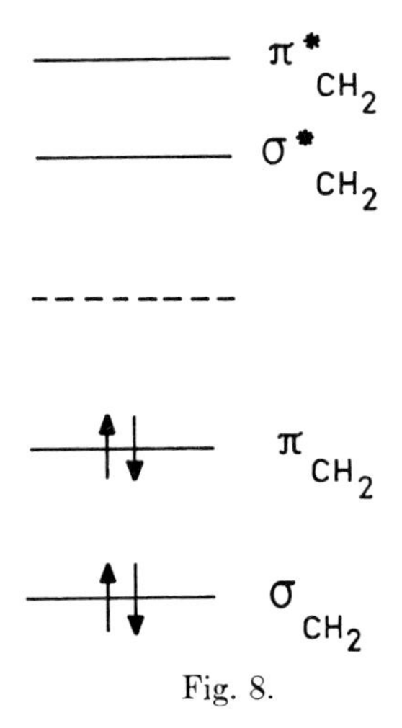

Fig. 8.

6. The Localized Orbitals of a CH_3 Group

The procedure for determining the six localized orbitals of a CH_3 group (Fig. 9), is similar to that for determining the four localized orbitals of a CH_2 group. This time the basic valence orbitals can be divided into three sets: σ orbitals with either cylindrical or pseudo-cylindrical symmetry about the x axis;[a] π_y orbitals, which

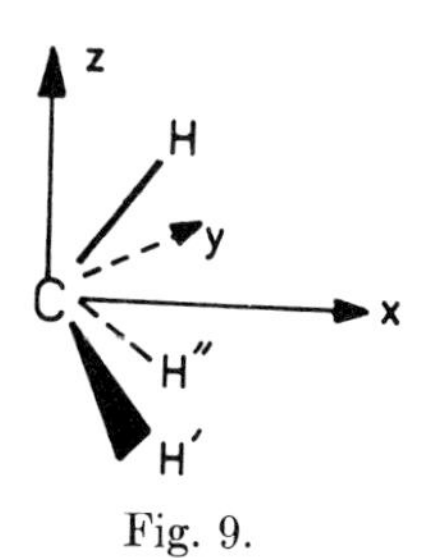

Fig. 9.

are antisymmetric with respect to the xz mirror-plane; and (pseudo) π_z orbitals, which approximate an antisymmetric behavior with respect to the xy plane.[b] Since this plane is *not* a mirror plane for the fragment, the terminology π_z is again an extension of the conventional terminology. The basic orbitals are shown in Fig. 10. The coefficients of the combinations over H, H′ and H″ are determined by symmetry;[c] even though one π combination is rigorously antisymmetric with respect to a mirror plane while the other is not, the two π combinations have the same energy.

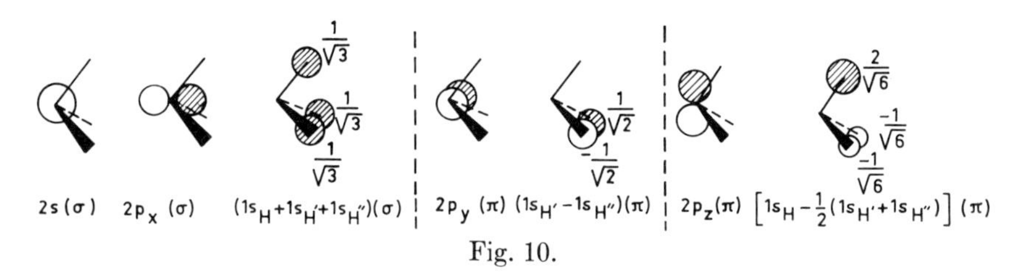

Fig. 10.

[a] These σ orbitals correspond to the orbitals of *a* symmetry in the C_{3v} point group of the CH_3 fragment.

[b] These π orbitals correspond to the degenerate orbitals of *e* symmetry in the C_{3v} point-group of the CH_3 fragment.

[c] The numerical coefficients in Fig. 10 correspond to neglecting the overlap between the hydrogen orbitals.

The manner in which we construct the σ_{CH_3} and π_{CH_3} orbitals parallels that for the CH_2 fragment. For the σ combinations we again set aside that (2s, $2p_x$) combination ("out") which points away from the CH_3 group. This orbital serves either to build the fourth bond or, for an isolated methyl group, remains as a non-bonding (n) orbital. In the anion it carries the lone pair of electrons; in the methyl radical the odd electron. The remaining "in" hybrid is then combined with the $\sigma(1s_H)$ combination to give two σ_{CH_3} group orbitals (Fig. 11). The bonding σ_{CH_3} combination is occupied while

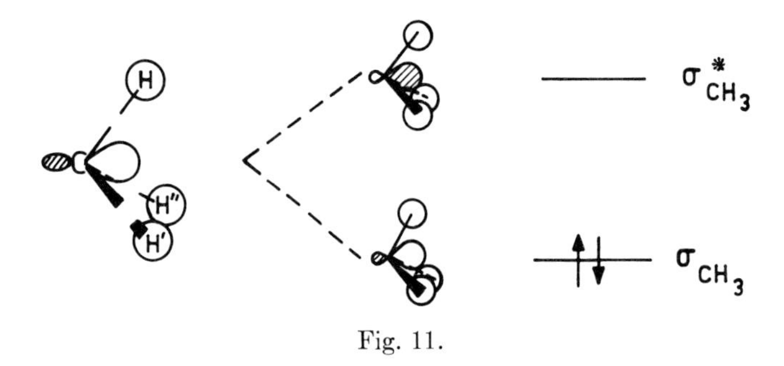

Fig. 11.

the antibonding $\sigma^*_{CH_3}$ combination is empty. For the π-type group orbitals, the overlap between the carbon 2p orbitals and the appropriate hydrogen counterparts leads to two pairs of degenerate orbitals (two $\pi^y_{CH_3}$ and two $\pi^z_{CH_3}$, indicated in Fig. 12. We have now obtained all six orbitals of a CH_3 group. The six valence electrons of the C—H bonds all occupy bonding orbitals, two π_{CH_3} localized orbitals and one σ_{CH_3} localized orbital.

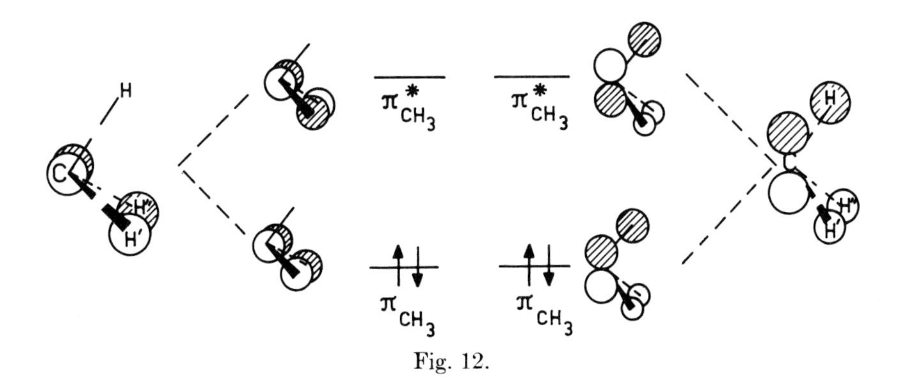

Fig. 12.

The reader will find far more elegant drawings of these orbitals in Section III, for instance by referring to the orbitals (III.6) of the pyramidal methyl radical. Similar orbitals exist, of course, for ammonia (III.8). Again, as in the CH_2 case, the energy ordering is

$$\sigma_{CH_3} < \pi_{CH_3} \ll \sigma^*_{CH_3} < \pi^*_{CH_3}$$

Now that we have established the major types of localized orbitals, we must inquire as to how they combine in a molecule to form delocalized combinations.

7. How Localized Orbitals Interact to Create Delocalized Combinations. General Rules for the Interaction between Orbitals of Different Energy

The mechanism which presides over the combination of localized bond or group orbitals to give birth to delocalized molecular orbitals is the same as that which leads from valence atomic orbitals to the localized orbitals. Overlap between *two* or *several* orbitals and interference of the orbital "waves," creates *two* or *several* new combinations in which the orbital amplitudes are either in-phase (bonding) or out-of-phase (antibonding). Often the symmetry of the molecule restricts the allowed interactions to sets of localized orbitals of given symmetry. Only those localized orbitals with the *same* symmetry can mix. In other cases, for instance in molecules with no symmetry elements, all localized orbitals can mix—although on geometrical grounds certain specific mixings will be preferred over others (See II.1).

The general rules for the interaction between two orbitals of different energy, typified by the scheme shown in Fig. 13, are as follows:

1. When two orbital levels interact they yield a lower, bonding combination and a higher antibonding combination. In the bonding level the lower orbital has mixed in some of the higher orbital in a bonding manner ($\lambda > 0$ if the overlap between ϕ_A and ϕ_B is positive, as in Fig. 13). In the antibonding level the higher orbital has mixed in some of the lower one in an antibonding manner ($\mu < 0$ for positive overlap).
2. The destabilization of ϕ_A is always slightly larger than the stabilization of ϕ_B.

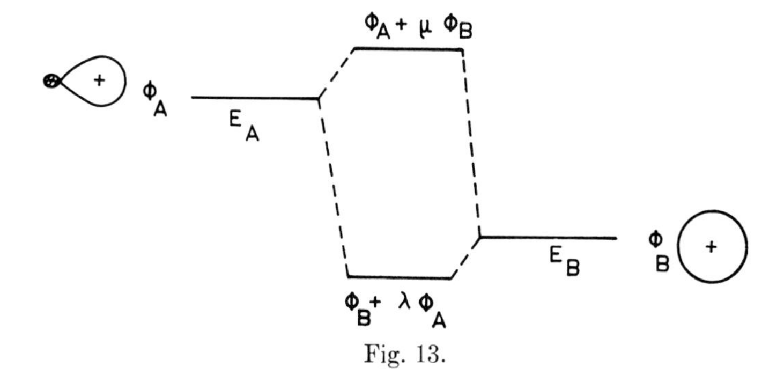

Fig. 13.

3. Only levels which are close to each other interact strongly; the closer the better.
4. Only orbitals which overlap significantly interact. In particular orbitals belonging to different symmetry representations of the molecular point-group have zero interaction.
5. If a given level (A) interacts with several others (B, C) of significantly different energy, the interactions are pairwise additive—level A is first lowered (or raised) by B, then by C, etc. The final energy of level A is the same irrespective of the order in which the interactions are accounted for. However, if one of the orbitals B, C, $\cdots$ has the *same* energy as A, and is allowed by the molecular symmetry to mix with A, it is important to take this interaction into account *first*.

These rules follow directly from the quantum-mechanical theory of perturbations and the resolution of the secular equations for the orbital interaction problem. The (small) interaction between orbitals of "significantly" different energy is the familiar "second order" type interaction, where the interaction energy is small relative to the difference between E_A and E_B. The (large) interaction between orbitals of "same" energy is the familiar "first order" type interaction between degenerate or nearly degenerate levels.

8. Ethylene

Let us now apply these results to the ethylene molecule (Fig. 14), for which we attempt to build the bonding molecular orbitals. Clearly there are three symmetry planes. Two of these are of special interest

Fig. 14.

to us—the *xy* molecular plane and the *xz* plane which bisects the CH_2 groups. Relative to these two planes we can in principal define four symmetry types—symmetric or antisymmetric with respect to *xy*, together with symmetric or antisymmetric with respect to *xz*. Let us list the available bonding localized orbitals with their symmetry classifications:

σ_{CC}, two σ_{CH_2}	Symmetric with respect to *xy*, *xz*
π_{CC}	Antisymmetric with respect to *xy* Symmetric with respect to *xz*
two π_{CH_2}	Symmetric with respect to *xy* Antisymmetric with respect to *xz*

One should also classify the antibonding orbitals. Since these are much higher in energy they mix only weakly with the bonding localized orbitals. We neglect here this small perturbation.

We are now in a position to draw qualitative pictures of the molecular orbitals. The simplest one is clearly the familiar π_{CC} orbital which mixes with no other orbital and becomes a molecular orbital in its own right (Fig. 15). Next the two π_{CH_2} group orbitals can interact to form a "more" bonding and a "less" bonding combination (Fig. 16). We do *not* call the upper combination *antibonding* since the bonding character due to positive overlap within each CH bond still outweighs the antibonding character across the CC bond or between CH bonds of a same CH_2 group. However the two antibonding features of the higher π_{CH_2} molecular orbital do raise its energy, and it lies just below the π_{CC} orbital (III.18).

Fig. 15.

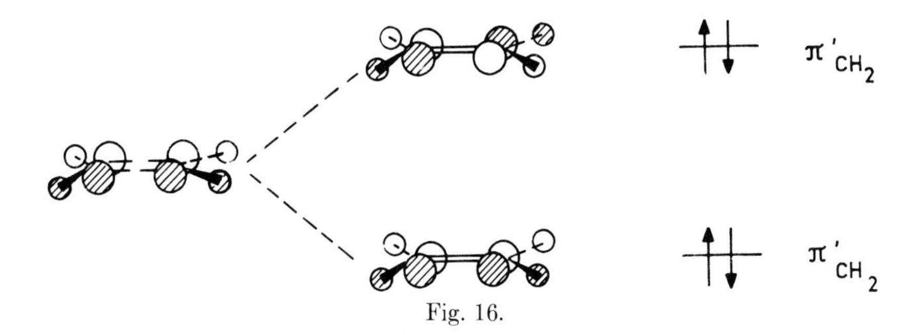

Fig. 16.

The reader may be confused to see in ethylene π orbitals which lie *in* the molecular plane. In order to avoid any confusion we label the two π_{CH_2} molecular combinations with a *prime* π'_{CH_2}. Similar orbitals occur in ketene (III.42), formaldehyde (III.20), butadiene (III.65), etc.

The interaction between the three σ localized orbitals is slightly more complex. As we have just shown, it is proper to start by combining orbitals of *same* energy (the σ_{CH_2} pair), and then to interact the new combinations with the remaining orbitals. The procedure is simple here because, by symmetry, only the in-phase combination of the σ_{CH_2} group orbitals mixes with the σ_{CC} bond orbital (see Fig. 17). The reader will notice that the σ_{CC} orbital has been placed,

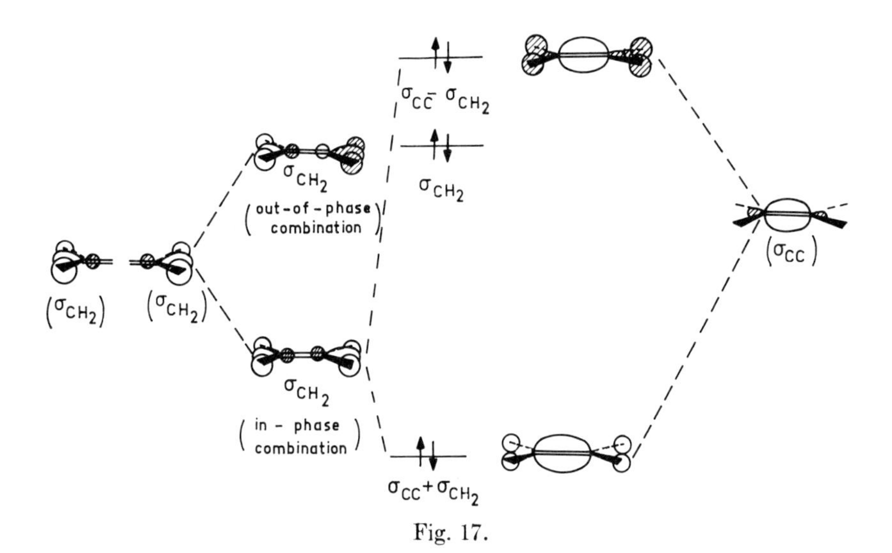

Fig. 17.

at the outset, at higher energy than the original pair of σ_{CH_2} orbitals. Generally, it is safe to apply the rule that a CC orbital lies higher than a CH orbital of similar type (Fig. 18). Indeed CH bond energies are nearly 100 kcal/mole, while CC single bond energies (83 kcal/mole) or incremental CC "double" bond energies (63 kcal/mole) are significantly smaller. The rule is, therefore, a rough illustration of the more sizable CH bonding. Again it is useful to compare accurate orbitals (III.18) with those drawn in Fig. 17.

σ_{CC} π_{CC}

σ_{CH_2} π_{CH_2}

Fig. 18.

9. Analogy between the σ Delocalized Orbitals of a Hydrocarbon Skeleton (Ethylene, Ethane, Allene, Propane, etc.) and the π Delocalized Orbitals of Conjugated Chains

The three σ-type molecular orbitals which we have just derived for ethylene can also be constructed by using a slightly different but equally instructive approach. We consider the left-hand σ_{CH_2} group orbital, the σ_{CC} bond orbital, and the right-hand σ_{CH_2} group orbital as three orbitals *in a chain*. They have relatively equal energies and in a good approximation only nearest-neighbor interactions are important. Thus the molecular orbitals can be constructed as one would the π molecular orbitals of a 3-atom conjugated skeleton, i.e. the allyl radical. Immediately we can write down the proper phase relationship between the localized orbitals to form the three molecular orbitals shown in Fig. 19. The comparison with the previous scheme, as well as with the exact pictures (III.18) is excellent. Of course, it would be incorrect to label the three combinations here as *bonding*, *nonbonding*, and *antibonding*—since all three molecular orbitals remain *overall bonding*. Thus the simple numbering σ_1, σ_2, σ_3 and its implications provides an alternative description to the qualitative labeling of the previous section.

The interaction pattern is exactly the same for ethane. Instead of constructing a pair of σ_{CH_3} orbitals and one σ_{CC} orbital (as is done

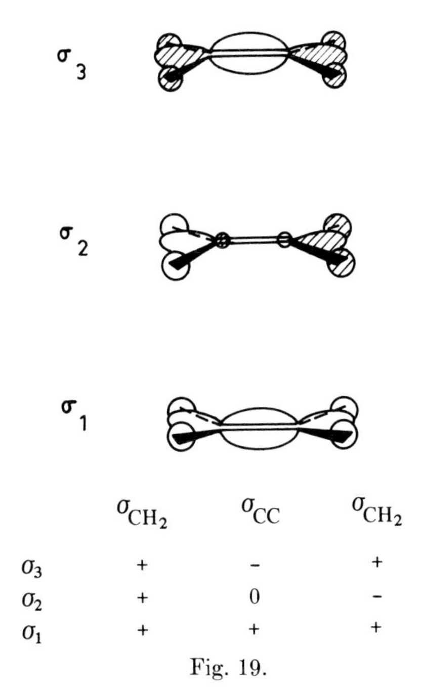

	σ_{CH_2}	σ_{CC}	σ_{CH_2}
σ_3	+	−	+
σ_2	+	0	−
σ_1	+	+	+

Fig. 19.

in the next section), one can use immediately the two σ_{CH_3} group orbitals and the CC bond orbital to build three σ molecular orbitals (see Fig. 20). Orbital σ_2 is of purely σ_{CH_3} type, while orbitals σ_1 and σ_3 involve both the σ_{CC} orbital and the in-phase combination of the σ_{CH_3} group orbitals. An interaction diagram of the type illustrated for ethylene would, of course, lead to very similar results. The comparison with the molecular orbitals $2A_1''$, $2A_2''$ and $3A_1''$ (III.30) is excellent.

Let us now extend the hydrocarbon chain to include an additional fragment. In allene, for example, there are four interacting units—CH_2 group, CC bond, CC bond and CH_2 group. Again we can construct the proper phase relationships in the σ molecular orbitals by analogy with the π orbitals of butadiene (see Fig. 21). The correct figures (III.41) show that not only are these phase relationships correct, but even the *relative amplitudes* of the local orbital components along the skeleton seem to reproduce extremely well the amplitudes of the π molecular orbitals of butadiene (large amplitudes at the termini on σ_2 and σ_3, large amplitudes on the central fragments in σ_1 and σ_4).

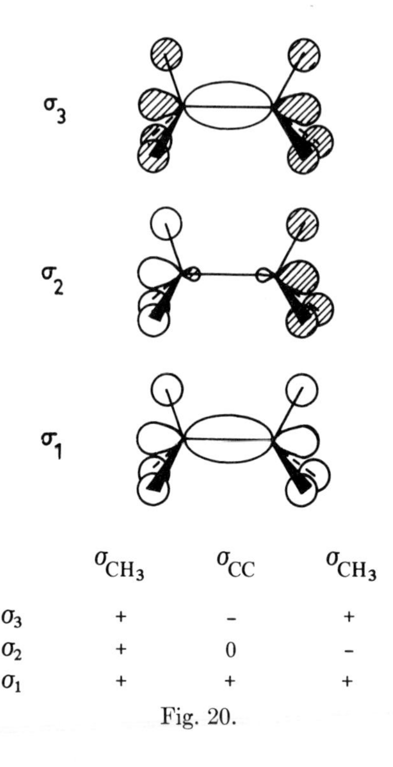

	σ_{CH_3}	σ_{CC}	σ_{CH_3}
σ_3	+	–	+
σ_2	+	0	–
σ_1	+	+	+

Fig. 20.

This analogy seems to be a very general one. For instance in methyl acetylene (III.38) molecular orbitals $4A_1$ to $7A_1$ are built up just like the four σ orbitals of allene. The same argument can be used to construct the σ orbitals of propane (five interacting fragments), propylene (see also II.1),[a] etc. The reader can verify by inspection of the figures for these molecules (III.61, III.49) that the analogy to the conjugated-chain π orbitals remains valid, and is independent of the number of interacting units.[b] The analogy

[a] In propane and propylene, the five-membered chain is branched rather than linear; a fair analogy would be given by the π orbitals calculated for a pentadienyl radical in which an additional matrix element would link atoms 2 and 4.

[b] Unpublished calculations on the linear C_n molecule show the σ_{CC} orbitals to combine like the π molecular orbitals of a linear conjugated molecule. Relative to the symmetry plane bisecting the chain the σ_{CC} orbitals are alternatively symmetric and antisymmetric. For even (odd) n the highest occupied orbital is antisymmetric (symmetric). Similarly for even (odd) n the lowest unoccupied σ^*_{CC} orbital is symmetric (antisymmetric). "Through-bond" interactions (see I.15) then make the ordering of the lone-pair orbitals alternate with n.

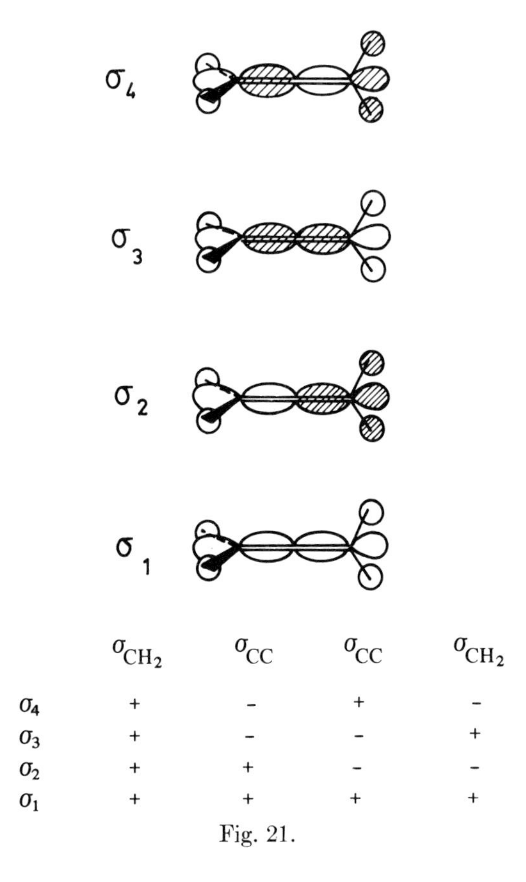

	σ_{CH_2}	σ_{CC}	σ_{CC}	σ_{CH_2}
σ_4	+	−	+	−
σ_3	+	−	−	+
σ_2	+	+	−	−
σ_1	+	+	+	+

Fig. 21.

also holds for cyclic molecules (see further cyclopropane, III.56 and benzene, III.94).[a]

10. The "π" Orbitals of Ethane

A beautiful illustration of the interaction between local group orbitals is given by the construction of ethane from two CH_3 groups. For the sake of argument we will consider the *eclipsed* configuration of the molecule. Since the local symmetry properties of the pyramidal CH_3 groups are preserved in the molecule, the interaction diagram is easy to construct—σ orbitals of one CH_3 group will mix only with σ orbitals of the other, π^y orbitals with π^y orbitals, π^z orbitals

[a] When interacting *antibonding* σ^* orbitals, the reader should beware that the *sign* of the overlap between adjacent σ^* bond orbitals depends on the bond angle.

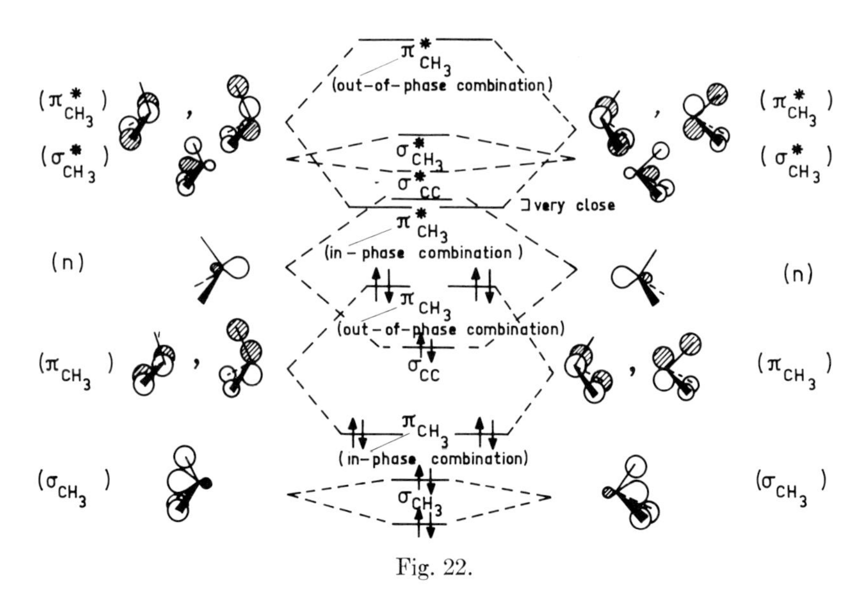

Fig. 22.

with π^z orbitals. Furthermore only mixings between orbitals in the same manifold (bonding, or antibonding) need be considered. The unrefined interaction diagram which arises after the "first-order" mixing between orbitals of same energy is shown in Fig. 22. Rather than obscure the diagram by drawing the molecular orbitals (the reader should refer to III.30) we have simply given their symmetry. We have also inserted the σ_{CC} and σ^*_{CC} orbitals of the central CC bond (in a refinement they would mix with their CH_3 counterparts). These two orbitals arise from the overlap of the "out" (2s, $2p_x$) n-type hybrids on the two CH_3 groups.

There are several striking features in this diagram:

1. The interaction between σ_{CH_3} (or $\sigma^*_{CH_3}$) orbitals is small relative to that between π_{CH_3} (or $\pi^*_{CH_3}$) orbitals. The reason is simply that the local σ_{CH_3} orbitals point away from the carbon-carbon bond. This feature is not so acute for the π orbitals, which involve a $2p_y$ or $2p_z$ orbital on carbon rather than a (2s, $2p_x$) hybrid pointing away.
2. The highest occupied orbitals in ethane are a pair of degenerate π-type orbitals, quite strongly localized on the CH_3 groups. They have been obtained by combining the local π_{CH_3} orbitals in an out-of-phase manner, with opposing amplitudes.

Thus the top occupied orbitals in ethane are CH bonding but CC antibonding.[a]

3. In a similar vein the in-phase combination of the $\pi^*_{CH_3}$ localized pairs, which is CH antibonding but CC bonding, is possibly the lowest unoccupied orbital; it lies very close to the σ^*_{CC} orbital and calculations are at variance as to their relative energy.[b]

Thus the π orbitals and π electrons lie in the outermost part of the valence shell of ethane. They should play a critical role in determining the chemical properties of the molecule. Some theories have ascribed the barrier to internal rotation to these orbitals. It should be noted that the existence of π electrons in ethane is not a novelty, and was first pointed out by Mulliken in 1935.

The σ orbitals of ethane will not, of course, remain pure σ_{CH_3} or pure σ_{CC}. In a further but important refinement the σ_{CC} bonding orbital and the σ_{CH_3} in-phase bonding combination will mix together. The manner in which this mixing occurs has been illustrated for ethylene. The resulting ethane orbitals have been constructed, by a different technique, in the previous section.

11. Cyclopropane

The manner in which orbitals belonging to CC bonds and to CH_n groups combine together in a linear chain was shown above to be very similar to the manner in which the $2p_z$ atomic orbitals of a conjugated molecule combine to form conventional π molecular orbitals. The analogy is also valid for cyclic molecules. For instance in the cyclopropenyl system there is a low-lying three-fold symmetric type π molecular orbital and a degenerate pair of antibonding π orbitals (see Fig. 23 and III.37). The coefficients which determine the relative amplitude of the atomic orbitals are determined by symmetry.

[a] Unfortunately the interesting consequence of a CC bond *shortening* upon excitation or ionization does not follow immediately. Indeed, in the ethane positive ion at least, as calculations by Pople and collaborators have shown, the ordering of the σ_{CC} orbital and of the highest π orbital pair seems reversed, with the hole lying in the σ_{CC} orbital. In the parent molecule both orbitals are very close.

[b] In III.30 (and III.29) the π_{CH_3} molecular combinations which lie in the vertical symmetry plane are labelled with a *prime*, π'_{CH_3} and $\pi^{*\prime}_{CH_3}$.

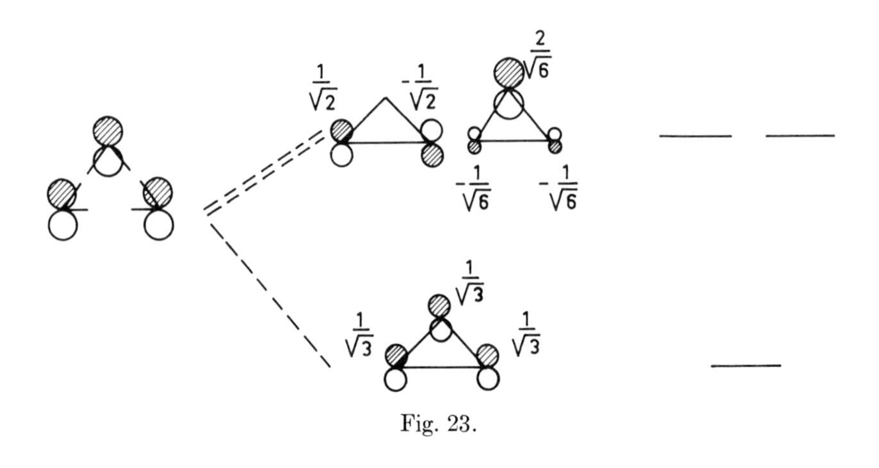

$\frac{1}{\sqrt{2}}$
$-\frac{1}{\sqrt{2}}$
$\frac{2}{\sqrt{6}}$
$-\frac{1}{\sqrt{6}}$
$-\frac{1}{\sqrt{6}}$
$\frac{1}{\sqrt{3}}$
$\frac{1}{\sqrt{3}}$
$\frac{1}{\sqrt{3}}$

Fig. 23.

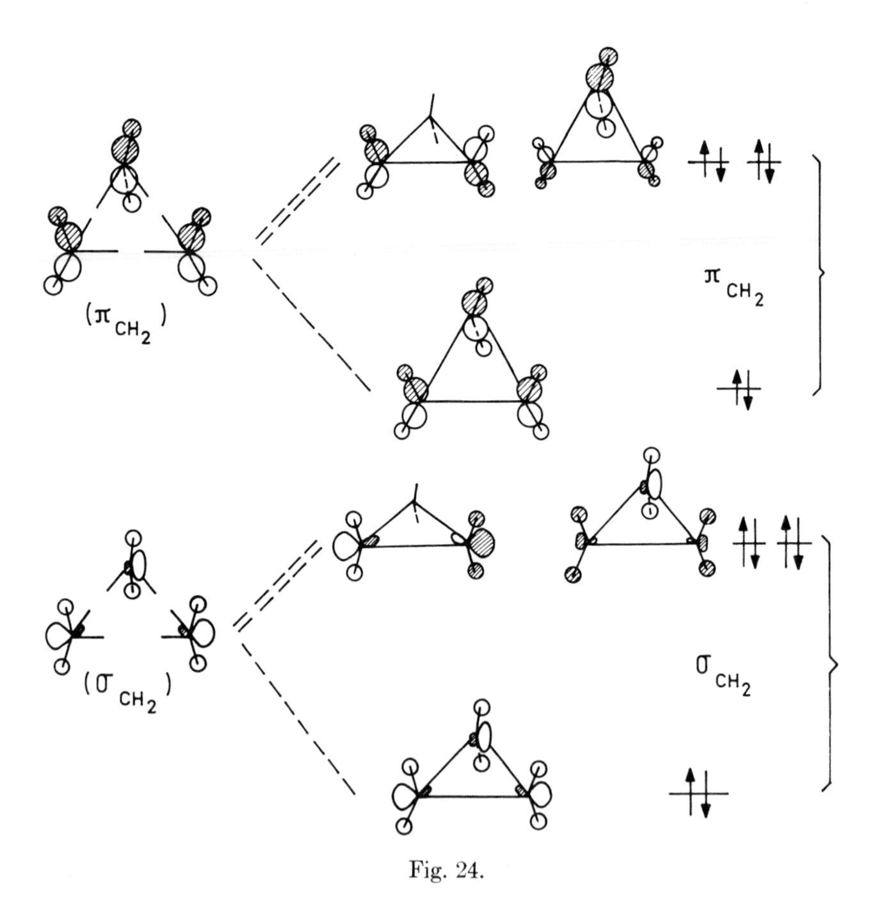

(π_{CH_2})
π_{CH_2}
(σ_{CH_2})
σ_{CH_2}

Fig. 24.

In cyclopropane, the molecular symmetry is the same (D_{3h} point group) as in cyclopropenyl, so that not only the π_{CH_2} and $\pi^*_{CH_2}$ orbitals—but also the σ_{CH_2} and σ_{CC} orbitals—combine just like the 2p orbitals of Fig. 23. This is shown in Fig. 24, for the π_{CH_2} orbitals and σ_{CH_2} orbitals. Since each group orbital has two electrons to begin with, all three π_{CH_2} molecular orbitals and all three σ_{CH_2} molecular orbitals are full. The molecular orbitals are overall bonding, the the C $\leftrightarrow$ H bonding character overcoming any CH $\leftrightarrow$ CH antibonding or $CH_2 \leftrightarrow CH_2$ antibonding character.

The construction of the σ_{CC} molecular orbitals is solved in exactly the same manner: each σ_{CC} bond orbital has a positive overlap with its two neighbors; via this overlap, the bond orbitals mix and form three typical combinations (see Fig. 25). A glance at

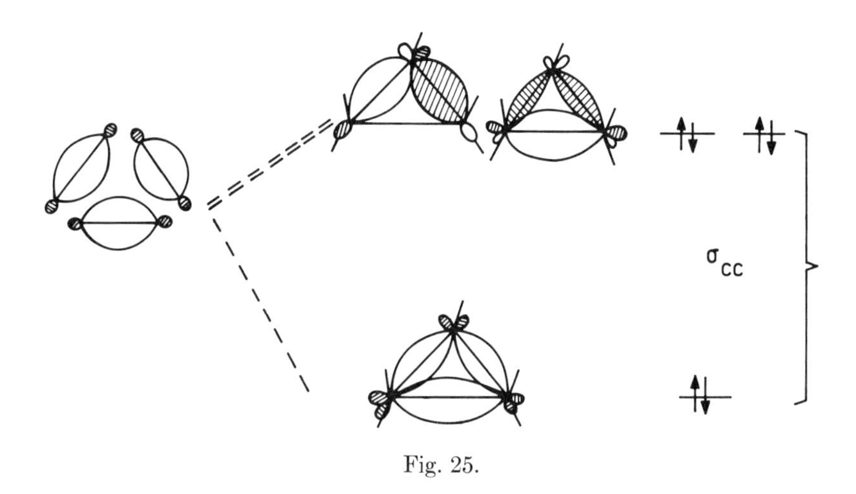

Fig. 25.

the more exact drawings for cyclopropane (III.56) should convince the reader that, although the delocalized combinations of σ_{CC} bond orbitals do reproduce the qualitative features of the highest occupied molecular orbitals of cyclopropane ($3A_1'$ and $3E'$), they are still approximations to the true picture. In particular the σ_{CC} orbitals of III.56 have a small admixture of σ_{CH_2} character. Yet it is rewarding that this simple description can reproduce such significant features as two successive sign changes in orbital amplitude across the molecule (right-hand side $3E'$ orbital, III.56; compare with top right-hand side σ_{CC} orbital).

12. An Alternative Description: Walsh Orbitals

It may be proper at this stage to lead the reader back to the stage where we constructed the localized orbitals of a CH_2 group. At that time two valence orbitals were set aside—the $2p_y$ orbital, and the outer (2s, $2p_x$) hybrid. Both of these orbitals lie in the x, y plane. Now in our description of cyclopropane, we used bond orbitals to describe the CC bonding; these bond orbitals are derived from in-plane (x, y) hybrids on each carbon. The two hybrids which are required on each carbon atom—in order to participate in two bond orbitals—are built precisely from the $2p_y$ orbital and the (2s, $2p_x$) "out" combination on each CH_2 group.

In the Walsh description, these very same valence orbitals are used on each CH_2 group, but one does not go to the trouble of combining them to make new orbitals pointing approximately along the bond directions.[a] One uses directly the three local $2p_y$-type orbitals of the three CH_2 groups to build one set of three molecular orbitals, and the three local (2s, $2p_x$) "out"-type hybrids to build a second set of molecular orbitals. The procedure is illustrated in Fig. 26.

In the interaction of the local $2p_y$ orbitals, two "more bonding" molecular orbitals are formed against one "less bonding." In all previous cases the opposite occurred. This is due to the negative overlap between adjacent $2p_y$ orbitals—whether, by convention, all positive lobes point in the clockwise direction, or whether all positive lobes point in the anticlockwise direction. The two bonding $2p_y$ combinations in fact fall below the two antibonding (hybrid 2s, $2p_x$) combinations. The former each have two electrons while the latter are empty. The six electrons of the three C—C bonds are nicely accounted for. The method creates simultaneously the σ_{CC} and σ^*_{CC} molecular orbitals of cyclopropane (note that the latter three lie relatively close in energy).

The reader may now wish to compare the three bonding molecular orbitals derived in this manner with the three σ_{CC} molecular orbitals shown at the end of the previous section. There is a strong resemblance. This similarity increases if, in the Walsh method, the $2p_y$-derived molecular orbitals are allowed to mix with the (2s, $2p_x$)-

[a] In practice the angle between the two in-plane hybrids still exceeds the bond angle (60°)—a defect which justifies the Walsh approach.

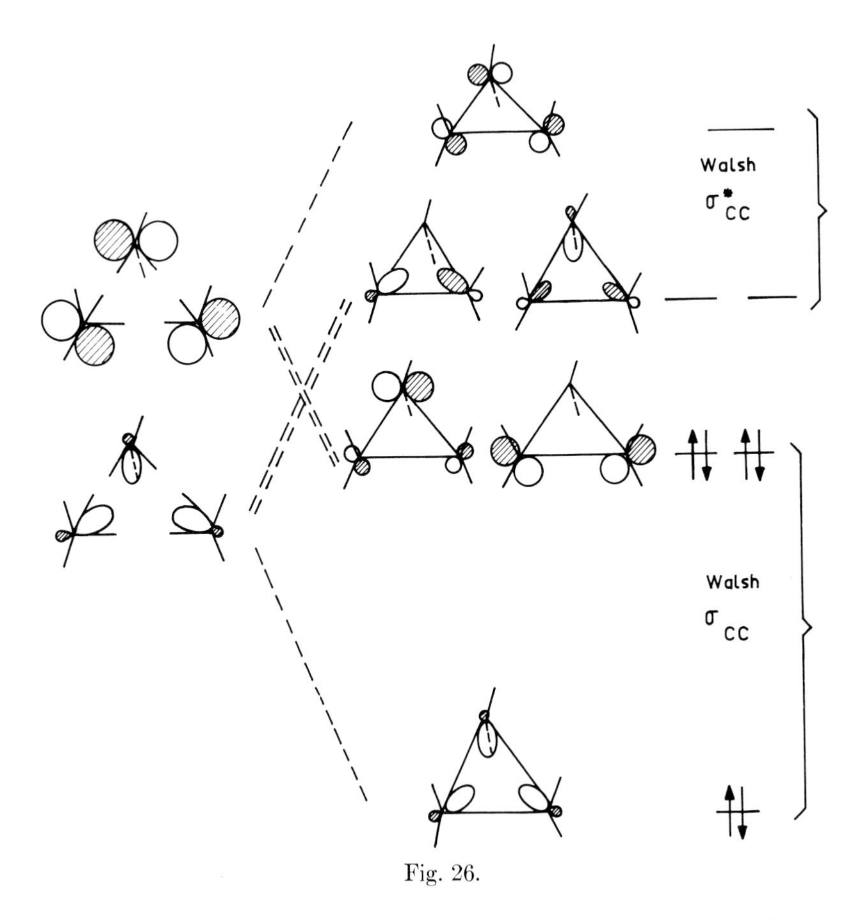

Fig. 26.

derived molecular orbitals. For instance the right-hand Walsh molecular orbital of the degenerate σ_{CC} pair acquires a second nodal surface by slightly mixing with the Walsh σ_{CC}^* orbital which lies just above it. Then the Walsh description and the bond-orbital description become essentially equivalent.

13. The σ Orbitals of Benzene

Another example which illustrates beautifully the mixing of σ group orbitals to form delocalized molecular orbitals is benzene. First of all the six σ_{CC} bond orbitals interact to give six linear combinations which are delocalized over the entire carbon skeleton. The amplitudes of the various bond orbitals in each σ_{CC} molecular orbital are identical with the well-known atomic orbital amplitudes in the molecular orbitals of a six-atom ring (for instance the π molecular orbitals of benzene). They are shown in Fig. 27 with simple + or − signs. Similarly the six σ_{CH} bond orbitals can be combined to give six molecular orbitals which cover the six CH bonds alone. Again the amplitude maps follow familiar patterns. The orbitals shown in Fig. 27 can be called *semilocalized* orbitals, since their delocalization is restricted to a given type of chemical group (CH or CC).

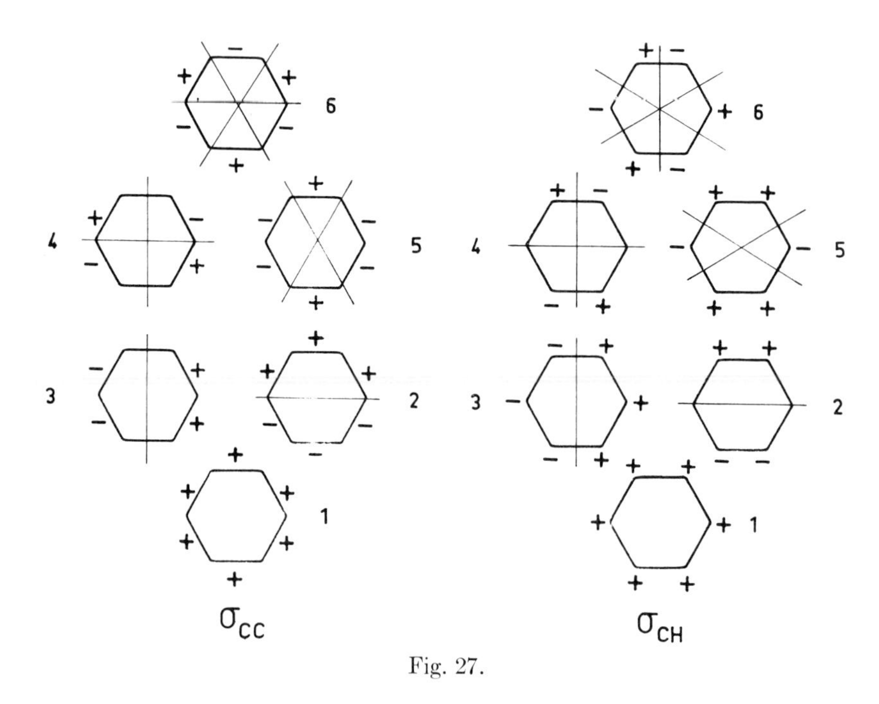

Fig. 27.

Since the CH bond orbitals are relatively distant, their interaction is very weak and the six σ_{CH} semilocalized orbitals are very close in energy. On the other hand the σ_{CC} bond orbitals interact strongly so that the energies of the six σ_{CC} semilocalized orbitals are spread out, with the lower ones falling well below the group of σ_{CH} orbitals.

We now go one step further and allow for the mixing of σ_{CC} semilocalized orbitals with the σ_{CH} semilocalized orbitals of appropriate symmetry. This mixing occurs just like that of the σ_{CC} bond orbital

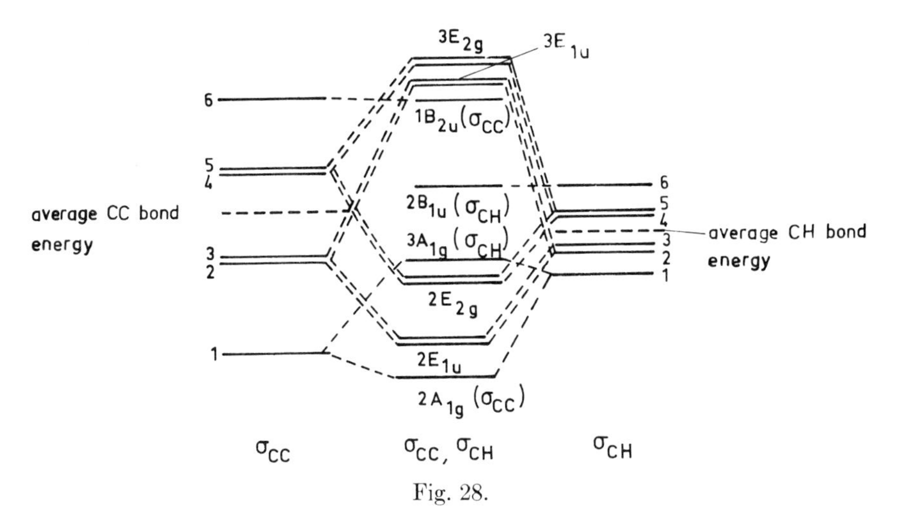

Fig. 28.

with the σ_{CH_2} in-phase orbital of ethylene, which we discussed in I.8. In Fig. 28 we simply sketch out the interaction diagram between the two sets of molecular orbitals, which leads to the benzene σ molecular orbitals (III.94). The interaction between the two totally symmetric orbitals, numbered 1 in each sequence, is very small because of their large separation. This leads to a $2A_{1g}$ orbital of benzene which is nearly pure σ_{CC} in character, and a $3A_{1g}$ orbital which is essentially a pure σ_{CH} molecular orbital. The degenerate (2, 3) pairs of molecular orbitals interact to give a more bonding $2E_{1u}$ pair and a less bonding $3E_{1u}$ pair. Both of these pairs have mixed σ_{CC}, σ_{CH} character, although the σ_{CH} character is not very apparent in the $2E_{1u}$ orbitals (III.94). Similarly, the $2E_{2g}$ and $3E_{2g}$ pairs of molecular orbitals of benzene are formed from the mixing of orbitals 4, 5 (σ_{CC}) with orbitals 4, 5 (σ_{CH}). Finally the top orbital, number 6, of each semilocalized series remains untouched when all interactions are accounted for. Indeed orbital 6 (σ_{CC}) has nodes running axially along the CH bonds and cannot borrow any σ_{CH} character, while orbital 6(σ_{CH}) has nodal planes across the CC bonds and cannot mix in any σ_{CC} character. They lead respectively to the $1B_{2u}$ and $2B_{1u}$ σ molecular orbitals of benzene.

Again this example shows how convenient bond orbitals can be as a starting point for classifying the molecular orbitals of large molecules.

14. Does Distance Affect the Interaction between Localized Orbitals? Cyclobutane

The direct, through-space interaction between two localized orbitals is extremely sensitive to distance. Atomic orbital overlaps decrease as inverse exponential powers of the distance, so that an overlap which is significant (~0.2) at 1.4 Å may be negligible (~0.01) at 2.8 Å. Furthermore, since the shape of a group orbital—for instance a σ_{CH_2} orbital or a π_{CH_2} orbital—is essentially determined by the optimization of bonding interactions *within* the group, the orbital generally offers little in terms of highly directional overlap for interaction with non-nearest neighbors. Thus only the overlap between *neighboring* localized orbitals is responsible for the delocalization of the orbitals in saturated hydrocarbons. Next-nearest-neighbor groups or bonds, unless the molecular geometry makes them unusually close, generally have a negligible through-space mixing.

The "1—3" CC bond-orbital interactions between opposite CC bonds in a cyclobutane ring are an interesting exception. In this system there is significant mixing between the σ_{CC} (and σ^*_{CC}) orbitals on opposite bonds (Fig. 29). The two σ_{CC} molecular orbitals are both occupied. The reader will recognize these orbitals, and the

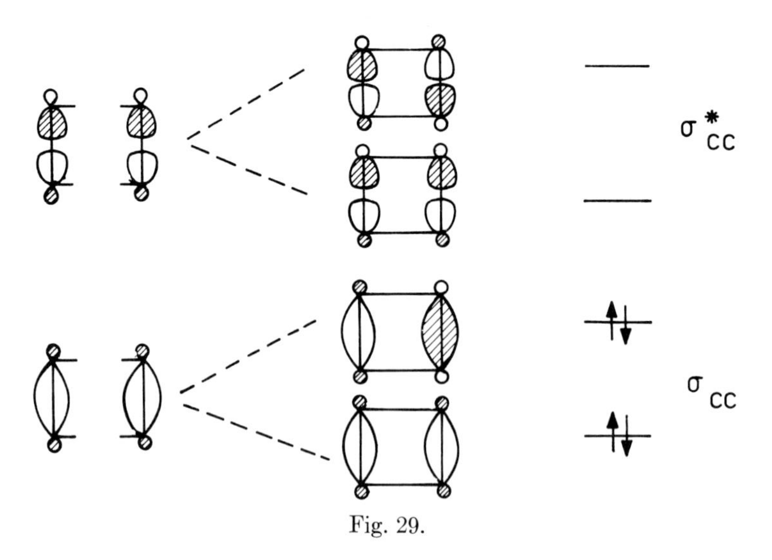

Fig. 29.

σ^*_{CC} orbitals, as identical with those used by Woodward and Hoffmann in their correlation diagram for the ethylene dimerization reaction. In 1965 it was somewhat unusual to draw out, as these authors did, delocalized σ_{CC} and σ^*_{CC} orbitals for the product cyclobutane, but the significant overlap between opposite bond orbitals (0.20 for σ_{CC}, 0.13 for σ^*_{CC}) justifies this realistic approach. The $\sigma_{CC} \leftrightarrow \sigma_{CC}$ cross-overlap is in fact so large that it exceeds that between adjacent σ_{CC} bond orbitals and thereby controls the energy ordering of the four σ_{CC} molecular orbitals of cyclobutane (Fig. 30). The reader can compare these approximate drawings with those of

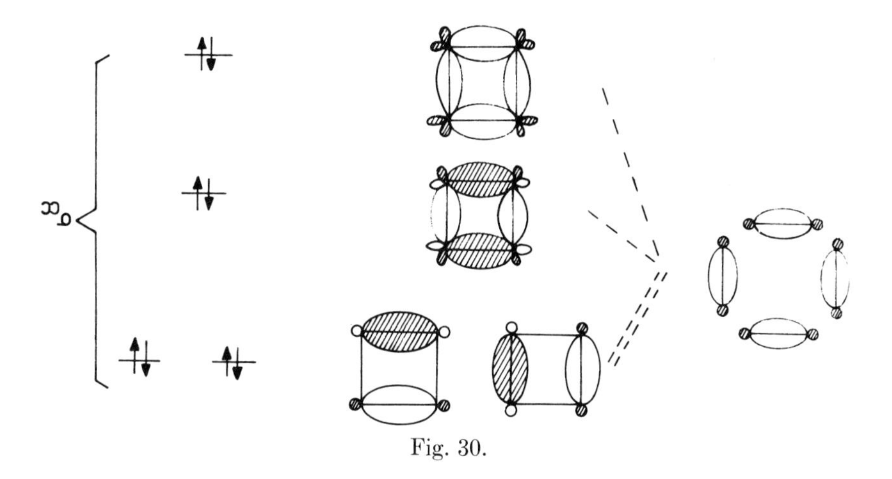

Fig. 30.

III.82. The degeneracy of the top occupied orbitals of cyclobutane has been verified by photoelectron spectroscopy. These orbitals are still overall bonding because their local bonding character dominates their "interbond" antibonding character.[a]

15. Through-Bond Interactions: *para*-Benzyne and Pyrazine

Since through-space interactions die out for groups more than one bond apart, is it true that two distant local orbitals have no interaction? The answer is that they do not overlap directly, but they may interact *indirectly via* the groups which lie between them.

[a] The choice of degenerate orbitals is not unique; it is also possible to construct Walsh orbitals for cyclobutane. These are precisely the $3E_u$ and $4E_u$ pairs of III.82.

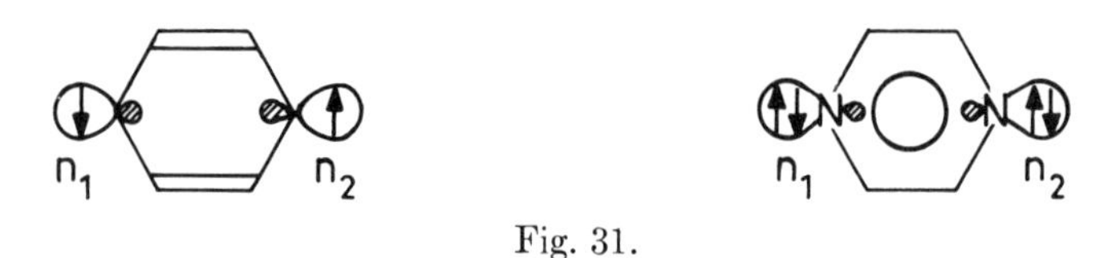

Fig. 31.

This is first illustrated for the two nonbonding *n*-type orbitals n_1 and n_2 of *para*-benzyne and pyrazine (Fig. 31). These nonbonding orbitals are derived from "outer" (2s, 2p) sp^2 type hybrids which have not been used in any bonding interaction. Although the overlap between n_1 and n_2 is zero each one overlaps with the central CC bond orbitals. All told, there will arise two distinct molecular orbitals in which n_1 and n_2 enter as combinations (symmetric or antisymmetric) and which have *different energies*, because of selective interactions with the central bonds.

Let us now look at the details of this *through-bond* interaction mechanism. We consider the four-orbital system shown in Fig. 32.

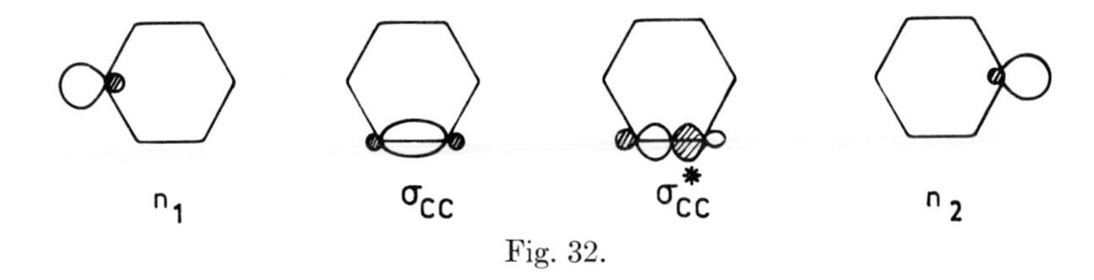

Fig. 32.

At the start, the relative energies of the four orbitals are represented in Fig. 33. The correct procedure requires that we first interact the degenerate orbitals n_1 and n_2, and then mix the resulting combinations with σ_{CC} and σ^*_{CC} which lie at significantly different energies. This is true even though the overlap between n_1 and n_2 is extremely small, because n_1 and n_2 are symmetry-equivalent orbitals. The

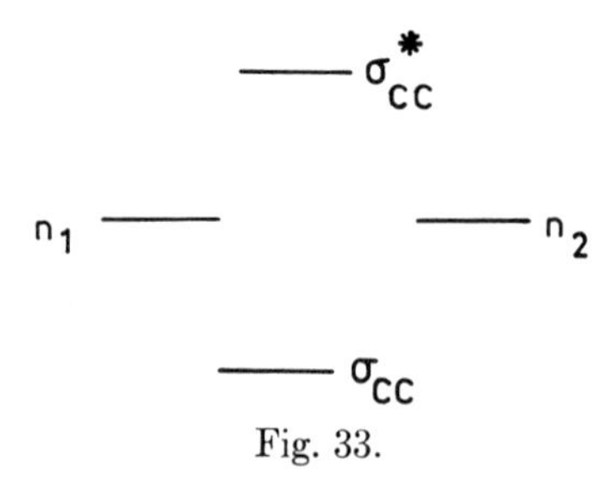

Fig. 33.

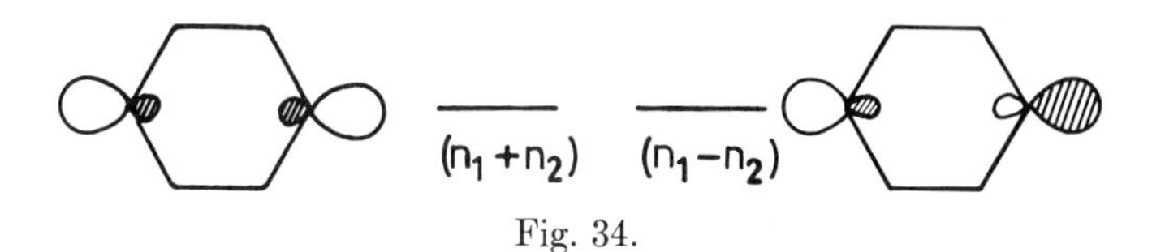

Fig. 34.

symmetry of the molecules requires that the proper wave functions be the combinations $(n_1 + n_2)$ and $(n_1 - n_2)$ (see Fig. 34). The reader should be aware that in this process we have *not* introduced any electronic delocalization; in either of the orbitals $(n_1 + n_2)$ and $(n_1 - n_2)$ the density vanishes strictly[a] throughout the region intermediate between the two para atoms.

But now let us interact $(n_1 + n_2)$ with σ_{CC} (clearly both are symmetric with respect to a plane bisecting the two central bonds), and $(n_1 - n_2)$ with σ^*_{CC} (both are antisymmetric). The results are shown in Fig. 35. The overlap between $(n_1 + n_2)$ and σ_{CC}, as well as that

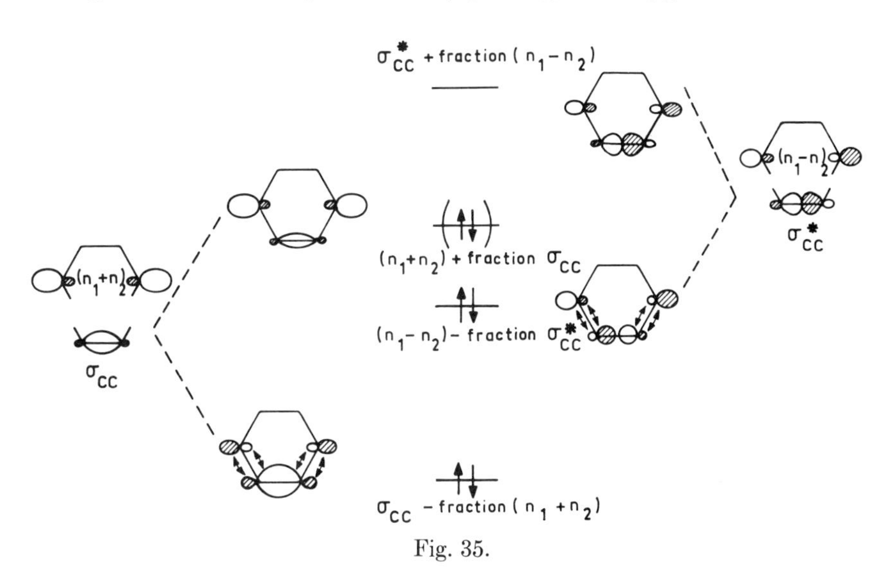

Fig. 35.

between $(n_1 - n_2)$ and σ^*_{CC}, is negative. It is dominated by the overlaps between the inner (small) lobes of the n orbitals and the inner (large) lobes of either σ_{CC} or σ^*_{CC} and between the outer (large) lobes of n and the outer (small) lobes of σ_{CC} or σ^*_{CC}. Thus a *positive*

[a] As strictly as the overlap between n_1 and n_2 vanishes.

overlap is indicated when adjacent inner or outer lobes have the same sign (see arrows).

Bearing this relationship in mind, we see that σ_{CC} is lowered by mixing in a small amount of $(n_1 + n_2)$ orbital in a bonding manner, while $(n_1 + n_2)$ is raised by the admixture of a small amount of σ_{CC} orbital in an antibonding manner. Similarly orbital $(n_1 - n_2)$ is lowered by admixture of a small fraction of σ_{CC}^* in a bonding way, while σ_{CC}^* itself is raised as it mixes in some $(n_1 - n_2)$ character in antibonding fashion. These results are an example of the general rules of behavior for the interaction of orbitals with different energy, which we gave earlier in Section I.7. We have already encountered an example with the mixing between the σ_{CC} orbital and the in-phase σ_{CH_2} combination in ethylene.

Now, after these interactions, two purely nonbonding orbitals $n_1 \pm n_2$ are replaced by two molecular orbitals which are delocalized over four atoms, one with slightly bonding character, the other with slightly antibonding character. The orbital in which n_1 and n_2 combine in an antisymmetric manner is actually stabilized relative to the symmetric combination, wholly because of *through-bond* interactions. When the electrons are distributed amongst the various orbitals, the two "odd" electrons of *para*-benzyne tend to occupy the antisymmetric delocalized molecular orbital,[a] while in pyrazine the electrons of highest energy occupy the symmetric delocalized molecular orbital. It is hardly suitable to speak any more of "lone" electron pairs.

The correct "nonbonding" molecular orbitals for these two systems are shown in III.92 and III.97. Of course the through-bond effect also involves the orbitals of the other central CC bond, as well as some mixing-in of the orbitals from the four other CC bonds. In pyrazine, orbital $6A_g$ (III.97) has the main features of the $(n_1 + n_2)$ orbital drawn above; orbital $5B_{1u}$, however, shows little mixing-in of the central σ_{CC}^* orbitals—possibly because the σ_{CC}^* orbitals are very high in energy so that the $\sigma_{CC}^* \leftrightarrow (n_1 - n_2)$ energy difference is too large to begin with.[b]

[a] We assume for the sake of argument that the splitting induced by through-bond interactions between the two previously nonbonding orbitals is large enough. If not, the ground state will be a triplet with one electron in each of the two orbitals.

[b] Epiotis has explained the relative stability of *cis* and *trans* 1,2-dihalo-ethylenes by invoking through-bond interactions between the lone-pair orbitals *via* the antibonding σ_{CC}^* and π_{CC}^* bond orbitals.

16. Are the Interactions between Localized Group Orbitals Useful in Predicting the Chemical Properties of Molecules? The Ethyl Cation and *n*-Propyl Cation

The most attractive feature of the various types of localized orbitals which have been defined—σ_{CC}, $\overset{*}{\sigma}_{CC}$, π_{CC}, $\overset{*}{\pi}_{CC}$, σ_{CH}, $\overset{*}{\sigma}_{CH}$, σ_{CH_2}, $\overset{*}{\sigma}_{CH_2}$, π_{CH_2}, $\overset{*}{\pi}_{CH_2}$, σ_{CH_3}, $\overset{*}{\sigma}_{CH_3}$, $2\pi_{CH_3}$, $2\overset{*}{\pi}_{CH_3}$—is the ease with which one can infer, from their interaction, the conformational properties and geometries of small hydrocarbon molecules. We have a handy tool for analyzing the discrete orbital entities which comprise any molecule and then synthesizing its overall delocalized molecular orbitals.

Take, for example, the ethyl cation. It is interesting to compare the eclipsed (E) and the bisected (B) conformations (see Fig. 36).

Fig. 36.

The methylene group carries an empty p orbital. How will the methyl group interact with this empty orbital? Clearly the π-type orbitals of the methyl group (π_y in E, π_z in B) have the appropriate symmetry to mix with the p orbital. A typical orbital interaction diagram (for E) is shown in Fig. 37. Several conclusions emerge immediately from this diagram:

1. The total energy is *lowered* by the interaction. The two electrons originally in π_{CH_3} now occupy an orbital in which there is a slight admixture of p_y character. The stabilization can be thought of as arising from the electronic delocalization.
2. The lowest molecular orbital has less than 100% π_{CH_3} character and a small percentage of p_y character. Therefore, a *transfer of electron density* occurs from the methyl group to the methylene group, a conclusion in accordance with the chemical viewpoint. The slight concomitant elongation of the $\pi^y_{CH_3}$ orbital shows up nicely in III.25 (orbital 1A″).

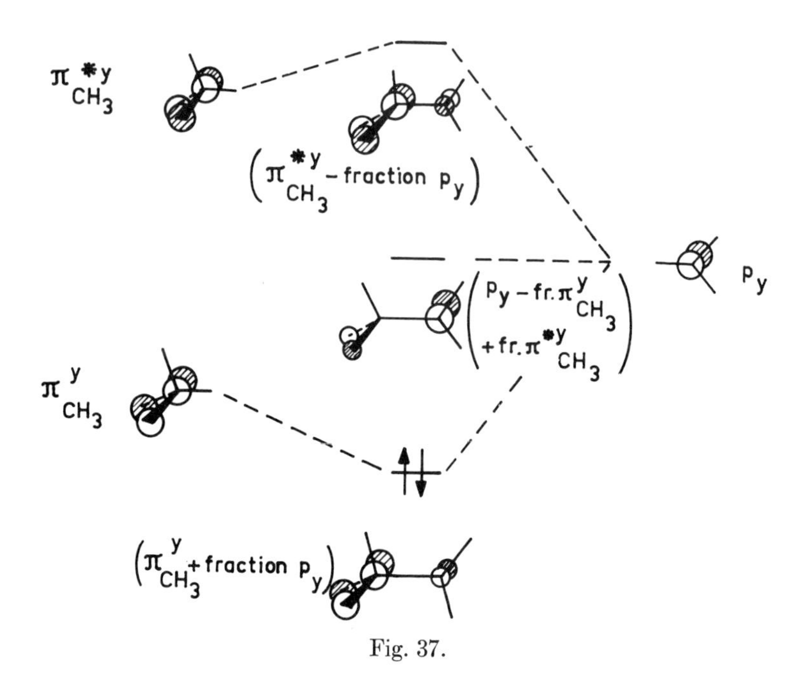

Fig. 37.

3. As a consequence of the electron transfer, the electron fraction on the π_{CH_3} orbital decreases. Since this orbital is C ↔ H bonding along two of the methyl CH bonds, we expect these bonds to *lengthen*. Also, it is antibonding between the two bonds, so that the interbond angle should *decrease*. The predicted geometrical consequences are shown in Fig. 38 for both E and B [a] (the bond lengths and bond angles from *ab initio* calculations are also shown—by comparison the calculated bond length in an isolated CH_3 group is 1.080 Å while a normal HCH

[a] The reader will obtain the geometrical consequences for B by carefully following a similar procedure for the interaction between $\pi^z_{CH_3}$ and p_z. Again all CH bonds of the CH_3 group should lengthen, but the top CH bond, with the largest hydrogen amplitude $(2/\sqrt{6})$ in the initial hydrogen combination (see III.24), is expected to lengthen most. There is indeed an extremely good correlation between calculated bond lengths in B and E and these amplitudes: 1.088(0), 1.091 $(1/\sqrt{6})$, 1.101 $(1/\sqrt{2})$, 1.110 $(2/\sqrt{6})$. In B also the angle between the lower two CH bonds should *open* ($\pi^z_{CH_3}$ is bonding between them), while that between their bisector and the top CH bond should *close* ($\pi^z_{CH_3}$ is antibonding along the z direction). This last prediction is the only one not to be borne out, possibly because of the dominant effect of the tilting of the upper CH bond towards the methylene group.

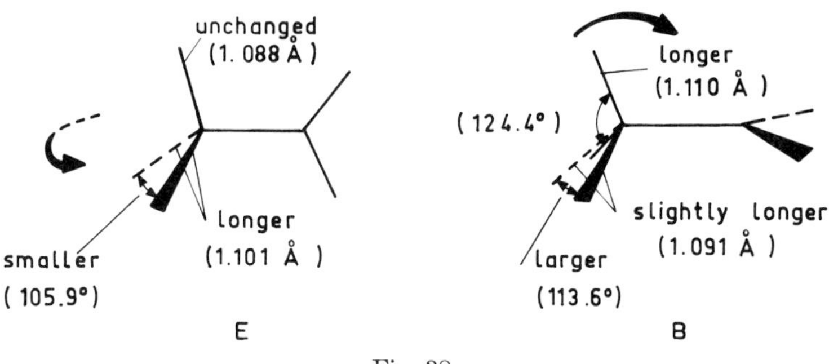

Fig. 38.

angle should be near 109.5°). There is a full symphony of geometrical changes. The molecule also distorts so as to increase the overlap between orbitals π_{CH_3} and p—and thereby the extent of the energy gain. Calculations have shown that the methyl group tilts sideways in E and forwards in B, along the directions shown by the arrows. The methylene group also tilts slightly, and the CC bond shortens.

4. The antibonding orbital $\pi^{*y}_{CH_3}$ is raised by the interaction with p_y; here the mixing occurs in a destabilizing manner. Finally, the p_y orbital stays relatively constant in energy, being pushed up by $\pi^{y}_{CH_3}$ and pushed down by $\pi^{*y}_{CH_3}$ (although possibly less, since the overlap between $\pi^{*y}_{CH_3}$, with an additional nodal surface cutting through the two CH bonds, and p_y, should be slightly smaller). The corresponding molecular orbital is now delocalized on a pair of hydrogens of the CH_3 group (the fractional contributions of $\pi^{y}_{CH_3}$ and $\pi^{*y}_{CH_3}$ on the methyl carbon atom roughly cancel out). Comparison can be made with orbital 2A″ in III.25 (Note that the $\pi^{z}_{CH_3}$ orbital is labelled with a *prime*, π'_{CH_3}, because it lies in the symmetry plane).

The phenomenon which we have just described—formation of delocalized π-type orbitals between methyl group and methylene group, and electron transfer from one to the other—is none other than the familiar hyperconjugative effect. One can compare the energy gain due to hyperconjugation in B and E. Since the orbitals $\pi^{y}_{CH_3}$ and $\pi^{z}_{CH_3}$ are rigorously degenerate, and since their respective overlaps with p^y and p^z are identical, the stabilization incurred through hyperconjugation, in E and B—barring differences due to molecular distortions—should be the same. Indeed the six-fold barrier to internal rotation is calculated to be tiny.

Let us now substitute a methyl group for one of the methyl hydrogens. In the resulting normal-propyl cation there should be a differentiation between E and B (Fig. 39). Indeed while the $\pi^y_{\mathrm{CH_3}}$ orbital is unaffected by the substitution, the $\pi^z_{\mathrm{CH_3}}$ orbital, which has a large concentration along the new carbon-carbon bond, should be raised.[a] It will lie closer in energy to the empty p orbital on the methylene group, and their mixing, with the concomitant stabilization should be greater. Thus the *n*-propyl cation should adopt a *bisected conformation.* We also expect geometrical consequences similar to those predicted for the ethyl cation, in particular a significant lengthening of the CC bond in the donating ethyl group of B.

Fig. 39.

Cations are by no means the only species where the effects of hyperconjugative delocalization reveal themselves in such a striking manner. Similar effects exist in neutral systems or in anions. For instance, the normal propyl anion should tend to be eclipsed (E) since in this manner the molecule would optimize the *4-electron* interactions between the ethyl group π orbital and the p orbital which carries the electron pair. In the bisected conformation, where $\pi^z_{\mathrm{CH_3}}$ and $\pi^{*z}_{\mathrm{CH_3}}$ have both been raised in energy, the four–electron, *destabilizing* (see Section I.7, rule 2) $p \leftrightarrow \pi$ interaction is stronger than in the eclipsed conformation. At the same time the two–electron, *stabilizing* $p \leftrightarrow \pi^*$ interaction is weaker than in the eclipsed conformation. Both effects favor the eclipsed conformation.

[a] We assume again that an orbital which is CC bonding is less stable than an orbital which—all other things being equal—is CH bonding (CC bond energy 83 kcal/mole, CH bond energy 99 kcal/mole).

17. The Vinyl Cation

The vinyl cation (Fig. 40 and III.13) in which the empty p orbital on the positive center is coplanar with the terminal methylene group, is clearly another candidate for hyperconjugative donation. The in-plane π'_{CH_2} orbital readily overlaps with the odd p orbital, with formation of two delocalized orbitals and an energy gain

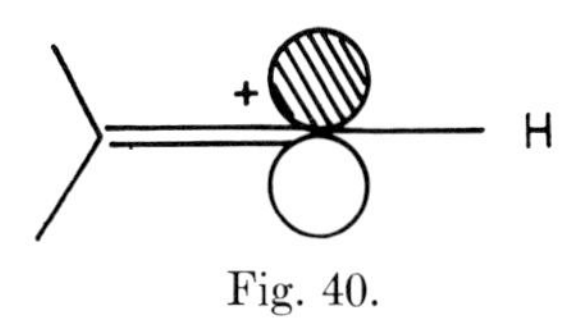

Fig. 40.

(Fig. 41). We predict a lengthening of the two CH bonds (calculations give 1.106 Å versus an average of 1.084 Å in the radical) and a closing of the terminal HCH angle (however, the calculated value 118.6° is slightly larger than in the radical, 116.3°—probably because the overlap increases if the angle opens and the CH bonds tilt towards the positive center).

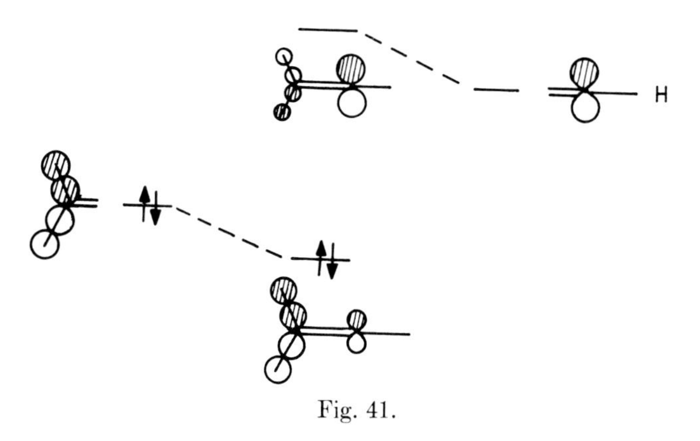

Fig. 41.

18. The Cyclopropylcarbinyl Cation

The cyclopropylcarbinyl cation is characterized, like the previous molecules, by the existence of a high labile 2p orbital which can interact with orbitals of appropriate symmetry in the ring. We can single out the important interactions in the two extreme configura-

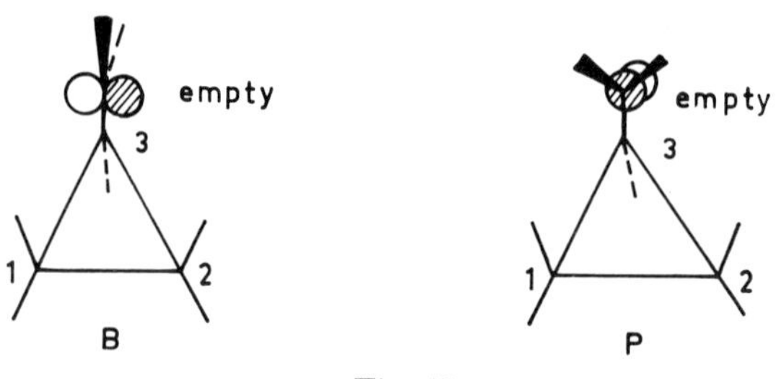

Fig. 42.

tions B (bisected) and P (perpendicular) illustrated in Fig. 42. The top bonding orbitals of cyclopropane—well approximated by the Walsh σ_{CC} orbitals—have the largest effect because they lie nearest in energy to the empty nonbonding p orbital. These Walsh orbitals have been previously schematized in I.12. The appropriate interactions are shown in Fig. 43, and lead in both systems to the forma-

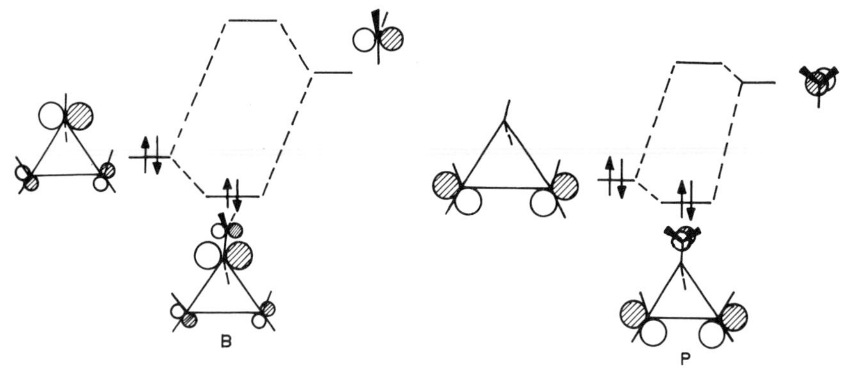

Fig. 43.

tion of a new pair of delocalized orbitals, one occupied and one empty. The essential difference between conformations B and P is the much larger overlap of the odd orbital with the Walsh orbital in B, thanks to a sizable component of this Walsh orbital on carbon atom 3. In conformation P, the appropriate Walsh orbital, with essentially zero amplitude on carbon 3,[a] overlaps only weakly with the empty p orbital. Hence the stabilization should be far

[a] The reader will remember that in the slightly more accurate bond–orbital description of cyclopropane, there is a small amplitude on this atom (III.56). So our statement should be somewhat qualified.

greater in the bisected conformation B. Proof of this has been obtained both experimentally and by *a priori* calculations. The reader should also refer to III.76 and III.77 for a correct view of the orbitals of B and P.

By now the reader will already have been alerted to an important geometrical consequence of the hyperconjugative stabilization of B. Bond 12 should shorten, due to a decreased electron density in the (12 antibonding) Walsh orbital.

19. Methylenecyclopropane and Cyclopropanone

What happens now if we attach to the cyclopropane ring an electron-donating group instead of an electron-accepting group? If this group has a high-energy, filled *in-plane* p-type orbital, as in cyclopropanone (n_o) or in methylenecyclopropane (π'_{CH_2}), the relevant interaction occurs with the highest empty Walsh σ^*_{CC} orbital and causes electron donation from the substituent group into the ring, as indicated by the arrows in Fig. 44. The main difference be-

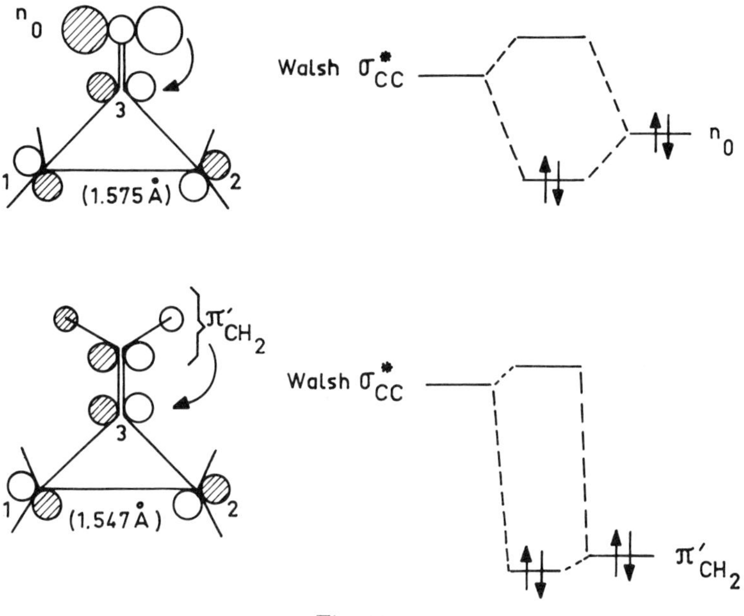

Fig. 44.

tween the two molecules lies in the much higher energy of the lone-pair orbital on oxygen relative to the in-plane π'_{CH_2} orbital of the exo methylene group. Hence a better interaction and larger electron donation in cyclopropanone than in methylenecyclopropane. A measurable consequence of this stronger electron transfer is the much longer C_1—C_2 bond length in cyclopropanone: there is more electron density in the Walsh σ^*_{CC} orbital, and thus more C_1—C_2 antibonding character.[a] Another consequence of the conjugative effect between the antibonding Walsh orbital and the adjacent in-plane p orbital is the lowering of the barrier to rotation about the CC double bond in methylenecyclopropane relative to ethylene.[b]

More exact drawings of the orbitals are shown in III.72 and III.73.

20. Introduction of Heteroatoms: The Effect of an Electronegativity Perturbation

Earlier in this chapter we considered the effect of orbital interactions on a previously noninteracting system. But suppose now we take as starting point two interacting orbitals ϕ_A and ϕ_B of equal energy and we introduce a change in electronegativity at centers A and B. The qualitative results of such a perturbation are again well known from elementary quantum chemistry:

1. When an electronegativity perturbation occurs, two molecular orbitals originally containing equal components of two atomic orbitals *localize* so that the lower one is primarily on the more electronegative atom, the upper one on the less electronegative atom.

[a] Note that in both molecules the C_1—C_2 bond length (cyclopropanone, 1.58 Å; methylenecyclopropane, 1.55 Å) is longer than in cyclopropane itself (1.51 Å). This is due in part to the hyperconjugative effect, but also arises partly from the change in hybridization at C_3.

[b] The donation of oxygen, sulfur or halogen lone pairs into the antibonding orbital of a neighboring bond seems to be a widespread effect. Suggestions have been made that it is responsible, via donation from the axial lone-pair of the oxygen atom into the empty σ^*_{CCl} orbital, for the long-puzzling *anomeric* effect for electronegative substituents (Cl, etc. . . .) on carbon atoms adjacent to oxygen in pyranose rings. The effect has already been used to interpret the nuclear quadrupole resonance experiments in these systems.

2. Both orbitals are stabilized if the average electronegativity increases; both destabilized if it decreases. If the average electronegativity remains constant, one orbital rises while the other one is lowered.

Let us consider, as an example, the orbital changes which occur in the "transformation" $N_2 \rightarrow CO$ (see III.15 and III.16). In the nitrogen molecule we have two localized (2s) electron pairs, one σ bond and two π bonds. The symmetric ($2\sigma_g$) and antisymmetric ($2\sigma_u$) molecular orbitals are essentially combinations of the nitrogen 2s atomic orbitals, while an in-phase combination of the 2pσ atomic orbitals yields the ($3\sigma_g$) orbital responsible for the σ bond.[a] The major orbital energy and amplitude changes when the two nitrogen atoms are replaced respectively by carbon and oxygen are shown in Fig. 45.

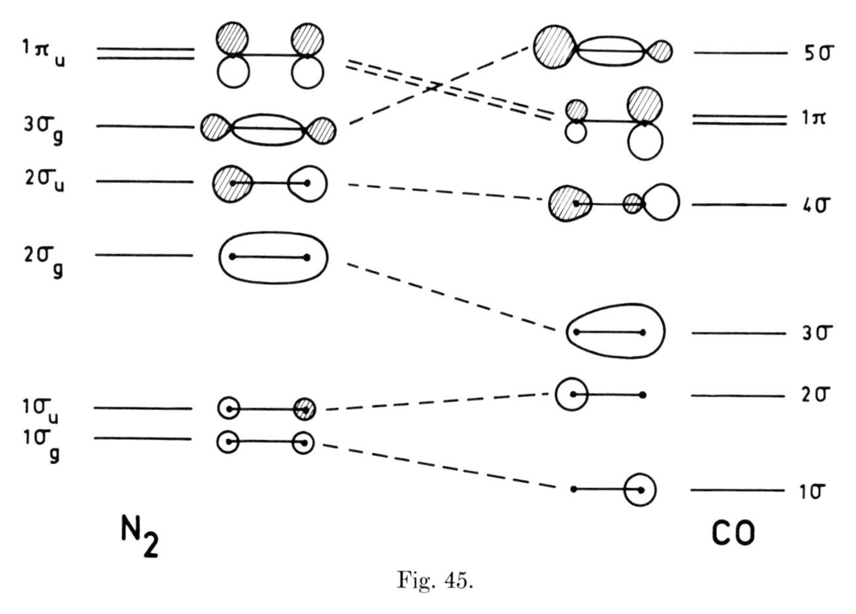

Fig. 45.

[a] An alternative point of view considers the $2\sigma_u$ and $3\sigma_g$ molecular orbitals as nitrogen "lone-pair orbitals," the $2\sigma_g$ orbital as a σ-bond orbital. Because of the through-bond interaction with the σ-bond orbital, the symmetric ($3\sigma_g$) lone-pair combination is pushed up above the antisymmetric combination. In this description the $2\sigma_u$ and $3\sigma_g$ lone-pair orbitals of N_2 correlate appropriately with the 4σ and 5σ lone-pair orbitals of CO.

The changes in the inner shell 1s-type orbitals are characteristic of all the molecular orbital modifications: the lower, symmetric combination of N_2 localizes on the oxygen atom in CO and becomes essentially the oxygen 1s orbital—while the higher antisymmetric combination correlates with the carbon 1s orbital. As the average electronegativity is roughly unchanged, one orbital goes up while the other goes down. Similarly the $2\sigma_g(N_2)$ orbital becomes a $3\sigma(CO)$ orbital localized mainly on oxygen: here the larger carbon amplitude is carried into a σ^* orbital of CO. The two π orbitals of nitrogen are lowered in carbon monoxide while simultaneously localizing on the oxygen atom. Finally the $2\sigma_u(N_2)$ orbital correlates with a $4\sigma(CO)$ orbital, to which $2p_\sigma$ admixture on the oxygen atom gives some poorly localized oxygen "lone-pair" character, while the $3\sigma_g$ orbital becomes a high "lone-pair" orbital localized on carbon. This high-lying orbital localized on the carbon side of the molecule is responsible for the small dipole moment of CO; its strong C^-O^+ character counterbalances the C^+O^- character of the 3σ, 4σ and 1π orbitals. The labile lone-pair orbital on carbon is also responsible for making CO a good ligand in inorganic systems: it has the high energy and the adequate directional character required to donate electrons into empty metal orbitals. (The antibonding π^* orbitals, with large amplitude on carbon, also facilitate back-donation from the metal to the ligand).

The general features introduced by an electronegativity perturbation are present in polyatomic molecules as well. For example, in the series ethylene (III.18), methylenimine (III.19), and formaldehyde (III.20), the lowest valence molecular orbital which is primarily the in-phase combination of the 2s orbitals, is progressively more localized on the more electronegative atom. Similarly in the butadiene, acrolein, glyoxal series, the increased localization of the π electron density on the oxygen atom(s)—compare π_1 and π_2 (III.65 to III.70)—is reflected in the increased central CC bond length and lower rotational barrier about this bond as one goes from butadiene to glyoxal.

21. The Reactivity of Cyclopropene and Diazirine

It is generally ambitious to try and explain reactive pathways by sole consideration of the electron distribution in the reactant.

Correlation diagrams include the product orbitals while perturbation approaches require knowledge of the empty orbitals of the reactant. However, the occupied molecular orbitals of diazirine (II), compared with those of cyclopropene (I), do seem to give some indication of a preferred thermal decomposition of (II) compared with the rearrangement of (I). Moreover these molecular orbitals are a typical illustration of the localization obtained in the presence of an electronegativity perturbation.

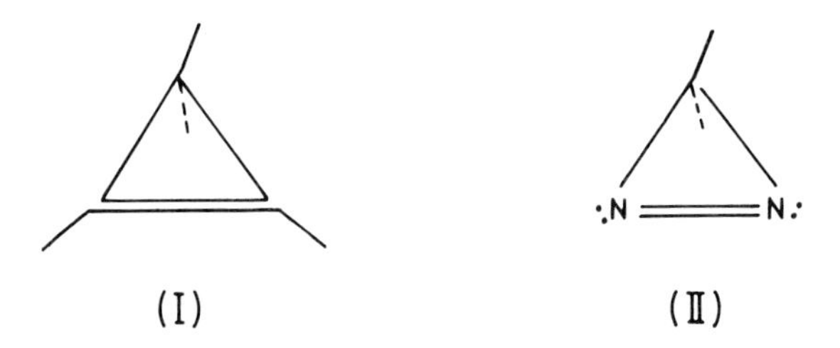

The relevant orbitals to consider are the molecular orbitals $5A_1$ and $6A_1$ of cyclopropene (III.46) which correlate with the lowest σ_{CC} orbital of cyclopropane ($3A_1'$, III.56) and with one of the higher degenerate σ_{CC} orbitals ($3E'$, III.56). The cyclopropene orbitals are very similar to their cyclopropane counterparts. In particular the $6A_1$ orbital is a linear combination of all three σ_{CC} bond orbitals and extends over the entire cyclopropene molecule. Its amplitude, shown in Fig. 46, is fairly accurately represented by the

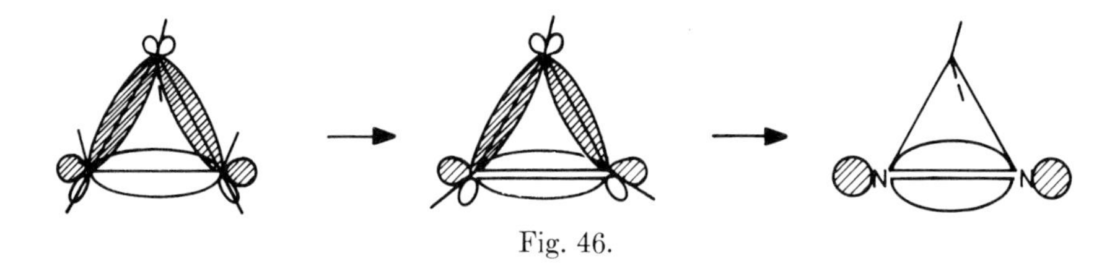

Fig. 46.

bond-orbital description of the cyclopropane orbitals given in Section I.11. However in diazirine the $6A_1$ orbital is localized solely in the N=N bond (the CN bond amplitudes are recovered in the σ^* orbitals) (III.47). It is strikingly akin to the $3\sigma_g$ orbital of the nitrogen molecule. Similarly the $5A_1$ orbital of diazirine resembles an (in-plane) π' orbital of the nitrogen molecule.

The facile thermal decomposition of the dimethyl and diethyl derivatives of (II) to nitrogen and carbene intermediates is emphasized by the readily discernible correlations between the reactant and product orbitals. On the other hand, the greater delocalization of the molecular orbitals of (I) may be a factor in its preference to rearrange, without decomposition, to methyl acetylene and allene.

22. Equivalence and Nonequivalence of Lone-Pairs. Water and Methyl Fluoride

In several previous Sections (I.15, I.19) we considered the interaction of lone-pair orbitals with orbitals localized in other regions of the molecules. In some cases an atom (ether oxygen or thioether sulfur, halogens) may carry several lone pairs. These lone pairs are generally described as localized orbitals pointing in tetrahedral directions on the atom to which they belong as in Fig. 47. The elec-

Fig. 47.

tronic density in these molecules agrees with such a description. However if the *energy* of the lone-pair electrons becomes an important factor—for instance, if we are seeking out favorable intramolecular interactions between lone-pairs and other orbitals—it is an absolute necessity to consider those lone-pair orbitals which have the proper local symmetry. In OH_2 for instance there is a π-type lone-pair molecular orbital and a less localized σ-type lone-pair molecular orbital (III.9). The latter, with s-character, lies below the former, as can be shown both by calculations and by photoelectron studies. The experimental energy difference is 2.1 eV.

In methyl fluoride the three lone-pair molecular orbitals divide into two degenerate, purely p-type orbitals and a deeper σ-type orbital (III.33) (see Fig. 48). The p-type orbitals on fluorine can be labeled "eclipsed" and "bisected", by analogy with the con-

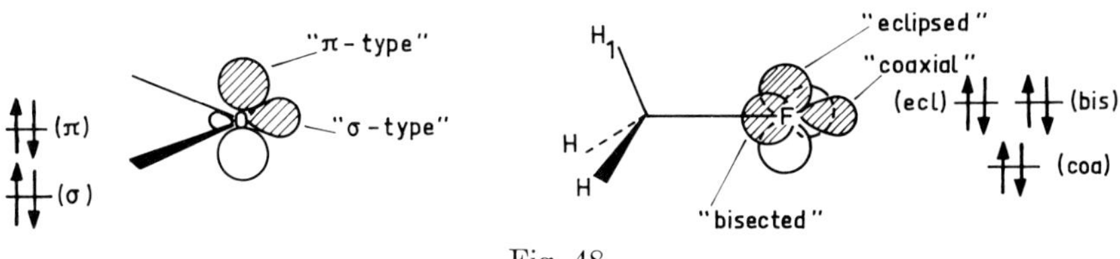

Fig. 48.

ventional conformational language; if H_1 is substituted by another atom their degeneracy is lifted.[a] They lie at significantly higher energy than the "coaxial" lone-pair orbital. The anomeric effect (Section I.19, footnote) appears to be a case where the energetic nonequivalence of the lone-pairs (on the ring oxygen) plays a significant role.

23. Conclusion Application to the Construction of Frontier Orbitals and to Correlation Diagrams

In this chapter we have described the various bond orbitals and group orbitals which are common to hydrocarbons and we have shown how they interact to form delocalized molecular orbitals. In molecules containing nitrogen or oxygen, similar group or bond orbitals exist; σ_{NH_2} or π_{NH_2} group orbitals, σ_{NH} or σ^*_{NH} bond orbitals, σ_{OH} or σ^*_{OH} bond orbitals. Their properties are identical to those of the corresponding orbitals involving carbon. We have also demonstrated how the properties of localized orbitals can be used to explain various phenomena such as the conformation of carbonium ions, the bond lengths in substituted cyclopropanes, the reactivity of cyclopropene analogs, etc. Finally we have considered the nature of lone-pair orbitals.

Many applications have not been dealt with. For instance, it now appears possible to construct, from elementary group orbitals, the

[a] In CH_3F the p-type lone-pair orbitals can actually occupy *any* set of mutually orthogonal directions in the plane perpendicular to the CF axis. But as soon as the C_{3v} symmetry is reduced to C_s symmetry (substitution by X at H_1) the "eclipsed" and "bisected" orientations become the only correct ones. Since the substitution allows the eclipsed and coaxial orbitals to mix, the eclipsed orbital will no longer be rigorously orthogonal to the CF bond; however it remains in the XCF plane. The coaxial lone-pair orbital also tilts in the plane.

important *frontier* orbitals of any small organic molecule. These frontier orbitals—the highest occupied molecular orbital and the lowest unoccupied molecular orbital—play a significant role in controlling the pathways of bimolecular reactions. It is also known that they are of paramount importance in determining optimal relaxation pathways for the nuclei in molecular distortions and unimolecular reactions. The *relaxability method* determines these pathways by seeking out a nuclear distortion which has the same symmetry as one of the lowest-lying electronic transitions, if possible that between frontier orbitals. Although a thorough study of the frontier method and of the relaxability method is beyond the scope of this Introduction, the frontier orbitals drawn out in Section III will provide the imaginative reader with a wealth of information and with strong interpretative power. He can, for example

1. choose to use the top occupied σ_{CC}-type and lowest unoccupied σ^*_{CC}-type semilocalized orbitals of cyclobutane to study the in-plane decomposition of this molecule;
2. compare the trans-rocking motion of the methyl groups in ethane with their cis-rocking by using the frontier π_{CH_3}-type and σ_{CC}-type orbitals in the molecule, and investigating whether either motion has a symmetry identical with a frontier transition;
3. check that the frontier orbitals of ethylene and butadiene interact favorably—highest occupied with lowest unoccupied and vice-versa—in the Diels-Alder addition, as shown many years ago by Fukui;
4. study the electrocyclic ring opening of cyclobutene by establishing the symmetry of the transition from highest occupied π_{CH_2}-type orbital to lowest unoccupied σ^*_{CC}-type orbital.

Another extremely important application, which deserves a book in its own right, is the construction of *correlation diagrams*. In these diagrams, the energetic variation of each orbital is obtained by joining its energy at the beginning, and at the end, of a given deformation or reaction. Orbitals of same symmetry are not allowed to cross. In the diagrams for the bending of a CH_2 group or for the pyramidalization of a CH_3 group, *the local nature of the molecular orbitals is conserved.*

Bond orbitals and group orbitals therefore seem to be a powerful tool for understanding organic phenomena. It is our hope that chemists, by using group orbitals to construct the molecular orbitals of organic systems, will thereby gain a better understanding of molecules and transition states.

24. Additional Reading

LOCALIZED ORBITALS, BOND ORBITALS

C. A. Coulson, *Trans. Far. Soc.*, **33,** 388 (1937).

J. Lennard-Jones, *Proc. Roy. Soc.*, **A 198,** 1, 14 (1949).

G. G. Hall, *Proc. Roy. Soc.*, **A 205,** 541 (1951).

G. G. Hall and J. Lennard-Jones, *Trans. Far. Soc.*, **48,** 581 (1952).

R. McWeeny, *Proc. Roy. Soc.*, **A 253,** 242 (1959); *Rev. Mod. Phys.*, **32,** 335 (1960).

C. Edmiston and K. Ruedenberg, *Rev. Mod. Phys.*, **35,** 4571 (1963); *J. Chem. Phys.*, **43,** S97 (1965).

R. Daudel, "Théorie Quantique de la Liaison Chimique." Presse Univ. France, Paris (1971).

O. J. Sovers, C. W. Kern, R. M. Pitzer, and M. Karplus, *J. Chem. Phys.*, **49,** 2592 (1968).

J. R. Hoyland, *J. Am. Chem. Soc.*, **90,** 2227 (1968); *J. Chem. Phys.*, **50,** 473 (1969).

I. T. Lyast, *Zh. Strukt. Khim.*, **12,** 1068 (1971).

H. A. Bent, *Topics in Current Chem.*, **14,** 1 (1970).

GROUP ORBITALS

(CH_2) L. Salem and J. S. Wright, *J. Am. Chem. Soc.*, **91,** 5947 (1969).

(CH_3) R. Hoffmann, *Pure and Applied Chem.*, **24,** 567 (1970).

RESULTS OF PERTURBATION THEORY

L. Pauling and E. B. Wilson, "Introduction to Quantum Mechanics." McGraw Hill, New York (1935).

S. R. La Paglia, "Introductory Quantum Chemistry." Harper & Row, New York (1971).

CONJUGATED MOLECULES

A. Streitwieser, Jr., "Molecular Orbital Theory for Organic Chemists." Wiley, New York (1961).

L. Salem, "The Molecular Orbital Theory of Conjugated Systems." W. A. Benjamin, New York (1966).

M. J. S. Dewar, "The Molecular Orbital Theory of Organic Chemistry." McGraw Hill, New York (1969).

C_n

A. Eschenmoser, P. Otto, and L. Salem, unpublished results.

ETHANE

R. Hoffmann, *Pure and Applied Chem.*, **24,** 567 (1970).

W. A. Lathan, L. A. Curtiss, and J. A. Pople, *Mol. Phys.*, **22,** 1081 (1971).

R. S. Mulliken, *J. Chem. Phys.*, **3,** 517 (1935).

W. E. Palke and W. N. Lipscomb, *J. Am. Chem. Soc.*, **88,** 2384 (1966).

W. H. Fink and L. C. Allen, *J. Chem. Phys.*, **46,** 2261 (1967).

W. L. Jorgensen and L. C. Allen, *J. Am. Chem. Soc.*, **93,** 567 (1971).

ORBITAL SYMMETRIES AND GROUP-THEORETICAL NOTATION

F. A. Cotton, "Chemical Applications of Group Theory." Wiley Interscience, 2nd Edition, Wiley, New York (1971).

WALSH ORBITALS

A. D. Walsh, *Nature*, **159,** 712 (1947); *Trans. Far. Soc.*, **45,** 179 (1949).

CYCLOPROPANE

H. Basch, M. B. Robin, N. A. Kuebler, C. Baker, and D. W. Turner, *J. Chem. Phys.*, **51,** 52 (1969).

R. M. Stevens, E. Switkes, E. A. Laws, and W. N. Lipscomb, *J. Am. Chem. Soc.*, **93,** 2603 (1971).

CYCLOBUTANE

L. Salem and J. S. Wright, *J. Am. Chem. Soc.*, **91,** 5947 (1969).

P. Bischof, E. Haselbach, and E. Heilbronner, *Angew. Chemie, Int. Ed. Engl.*, **9,** 953 (1970).

R. Hoffmann and R. B. Woodward, *J. Am. Chem. Soc.*, **87,** 2046 (1965).

THROUGH-BOND INTERACTIONS

R. Hoffmann, *Acc. Chem. Res.*, **4,** 1 (1971).

E. Heilbronner, *Pure and Applied Chem., Suppl.* (23rd Congress), **7,** 9 (1971).

LONE-PAIR INTERACTIONS

N. D. Epiotis, *J. Am. Chem. Soc.*, in press.

ETHYL CATION

W. A. Lathan, W. J. Hehre, and J. A. Pople, *J. Am. Chem. Soc.*, **93,** 808 (1971).

Hyperconjugation

R. Hoffmann, L. Radom, J. A. Pople, P. v.R. Schleyer, W. J. Hehre, and L. Salem, *J. Am. Chem. Soc.*, **94,** 6221 (1972).

n-Propyl Cation

L. Radom, J. A. Pople, V. Buss, and P. v.R. Schleyer, *J. Am. Chem. Soc.*, **93,** 1813 (1971).

Vinyl Cation

W. A. Lathan, W. J. Hehre, and J. A. Pople, *J. Am. Chem. Soc.*, **93,** 808 (1971).

Cyclopropylcarbinyl Cation

R. Hoffmann, *Pure and Applied Chem., Suppl.* (23rd Congress), **2,** 233 (1971).

R. Hoffmann and R. B. Davidson, *J. Am. Chem. Soc.*, **93,** 5699 (1971).

L. Radom, J. A. Pople, V. Buss, and P. v.R. Schleyer, *J. Am. Chem. Soc.*, **92,** 6380 (1970).

C. V. Pittmann and G. A. Olah, *J. Am. Chem. Soc.*, **87,** 2998 (1965).

Electronegativity Perturbation

R. Hoffmann, Lectures presented at the ETH, Zurich (1971).

Methylenecyclopropane, Cyclopropanone

V. W. Laurie and W. M. Stigliani, *J. Am. Chem. Soc.*, **92,** 1485 (1970).

J. M. Pochan, J. E. Baldwin, and W. H. Flygare, *J. Am. Chem. Soc.*, **91,** 1896 (1969).

DONATION FROM LONE-PAIRS INTO ADJACENT ANTIBONDING BOND ORBITALS, ANOMERIC EFFECT

C. Romers, C. Altona, H. R. Buys, and E. Havinga, *Topics in Stereochemistry*, **4**, 39 (1967), in particular pp. 76-77.

S. David, O. Eisenstein, W. J. Hehre, L. Salem, and R. Hoffmann, unpublished results.

P. Linscheid and E. A. C. Lucken, *Chem. Comm.*, 425 (1970).

L. Radom, W. J. Hehre, and J. A. Pople, *J. Am. Chem. Soc.*, **94**, 2371 (1972).

CYCLOPROPENE, DIAZIRINE

H. M. Frey and I. D. R. Stevens, *J. Chem. Soc.*, 3865 (1962).

H. M. Frey and A. W. Scaplehorn, *J. Chem. Soc.* **A**, 968 (1966).

K. B. Wiberg and W. J. Bartley, *J. Am. Chem. Soc.*, **82**, 6375 (1960).

T. Terao, N. Sakai, and S. Shida, *J. Am. Chem. Soc.*, **85**, 3919 (1963).

LONE-PAIRS, WATER

J. A. Pople, *Proc. Roy. Soc.*, **A202**, 329 (1950).

S. R. La Paglia, "Introductory Quantum Chemistry," pp. 277-287. Harper & Row, New York (1971).

S. Aung, R. M. Pitzer, and S. I. Chan, *J. Chem. Phys.*, **49**, 2071 (1968).

C. R. Brundle and D. W. Turner, *Proc. Roy. Soc.*, **A307**, 27 (1968).

LOCALIZATION, DELOCALIZATION

G. Berthier *In* "Aspects de la Chimie Quantique Contemporaine." (Editions du C.N.R.S., Paris, 1971), p. 49—see also the discussion, pp. 67–86.

Frontier Orbitals

K. Fukui, *Topics in Current Chem.*, **15,** 1 (1970).

L. Salem, *Chem. Brit.*, **5,** 449 (1969).

G. Klopman, *J. Am. Chem. Soc.*, **90,** 223 (1968).

R. F. Hudson, *Angew. Chemie, Int. Ed. Engl.*, **12,** 36 (1973).

Correlation Diagrams

A. D. Walsh, *J. Chem. Soc.*, 2260, 2266 (1953).

B. M. Gimarc, *J. Am. Chem. Soc.*, **93,** 593 (1971).

S. D. Peyerimhoff, R. J. Buenker, and L. C. Allen, *J. Chem. Phys.*, **45,** 734 (1966).

R. B. Woodward and R. Hoffmann, "The Conservation of Orbital Symmetry." Academic Press, New York (1970).

Relaxability Method

R. F. W. Bader, *Mol. Phys.*, **3,** 137 (1960); *Can. J. Chem.*, **40,** 1164 (1962).

R. Pearson, *Acc. Chem. Res.*, **4,** 152 (1971).

L. Salem, *Chem. Phys. Lett.*, **3,** 99 (1969).

L. S. Bartell, *J. Chem. Ed.*, **45,** 754 (1968).

J. K. Burdett, *J. Chem. Soc.* **A,** 1195 (1971).

II. Basic Data Concerning the Orbital Drawings in Chapter III

1. Notation

Below the graphic illustration of each molecular orbital the reader will find:

1. the group-theoretical symmetry notation of the orbital;
2. the orbital energy, calculated preferably by an *ab initio* method or alternatively by a semiempirical method (the relevant calculation is quoted in Section IV.1). The energy is quoted in atomic units (1 a.u. = 27.21 eV = 627 kcal/mole);
3. often, following the numerical energy, the *orbital type*. By orbital type we mean the nature of the localized orbital or orbitals which give the major contributions to the molecular orbital.

Several cases can be distinguished for the orbital-type label. Only group orbitals of a single type—for instance σ_{CH_2} or π_{CH_3}—may contribute significantly to the molecular orbital. This occurs either because orbitals of this type have a unique symmetry (π_{CC}, π'_{CH_2} in ethylene, π_{CH_2} in cyclopropane or cyclobutane) or because they are energetically well separated from other group-orbital types with the same symmetry. Then the notation is simple. The molecular orbital has the label of the particular group orbital: π_{CC} for the ordinary π orbital of ethylene, π'_{CH_2} for the in-plane molecular π orbitals of ethylene (III.18), π_{CH_2} for the four π molecular orbitals of cyclobutane (III.82), etc.

Sometimes several different types of localized group orbitals contribute significantly to the molecular orbital. The molecular orbital is then labeled by listing all the different types of group orbitals which have a large amplitude in the overall molecular orbital. Thus the lowest molecular orbital of ethylene is labeled σ_{CC}, σ_{CH_2} indicating the admixture of two different localized types. This orbital is the in-phase combination of σ_{CC} and of the bonding σ_{CH_2} combination (see I.8). The out-of-phase combination is similarly labeled σ_{CH_2}, σ_{CC}. If one type dominates, the other is written *in parentheses*: n, (σ_{CH_3}) for the half-filled orbital in pyramidal methyl radical (III.6). In methylenimine (III.19) three of the σ orbitals are built from the σ_{NH} bond orbital, the σ_{CN} bond orbital and the σ_{CH_2} group orbital. However, the lower one (3A′) is mainly localized on the CN and NH bonds and is labeled σ_{CN}, σ_{NH}. The middle one (4A′) is located at the ends and is labeled σ_{CH_2}, σ_{NH}. Finally the higher one (6A′) involves all three components and is labeled σ for the sake of simplicity. The manner in which the localized components combine to form the three delocalized combinations is very similar to the construction of the three π molecular orbitals of an allylic skeleton (see I.9).

The same notation applies to lone pair orbitals whose principal component is labeled n and to the unique 2p orbitals in some radicals and cations, e.g. methyl and ethyl, where the primary label is P.

As the molecule grows larger, the number of different localized orbital types which can enter the formation of the delocalized molecular orbitals increases and it may become difficult to assign dominant types. In propylene for instance (III.49), the σ orbitals, 4A′, 5A′, 6A′, 9A′ and 10A′ are built from two σ_{CC} bond orbitals, one σ_{CH} bond orbital, one σ_{CH_2} group orbital and one σ_{CH_3} group orbital. (The in-plane π'_{CH_2} and π'_{CH_3} group orbitals can also mix in, since there is no vertical plane of symmetry, but they separate fairly well and are concentrated in the 7A′ and 8A′ molecular orbitals.) The dominant character of 4A′ (σ_{CC}), 5A′ (σ_{CH_3}, σ_{CH_2}) and 6A′ (σ_{CH}) is relatively well defined, but 9A′ and 10A′ are really mixtures of all four types. We label them simply σ.

Mixed labeling involving both π' and σ orbitals occurs in certain molecules: the $5B_u$ molecular orbitals of *trans*-2-butene (III.78) and transoid 1,3-butadiene (III.65) are labeled π'_{CH_3}, (σ_{CC}) and π'_{CH_2}, (σ_{CC}) because one lobe of the π' orbital overlaps well with the adjacent CC bond-orbital to form a delocalized combination. In cisoid acrolein, orbitals 9A′ and 10A′ are labeled π'_{CH_2}, σ_{CH} because the nodal surfaces of the two localized orbitals coincide and allow for a delocalized combination (III.68).

In molecules with little or no symmetry, it may still be possible to recognize the main localized-orbital component of certain molecular orbitals. It is then convenient to adopt the label of this localized type as the label of the molecular orbital, even though the molecular symmetry does not coincide with the local symmetry. For instance, in methylenimine again, the 5A′ orbital is clearly built out of the in-plane π_{CH_2} group orbital, with a small NH component. We therefore label the orbital π'_{CH_2}, although the molecule does not have a vertical symmetry plane. Similarly, the orbitals 7A′ and 8A′ of propylene are labeled π'_{CH_3}, π'_{CH_2} (III.49).[a] Other examples where the local symmetry is sufficiently preserved and only weakly perturbed by the molecular environment are hydrazine (III.34) and methylamine (III.31). In some cases we have omitted the label as no unambiguous classification is possible.

2. Computational Details

This section provides explicit information concerning the construction of the molecular orbital drawings which comprise Chapter III. It is not essential for the appreciation and utilization of the drawings, but is included for completeness and for the theoretical chemists who would require such details.

The primary concern in producing the drawings was the selection of wavefunctions, orbital energies and of a contour level for the surfaces. Ideally, wavefunctions and orbital energies from accurate *ab initio* molecular orbital calculations would be employed. Since *ab initio* wavefunctions for complex molecules are rarely recorded in the literature and since we had to treat a large number of systems, it was necessary to use wavefunctions calculated from semiempirical molecular orbital theories. Several of these methods produce orbital coefficients and charge distributions in good agreement with *ab initio* values. In contrast, *ab initio* orbital energies for at least the occupied molecular orbitals may often be found in the literature. These values have been quoted whenever possible. In Section IV.1 references for

[a] The reason for the weak mixing of π' and σ orbitals in propylene is that there is a weak "pseudo" vertical symmetry plane in the molecule. It is the plane which would have existed if the carbon skeleton were linear, with no central hydrogen atom. One can also visualize its existence by joining the two *local* vertical symmetry planes of the CH_2 and CH_3 groups.

the orbital energies of each molecule are given. Preference was given to *ab initio* calculations using extended basis sets and to calculations performed on a series of molecules, for example, the molecules isoelectronic with ethane. Caution must always be exercised when comparing orbital energies from calculations with different basis sets. When *ab initio* values were not available, MINDO/2 orbital energies have been cited as they are comparably reliable.

For ions and for uncharged molecules with singlet ground states, Extended Hückel[1] wavefunctions were used. For radicals, the wavefunctions were provided by an open-shell calculation using the MINDO/2 method in an unrestricted, single-determinant approach. To compensate for the neglect of differential overlap, the MINDO/2 wavefunctions were orthonormalized with the overlap matrix of the molecular orbitals in accordance with a method proposed by Löwdin. This transformation is more successful in preserving the shapes of the orbitals than the more familiar Löwdin transformation which back-transforms the wavefunction to a basis set of orthogonal atomic orbitals. Dr. R. B. Davidson of Yale University suggested the procedure and kindly provided a computer program to execute it.

The Extended Hückel calculations employed Hoffmann's[2] most recent carbon and hydrogen parameters. The boron parameters were also due to Hoffmann,[3] while the nitrogen, oxygen, and fluorine parameters were chosen to reproduce charge distributions and population analyses of several *ab initio* calculations. The MINDO/2 method and parameters were those of Dewar and Lo[4] excluding the use of the half-electron method for the open-shell calculations. A comparison of several two-dimensional orbital maps for semiempirical and *ab initio* wavefunctions is presented on the following page. The agreement for occupied orbitals is generally excellent. However, considerable disagreement between semiempirical and *ab initio*, and between *ab initio* calculations with different basis sets, can occur in the shapes and energies of unoccupied orbitals except when the coefficients are constrained by symmetry, for example in pure π orbitals.

For molecular systems with up to thirty valence electrons, an amplitude of ± 0.1 a.u. was chosen for the contour level. For systems with more than thirty valence electrons it was necessary to reduce this value to 0.08 a.u. to maintain the orbital size at a comfortable visual level. The molecular orbitals were normalized to an occupancy

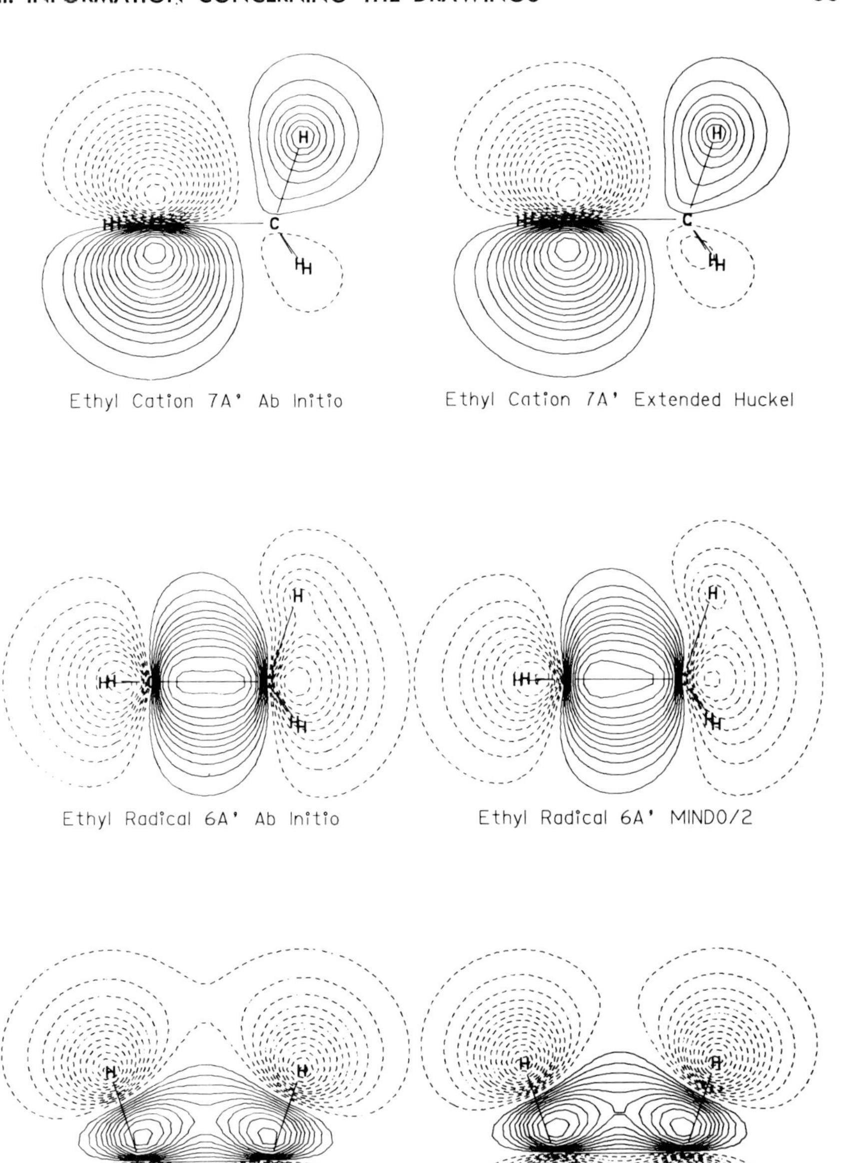

Ethyl Cation 7A' Ab Initio

Ethyl Cation 7A' Extended Huckel

Ethyl Radical 6A' Ab Initio

Ethyl Radical 6A' MINDO/2

Eclipsed Ethane 2E' Ab Initio

Eclipsed Ethane 2E' Extended Huckel

of one electron, so the contour levels correspond to charge densities per electron of 0.01 a.u. and 0.0064 a.u., respectively (1 a.u. = $1\ e^-/a_o^3 = 6.75\ e^-/\text{Å}^3$). The sections or slices through the molecular orbitals were taken, in all cases, at intervals of 0.2 angstroms and were spaced to coincide with any symmetry planes. The existence of symmetry planes should always be kept in mind when viewing the drawings. Full contours correspond to a positive orbital amplitude, dotted contours to a negative orbital amplitude.

References for the molecular geometries used in generating the drawings are given in Section IV.1. Effort was made to make these as up-to-date as possible. For this reason the geometry references are not always the same as those used to determine the *ab initio* orbital energies.

The drawings were produced using the computers and substantial array of graphical output devices at the Center for Research in Computing Technology at Harvard University.[5]

References

1. R. Hoffmann, *J. Chem. Phys.*, **39,** 1397 (1963).
2. R. Hoffmann, S. Swaminathan, B. G. Odell and R. Gleiter, *J. Amer. Chem. Soc.*, **92,** 7091 (1970).
3. R. Hoffmann, *J. Chem. Phys.*, **40,** 2474 (1964).
4. M. J. S. Dewar and D. H. Lo, *J. Amer. Chem. Soc.*, **94,** 5296 (1972).
5. The computer programs were written by W. L. Jorgensen and inquiries concerning them should be directed to him.

III. Three-Dimensional Molecular Orbitals

Listing of Molecules and Orbital Occupancies

Molecule name	No. valence electrons	No. filled orbitals	No. half-filled orbitals
1. Hydrogen	2	1	0
2. Triplet methylene	6	2	2
3. Singlet methylene	6	3	0
4. Amino cation (triplet)	6	2	2
5. Planar methyl radical	7	3	1
6. Pyramidal methyl radical	7	3	1
7. Methane	8	4	0
8. Ammonia	8	4	0
9. Water	8	4	0
10. Hydrogen fluoride	8	4	0
11. Protonated methane	8	4	0
12. Acetylene	10	5	0
13. Vinyl cation	10	5	0
14. Hydrogen cyanide	10	5	0
15. Carbon monoxide	10	5	0
16. Nitrogen	10	5	0
17. Nitric oxide	11	5	1
18. Ethylene	12	6	0
19. Methylenimine	12	6	0
20. Formaldehyde	12	6	0
21. Diimide	12	6	0
22. Diborane	12	6	0

Listing of Molecules (*Continued*)

Molecule name	No. valence electrons	No. filled orbitals	No. half-filled orbitals
23. Oxygen (triplet)	12	5	2
24. Ethyl cation, bisected	12	6	0
25. Ethyl cation, eclipsed	12	6	0
26. Ethyl cation, bridged	12	6	0
27. Ethyl radical, bisected	13	6	1
28. Ethyl radical, eclipsed	13	6	1
29. Ethane, staggered	14	7	0
30. Ethane, eclipsed	14	7	0
31. Methylamine	14	7	0
32. Methanol	14	7	0
33. Methyl fluoride	14	7	0
34. Hydrazine	14	7	0
35. Hydrogen peroxide	14	7	0
36. Fluorine	14	7	0
37. Cyclopropenium cation	14	7	0
38. Methyl acetylene	16	8	0
39. Acetonitrile	16	8	0
40. Methyl isocyanide	16	8	0
41. Allene	16	8	0
42. Ketene	16	8	0
43. Diazomethane	16	8	0
44. Carbodiimide	16	8	0
45. Carbon dioxide	16	8	0
46. Cyclopropene	16	8	0
47. Diazirine	16	8	0
48. Allyl cation	16	8	0
49. Propylene	18	9	0
50. Acetaldehyde	18	9	0
51. Formamide	18	9	0
52. Formic acid	18	9	0
53. Formyl fluoride	18	9	0
54. Nitrosomethane	18	9	0

Listing of Molecules (*Continued*)

Molecule name	No. valence electrons	No. filled orbitals	No. half-filled orbitals
55. Ozone	18	9	0
56. Cyclopropane	18	9	0
57. Aziridine	18	9	0
58. Ethylene oxide	18	9	0
59. Trimethylene, edge-to-edge	18	8	2
60. *n*-Propyl cation, bisected	18	9	0
61. Propane	20	10	0
62. Dimethylether	20	10	0
63. Ethyl fluoride	20	10	0
64. Cyclobutadiene (rectangular singlet)	20	10	0
65. 1,3-Butadiene, transoid	22	11	0
66. 1,3-Butadiene, cisoid	22	11	0
67. Acrolein, transoid	22	11	0
68. Acrolein, cisoid	22	11	0
69. Glyoxal, transoid	22	11	0
70. Glyoxal, cisoid	22	11	0
71. Methylazide	22	11	0
72. Methylenecyclopropane	22	11	0
73. Cyclopropanone	22	11	0
74. Cyclobutene	22	11	0
75. Bicyclobutane	22	11	0
76. Cyclopropylcarbinyl cation, bisected	22	11	0
77. Cyclopropylcarbinyl cation, perpendicular	22	11	0
78. *trans*-2-butene	24	12	0
79. Acetone	24	12	0
80. Isopropenol	24	12	0
81. Nitromethane	24	12	0
82. Cyclobutane, planar	24	12	0

Listing of Molecules (*Continued*)

Molecule name	No. valence electrons	No. filled orbitals	No. half-filled orbitals
83. Cyclopentadiene	26	13	0
84. (2.1.0)-Bicyclopentene-2	26	13	0
85. Pyrrole	26	13	0
86. Furan	26	13	0
87. Cyclopentadienyl anion	26	13	0
88. Pentadienyl radical*	27	13	1
89. Cyclopentene	28	14	0
90. (1.1.1)-Bicyclopentane	28	14	0
91. Spiropentane	28	14	0
92. *para*-Benzyne	28	14	0
93. Cyclopentane	30	15	0
94. Benzene	30	15	0
95. Dewar benzene	30	15	0
96. Pyridine	30	15	0
97. Pyrazine	30	15	0
98. Cyclopentadienone*	30	15	0
99. 1,3,5-Hexatriene*	32	16	0
100. (2.1.1)-Bicyclohexene-2	32	16	0
101. Cyclohexene, half–boat	34	17	0
102. Cyclohexane, chair	36	18	0
103. Norbornadiene	36	18	0
104. Maleic anhydride*	36	18	0

* Only π orbitals drawn.

1. Hydrogen

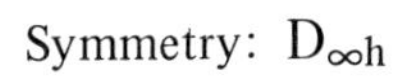

Symmetry: $D_{\infty h}$

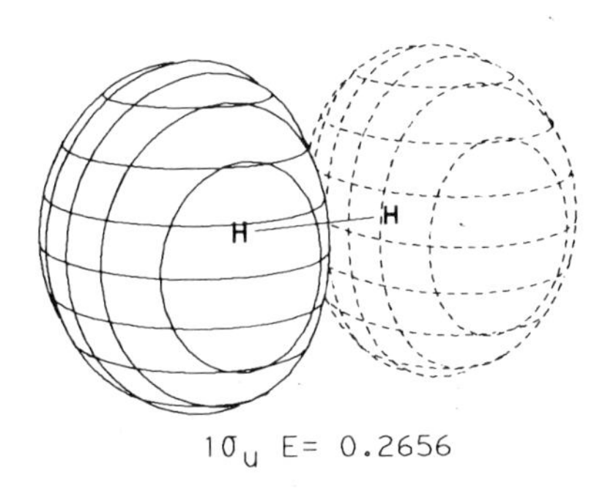

$1\sigma_u$ E= 0.2656

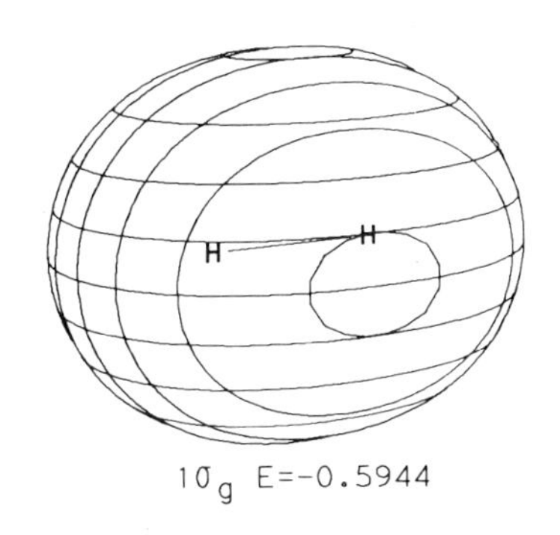

$1\sigma_g$ E=-0.5944

2. Triplet Methylene

Symmetry: C_{2v}

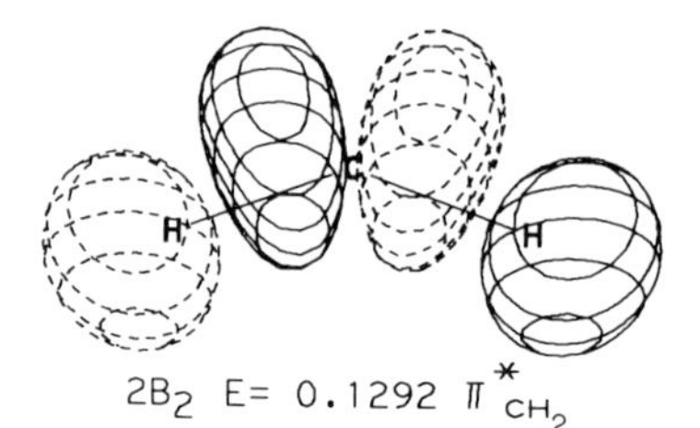

$2B_2$ E= 0.1292 $\pi^*_{CH_2}$

$4A_1$ E= 0.0497 $\sigma^*_{CH_2}$

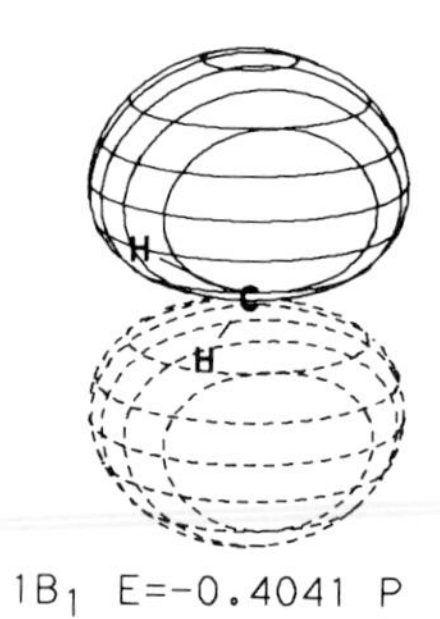

$1B_1$ E=-0.4041 P

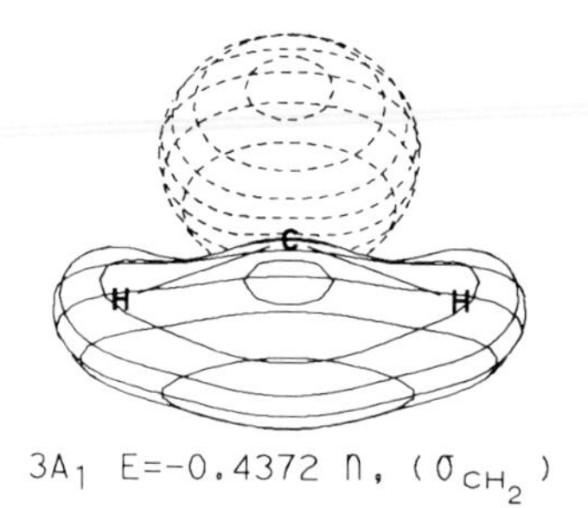

$3A_1$ E=-0.4372 n, (σ_{CH_2})

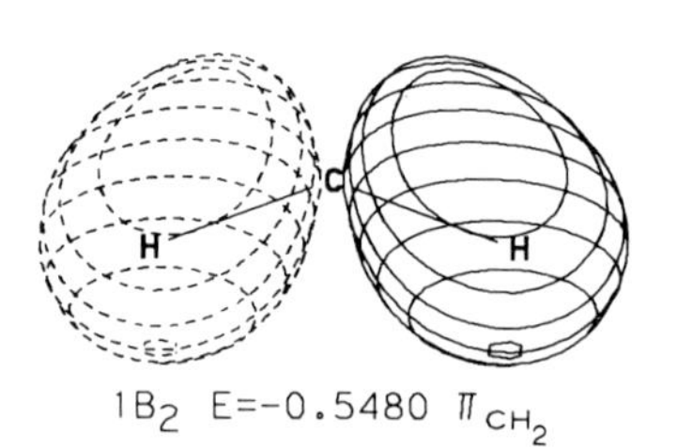

$1B_2$ E=-0.5480 π_{CH_2}

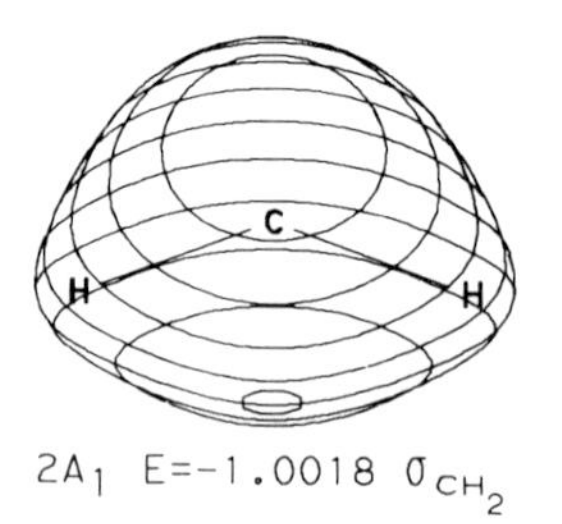

$2A_1$ E=-1.0018 σ_{CH_2}

3. Singlet Methylene

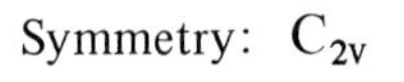
Symmetry: C_{2v}

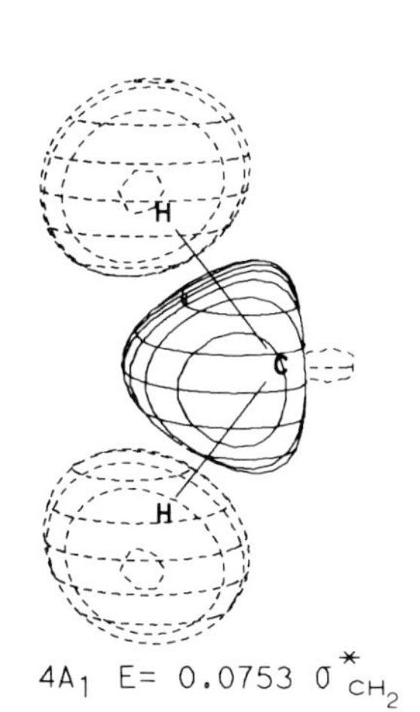

$4A_1$ E= 0.0753 $\sigma^*_{CH_2}$

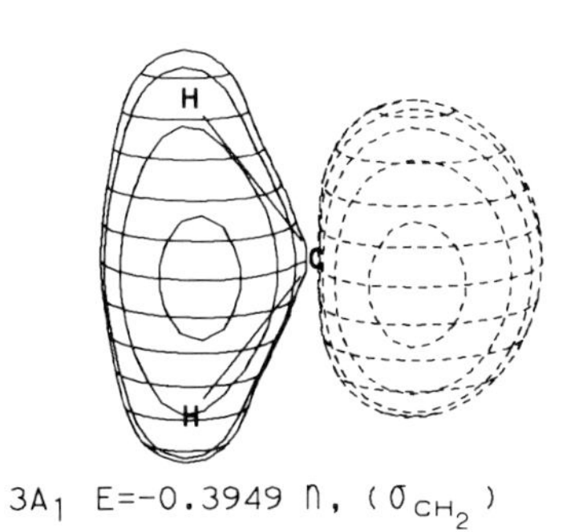

$3A_1$ E=-0.3949 n, (σ_{CH_2})

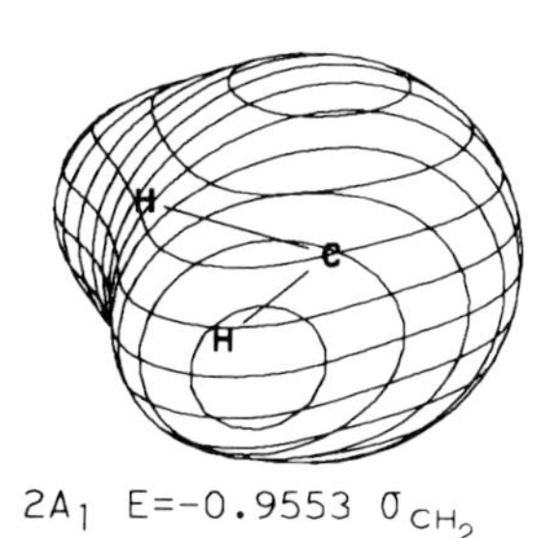

$2A_1$ E=-0.9553 σ_{CH_2}

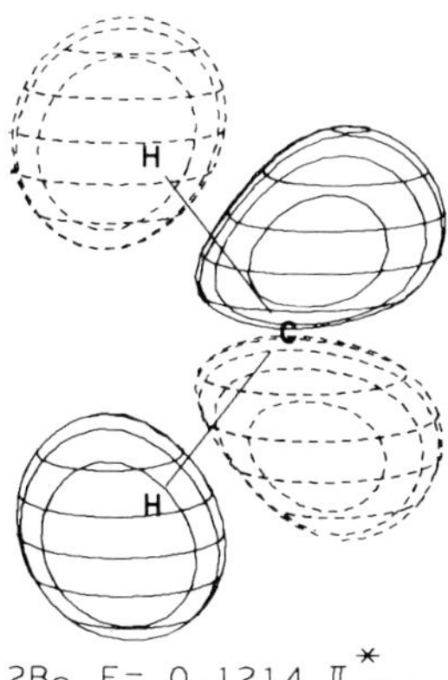

$2B_2$ E= 0.1214 $\pi^*_{CH_2}$

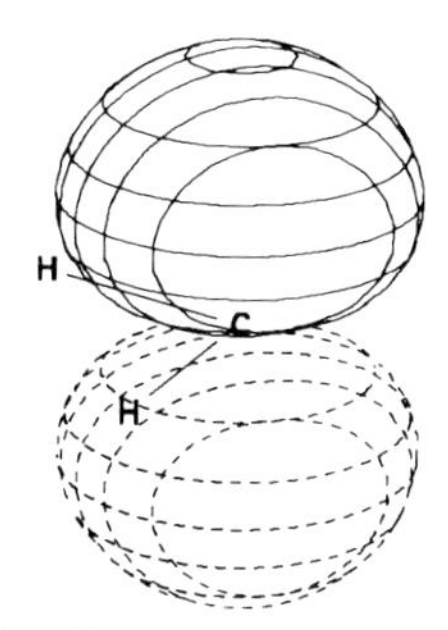

$1B_1$ E=-0.0369 P

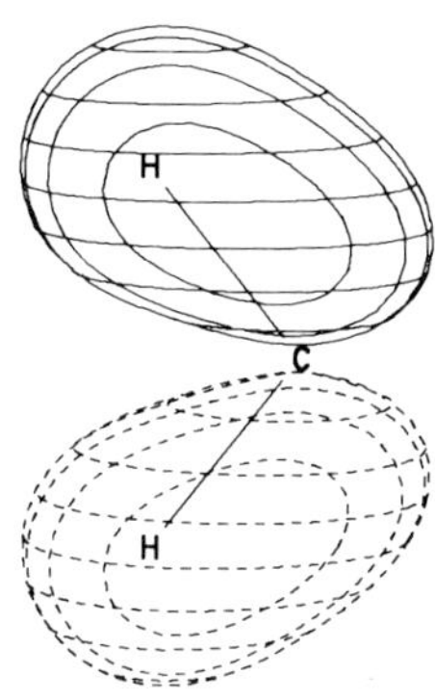

$1B_2$ E=-0.4860 π_{CH_2}

4. Amino Cation (Triplet)

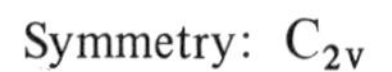
Symmetry: C_{2v}

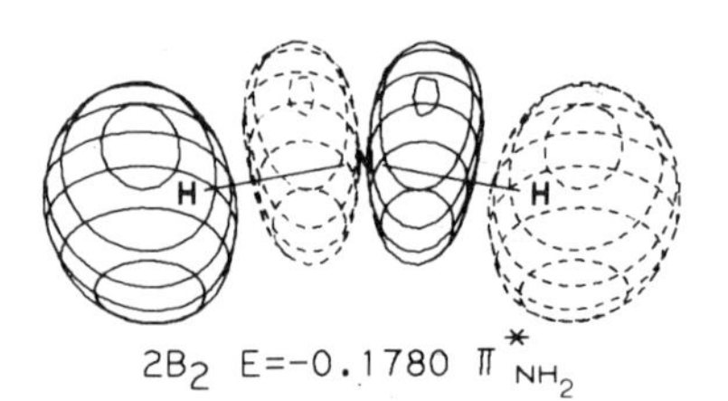

$2B_2$ E=-0.1780 $\pi^*_{NH_2}$

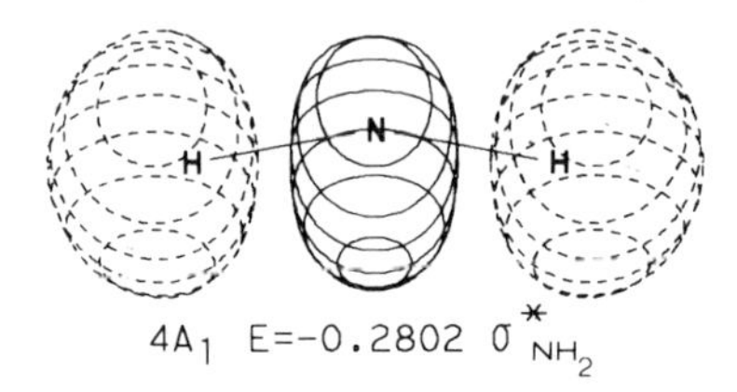

$4A_1$ E=-0.2802 $\sigma^*_{NH_2}$

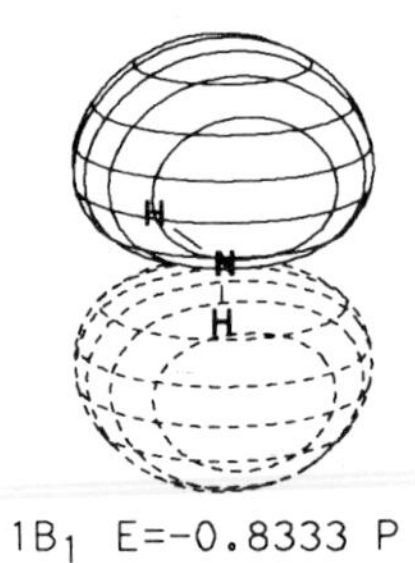

$1B_1$ E=-0.8333 P

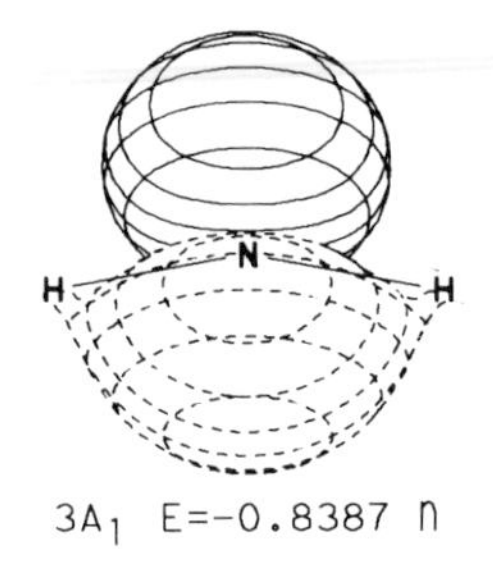

$3A_1$ E=-0.8387 n

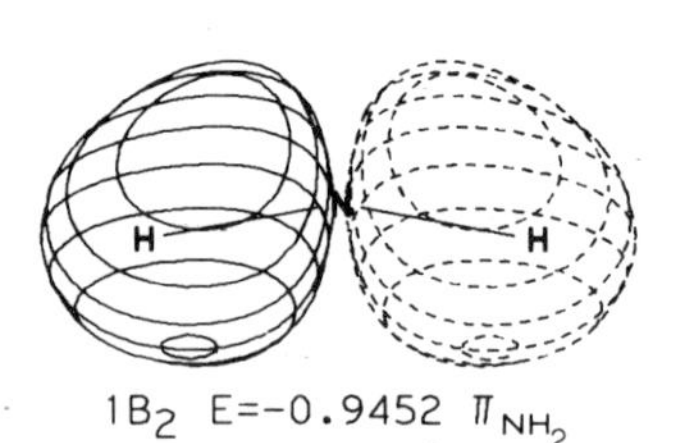

$1B_2$ E=-0.9452 π_{NH_2}

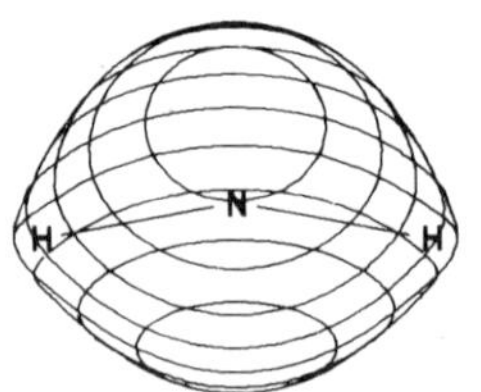

$2A_1$ E=-1.5890 σ_{NH_2}

5. Planar Methyl Radical

Symmetry: D_{3h}

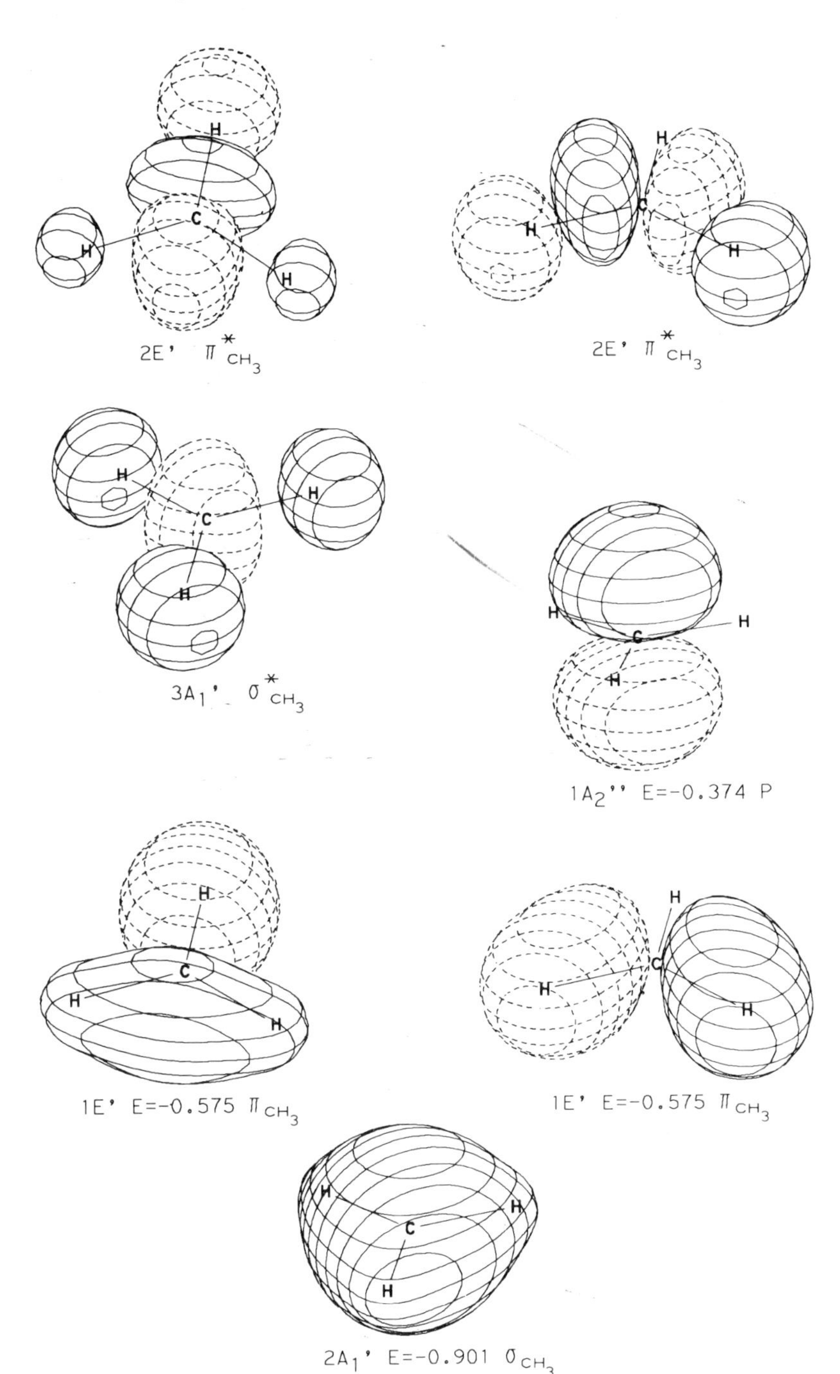

6. Pyramidal Methyl Radical

Symmetry: C_{3v}

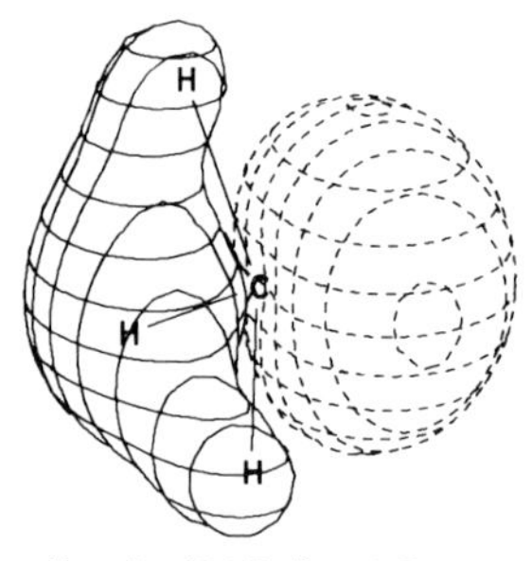

$3A_1$ E=-0.4315 n, (σ_{CH_3})

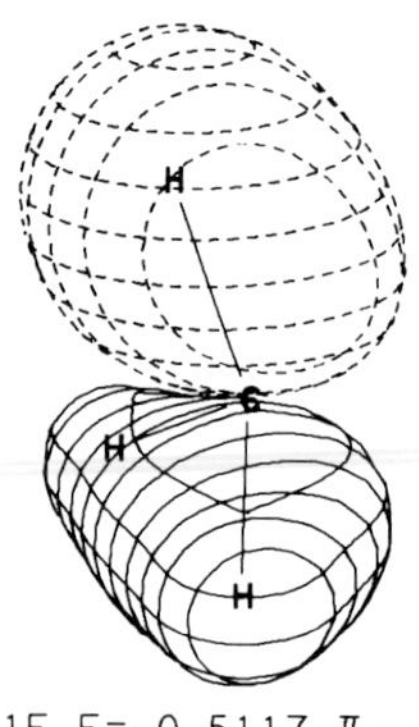

1E E=-0.5117 π_{CH_3}

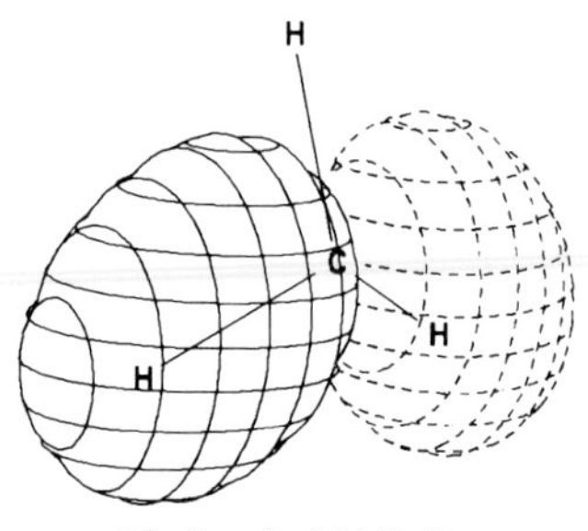

1E E=-0.5117 π_{CH_3}

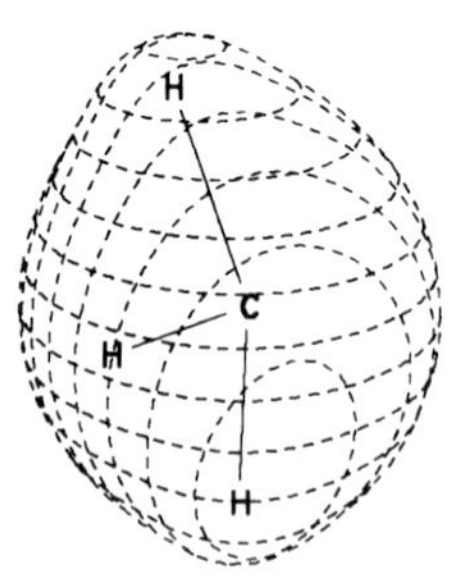

$2A_1$ E=-1.0653 σ_{CH_3}

Pyramidal Methyl Radical (Continued)

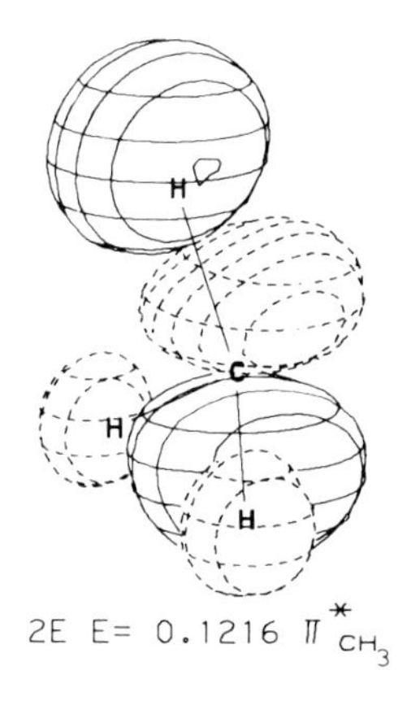

2E E= 0.1216 $\pi^{*}_{CH_3}$

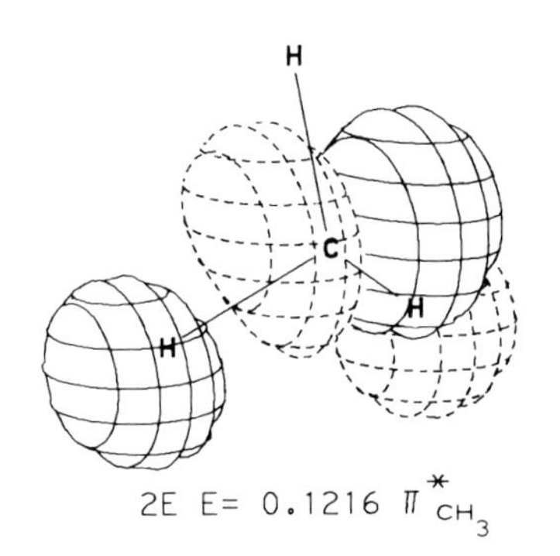

2E E= 0.1216 $\pi^{*}_{CH_3}$

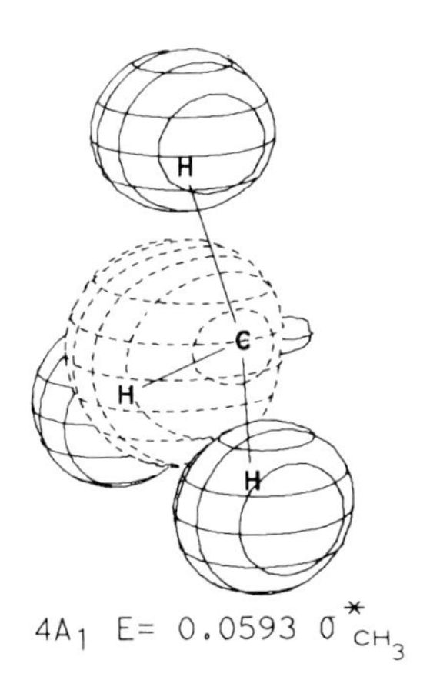

$4A_1$ E= 0.0593 $\sigma^{*}_{CH_3}$

7. Methane

Symmetry: T_d

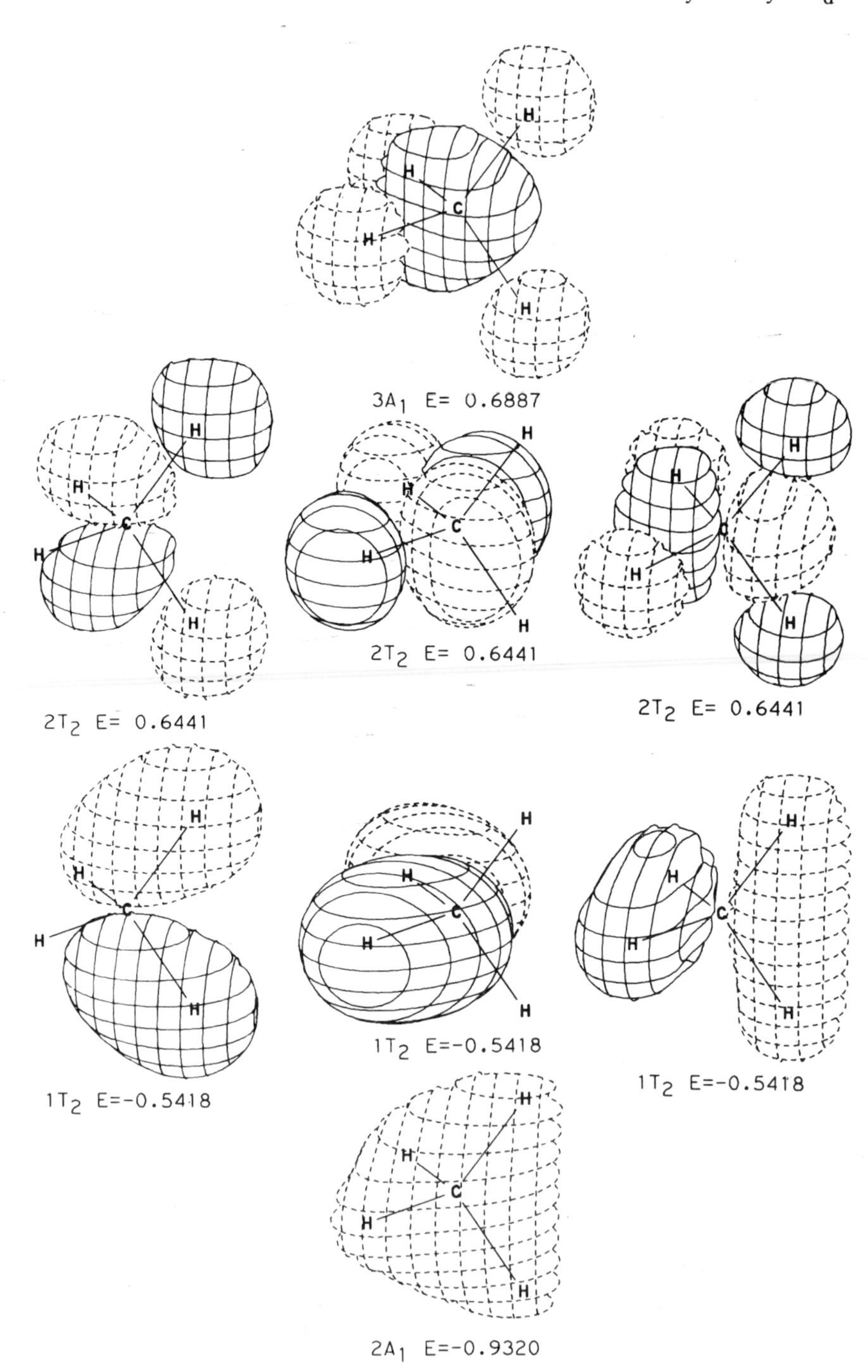

8. Ammonia

Symmetry: C_{3v}

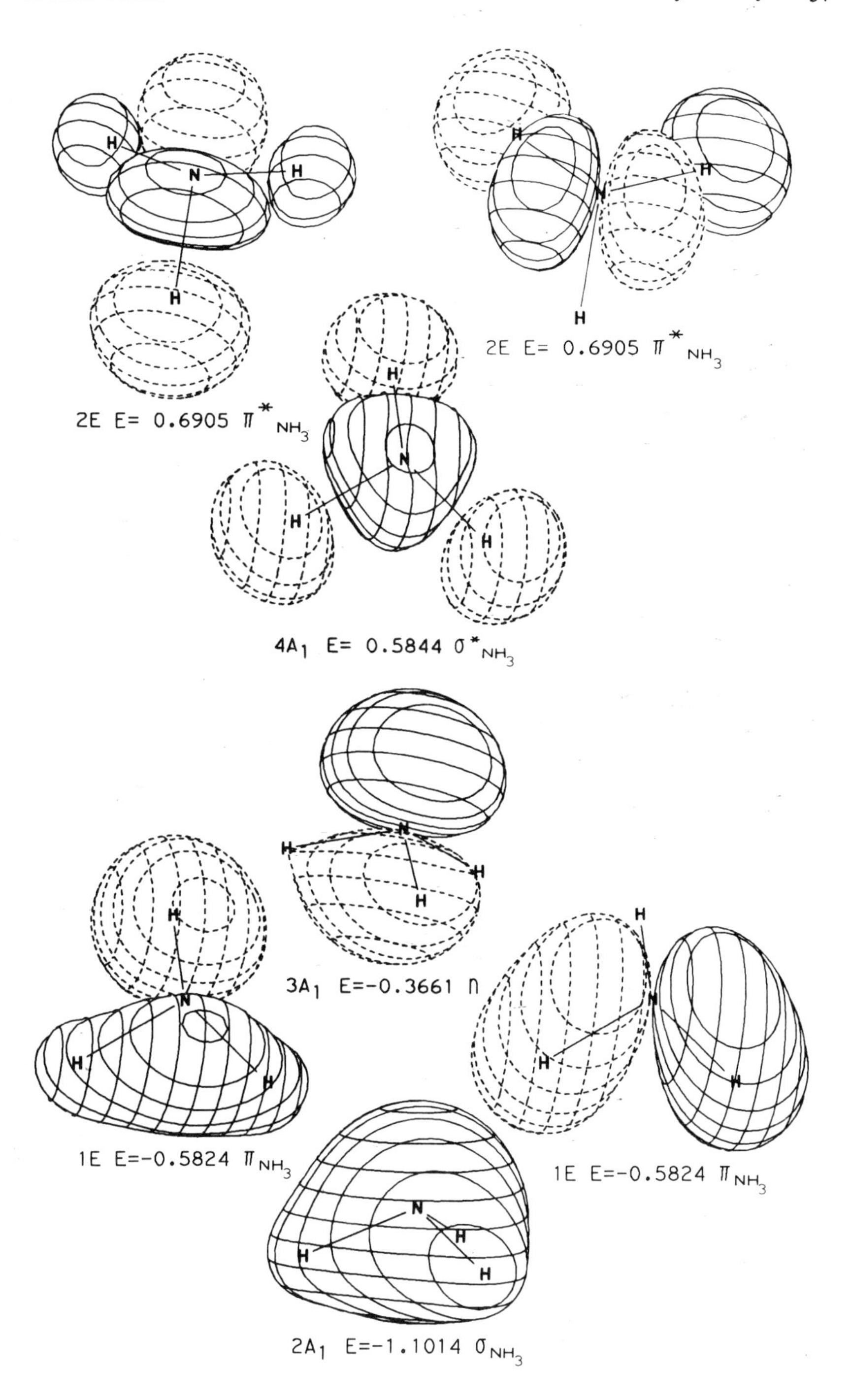

2E E= 0.6905 $\pi^*_{NH_3}$

2E E= 0.6905 $\pi^*_{NH_3}$

$4A_1$ E= 0.5844 $\sigma^*_{NH_3}$

$3A_1$ E=-0.3661 n

1E E=-0.5824 π_{NH_3}

1E E=-0.5824 π_{NH_3}

$2A_1$ E=-1.1014 σ_{NH_3}

9. Water

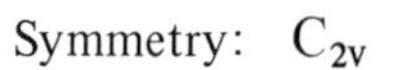
Symmetry: C_{2v}

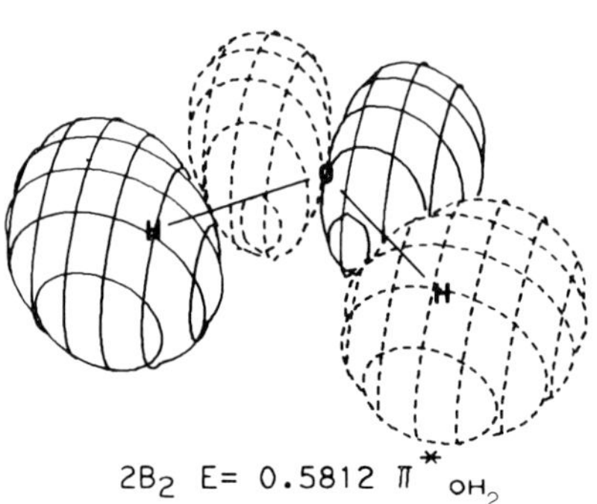

2B$_2$ E= 0.5812 $\pi^*_{OH_2}$

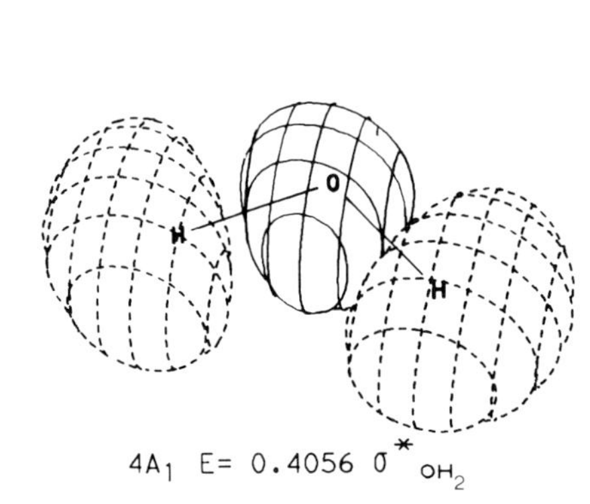

4A$_1$ E= 0.4056 $\sigma^*_{OH_2}$

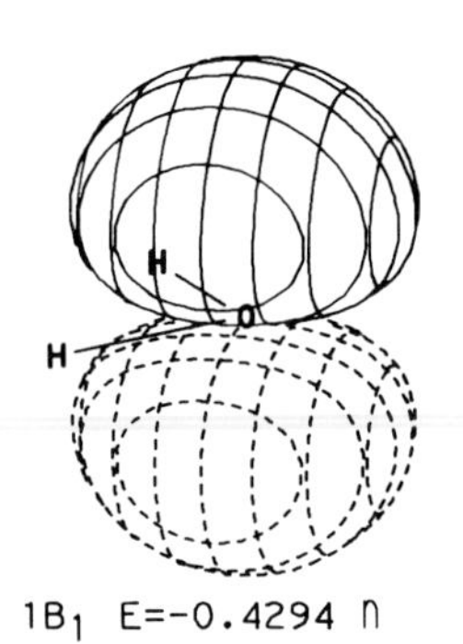

1B$_1$ E=−0.4294 n

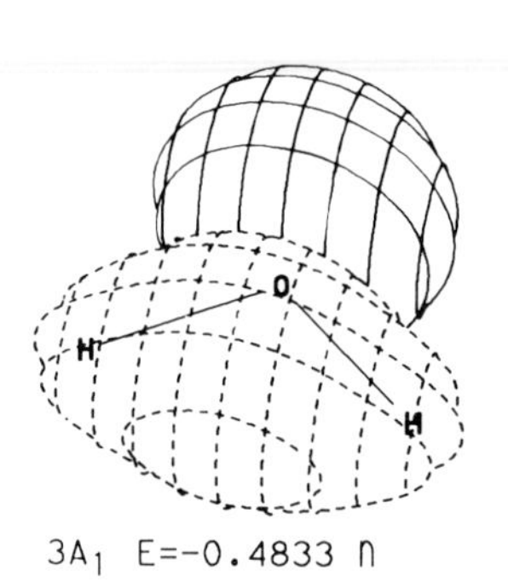

3A$_1$ E=−0.4833 n

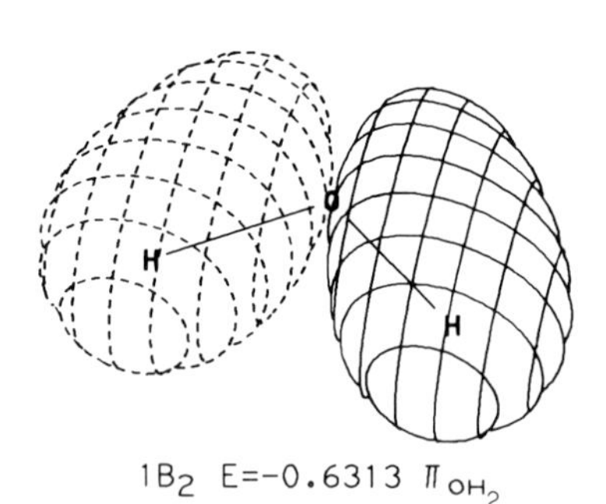

1B$_2$ E=−0.6313 π_{OH_2}

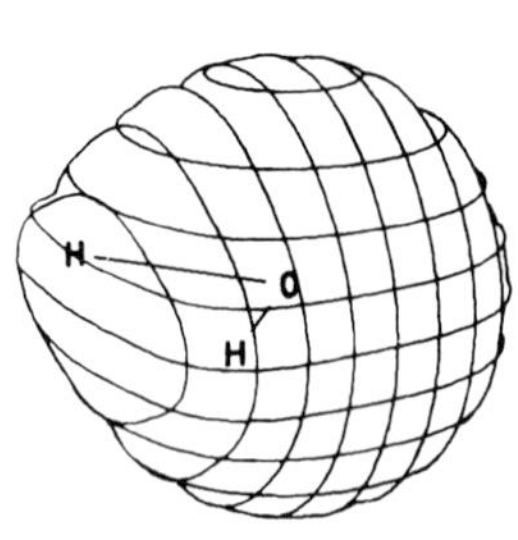

2A$_1$ E=−1.3049 σ_{OH_2}

10. Hydrogen Fluoride

Symmetry: $C_{\infty v}$

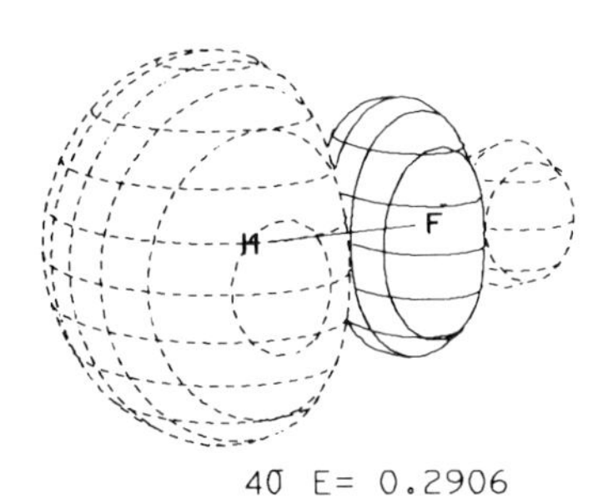

4σ E= 0.2906

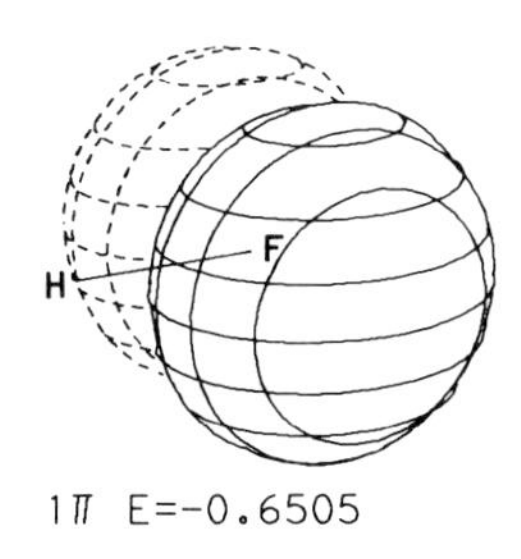

1π E=-0.6505

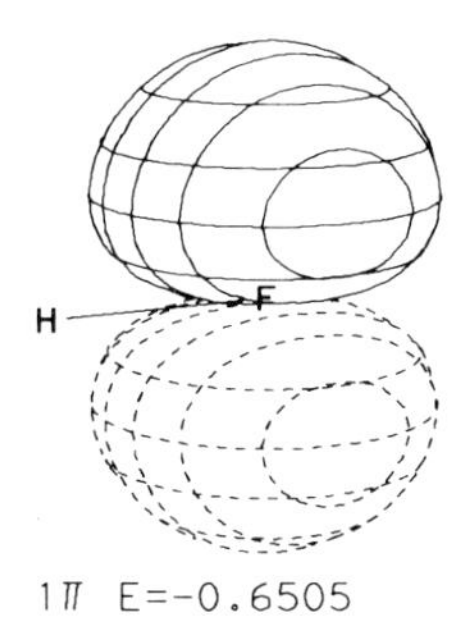

1π E=-0.6505

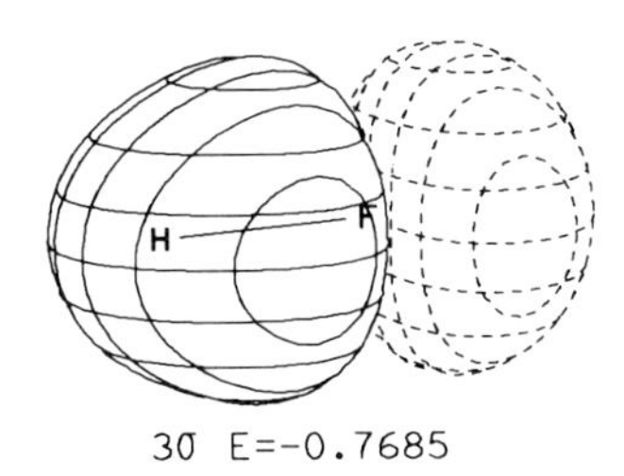

3σ E=-0.7685

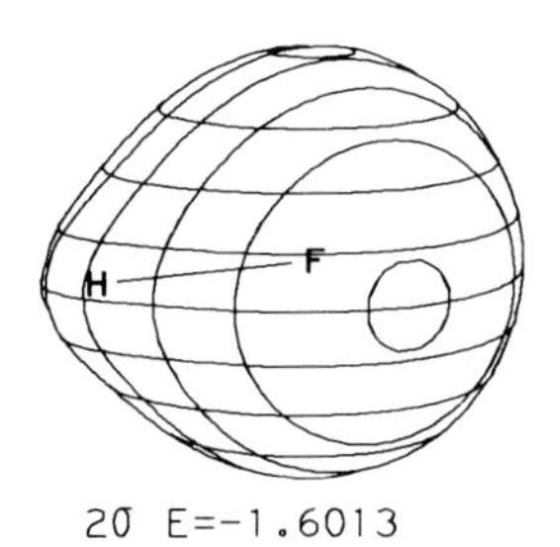

2σ E=-1.6013

11. Protonated Methane

Symmetry: C_s

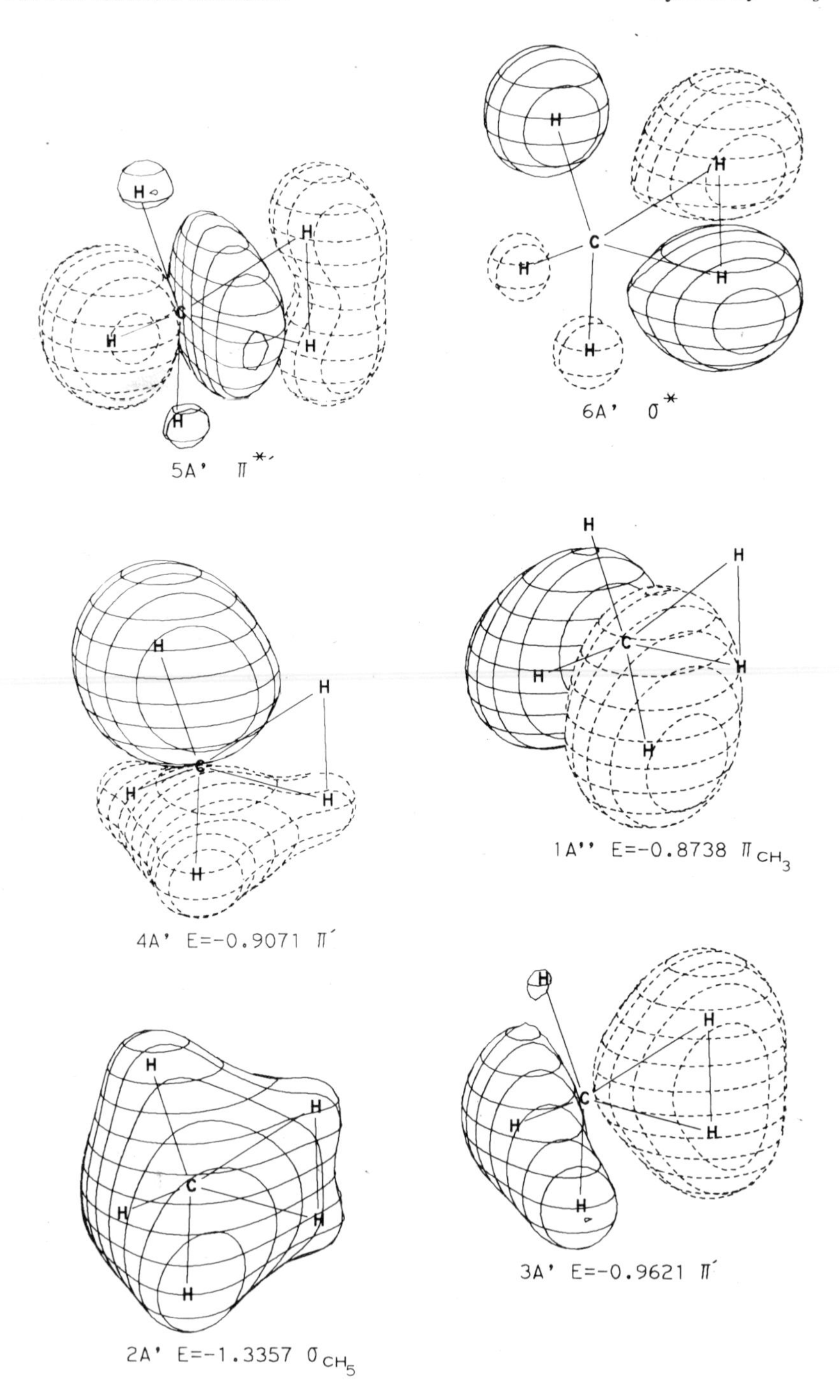

12. Acetylene

Symmetry: $D_{\infty h}$

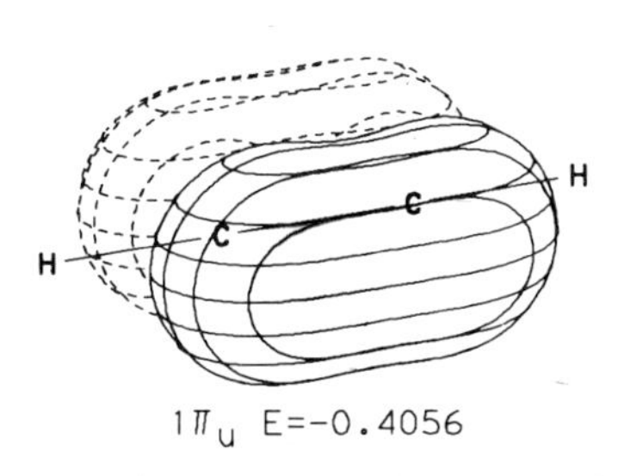

$1\pi_u$ E=-0.4056

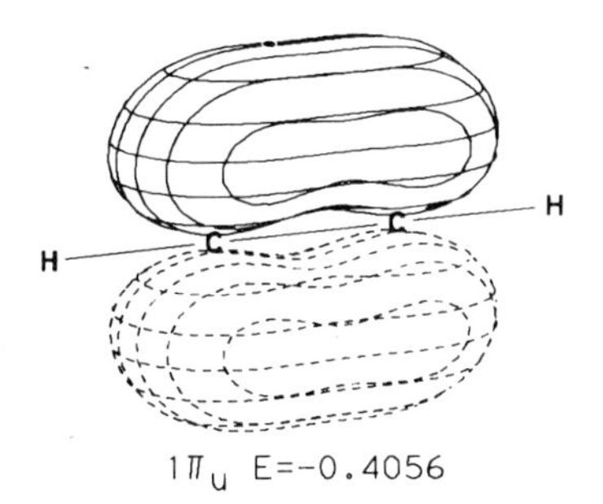

$1\pi_u$ E=-0.4056

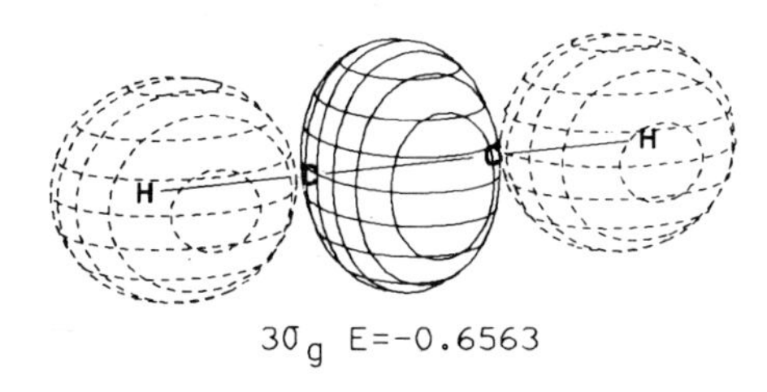

$3\sigma_g$ E=-0.6563

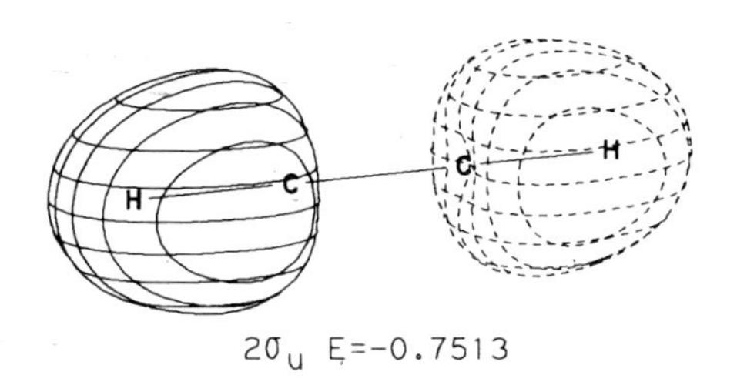

$2\sigma_u$ E=-0.7513

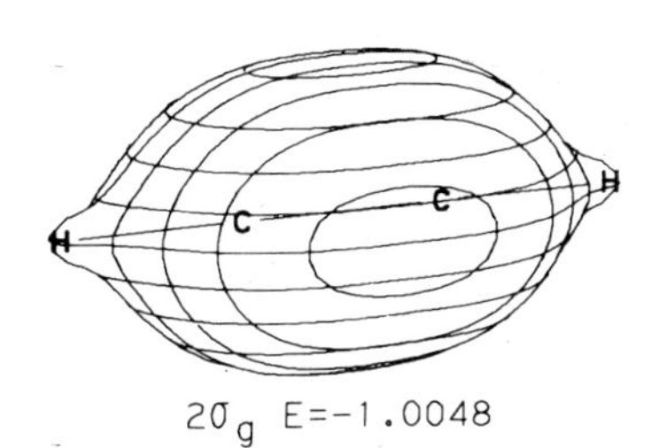

$2\sigma_g$ E=-1.0048

Acetylene (Continued)

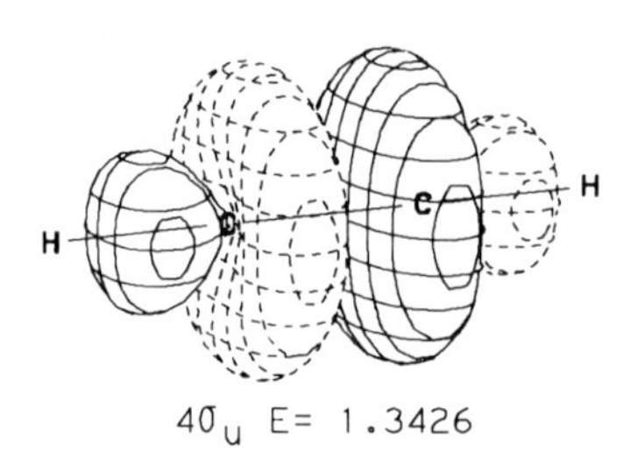

$4\sigma_u$ E= 1.3426

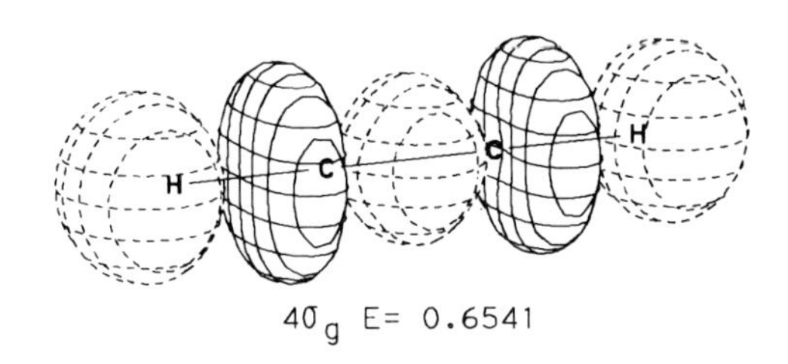

$4\sigma_g$ E= 0.6541

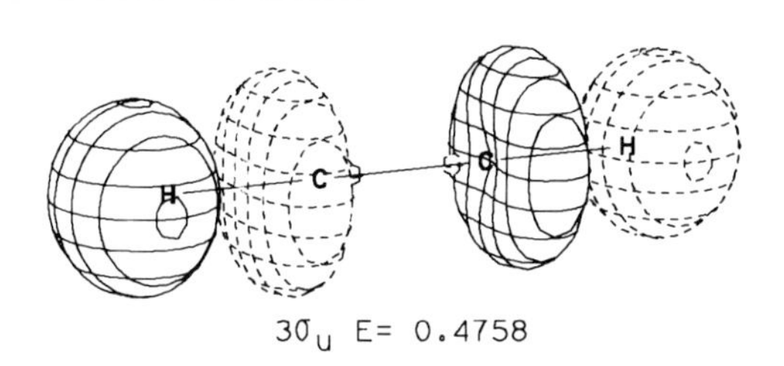

$3\sigma_u$ E= 0.4758

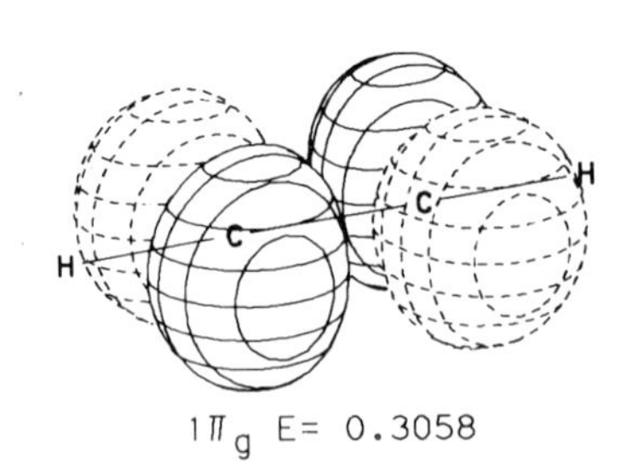

$1\pi_g$ E= 0.3058

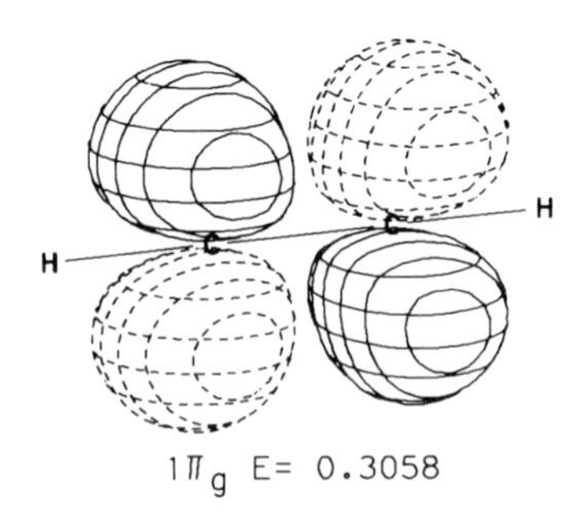

$1\pi_g$ E= 0.3058

13. Vinyl Cation

Symmetry: C_{2v}

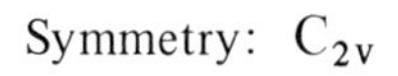

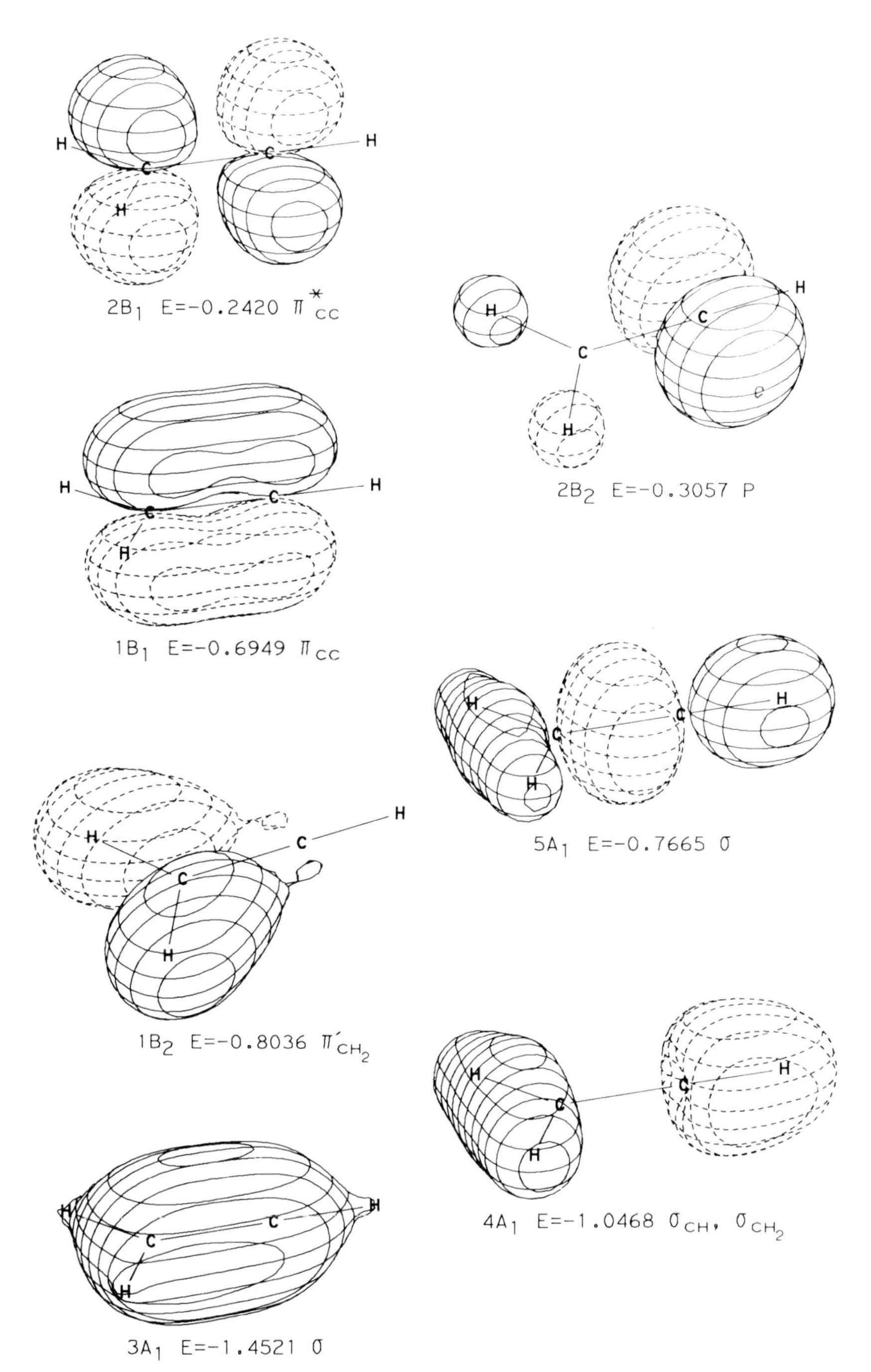

$2B_1$ E=−0.2420 π^*_{CC}

$2B_2$ E=−0.3057 P

$1B_1$ E=−0.6949 π_{CC}

$5A_1$ E=−0.7665 σ

$1B_2$ E=−0.8036 π'_{CH_2}

$4A_1$ E=−1.0468 σ_{CH}, σ_{CH_2}

$3A_1$ E=−1.4521 σ

14. Hydrogen Cyanide

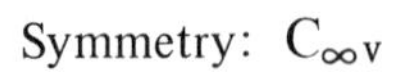
Symmetry: $C_{\infty v}$

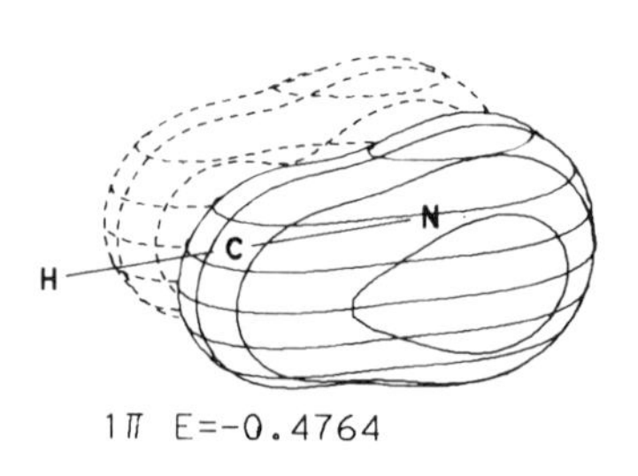

1π E=-0.4764

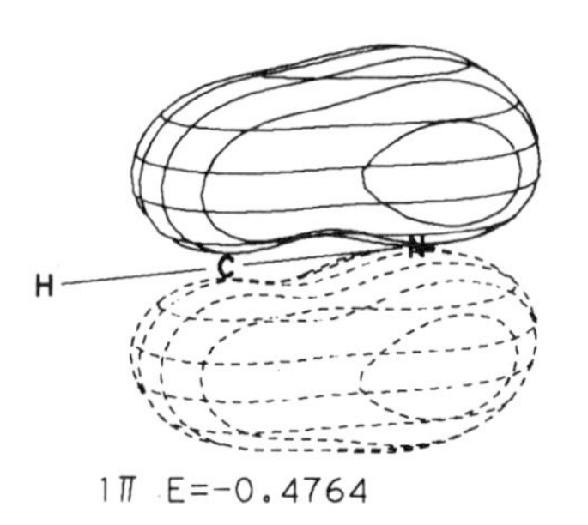

1π E=-0.4764

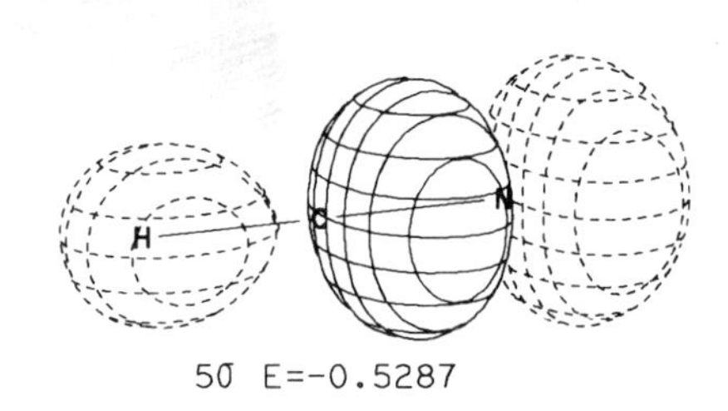

5σ E=-0.5287

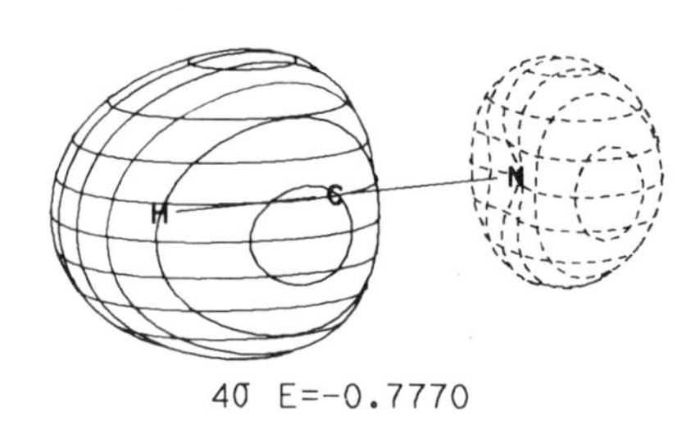

4σ E=-0.7770

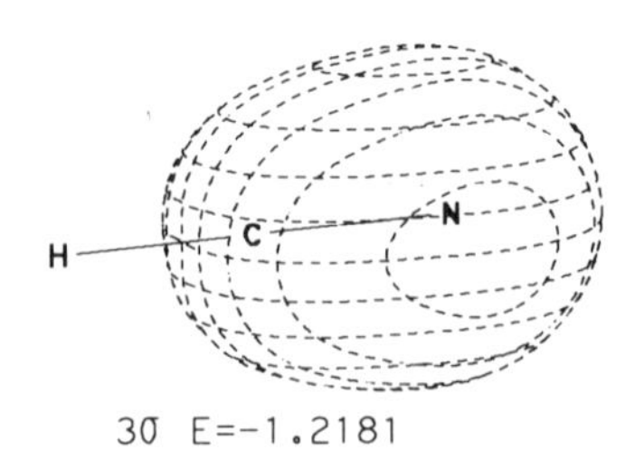

3σ E=-1.2181

Hydrogen Cyanide (Continued)

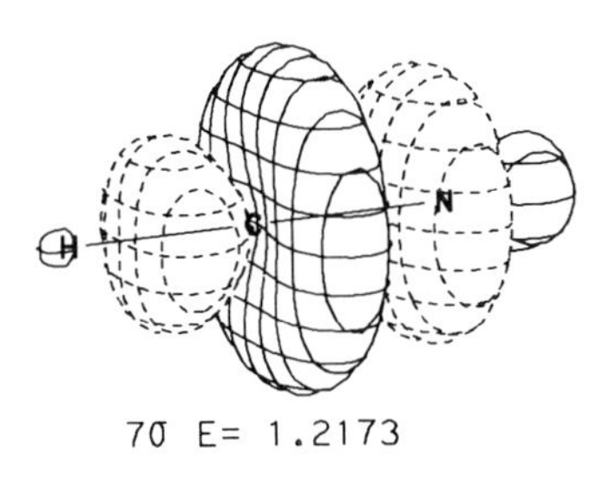

7σ E= 1.2173

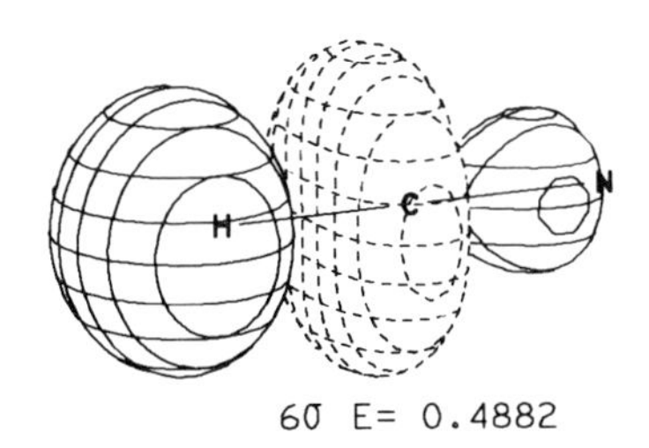

6σ E= 0.4882

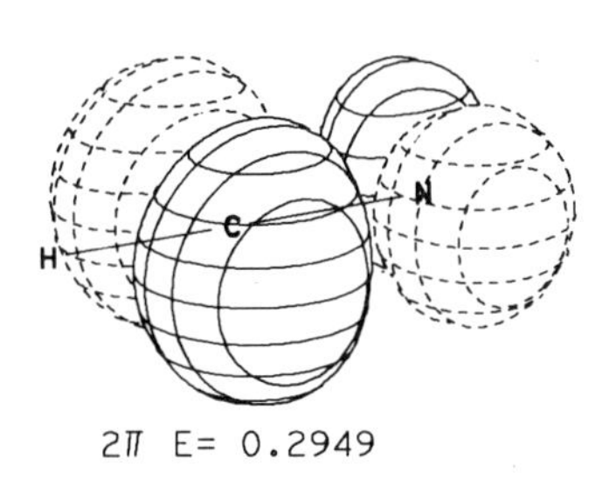

2π E= 0.2949

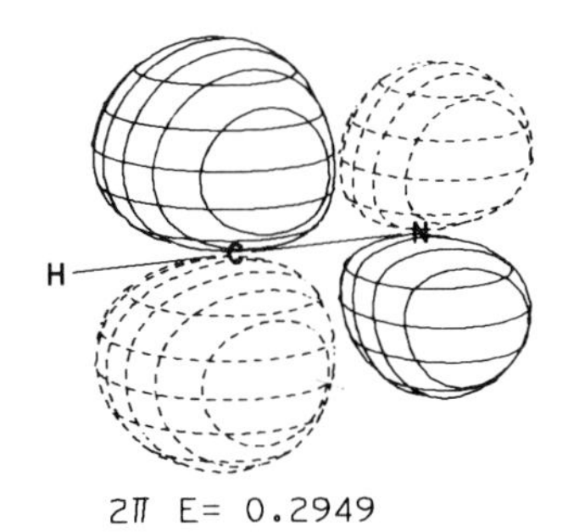

2π E= 0.2949

15. Carbon Monoxide

Symmetry: $C_{\infty v}$

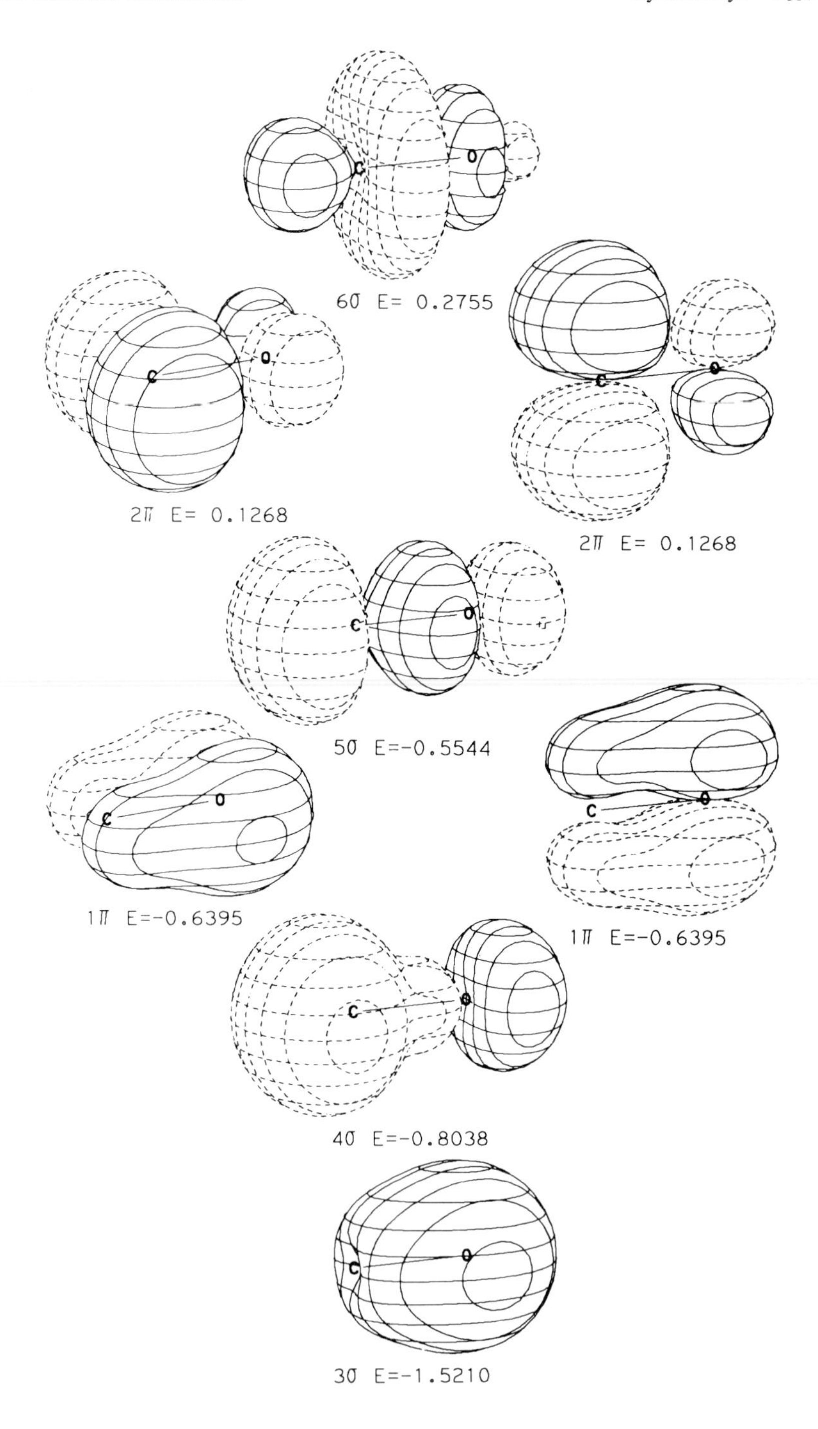

16. Nitrogen

Symmetry: $D_{\infty h}$

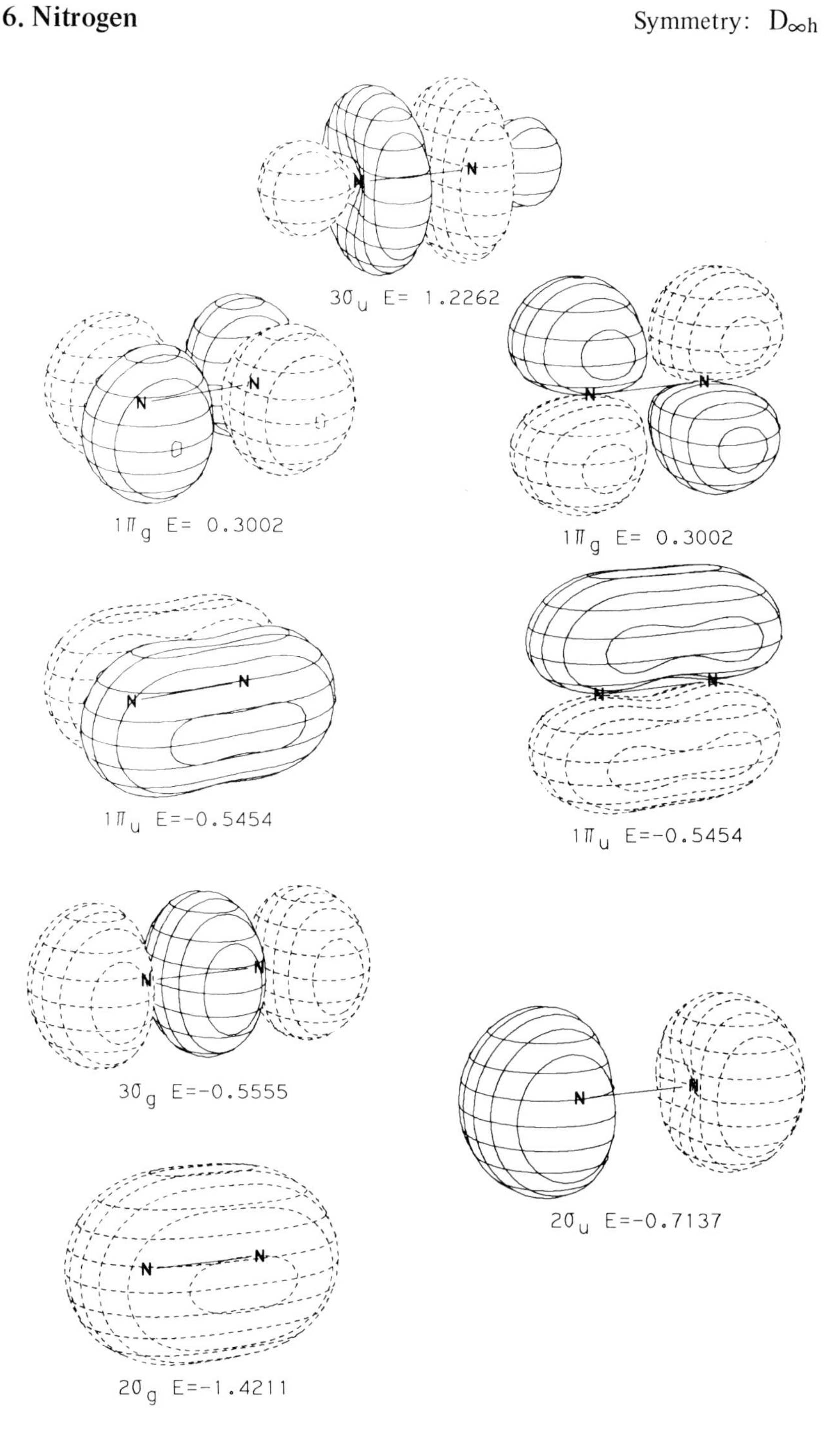

17. Nitric Oxide

Symmetry: $C_{\infty v}$

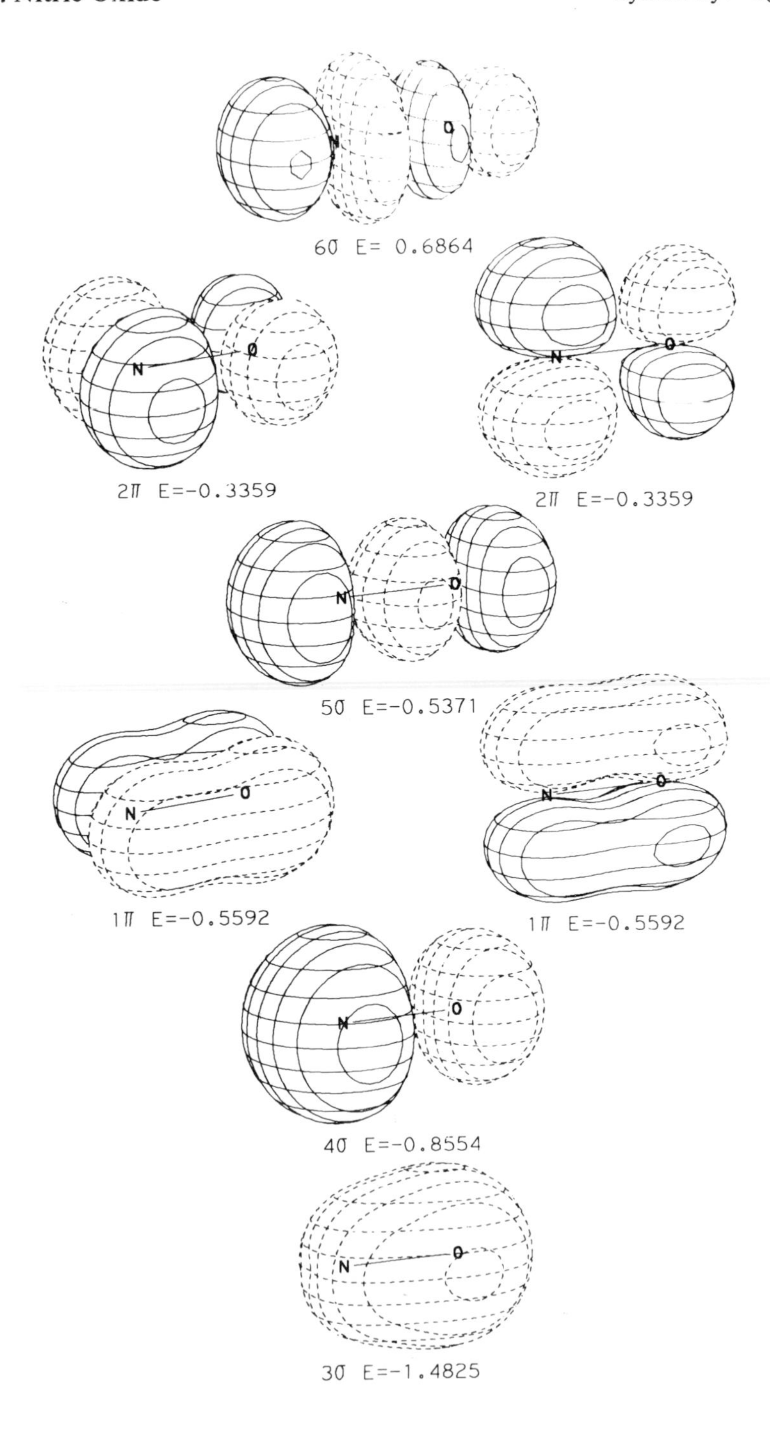

18. Ethylene

Symmetry: D_{2h}

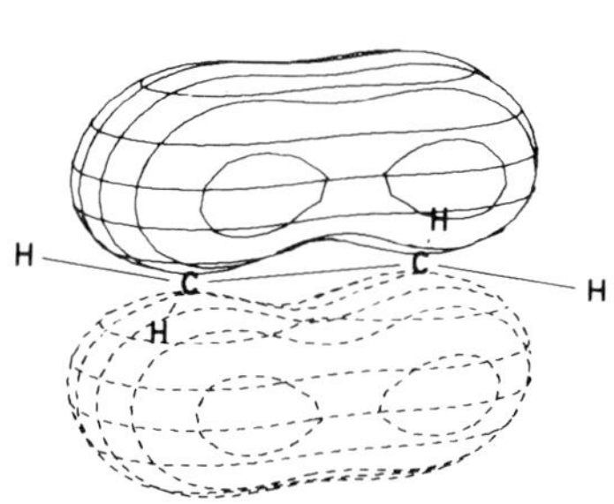

$1B_{2u}$ E=−0.3709 π_{CC}

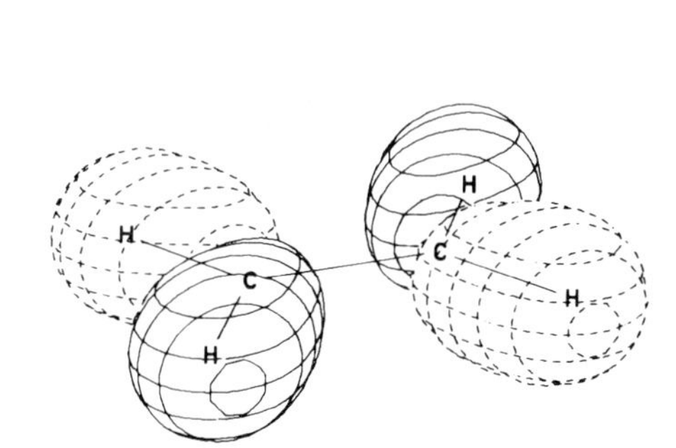

$1B_{2g}$ E=−0.5061 π'_{CH_2}

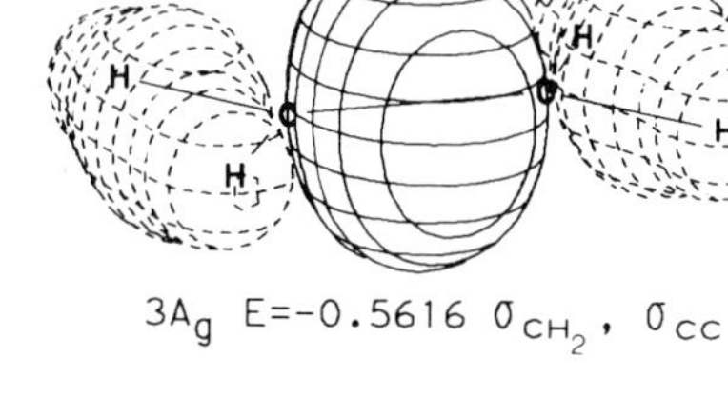

$3A_g$ E=−0.5616 σ_{CH_2}, σ_{CC}

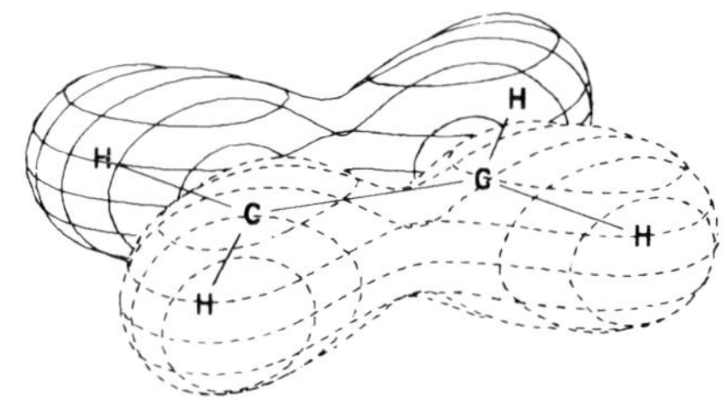

$1B_{3u}$ E=−0.6438 π'_{CH_2}

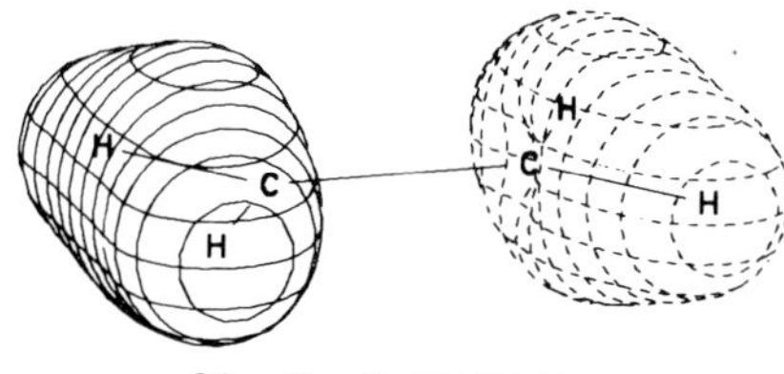

$2A_u$ E=−0.7823 σ_{CH_2}

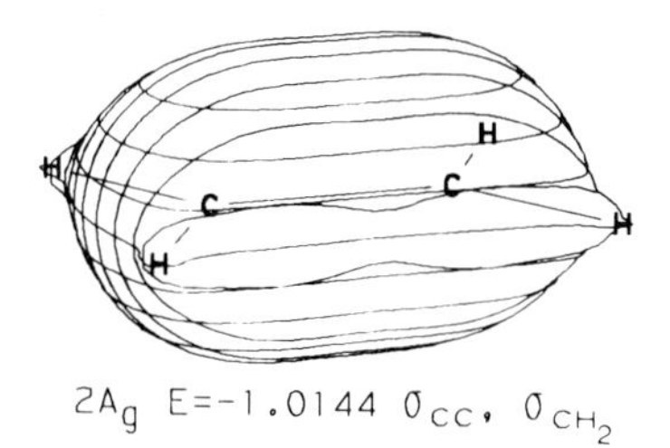

$2A_g$ E=−1.0144 σ_{CC}, σ_{CH_2}

Ethylene (Continued)

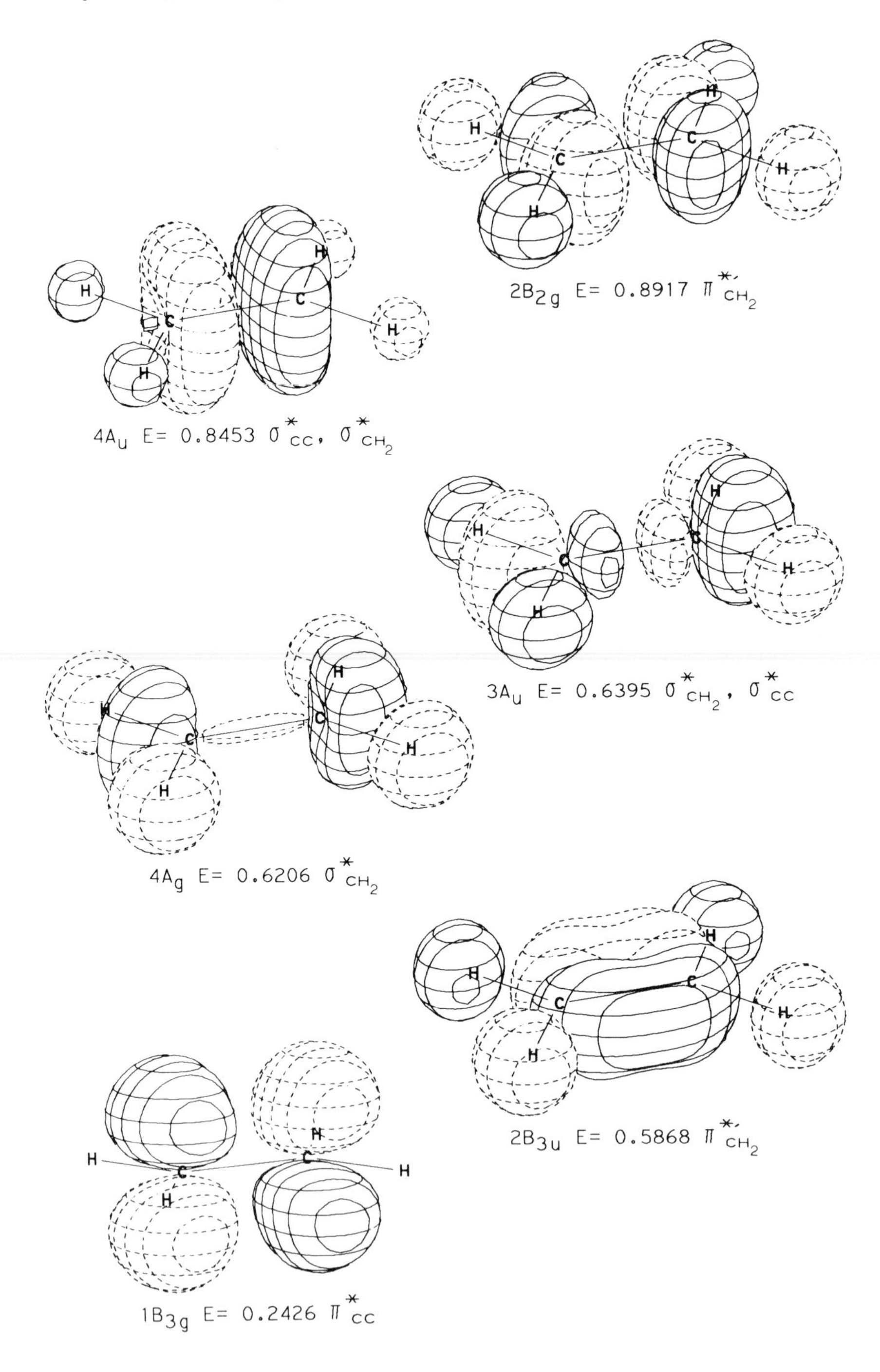

$2B_{2g}$ E= 0.8917 $\pi^{*}_{CH_2}$

$4A_u$ E= 0.8453 σ^{*}_{CC}, $\sigma^{*}_{CH_2}$

$3A_u$ E= 0.6395 $\sigma^{*}_{CH_2}$, σ^{*}_{CC}

$4A_g$ E= 0.6206 $\sigma^{*}_{CH_2}$

$2B_{3u}$ E= 0.5868 $\pi^{*}_{CH_2}$

$1B_{3g}$ E= 0.2426 π^{*}_{CC}

19. Methylenimine

Symmetry: C_s

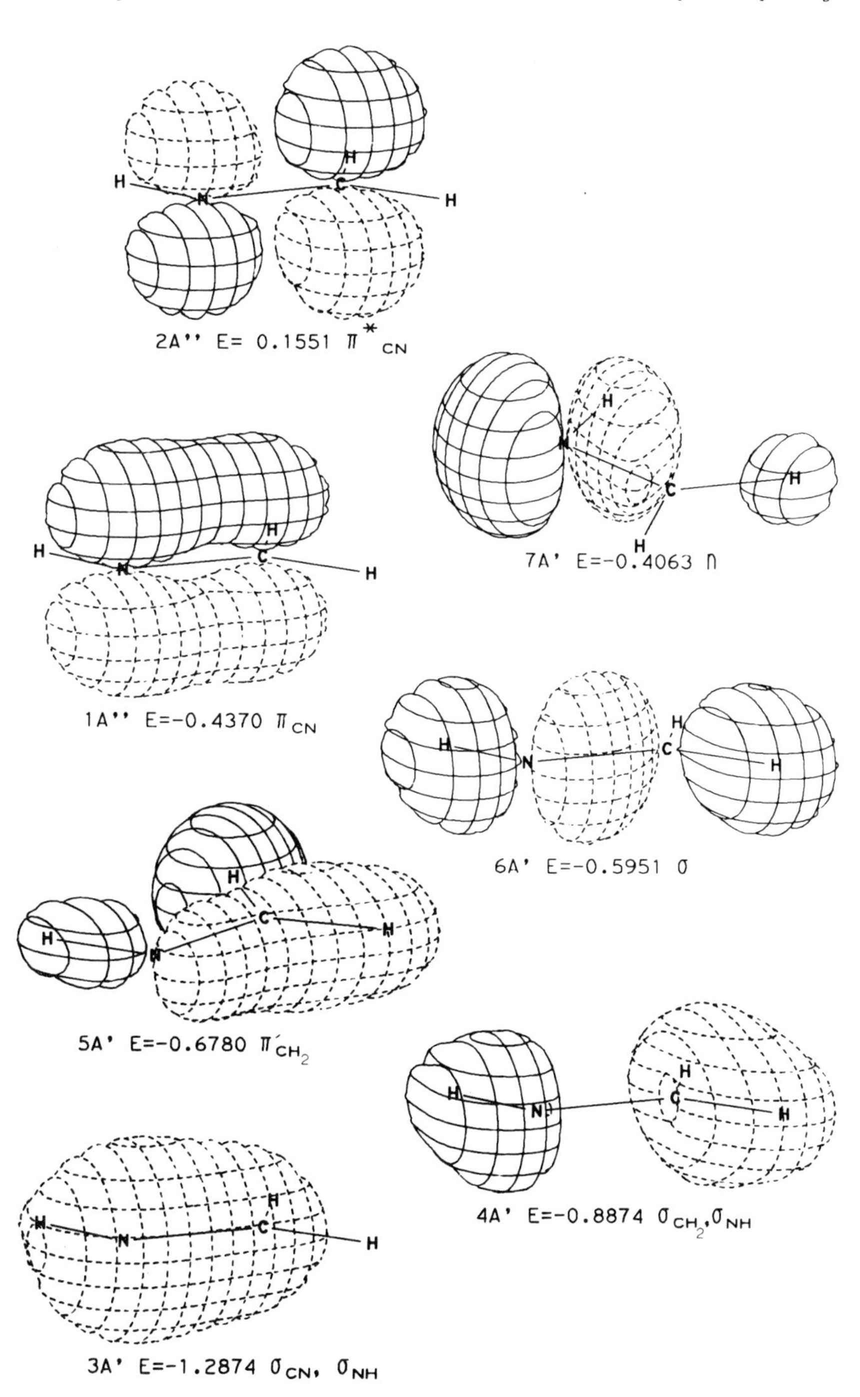

20. Formaldehyde

Symmetry: C_{2v}

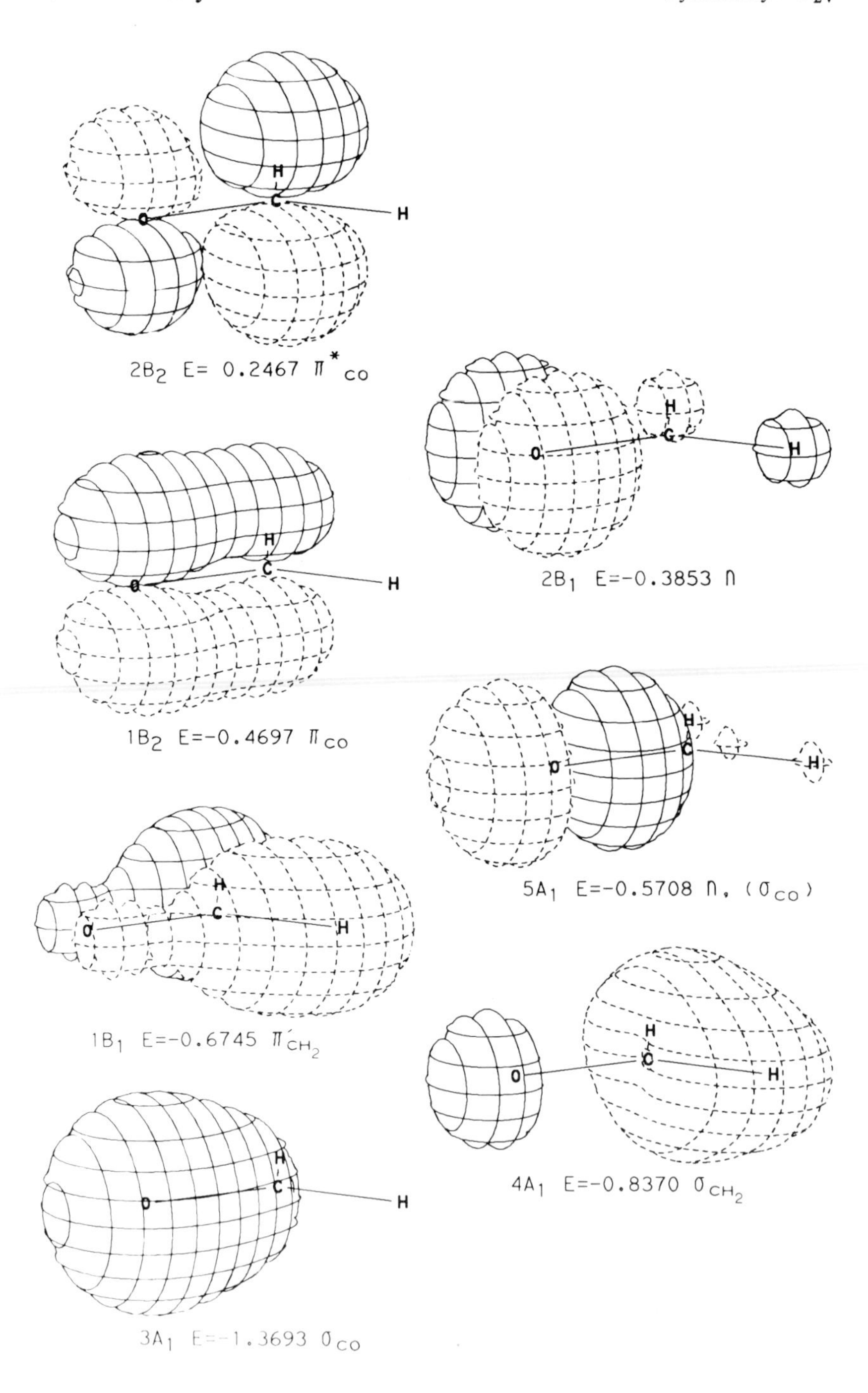

$2B_2$ E= 0.2467 π^*_{CO}

$2B_1$ E=-0.3853 n

$1B_2$ E=-0.4697 π_{CO}

$5A_1$ E=-0.5708 n, (σ_{CO})

$1B_1$ E=-0.6745 π'_{CH_2}

$4A_1$ E=-0.8370 σ_{CH_2}

$3A_1$ E=-1.3693 σ_{CO}

21. Diimide

Symmetry: C_{2h}

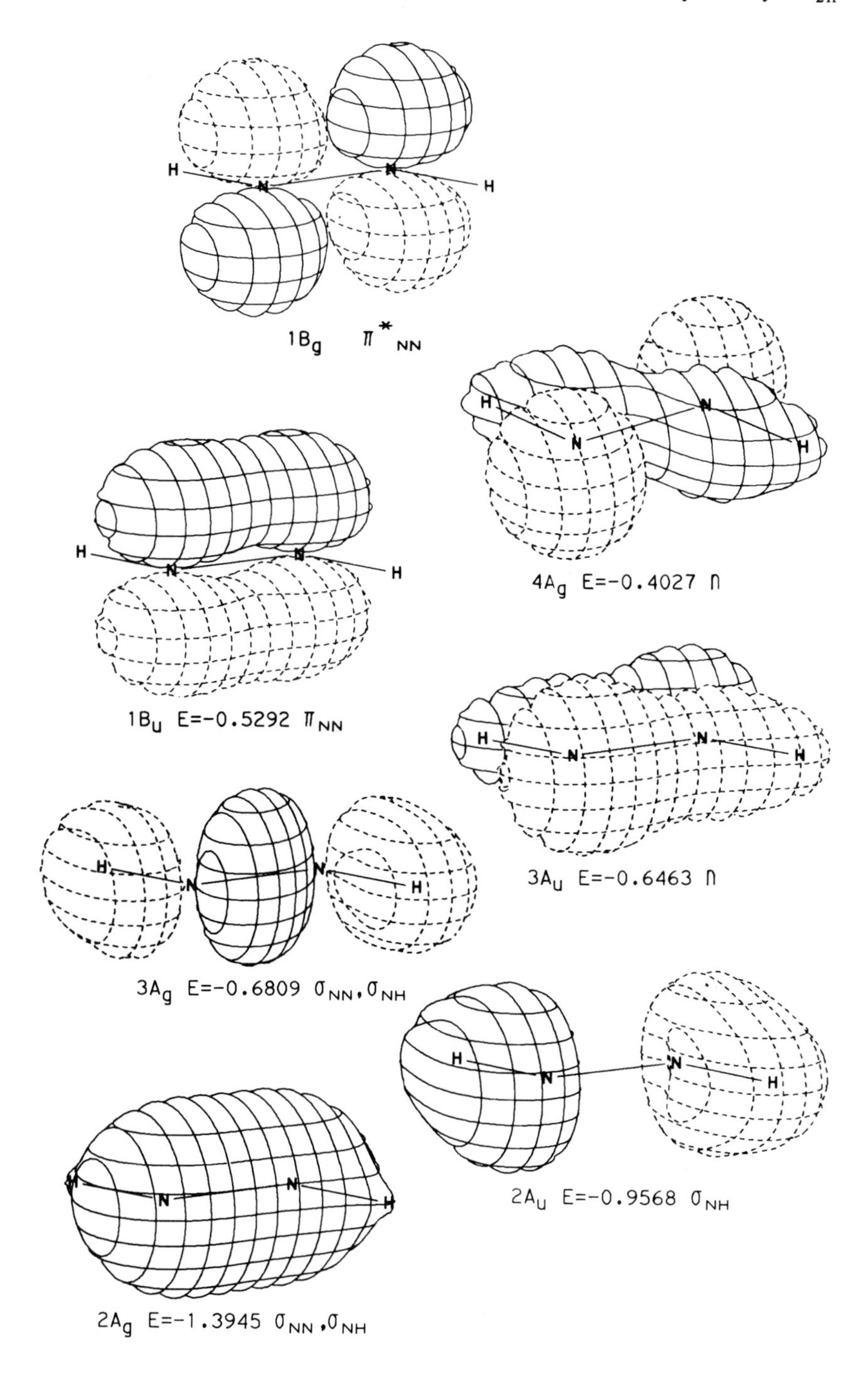

22. Diborane

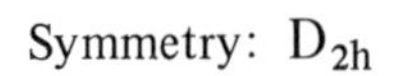
Symmetry: D_{2h}

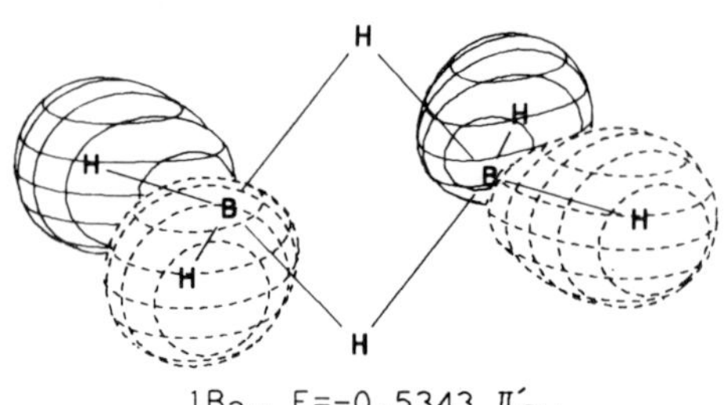

$1B_{3u}$ E=−0.5343 π'_{BH_2}

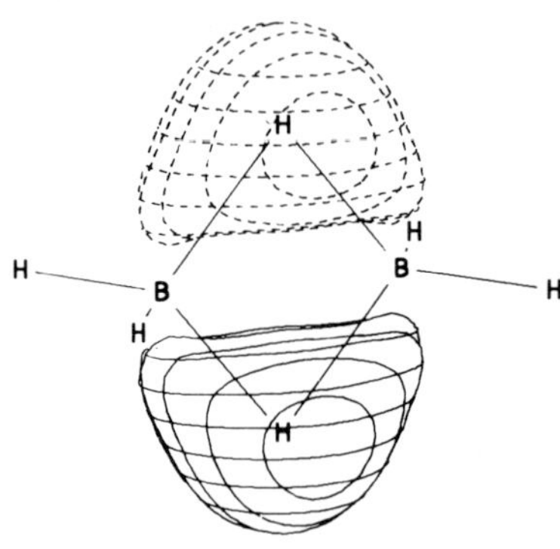

$1B_{2u}$ E=−0.5564 π

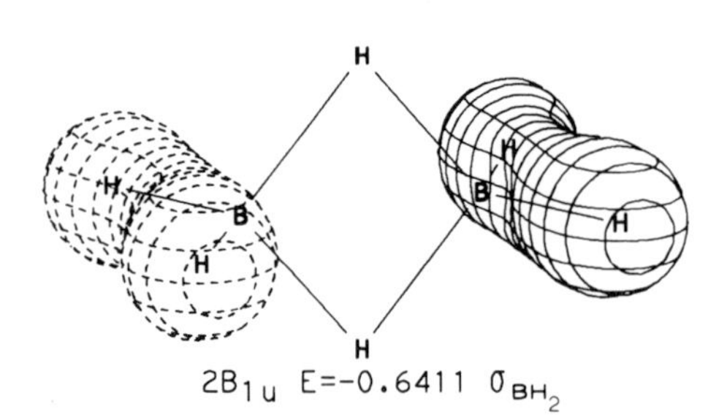

$2B_{1u}$ E=−0.6411 σ_{BH_2}

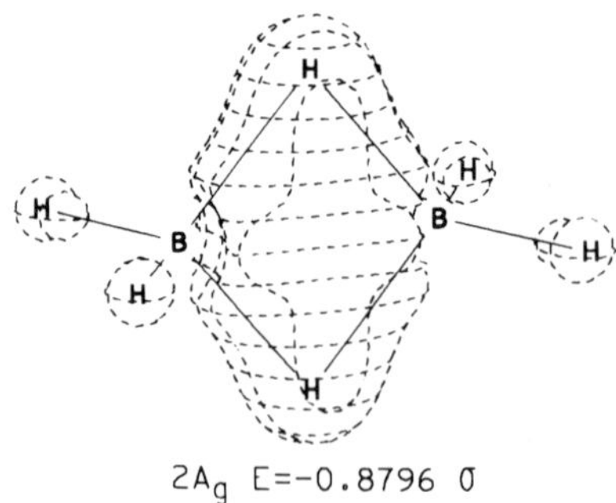

$2A_g$ E=−0.8796 σ

Diborane (Continued)

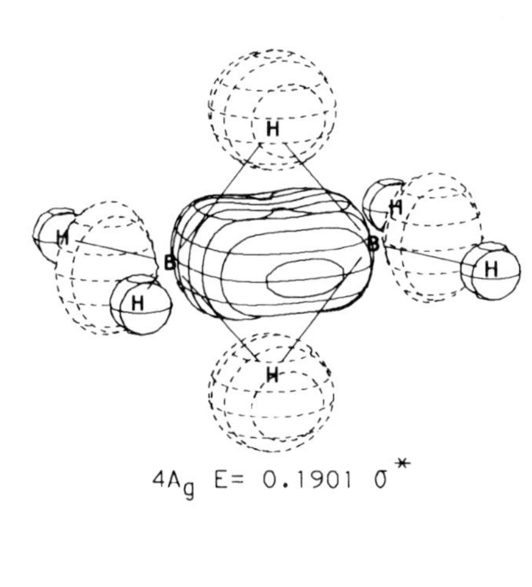

$4A_g$ E= 0.1901 σ^*

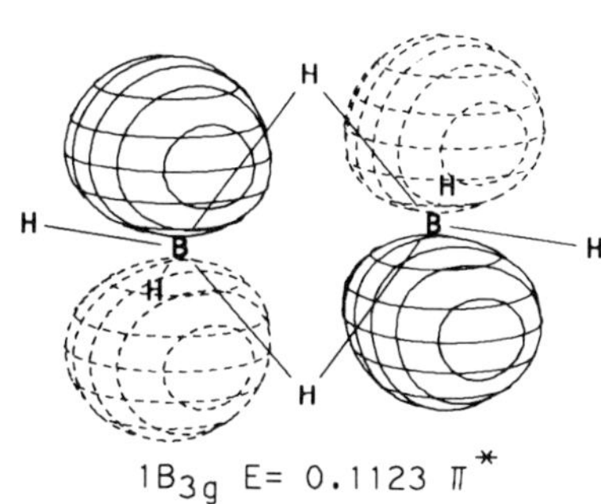

$1B_{3g}$ E= 0.1123 π^*

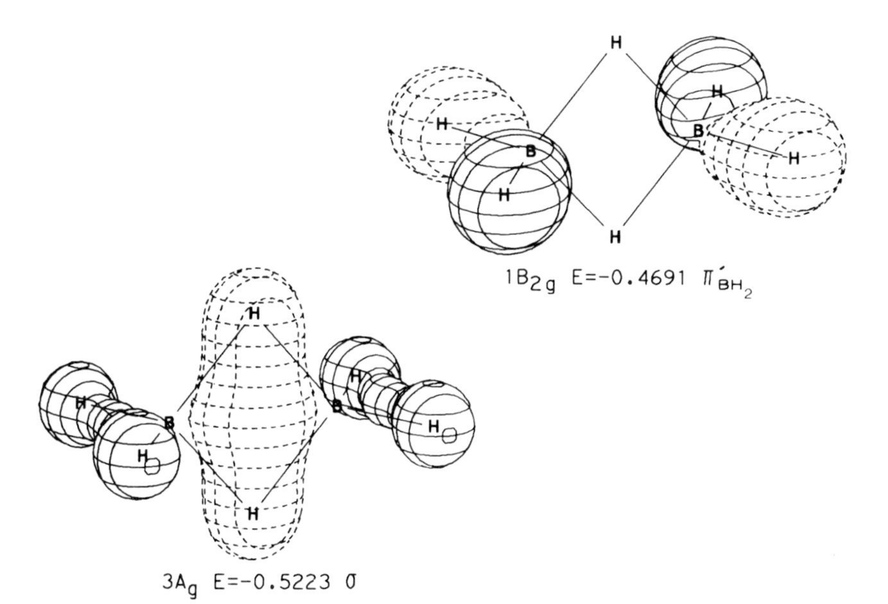

$1B_{2g}$ E=−0.4691 π'_{BH_2}

$3A_g$ E=−0.5223 σ

23. Oxygen (Triplet)

Symmetry: $D_{\infty h}$

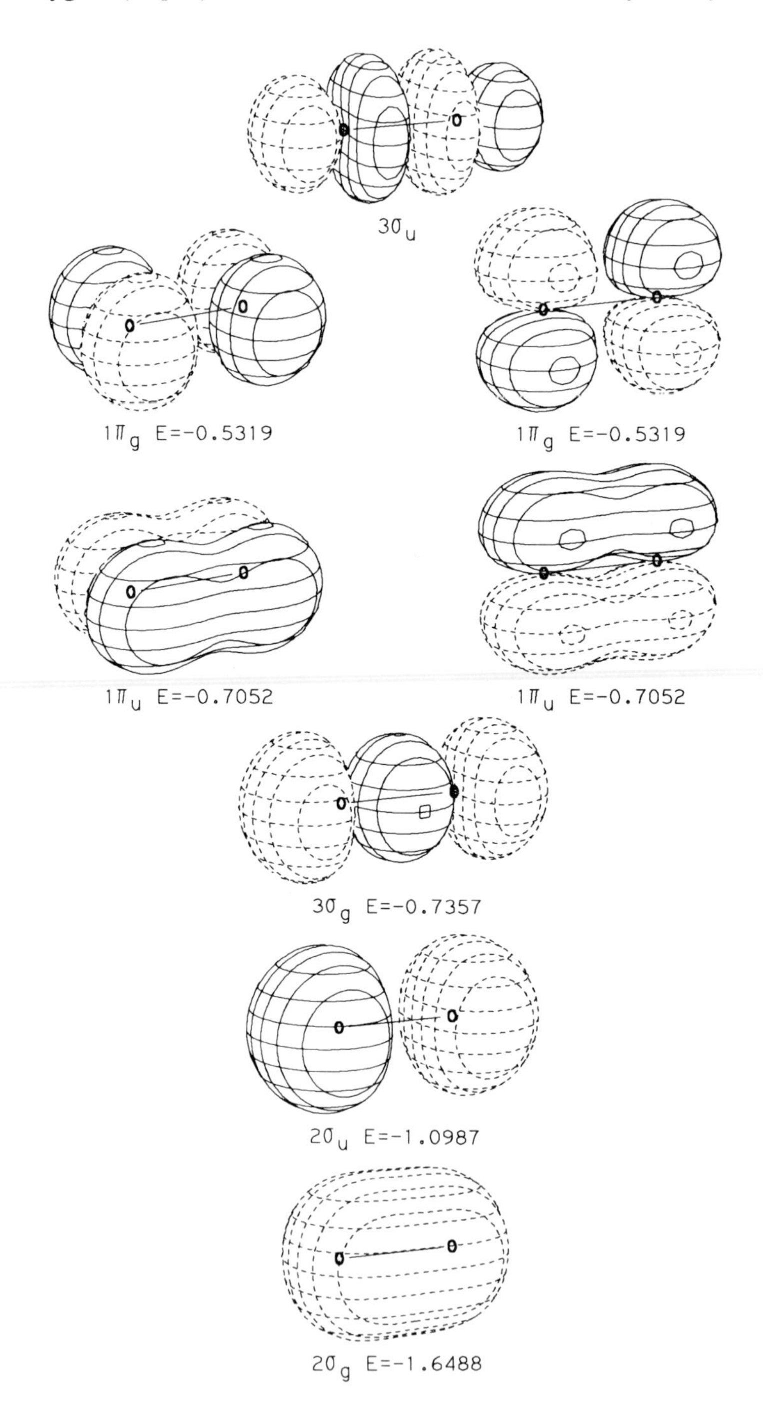

24. Ethyl Cation, Bisected

Symmetry: C_s

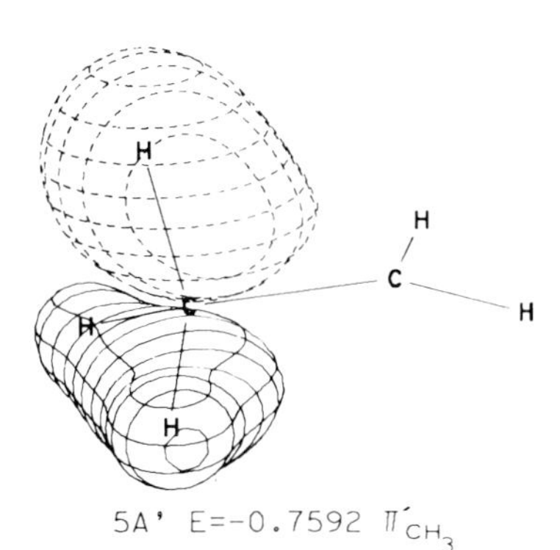

5A' E=-0.7592 π'_{CH_3}

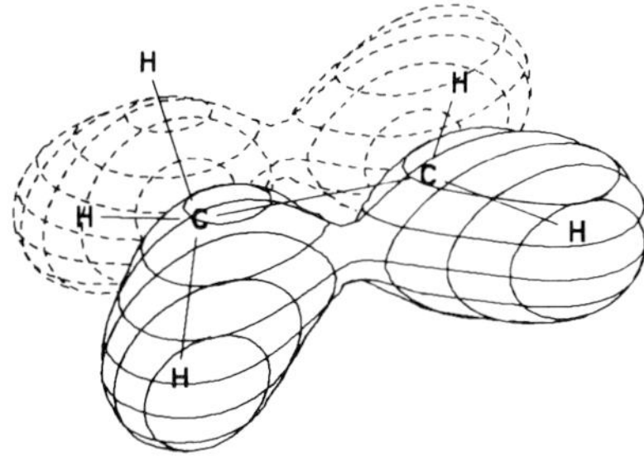

1A'' E=-0.8439 π_{CH_3}, π_{CH_2}

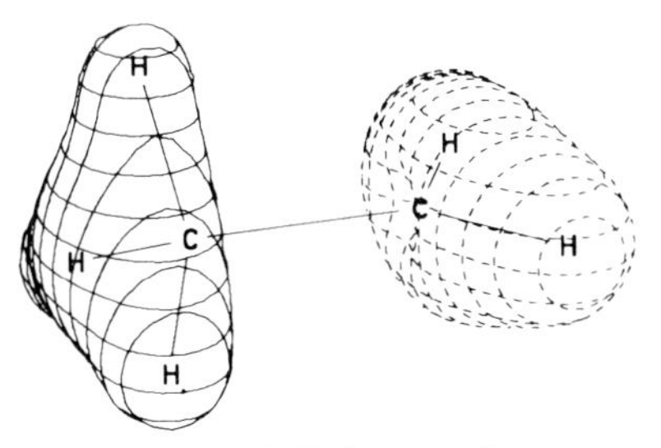

4A' E=-1.1067 σ_{CH_3}, σ_{CH_2}

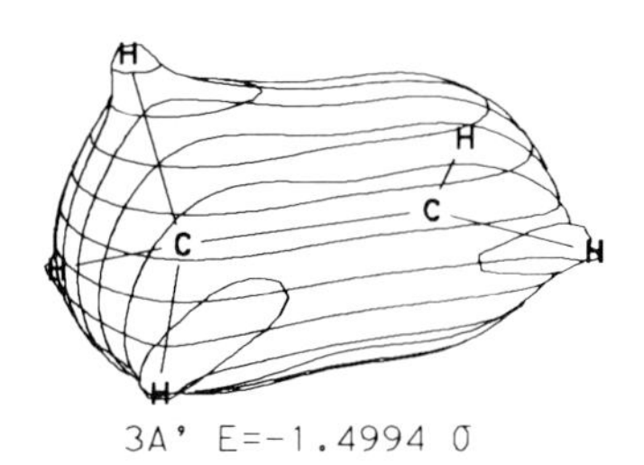

3A' E=-1.4994 σ

Ethyl Cation, Bisected (Continued)

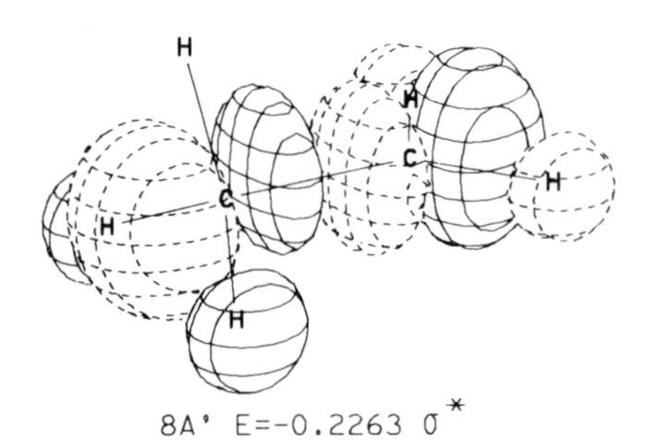

8A' E=-0.2263 σ^*

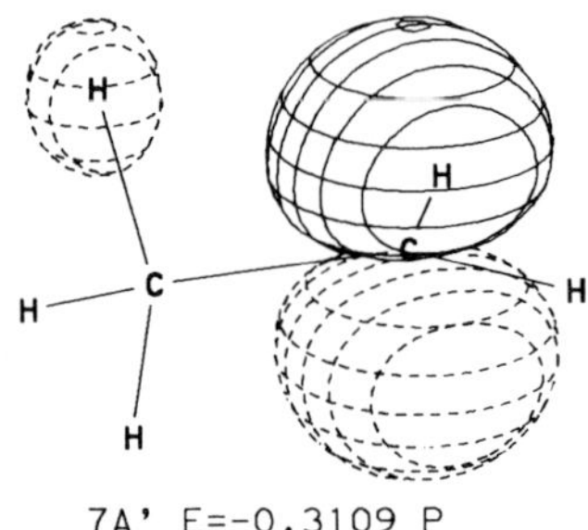

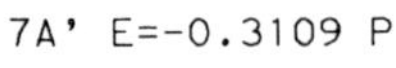

7A' E=-0.3109 P

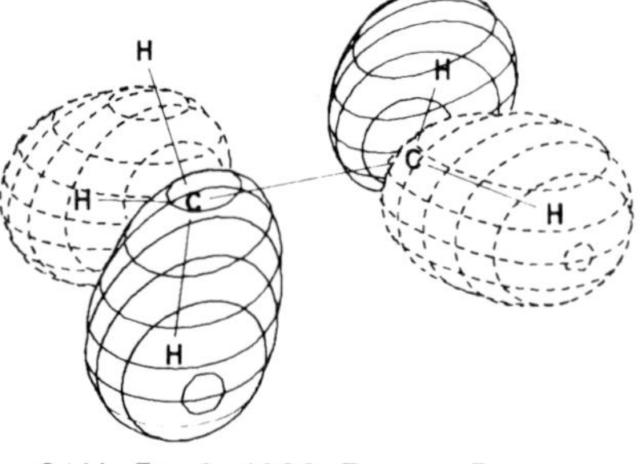

2A'' E=-0.6933 π_{CH_3}, π_{CH_2}

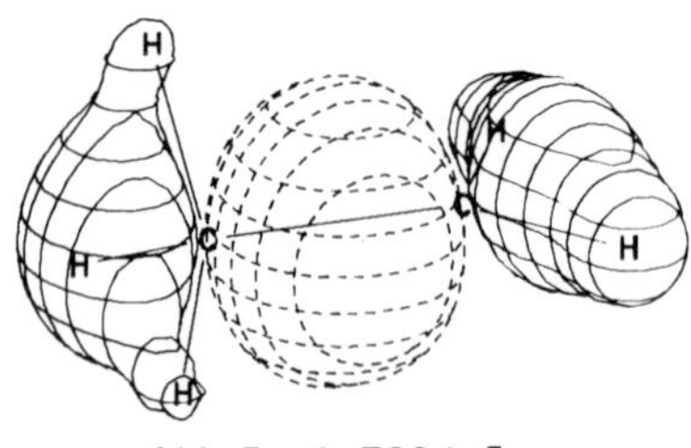

6A' E=-0.7394 σ

25. Ethyl Cation, Eclipsed

Symmetry: C_s

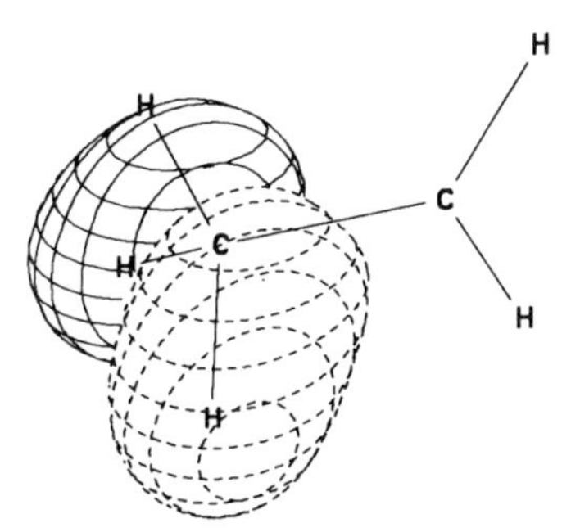

1A'' E=-0.7635 π_{CH_3}

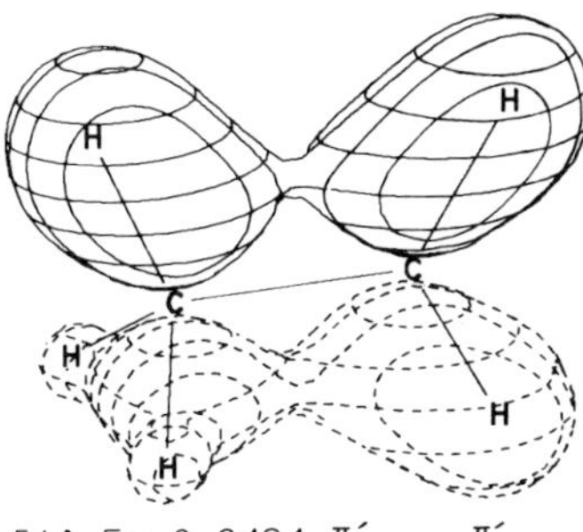

5A' E=-0.8424 π'_{CH_2}, π'_{CH_3}

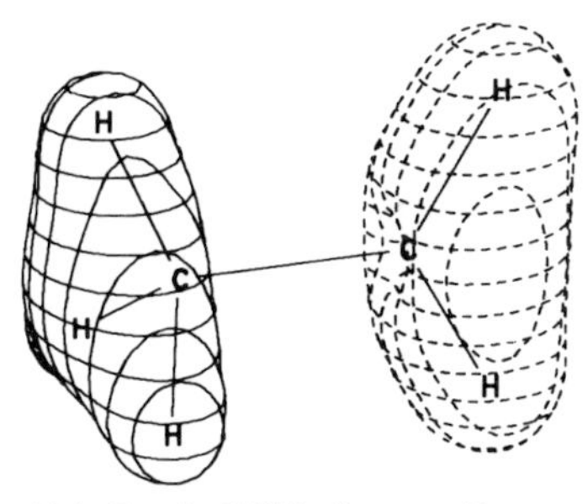

4A' E=-1.1070 σ_{CH_3}, σ_{CH_2}

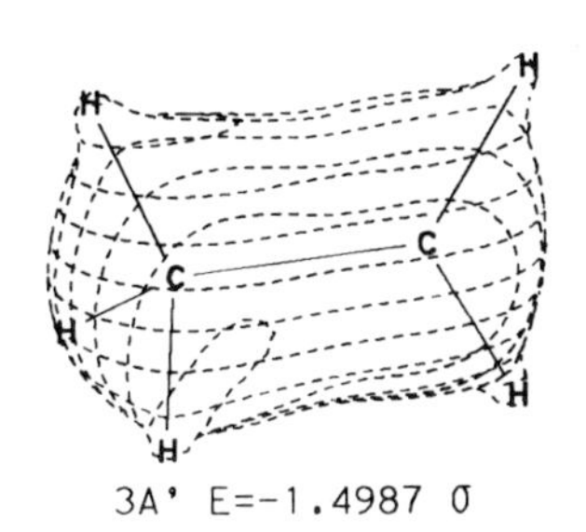

3A' E=-1.4987 σ

Ethyl Cation, Eclipsed (Continued)

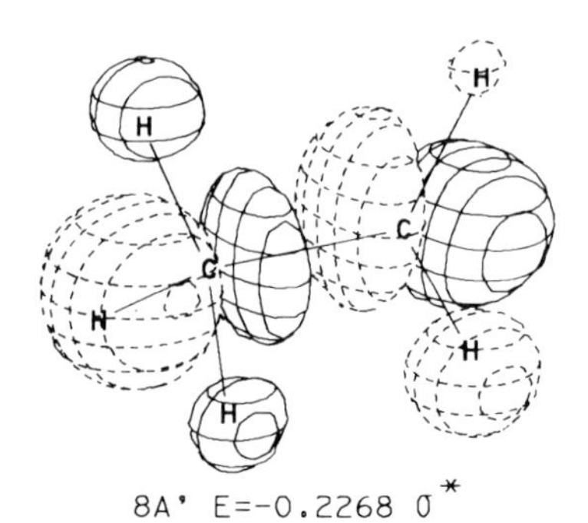

8A' E=-0.2268 σ^*

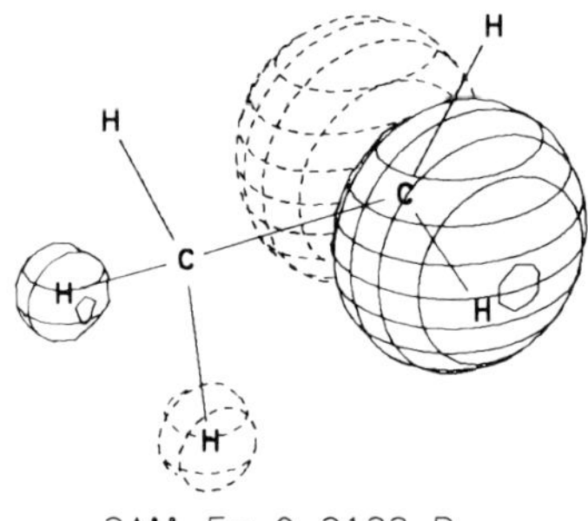

2A'' E=-0.3122 P

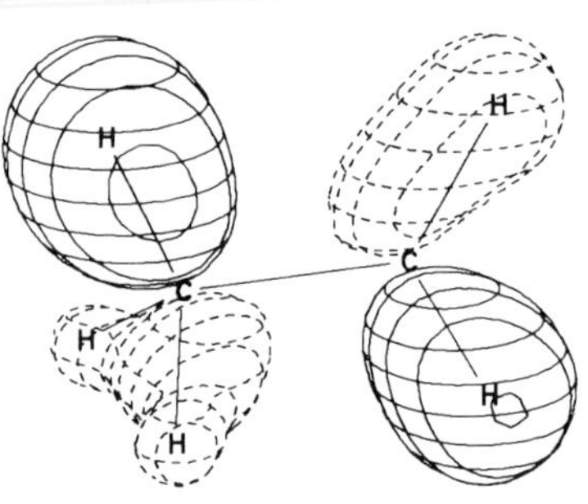

7A' E=-0.6912 π'_{CH_3}, π'_{CH_2}

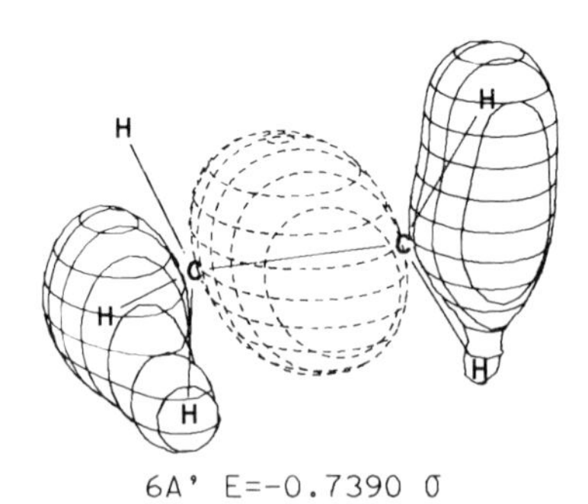

6A' E=-0.7390 σ

26. Ethyl Cation, Bridged

Symmetry: C_{2v}

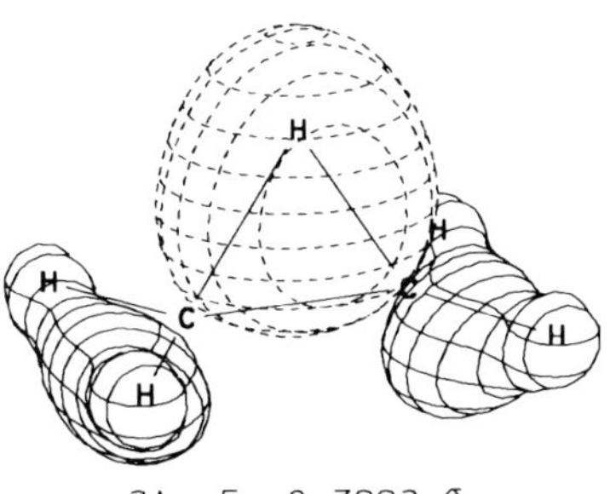

$3A_1$ E=-0.7993 σ

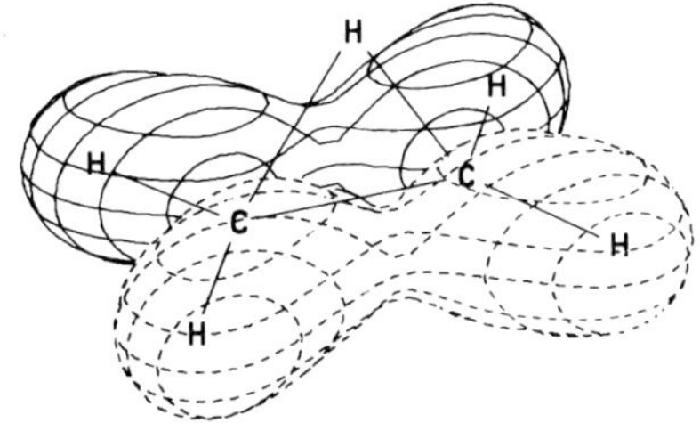

$1B_1$ E=-0.8595 π_{CH_2}

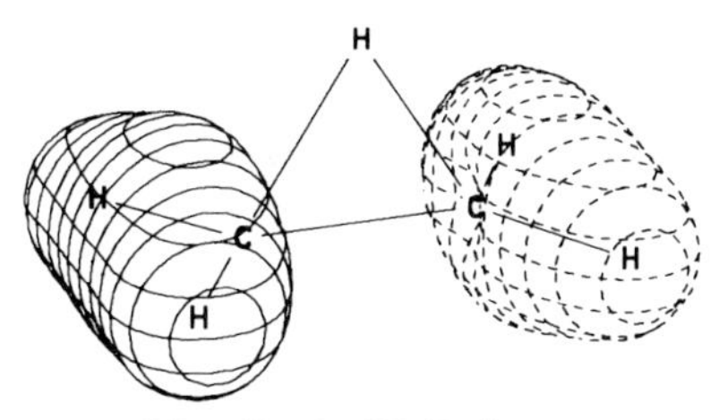

$2B_2$ E=-1.0797 σ_{CH_2}

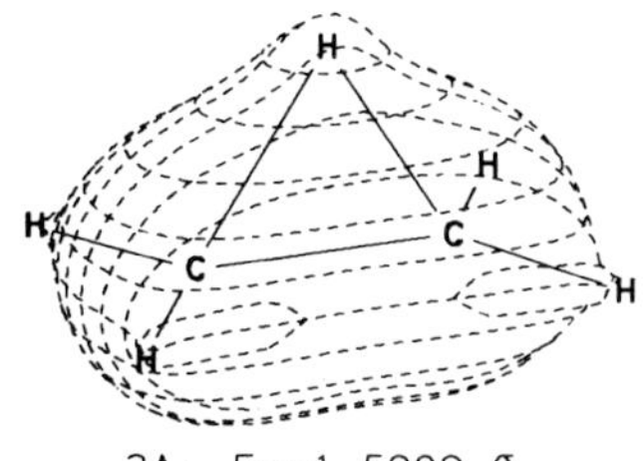

$2A_1$ E=-1.5989 σ

Ethyl Cation, Bridged (Continued)

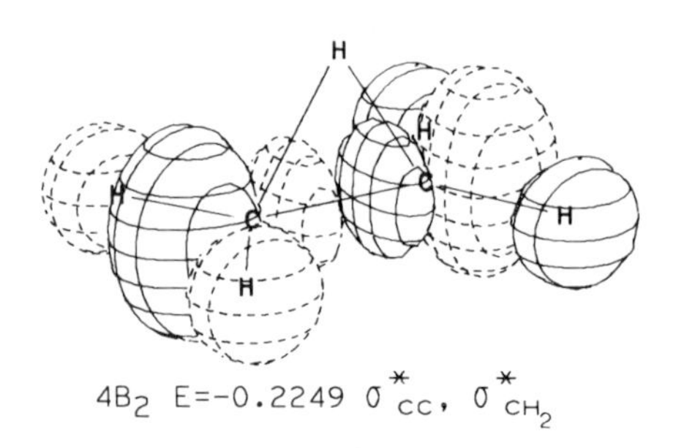

$4B_2$ E=−0.2249 σ^*_{CC}, $\sigma^*_{CH_2}$

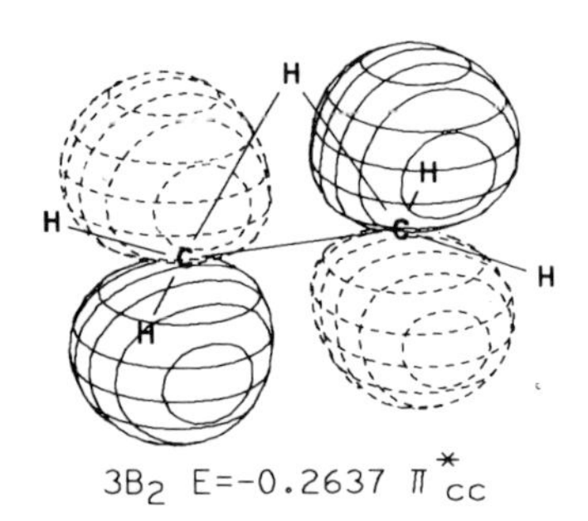

$3B_2$ E=−0.2637 π^*_{CC}

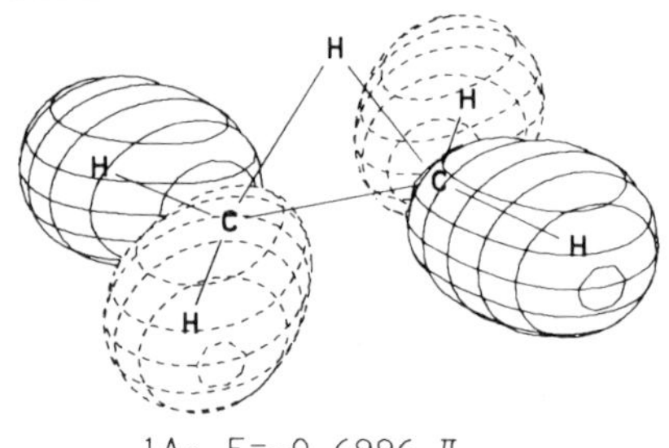

$1A_2$ E=−0.6996 π_{CH_2}

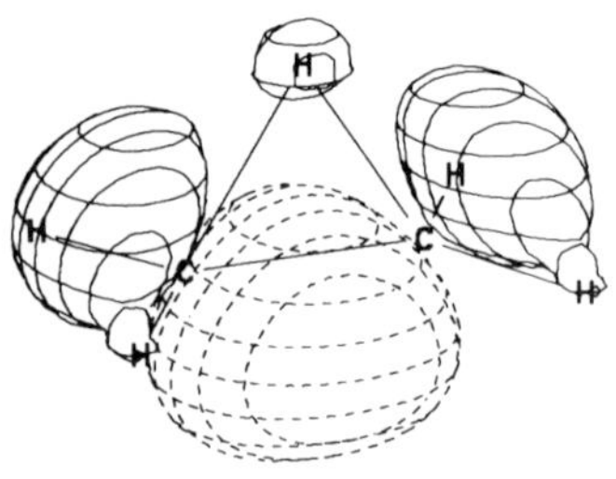

$4A_1$ E=−0.7381 σ_{CC}, π_{CC}

27. Ethyl Radical, Bisected

Symmetry: C_s

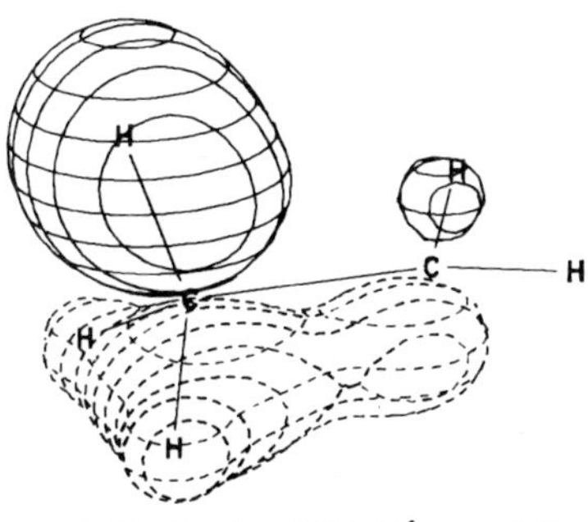

5A' E=-0.5296 π'_{CH_3}, (P)

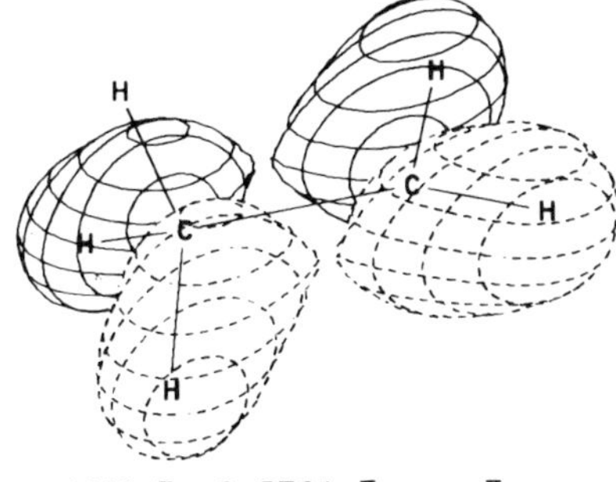

1A'' E=-0.5731 π_{CH_3}, π_{CH_2}

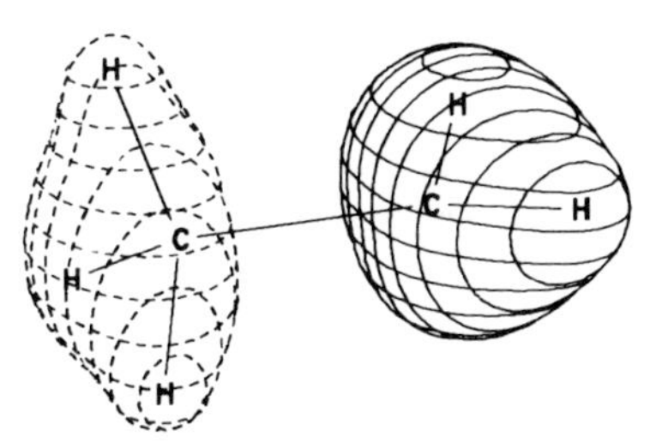

4A' E=-0.8623 σ_{CH_3}, σ_{CH_2}

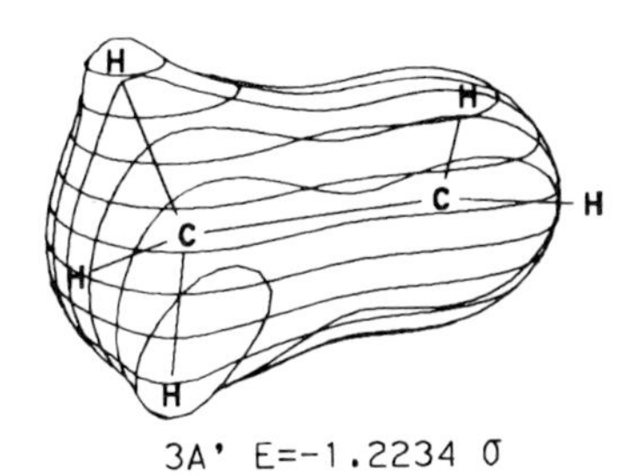

3A' E=-1.2234 σ

Ethyl Radical, Bisected (Continued)

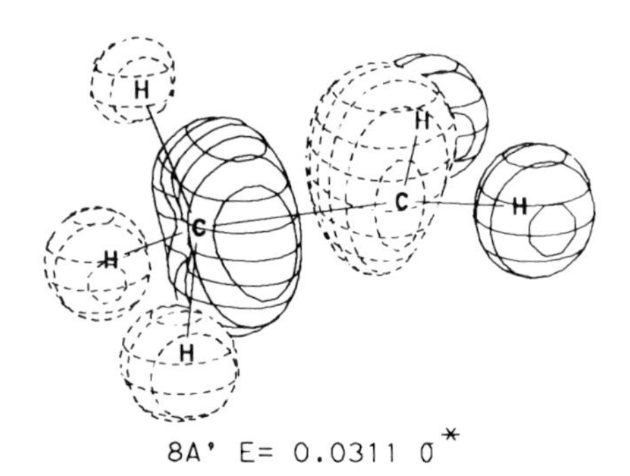

8A' E= 0.0311 σ^*

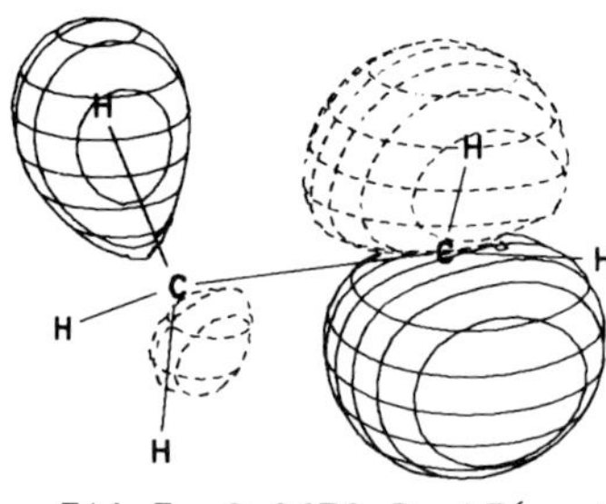

7A' E=-0.3670 P, (π'_{CH_3})

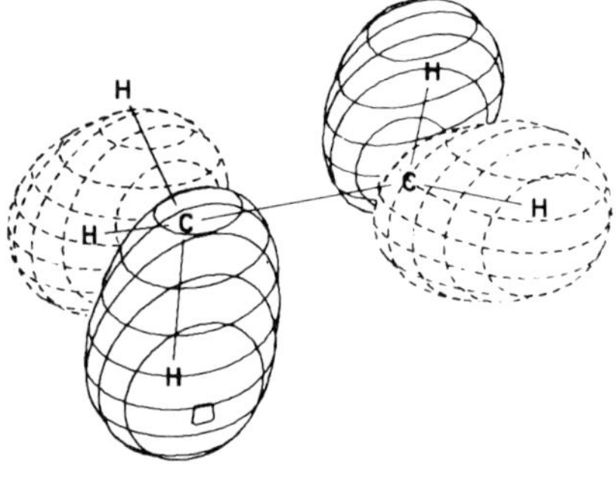

2A'' E=-0.4421 π_{CH_3}, π_{CH_2}

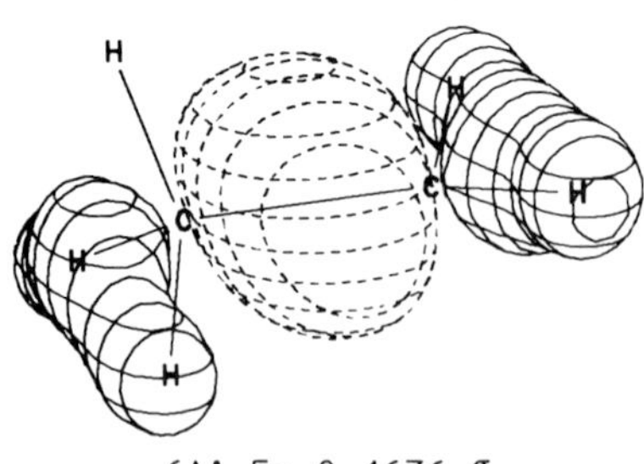

6A' E=-0.4676 σ

28. Ethyl Radical, Eclipsed

Symmetry: C_s

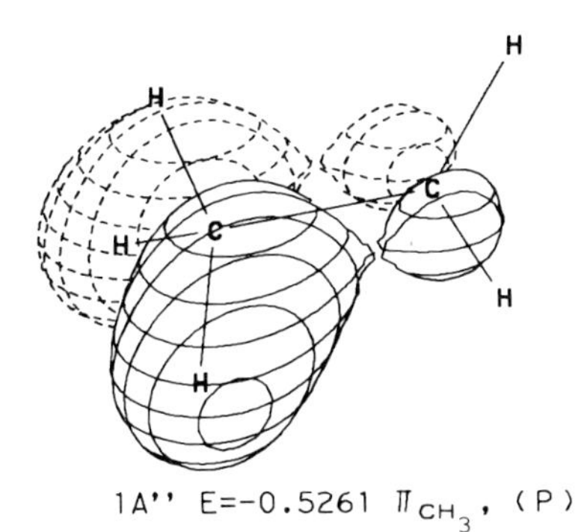

1A'' E=-0.5261 π_{CH_3}, (P)

5A' E=-0.5732 π'_{CH_2}, π'_{CH_3}

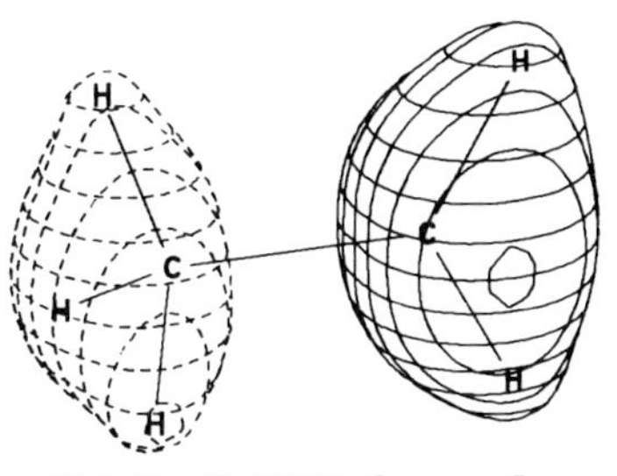

4A' E=-0.8630 σ_{CH_3}, σ_{CH_2}

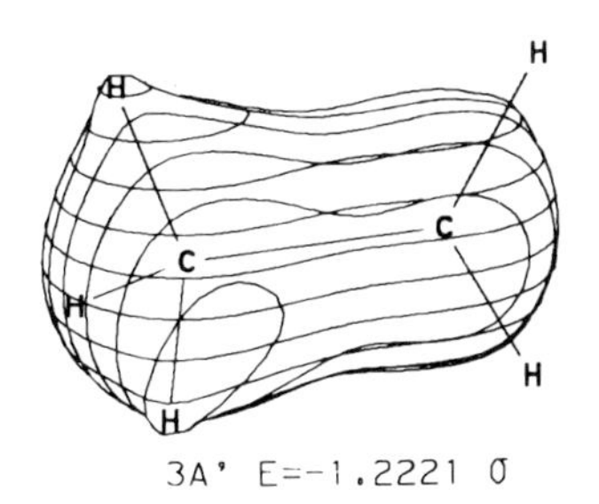

3A' E=-1.2221 σ

Ethyl Radical, Eclipsed (Continued)

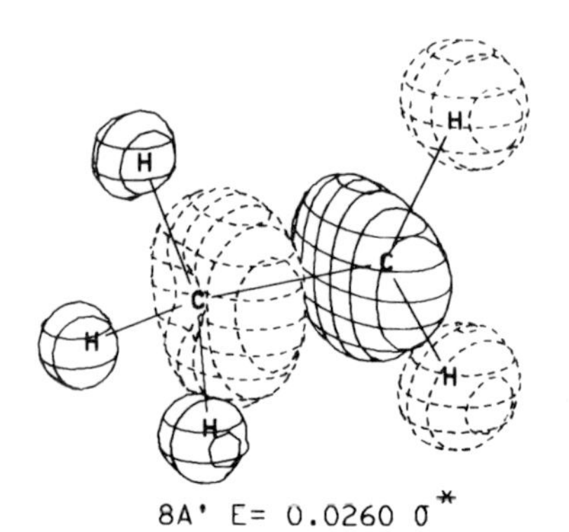

8A' E= 0.0260 σ^*

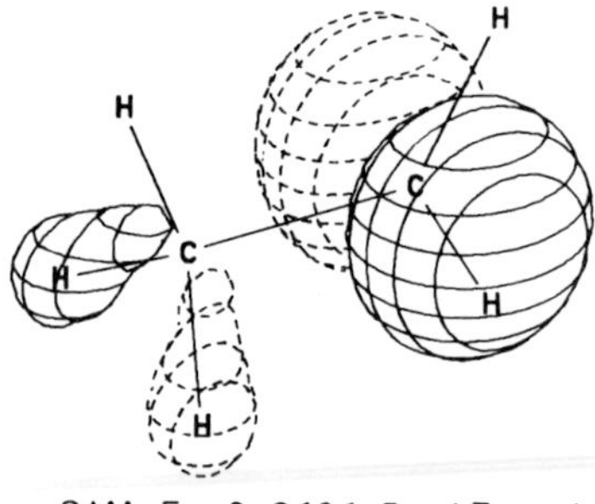

2A'' E=-0.3606 P, (π_{CH_3})

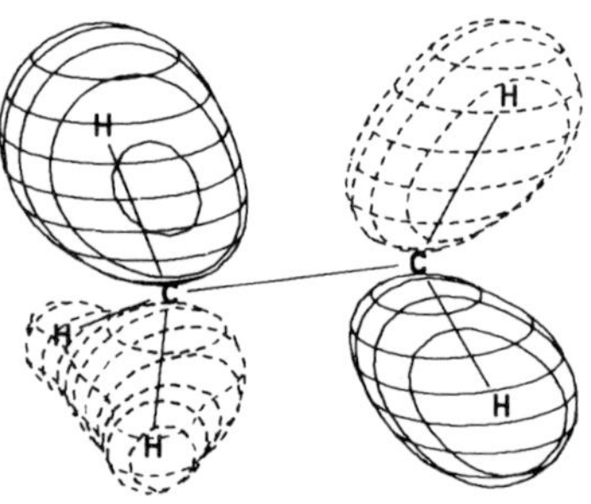

7A' E=-0.4420 π'_{CH_3}, π'_{CH_2}

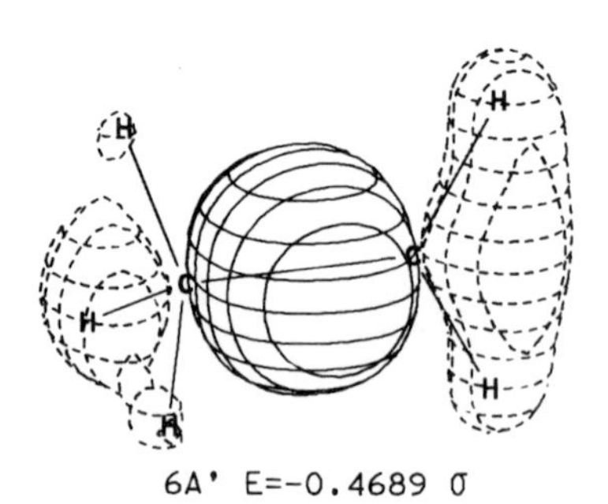

6A' E=-0.4689 σ

29. Ethane, Staggered

Symmetry: D_{3d}

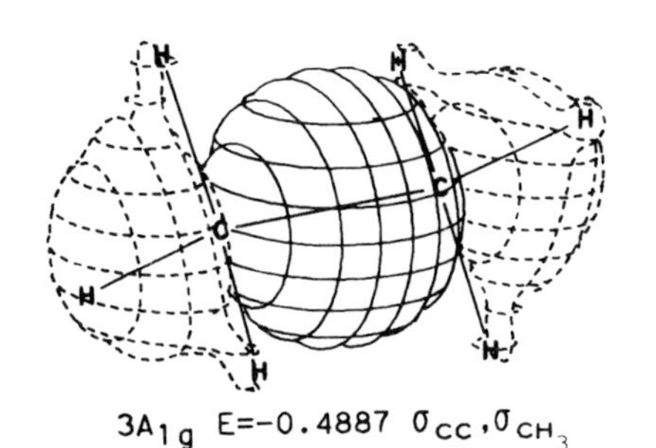

$3A_{1g}$ E=-0.4887 $\sigma_{CC}, \sigma_{CH_3}$

$1E_u$ E=-0.5955 π_{CH_3}

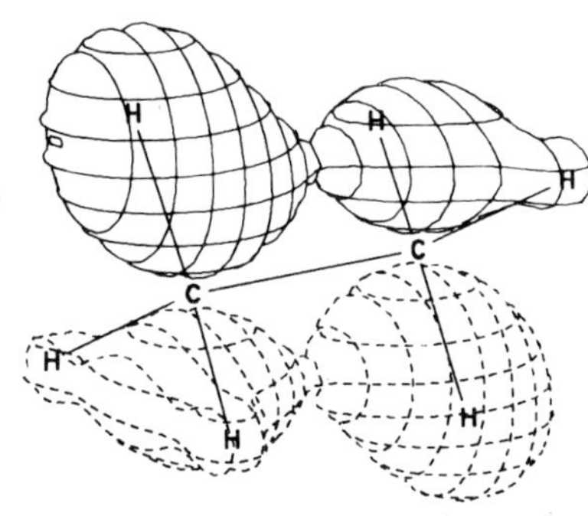

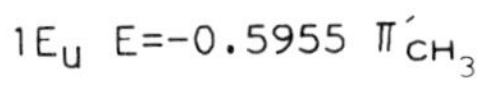

$1E_u$ E=-0.5955 π'_{CH_3}

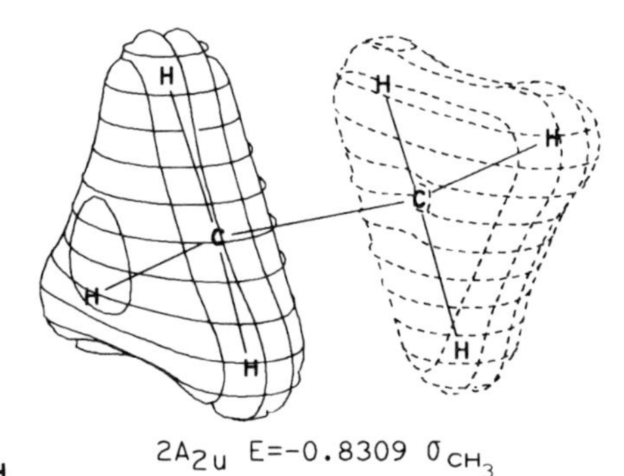

$2A_{2u}$ E=-0.8309 σ_{CH_3}

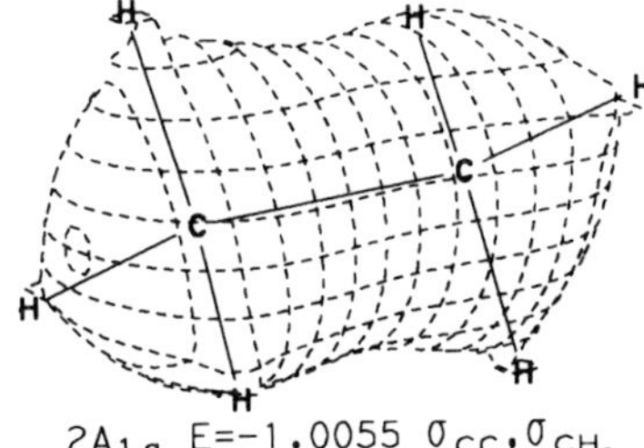

$2A_{1g}$ E=-1.0055 $\sigma_{CC}, \sigma_{CH_3}$

Ethane, Staggered (Continued)

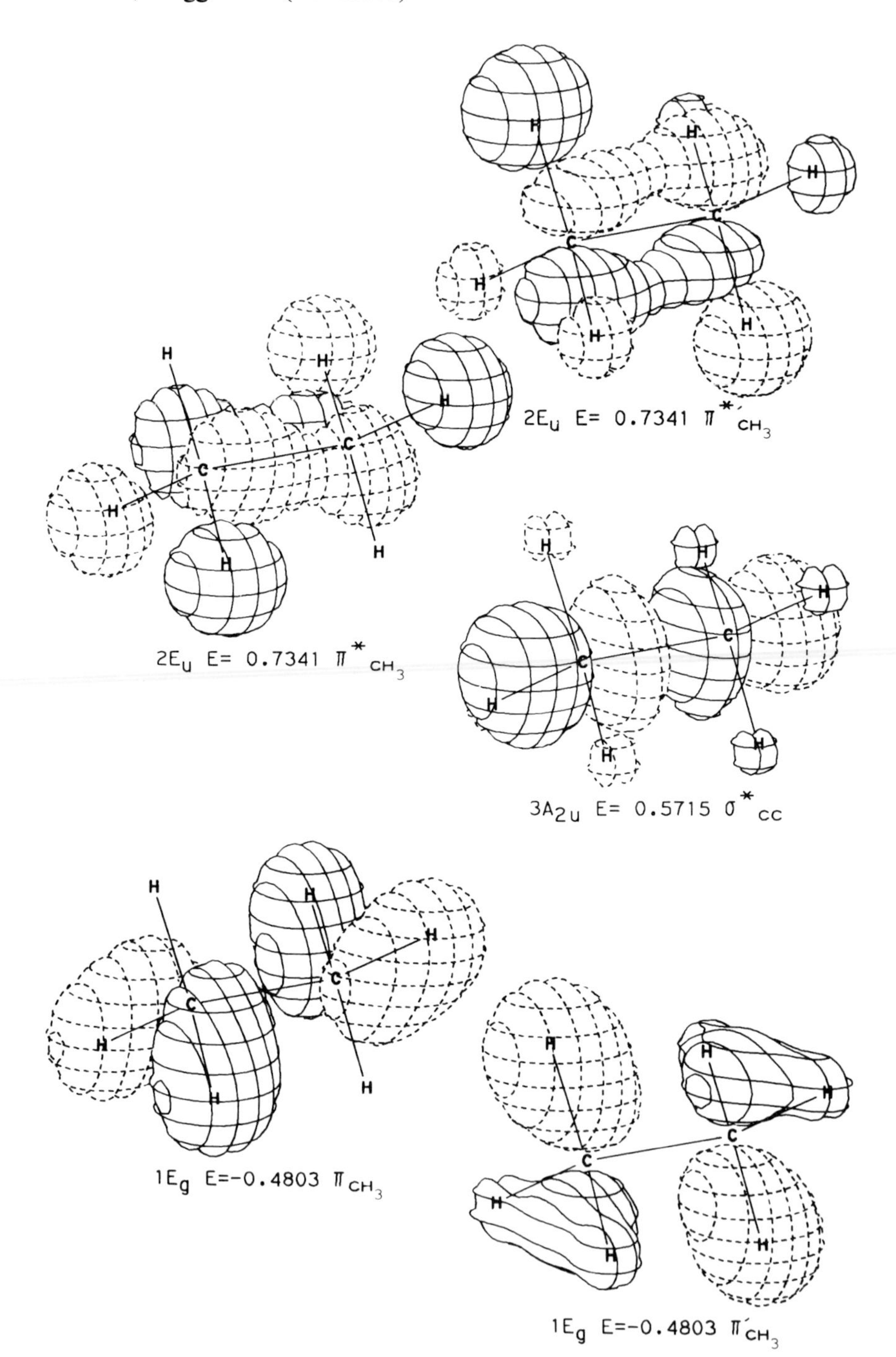

30. Ethane, Eclipsed

Symmetry: D_{3h}

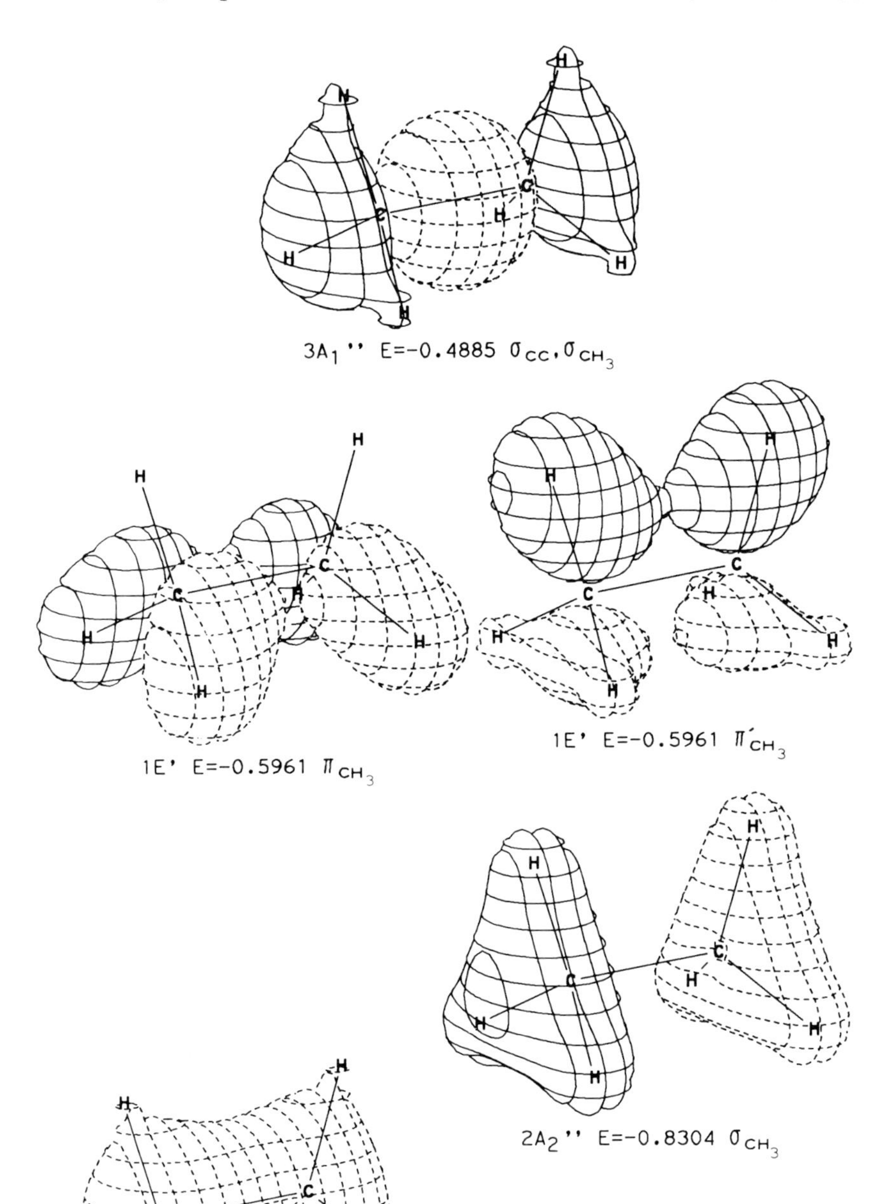

$3A_1''$ E=-0.4885 $\sigma_{CC}, \sigma_{CH_3}$

$1E'$ E=-0.5961 π_{CH_3}

$1E'$ E=-0.5961 π'_{CH_3}

$2A_2''$ E=-0.8304 σ_{CH_3}

$2A_1''$ E=-1.0052 $\sigma_{CC}, \sigma_{CH_3}$

Ethane, Eclipsed (Continued)

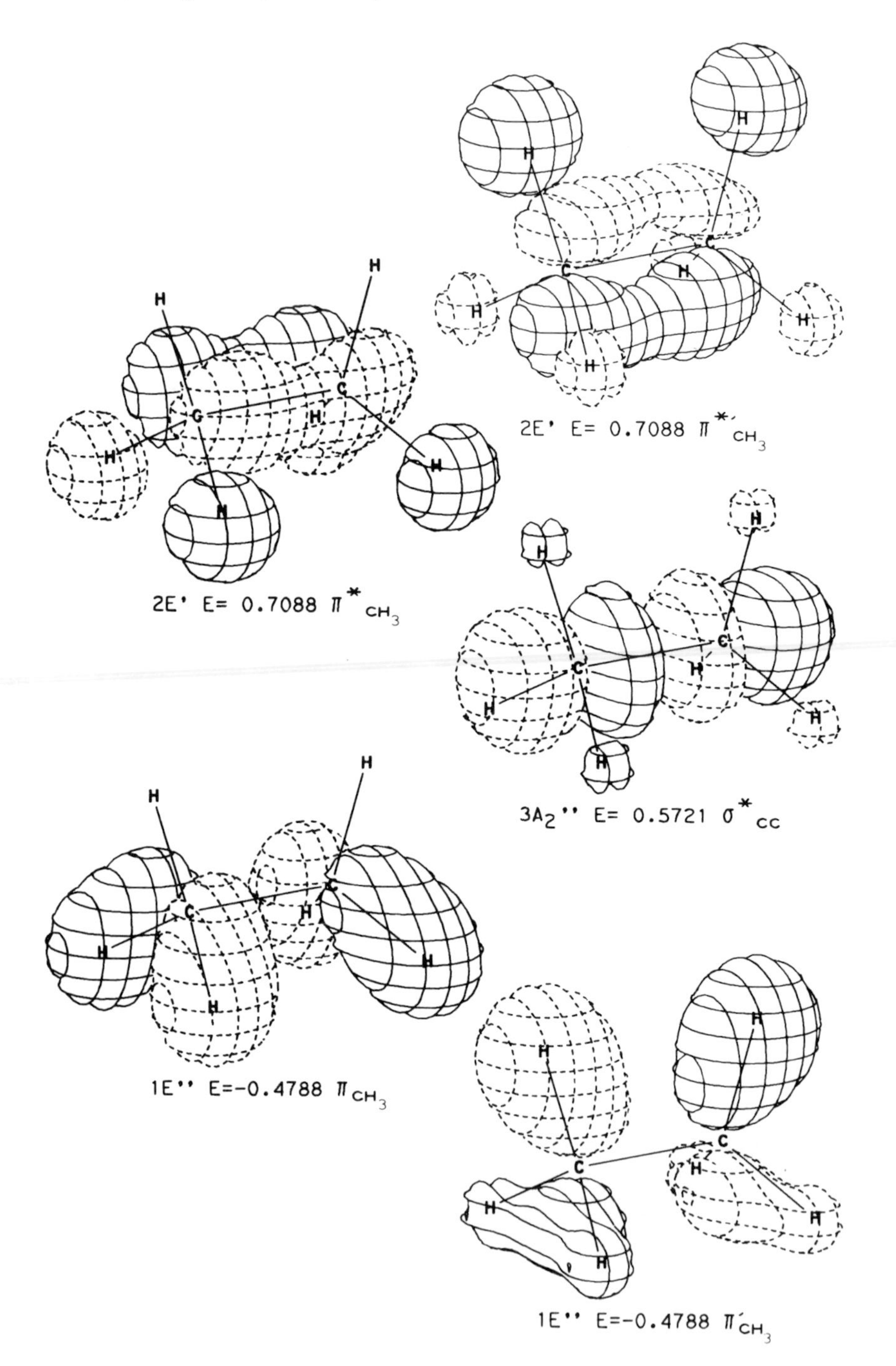

2E' E= 0.7088 $\pi^*_{CH_3}$

2E' E= 0.7088 $\pi^{*\prime}_{CH_3}$

3A$_2$'' E= 0.5721 σ^*_{CC}

1E'' E=-0.4788 π_{CH_3}

1E'' E=-0.4788 π'_{CH_3}

31. Methylamine

Symmetry: C_s

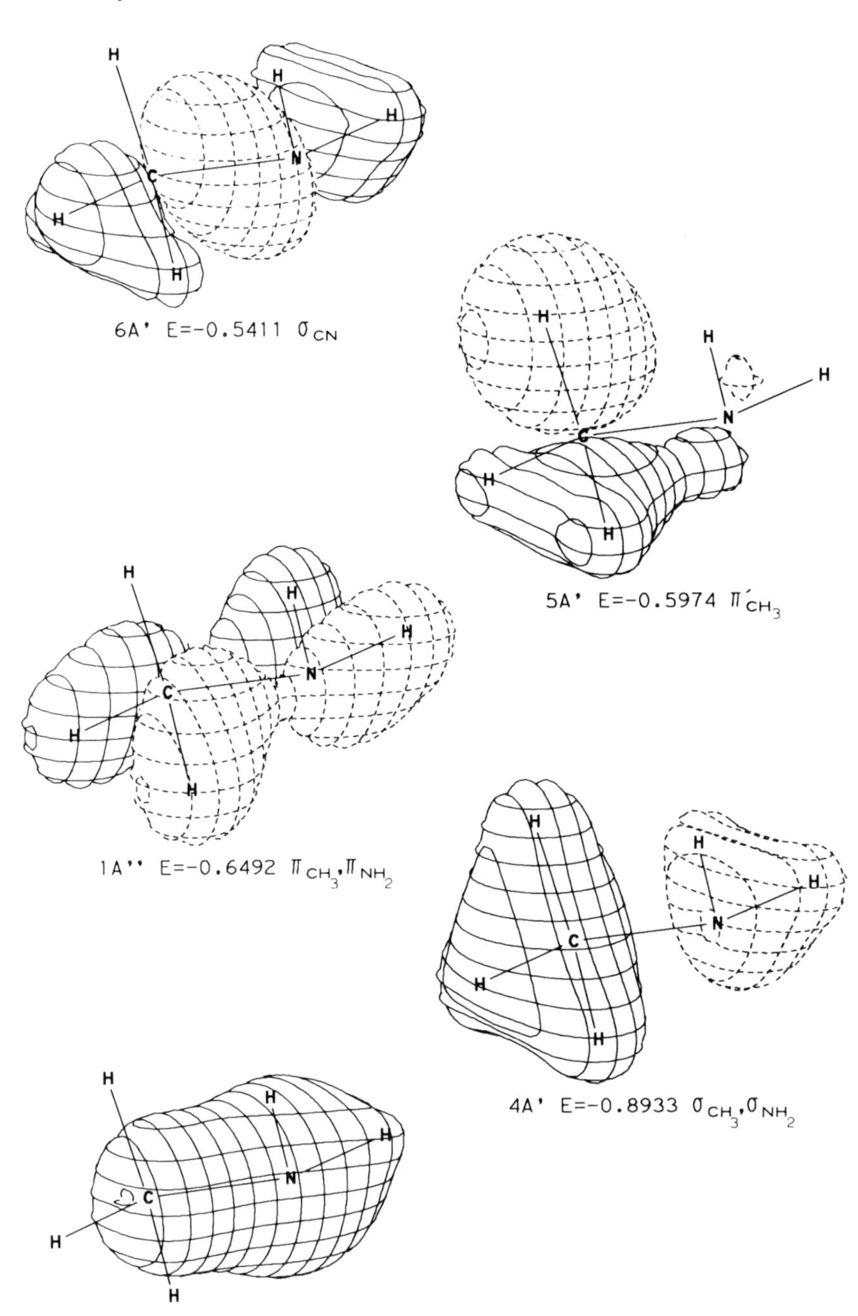

6A' E=-0.5411 σ_{CN}

5A' E=-0.5974 π'_{CH_3}

1A'' E=-0.6492 π_{CH_3}, π_{NH_2}

4A' E=-0.8933 $\sigma_{CH_3}, \sigma_{NH_2}$

3A' E=-1.1474 $\sigma_{CN}, \sigma_{NH_2}$

Methylamine (Continued)

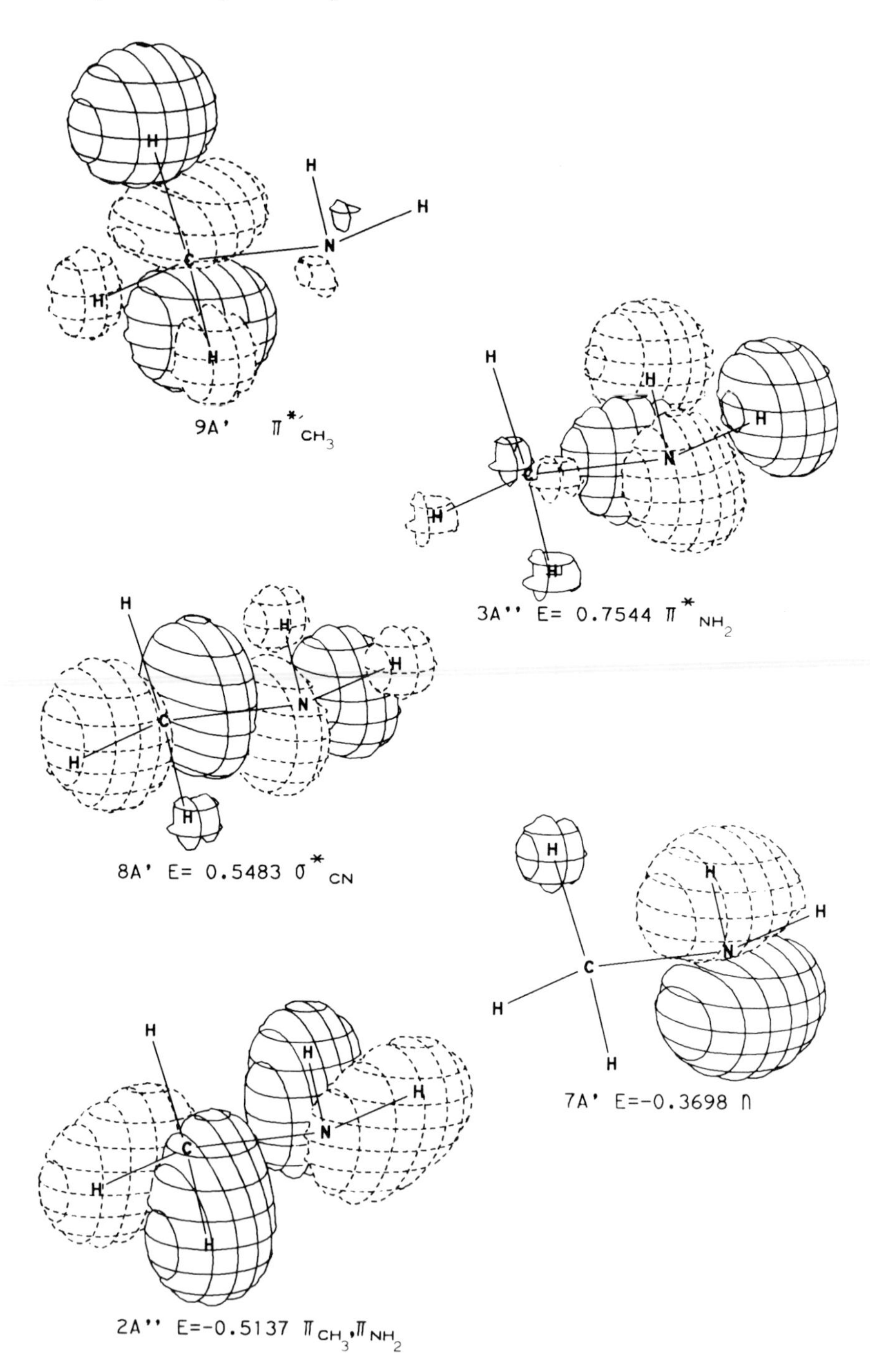

9A' $\pi^{*\prime}_{CH_3}$

3A'' E= 0.7544 $\pi^{*}_{NH_2}$

8A' E= 0.5483 σ^{*}_{CN}

7A' E=-0.3698 n

2A'' E=-0.5137 π_{CH_3}, π_{NH_2}

32. Methanol

Symmetry: C_s

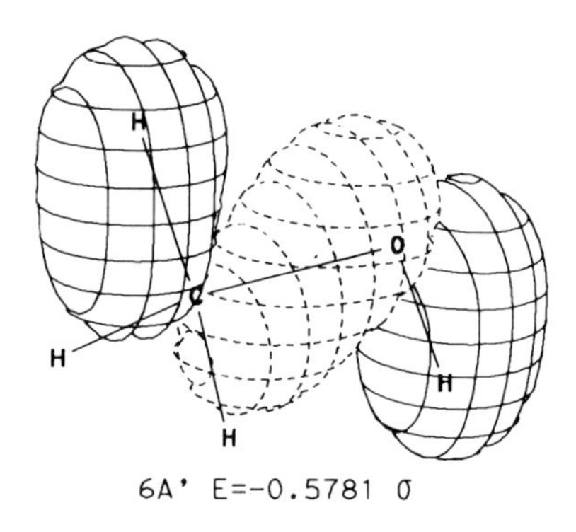

6A' E=-0.5781 σ

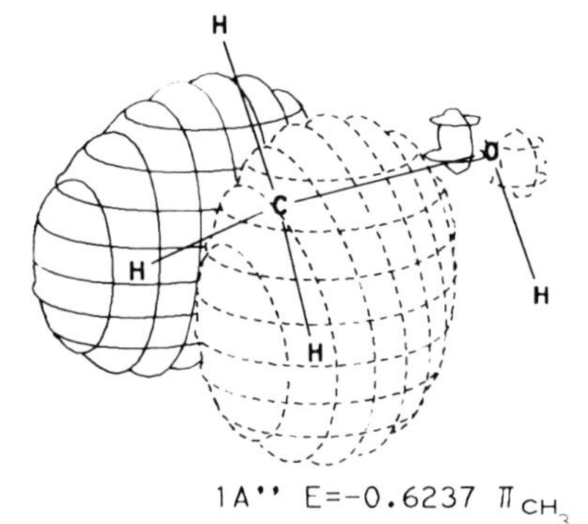

1A'' E=-0.6237 π_{CH_3}

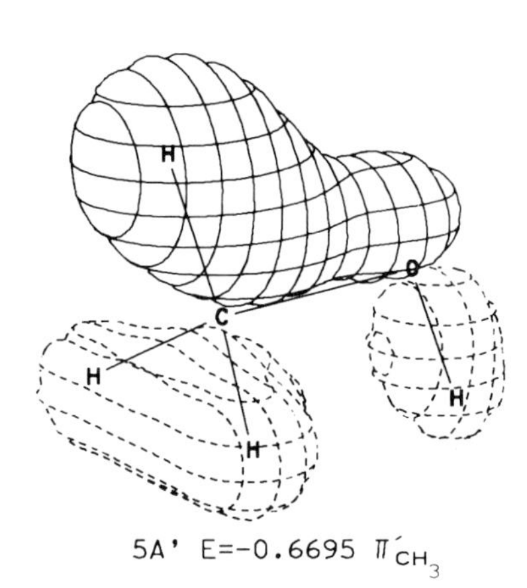

5A' E=-0.6695 π'_{CH_3}

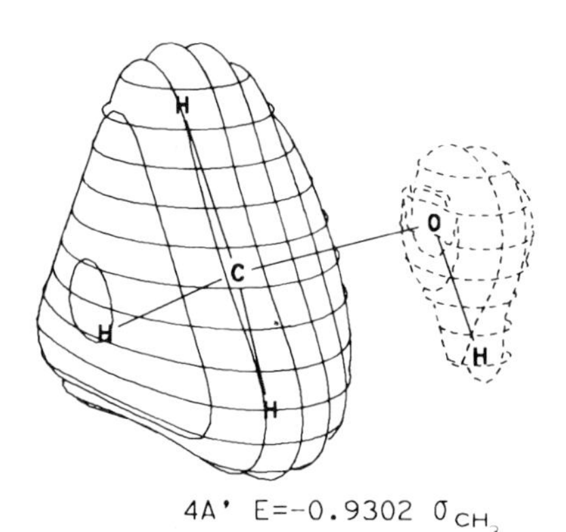

4A' E=-0.9302 σ_{CH_3}

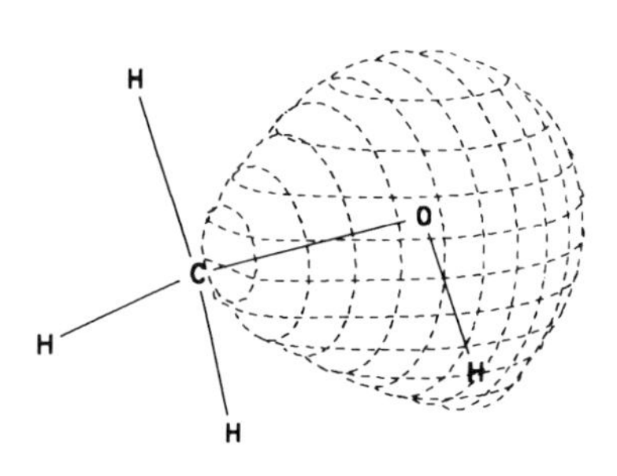

3A' E=-1.3293 σ_{OH}, σ_{CO}

Methanol (Continued)

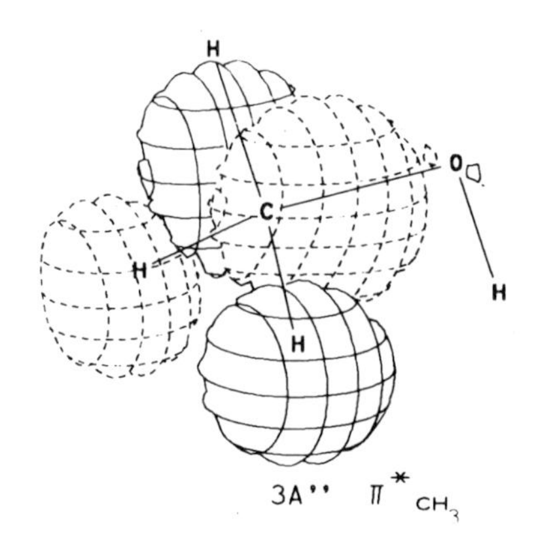

3A'' $\pi^{*}_{CH_3}$

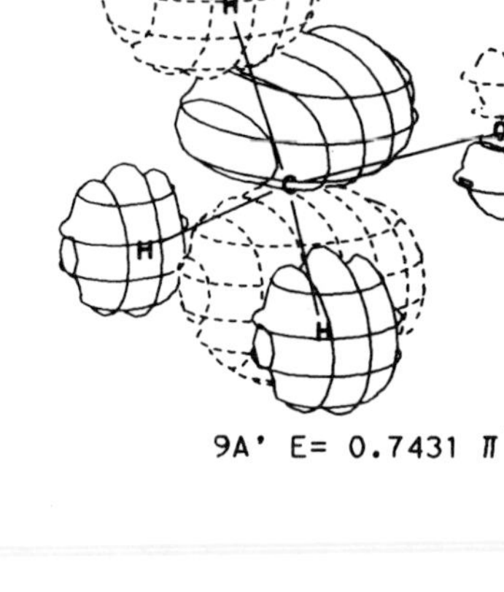

9A' E= 0.7431 $\pi^{*\prime}_{CH_3}$

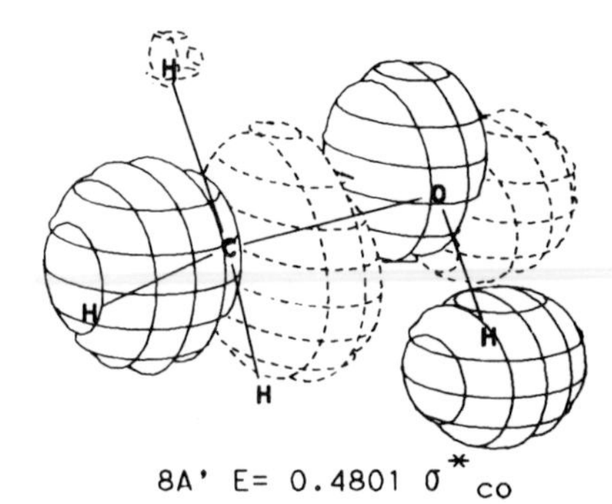

8A' E= 0.4801 σ^{*}_{CO}

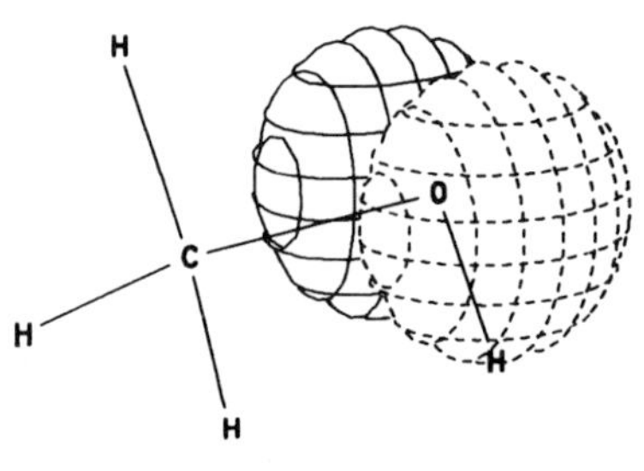

2A'' E=-0.4331 n

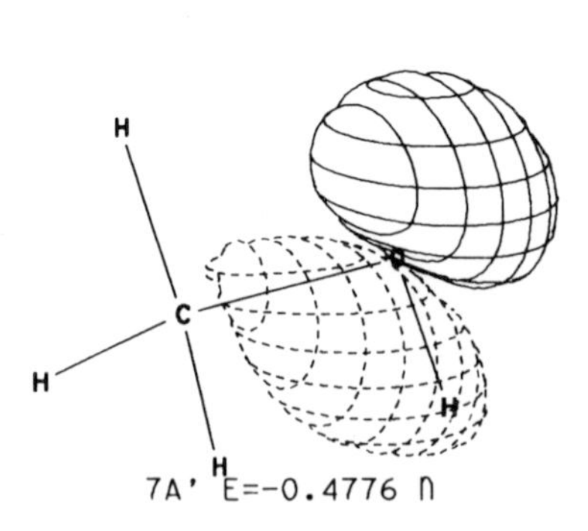

7A' E=-0.4776 n

33. Methyl Fluoride

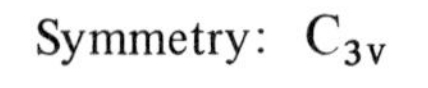
Symmetry: C_{3v}

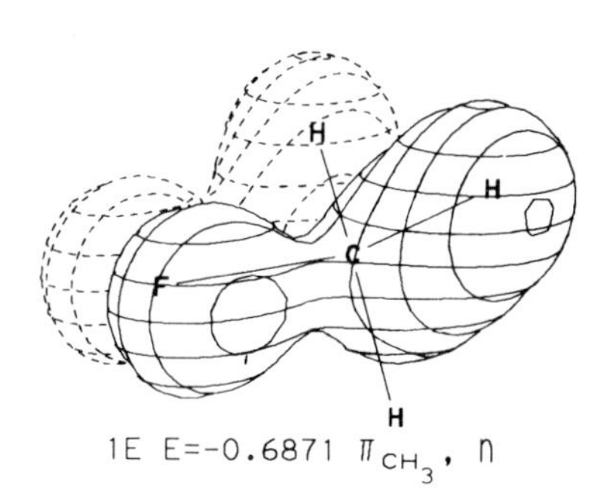

1E E=-0.6871 π_{CH_3}, n

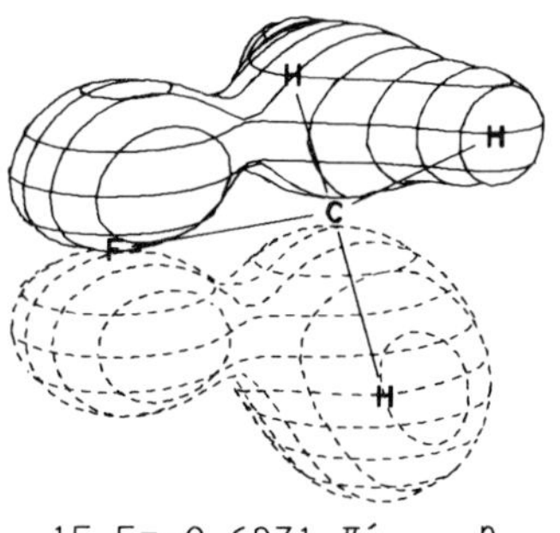

1E E=-0.6871 π'_{CH_3}, n

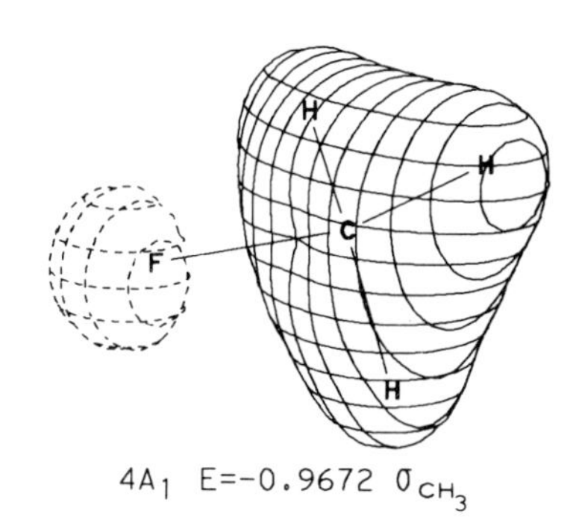

$4A_1$ E=-0.9672 σ_{CH_3}

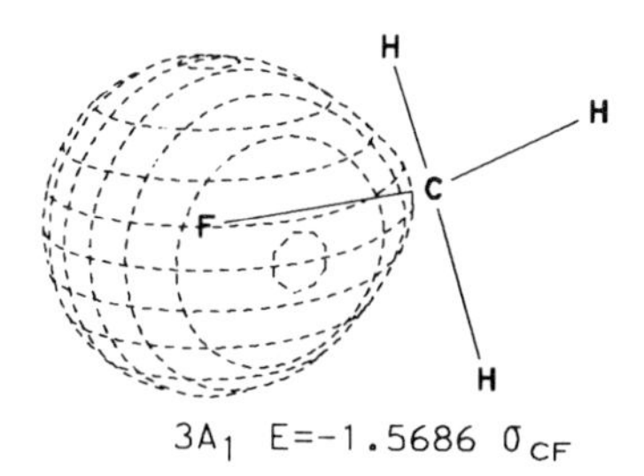

$3A_1$ E=-1.5686 σ_{CF}

Methyl Fluoride (Continued)

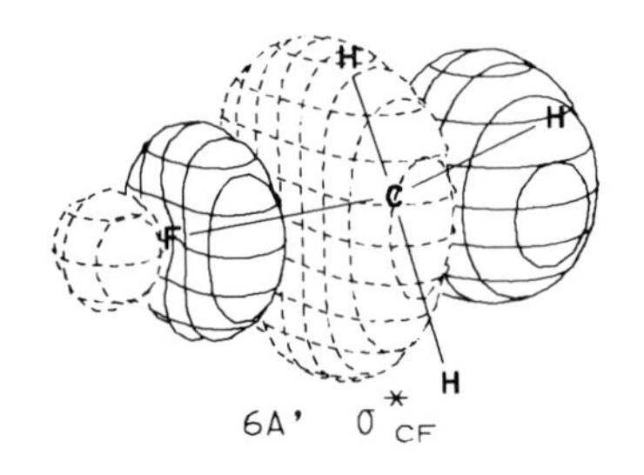

6A' σ^*_{CF}

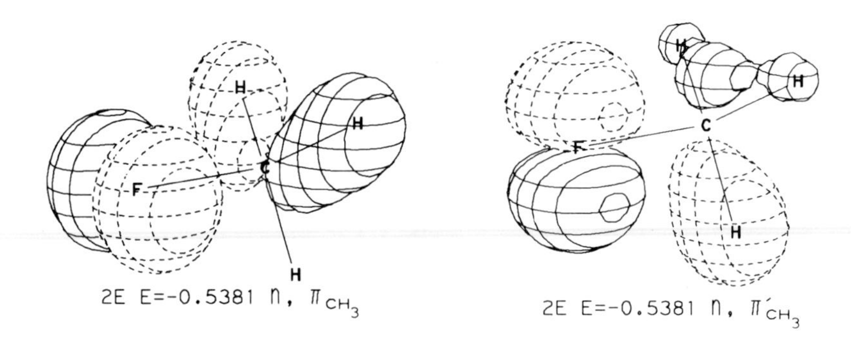

2E E=-0.5381 n, π_{CH_3}

2E E=-0.5381 n, π'_{CH_3}

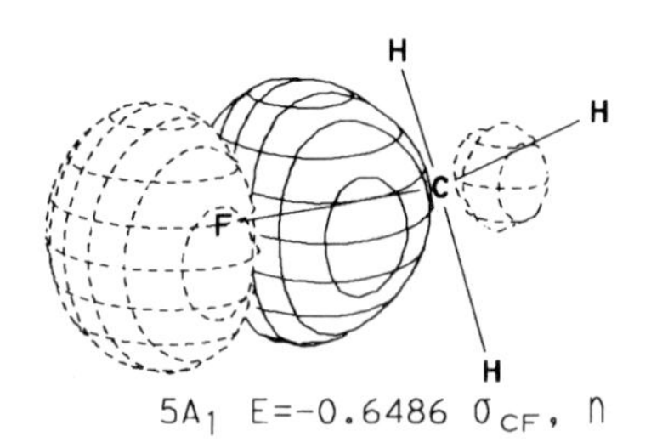

$5A_1$ E=-0.6486 σ_{CF}, n

34. Hydrazine

Symmetry: C_2

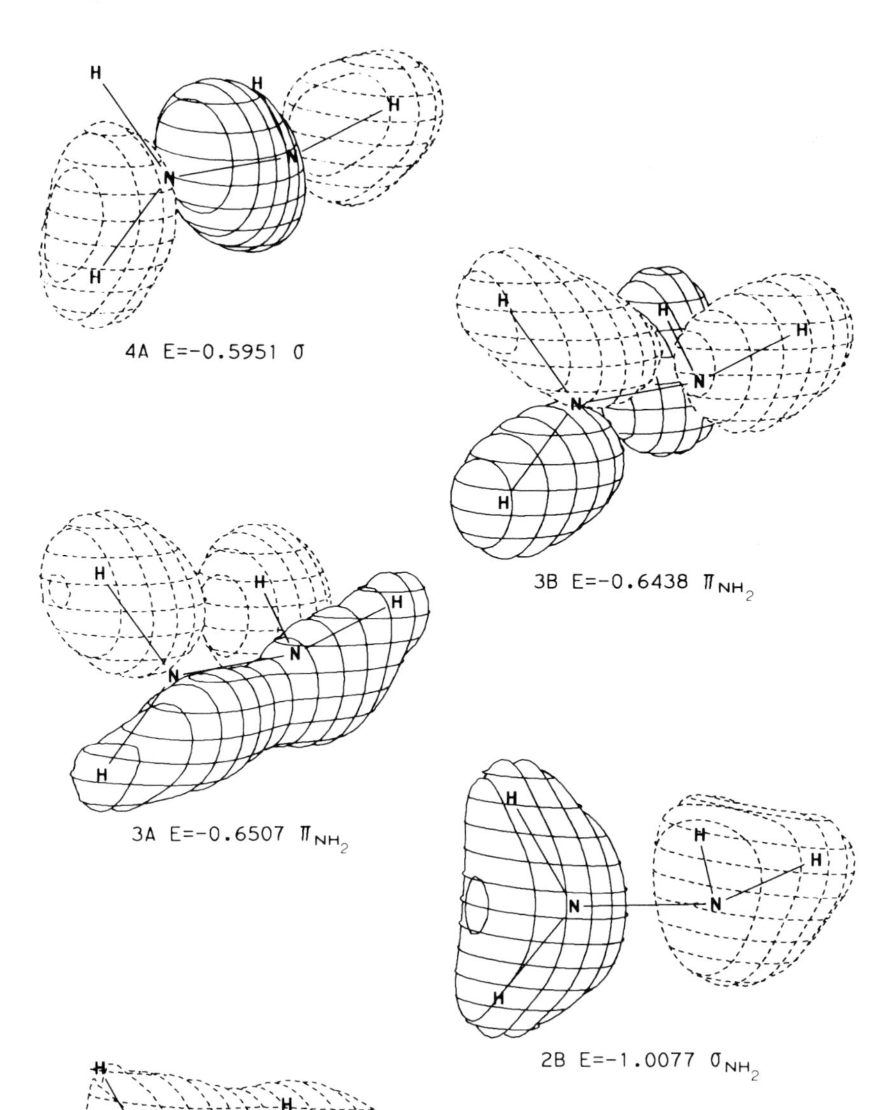

4A E=-0.5951 σ

3B E=-0.6438 π_{NH_2}

3A E=-0.6507 π_{NH_2}

2B E=-1.0077 σ_{NH_2}

2A E=-1.2257 σ_{NN}, σ_{NH_2}

Hydrazine (Continued)

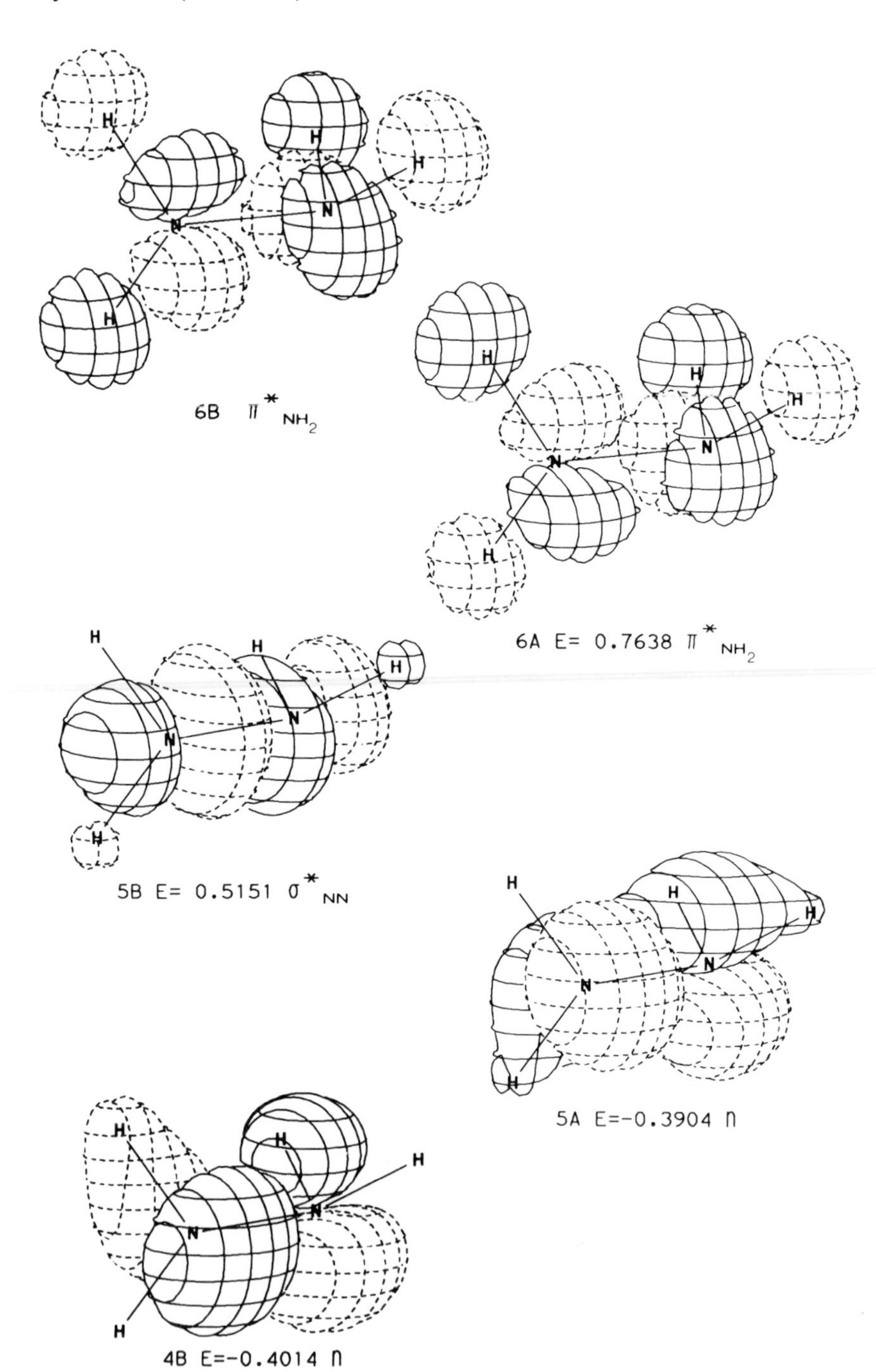

35. Hydrogen Peroxide

Symmetry: C_2

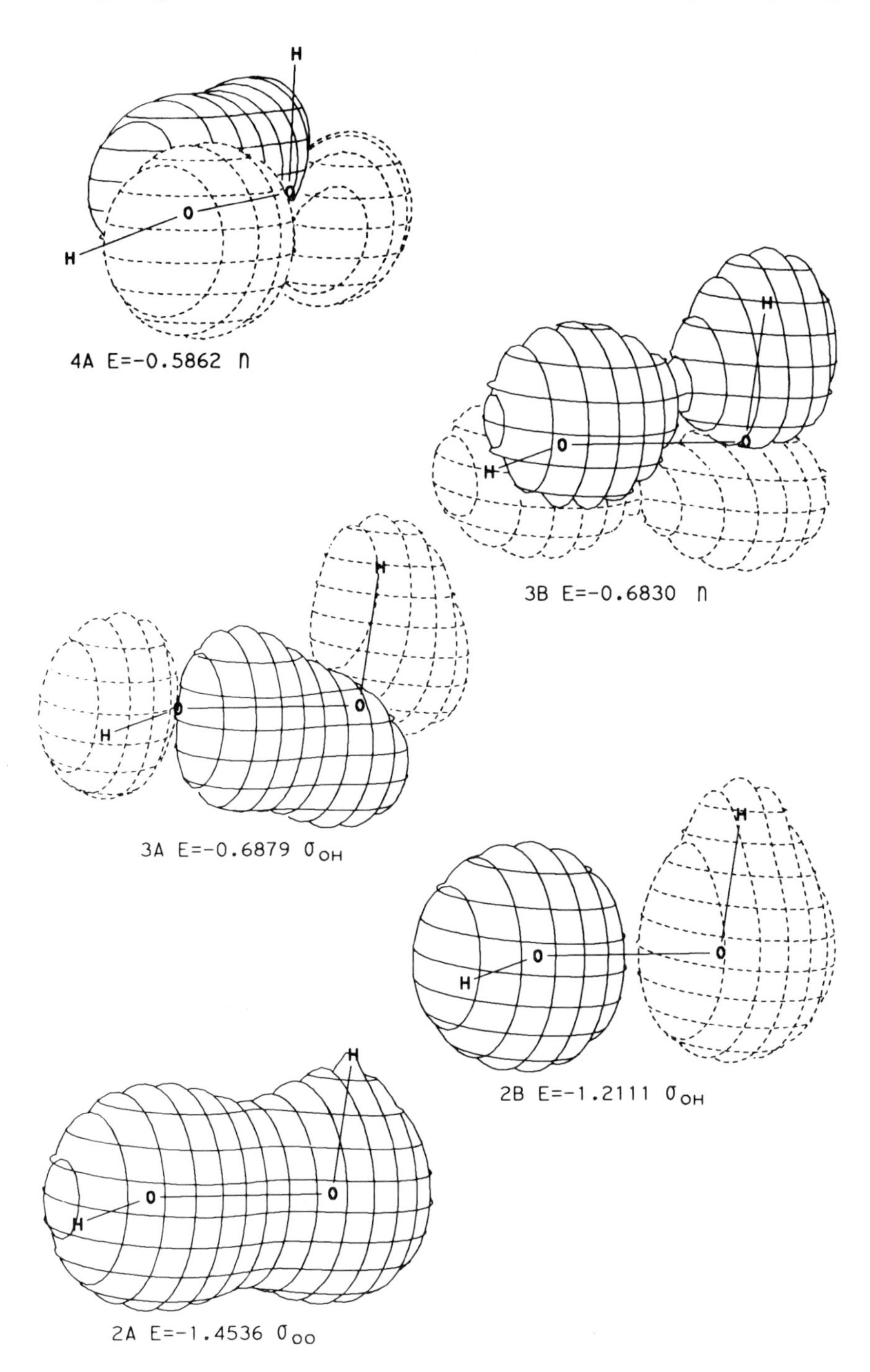

Hydrogen Peroxide (Continued)

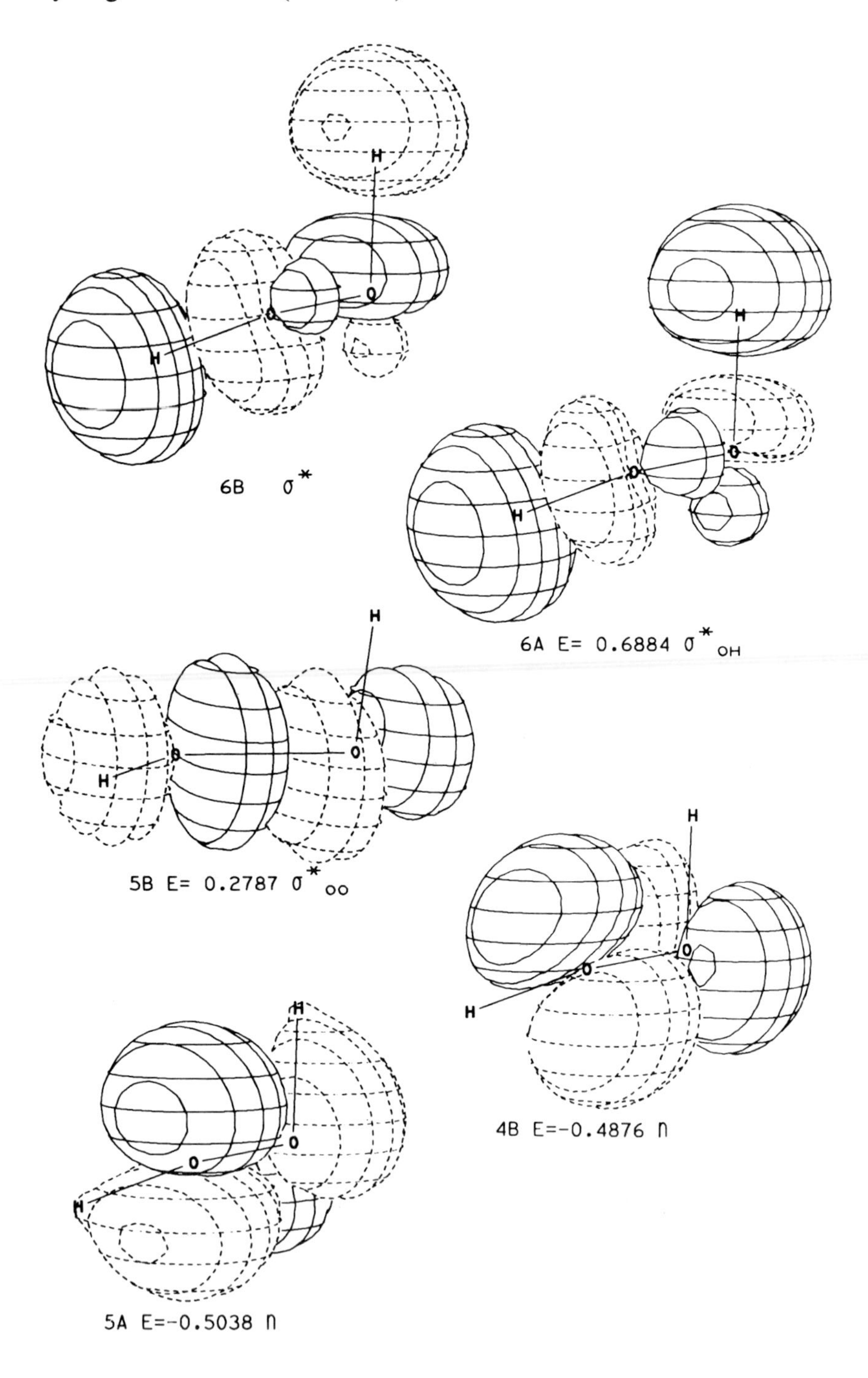

36. Fluorine

Symmetry: $D_{\infty h}$

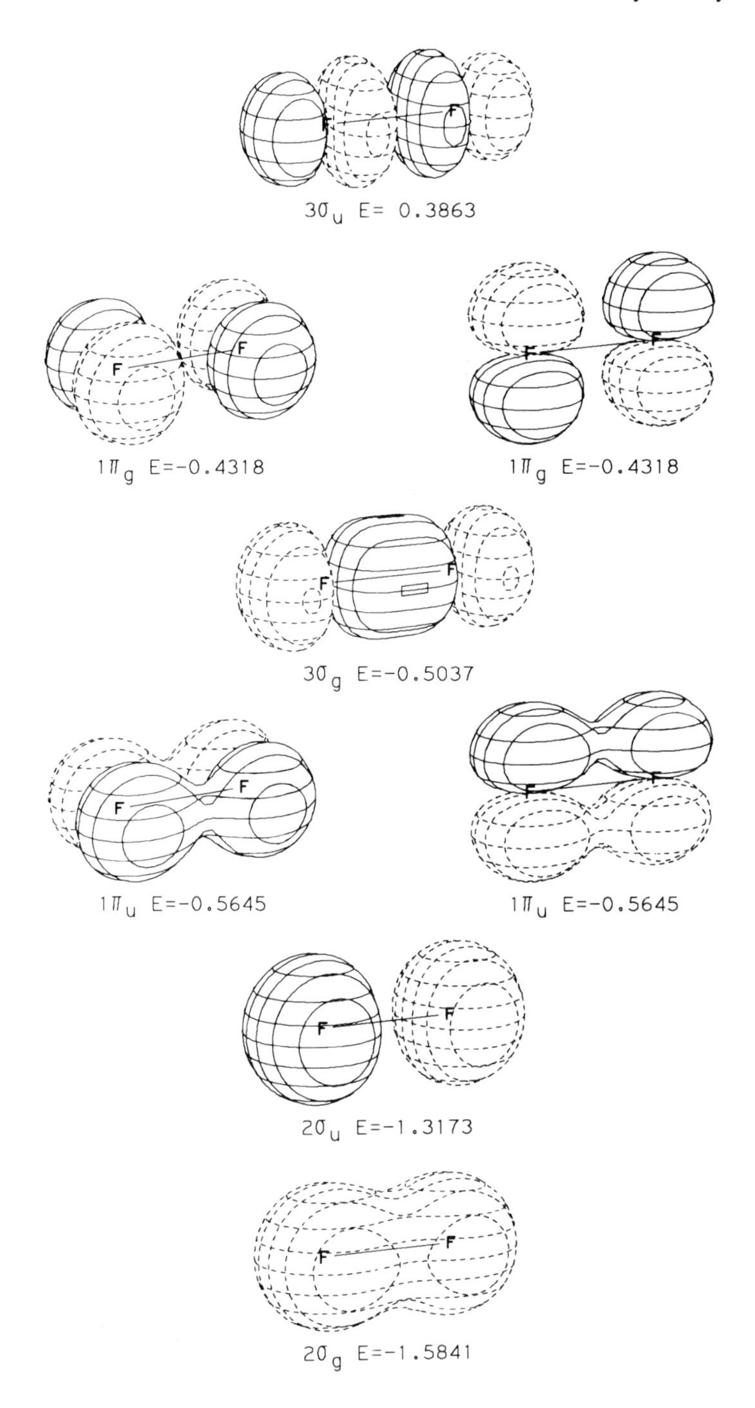

37. Cyclopropenium Cation

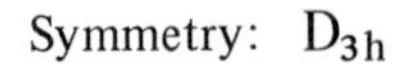
Symmetry: D_{3h}

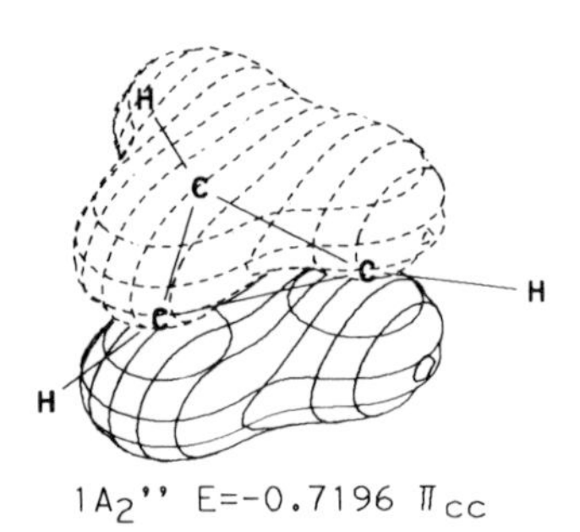

$1A_2''$ E=-0.7196 π_{CC}

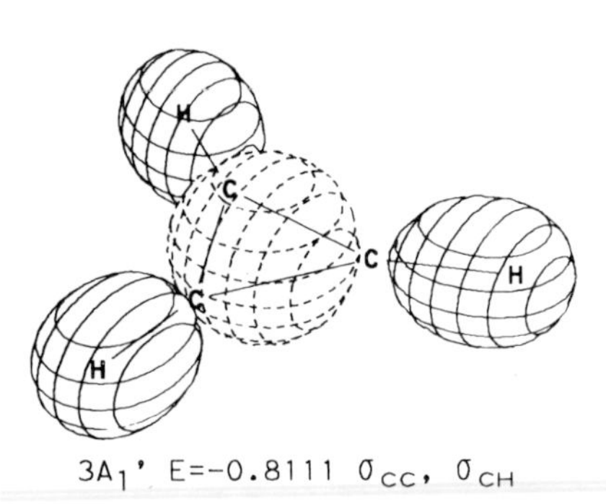

$3A_1'$ E=-0.8111 σ_{CC}, σ_{CH}

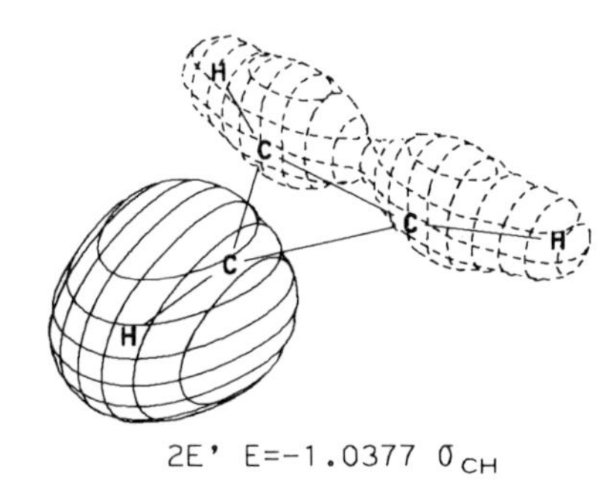

$2E'$ E=-1.0377 σ_{CH}

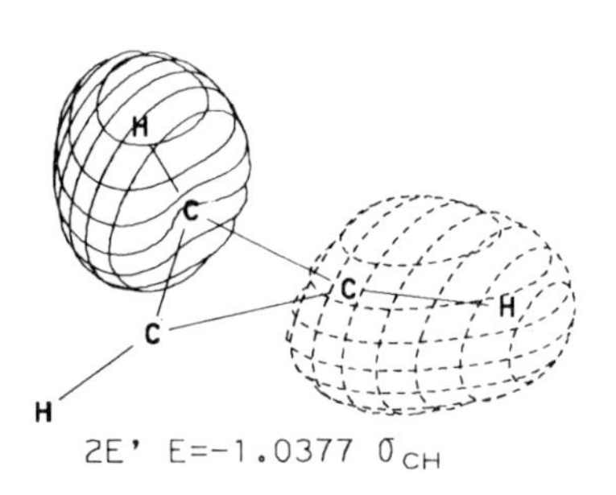

$2E'$ E=-1.0377 σ_{CH}

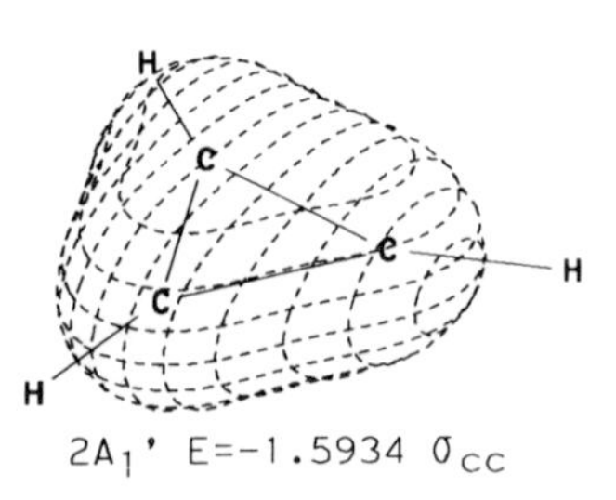

$2A_1'$ E=-1.5934 σ_{CC}

Cyclopropenium Cation Continued)

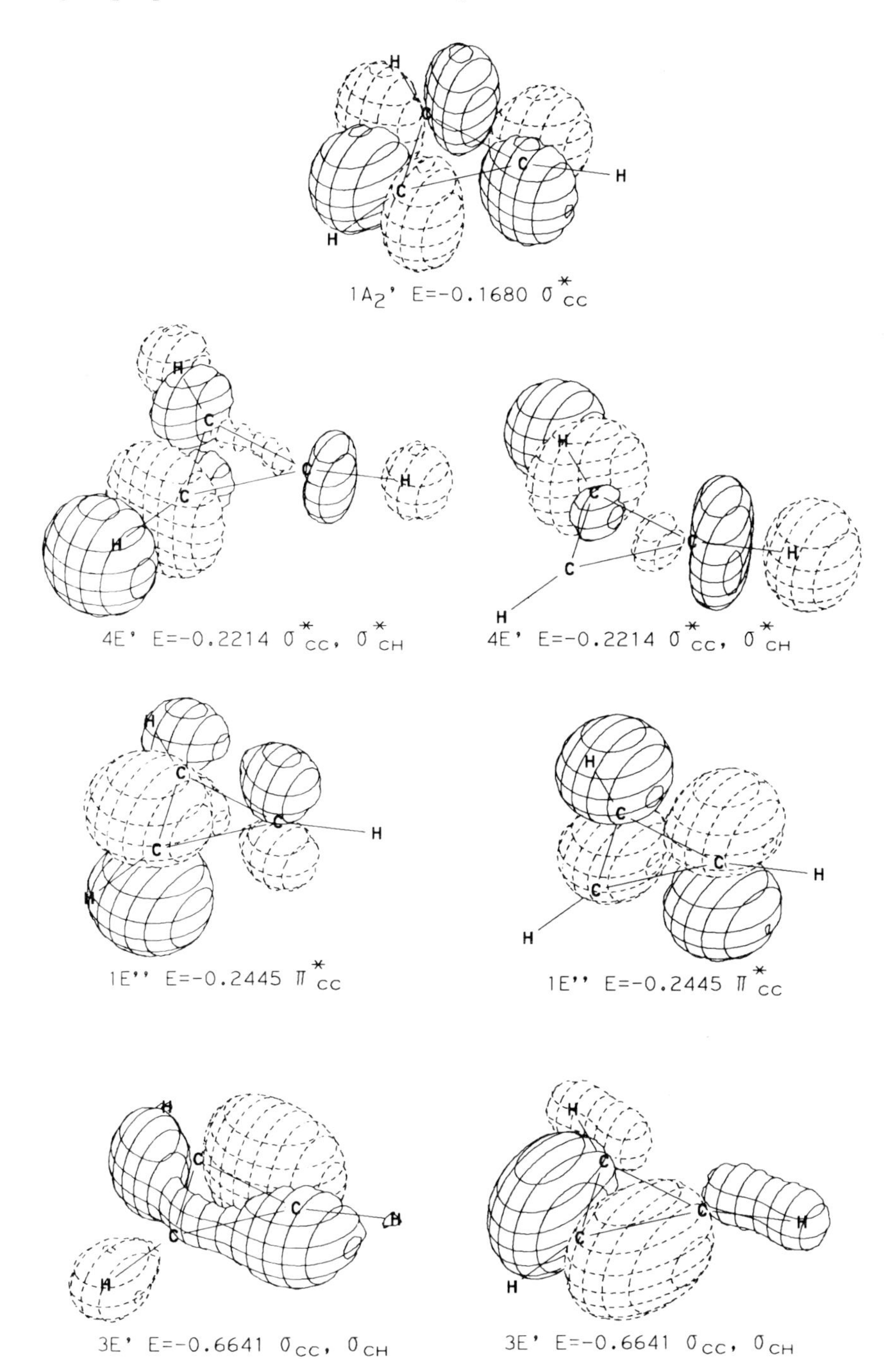

38. Methyl Acetylene

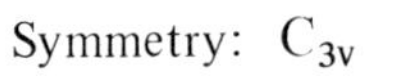

Symmetry: C_{3v}

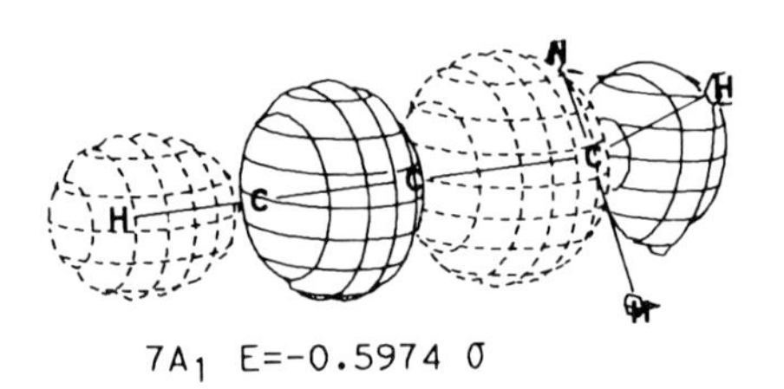

$7A_1$ E=−0.5974 σ

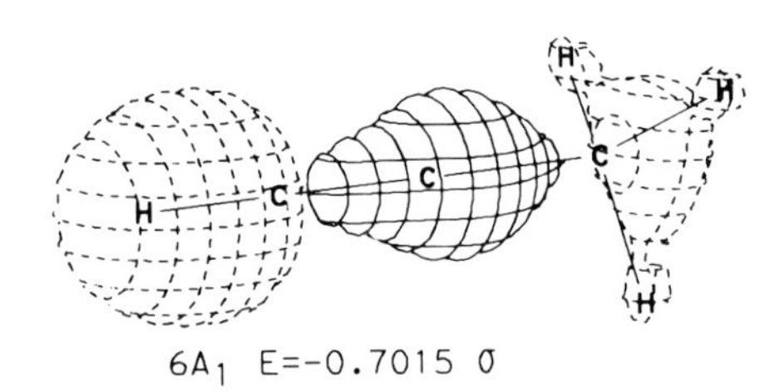

$6A_1$ E=−0.7015 σ

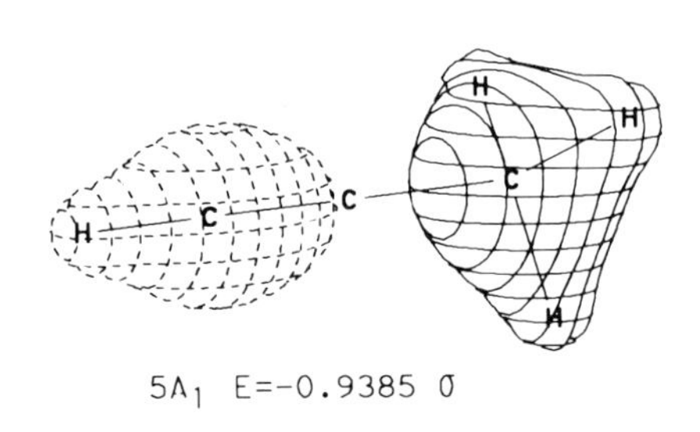

$5A_1$ E=−0.9385 σ

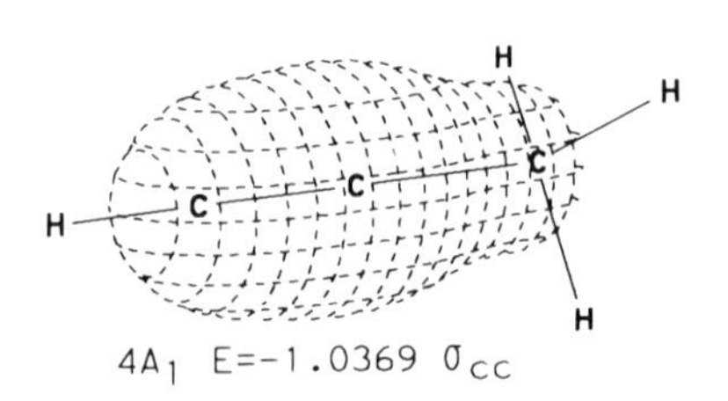

$4A_1$ E=−1.0369 σ_{CC}

Methyl Acetylene (Continued)

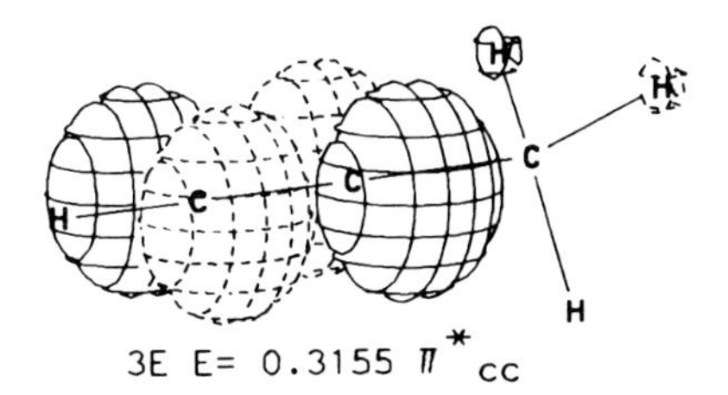

3E E= 0.3155 π^*_{CC}

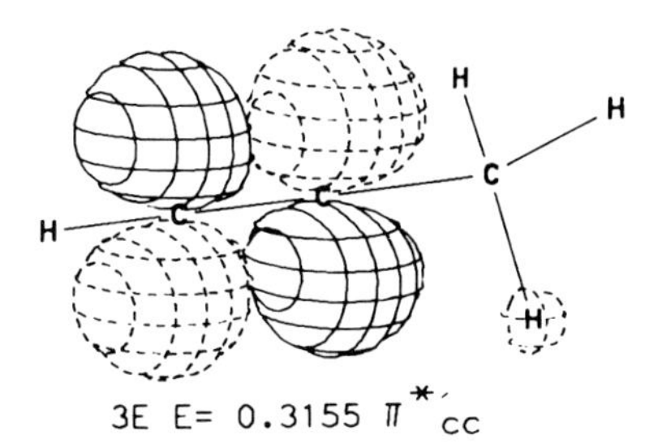

3E E= 0.3155 $\pi^{*\prime}_{CC}$

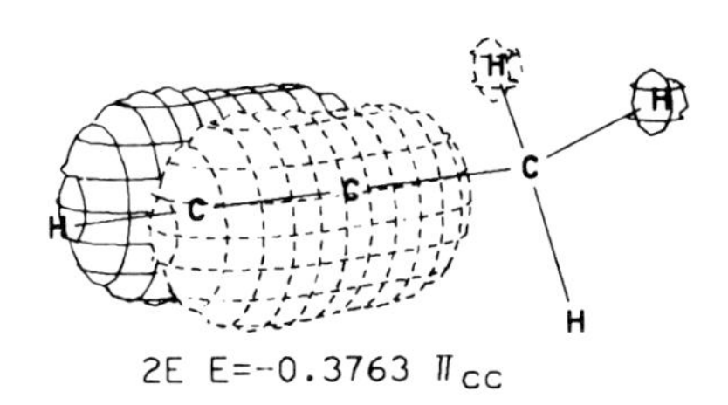

2E E=-0.3763 π_{CC}

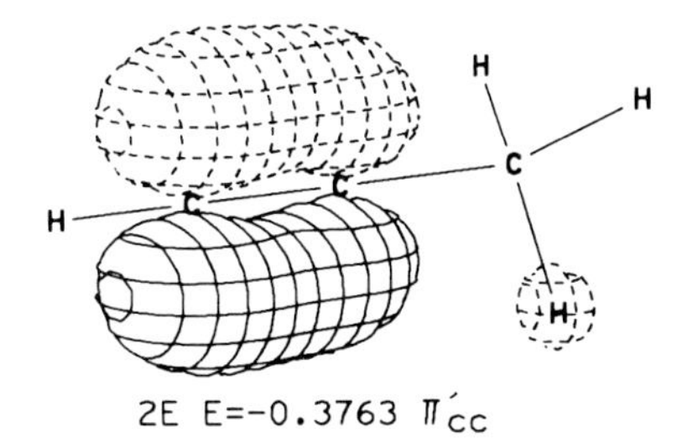

2E E=-0.3763 π'_{CC}

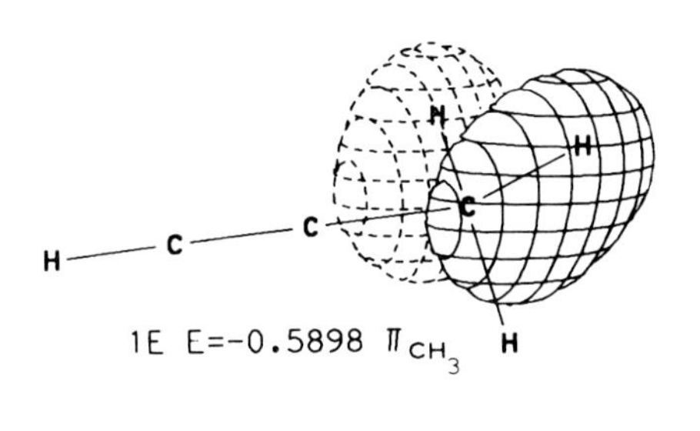

1E E=-0.5898 π_{CH_3}

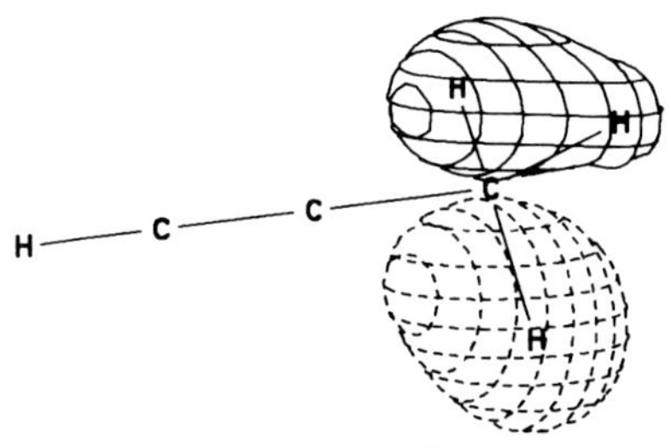

1E E=-0.5898 π'_{CH_3}

39. Acetonitrile

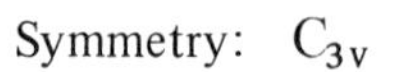

Symmetry: C_{3v}

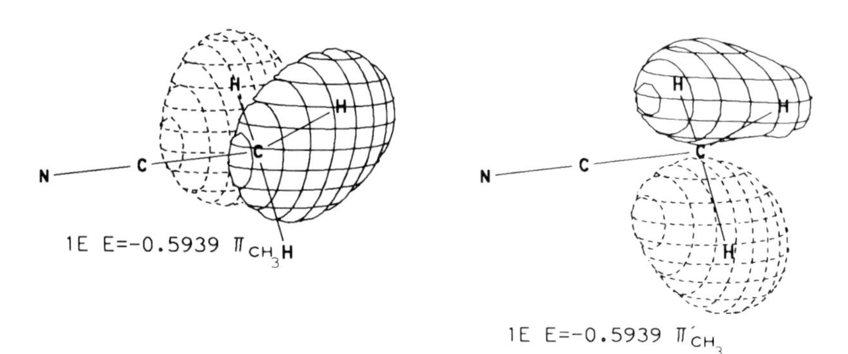

1E E=−0.5939 π_{CH_3}

1E E=−0.5939 π'_{CH_3}

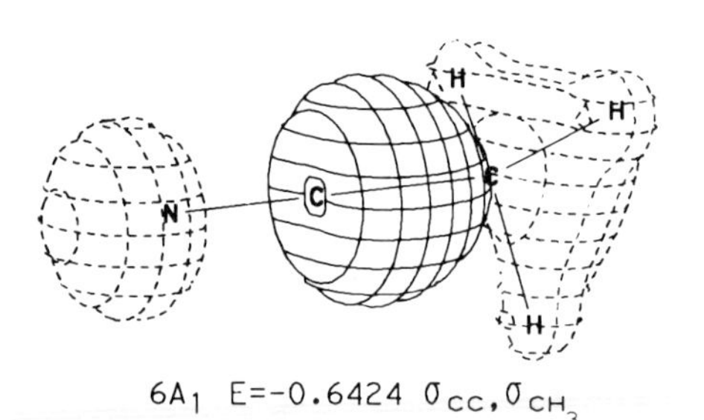

$6A_1$ E=−0.6424 $\sigma_{CC}, \sigma_{CH_3}$

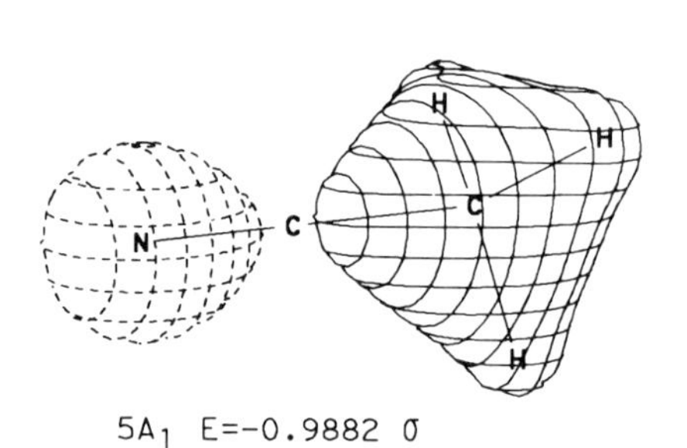

$5A_1$ E=−0.9882 σ

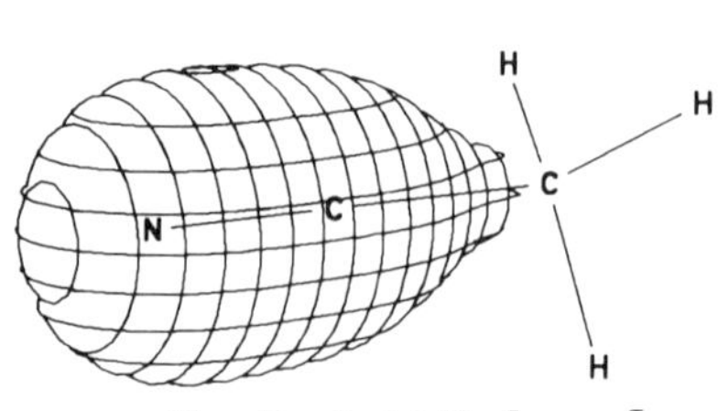

$4A_1$ E=−1.1641 σ_{CN}, σ_{CC}

Acetonitrile (Continued)

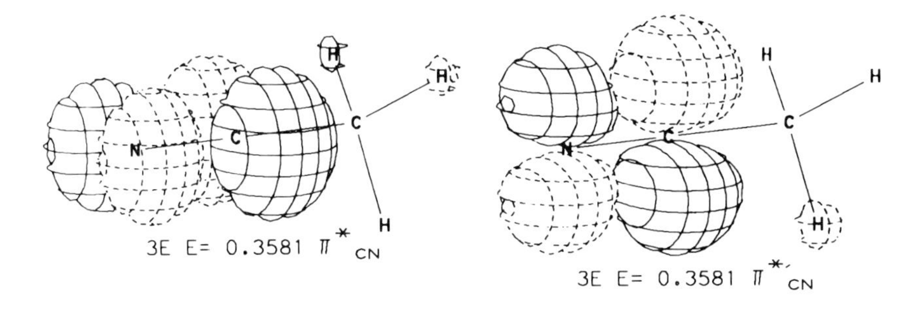

3E E= 0.3581 π^{*}_{CN}

3E E= 0.3581 $\pi^{*\prime}_{CN}$

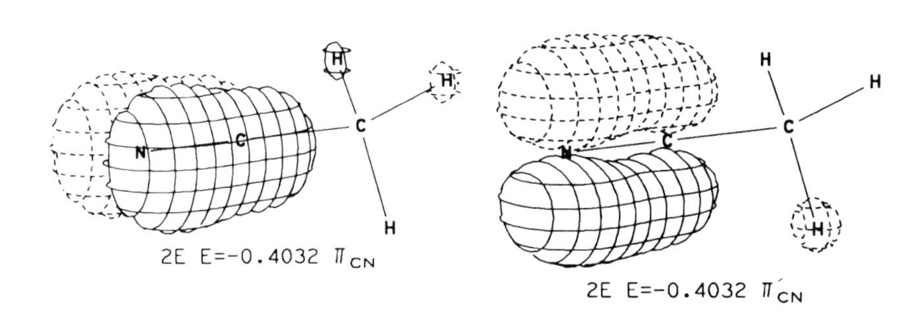

2E E=−0.4032 π_{CN}

2E E=−0.4032 π'_{CN}

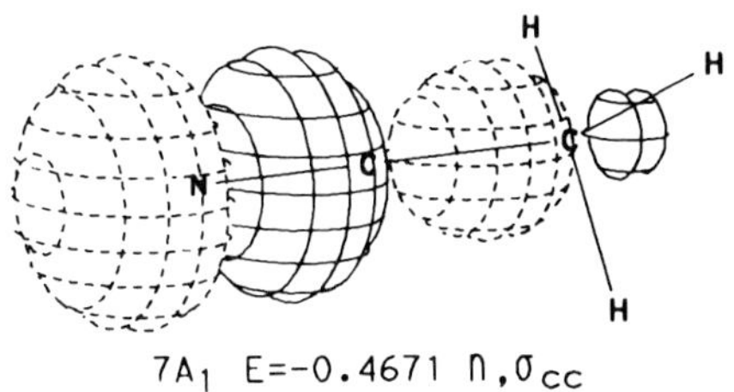

$7A_1$ E=−0.4671 n, σ_{CC}

40. Methyl Isocyanide

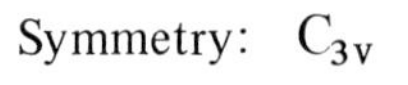

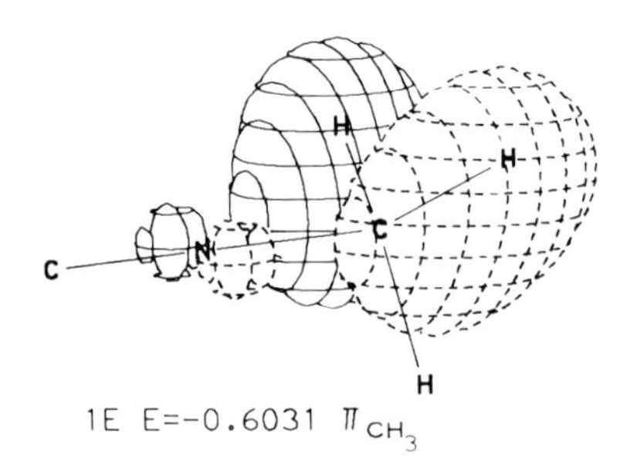

1E E=-0.6031 π_{CH_3}

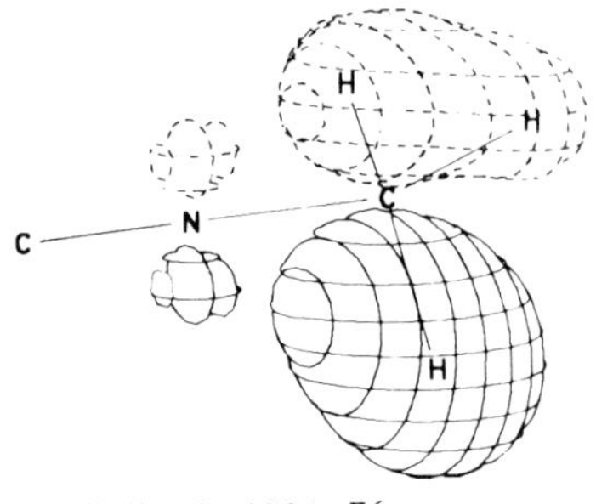

1E E=-0.6031 π'_{CH_3}

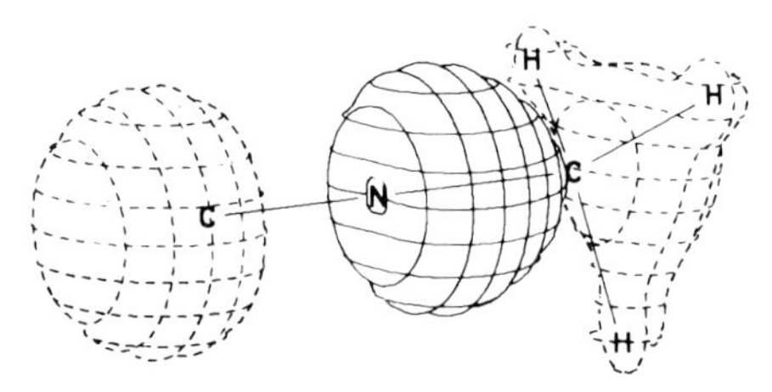

$6A_1$ E=-0.6702 σ_{CN}, σ_{CH_3}, n

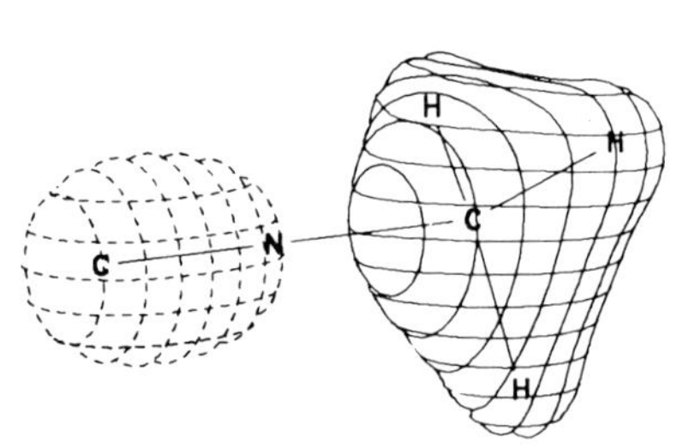

$5A_1$ E=-0.9677 σ_{CH_3}, σ_{CN}

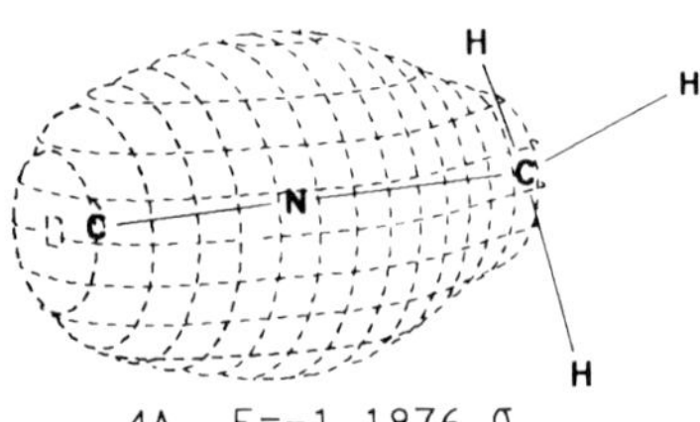

$4A_1$ E=-1.1876 σ_{CN}

Methyl Isocyanide (Continued)

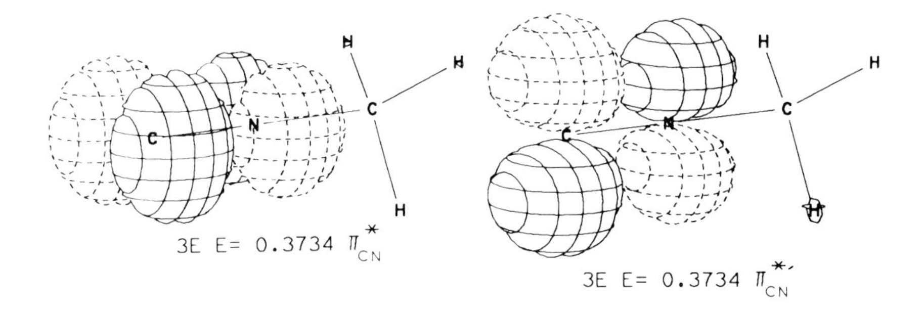

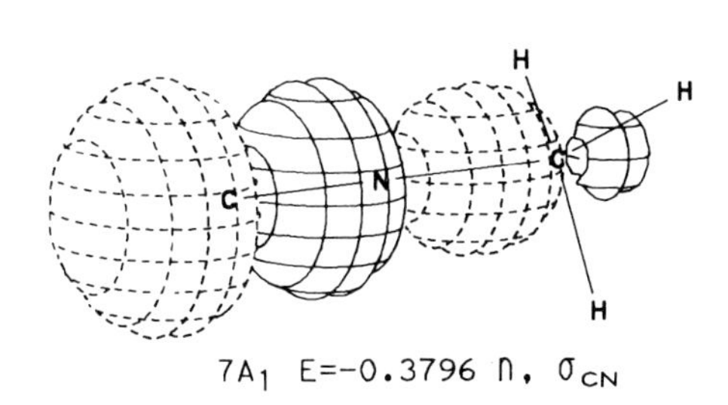

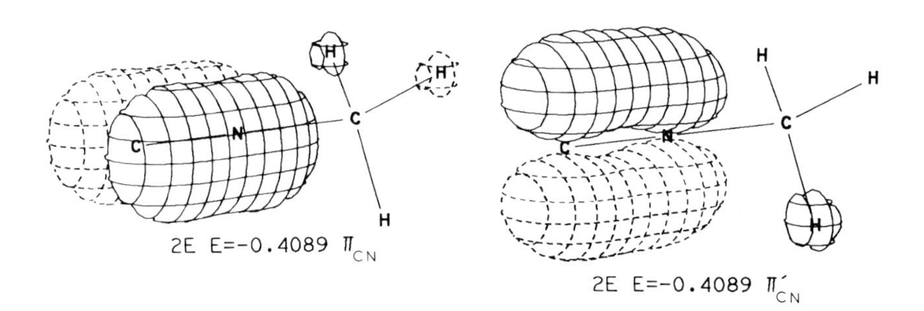

41. Allene

Symmetry: D_{2d}

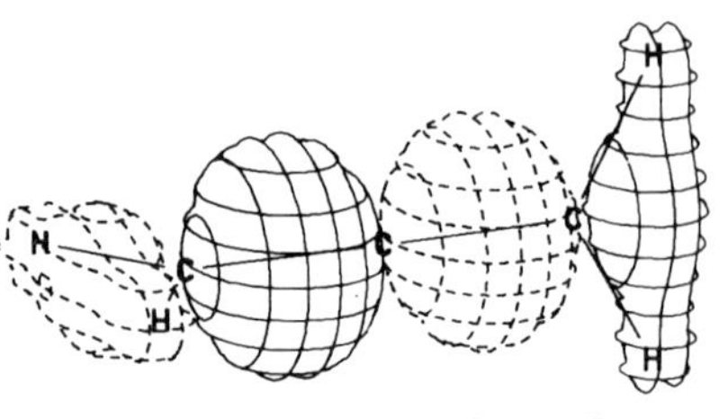

$3B_2$ E=−0.6067 σ_{CC}, σ_{CH_2}

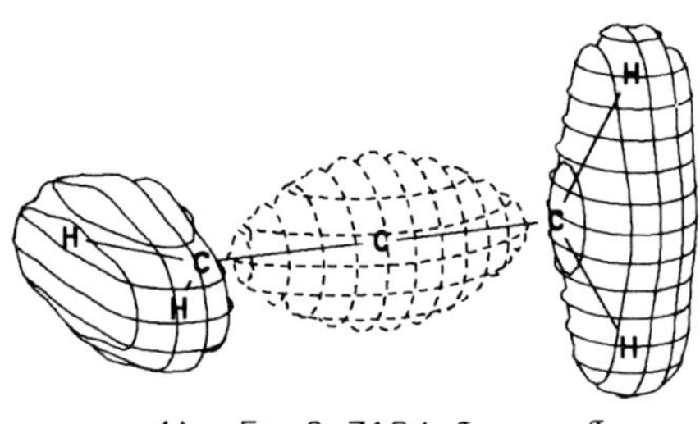

$4A_1$ E=−0.7124 σ_{CH_2}, σ_{CC}

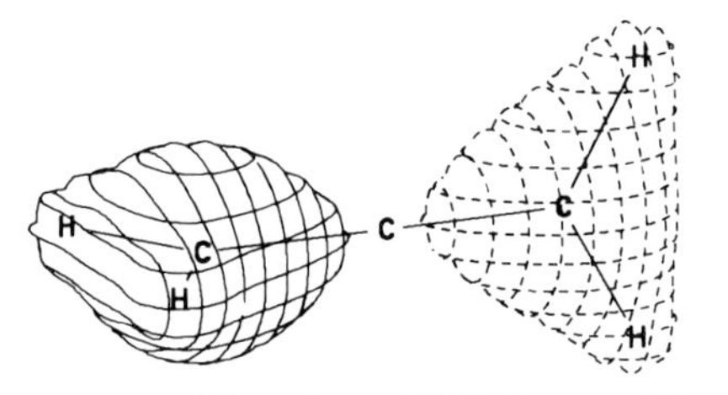

$2B_2$ E=−0.9822 σ_{CH_2}, (σ_{CC})

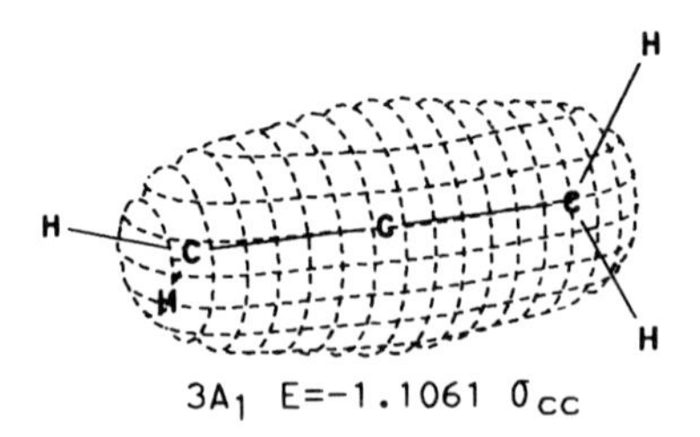

$3A_1$ E=−1.1061 σ_{CC}

Allene (Continued)

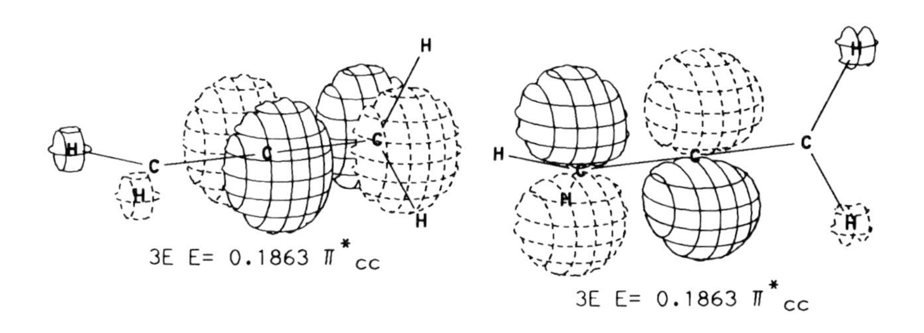

3E E= 0.1863 π^*_{CC} 3E E= 0.1863 π^*_{CC}

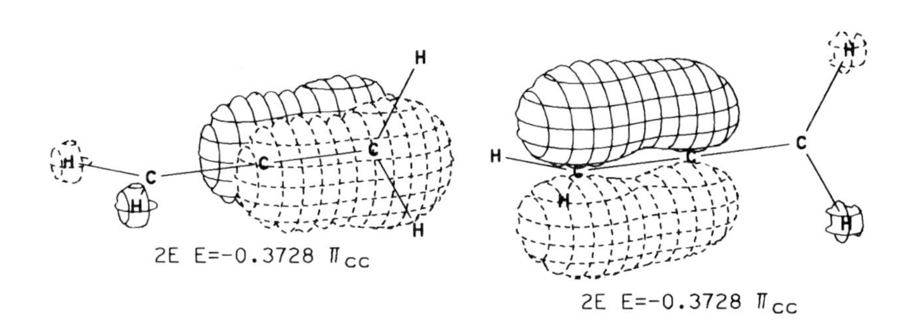

2E E=-0.3728 π_{CC} 2E E=-0.3728 π_{CC}

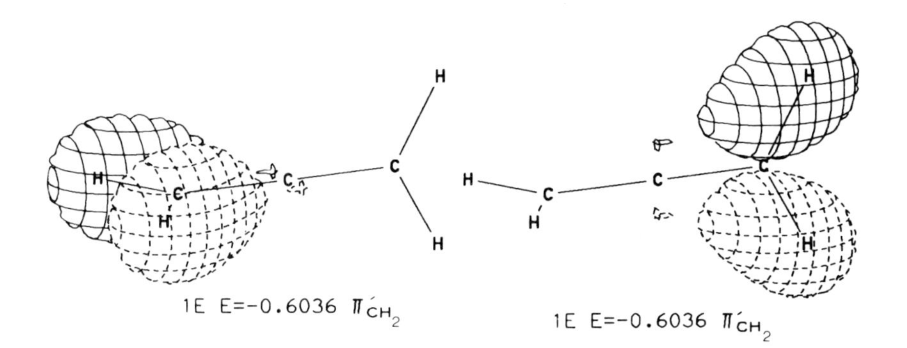

1E E=-0.6036 π'_{CH_2} 1E E=-0.6036 π'_{CH_2}

42. Ketene

Symmetry: C_{2v}

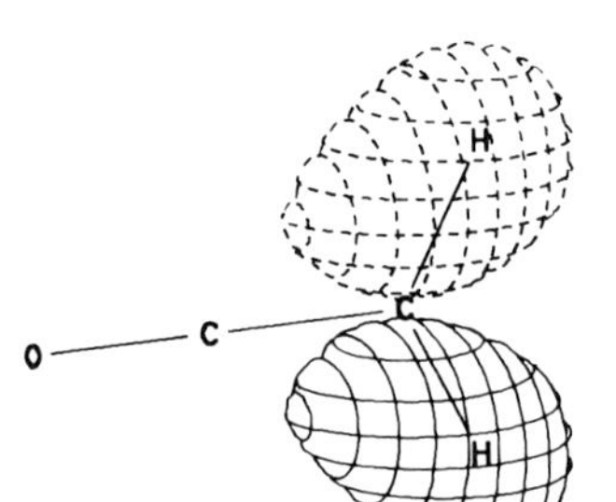

$1B_2$ E=-0.6536 π'_{CH_2}

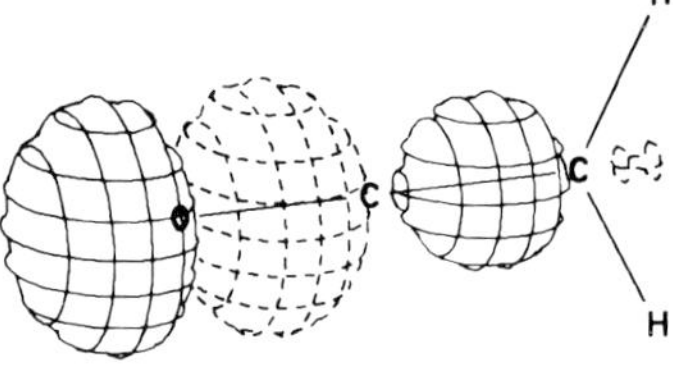

$7A_1$ E=-0.6611 n, σ_{CC}

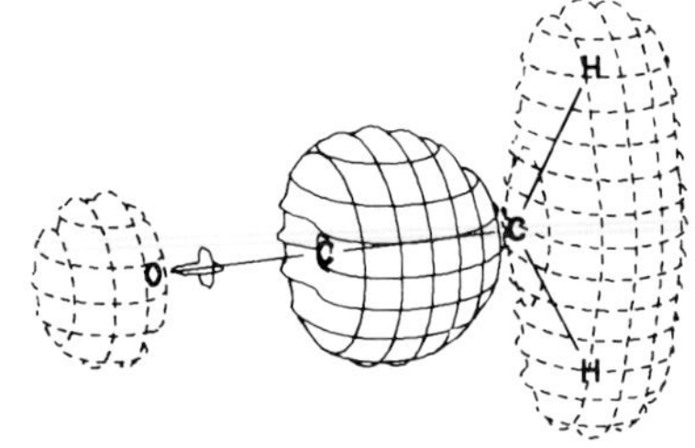

$6A_1$ E=-0.7403 $\sigma_{CH_2}, \sigma_{CC}$

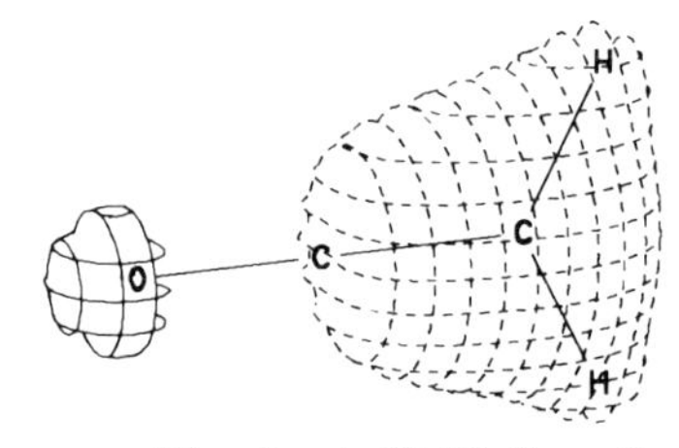

$5A_1$ E=-1.0532 $\sigma_{CC}, \sigma_{CH_2}$

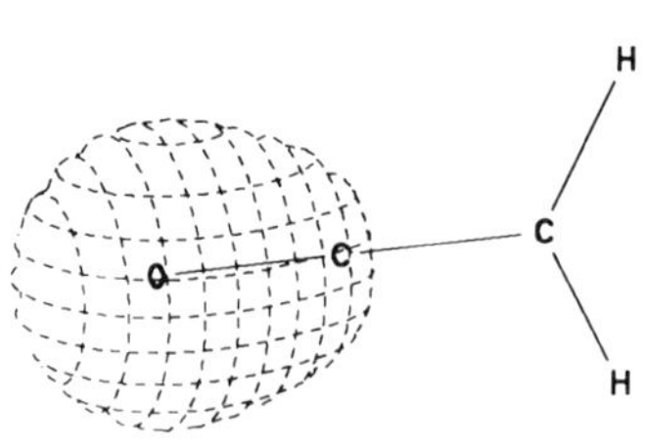

$4A_1$ E=-1.4992 σ_{CO}

Ketene (Continued)

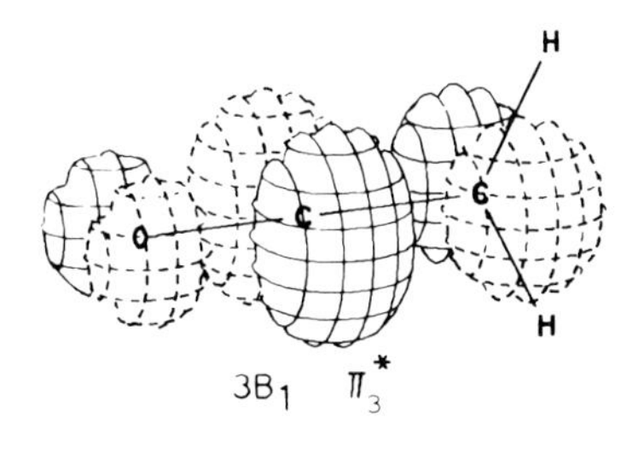

$3B_1$ π_3^*

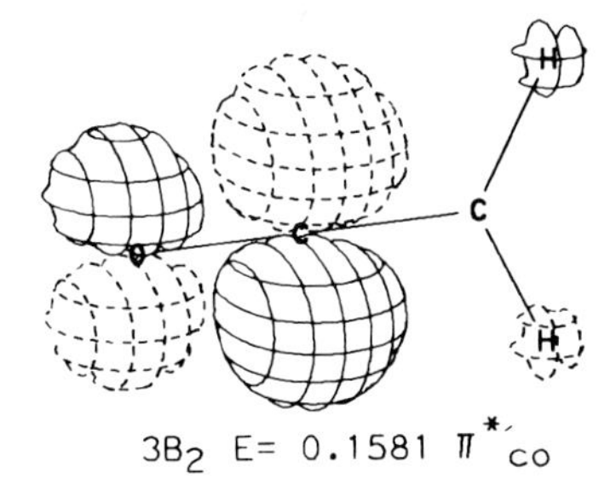

$3B_2$ E= 0.1581 $\pi^{*\prime}_{CO}$

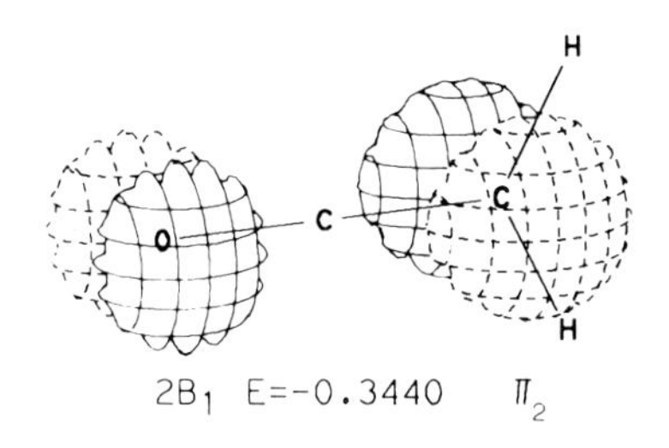

$2B_1$ E=-0.3440 π_2

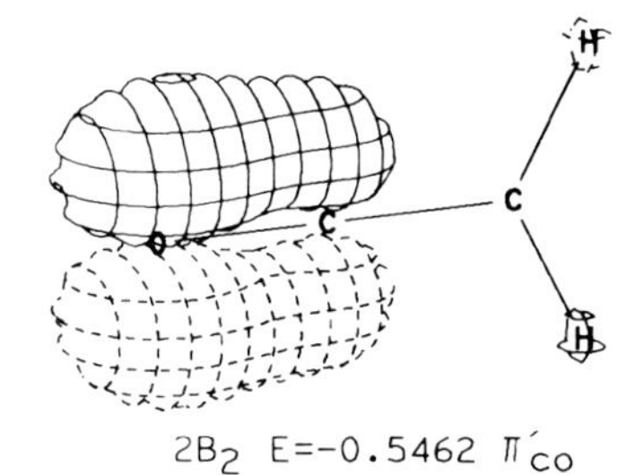

$2B_2$ E=-0.5462 π'_{CO}

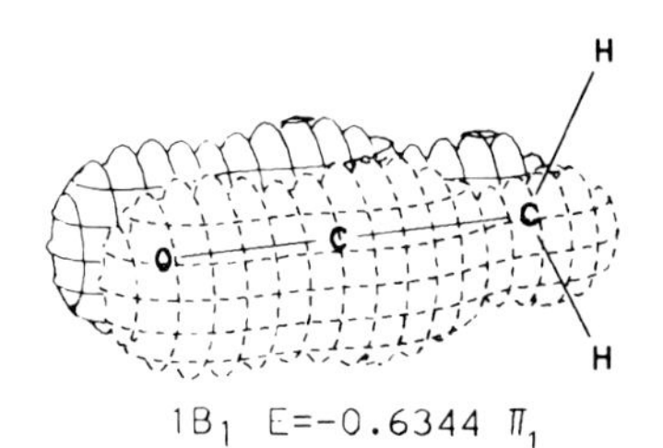

$1B_1$ E=-0.6344 π_1

43. Diazomethane

Symmetry: C_{2v}

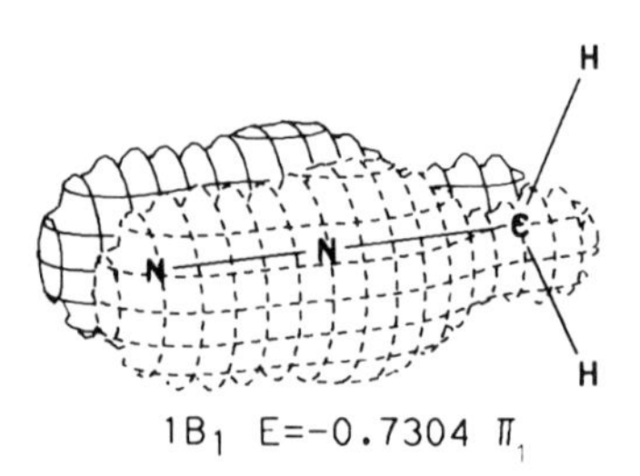

1B1 E=-0.7304 π_1

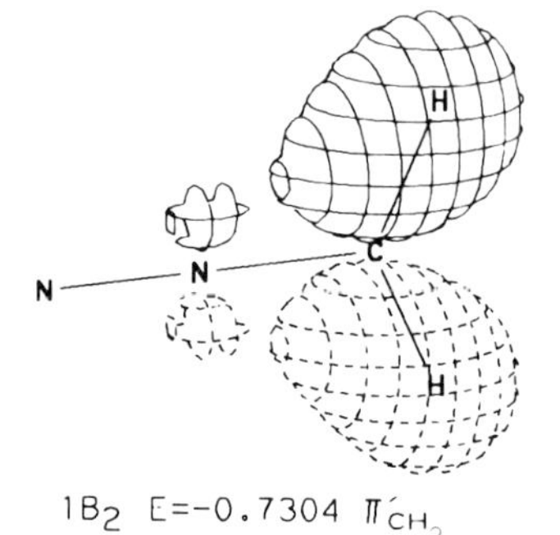

1B2 E=-0.7304 π'_{CH_2}

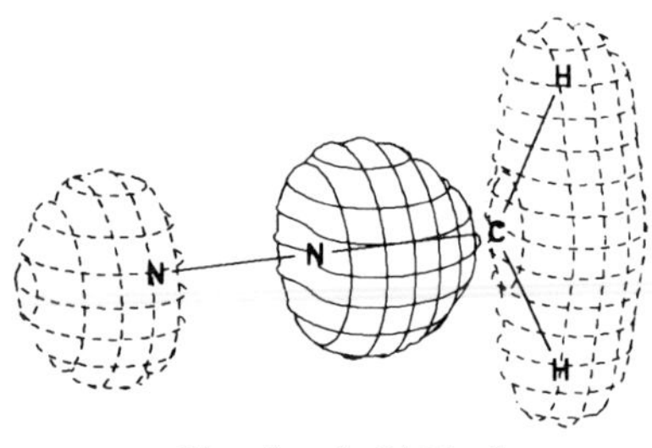

6A1 E=-0.9161 σ

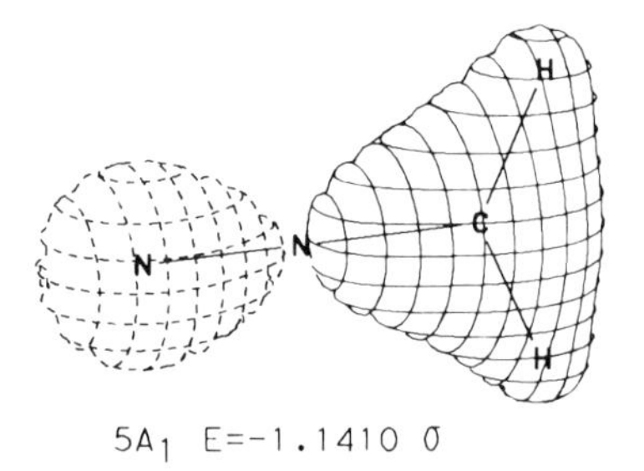

5A1 E=-1.1410 σ

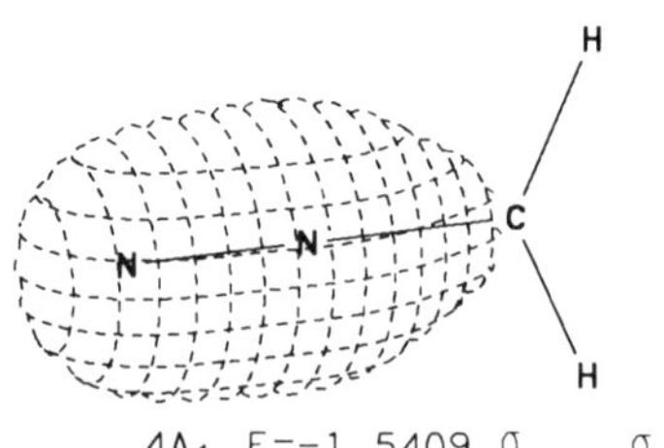

4A1 E=-1.5409 σ_{NN}, σ_{CN}

Diazomethane (Continued)

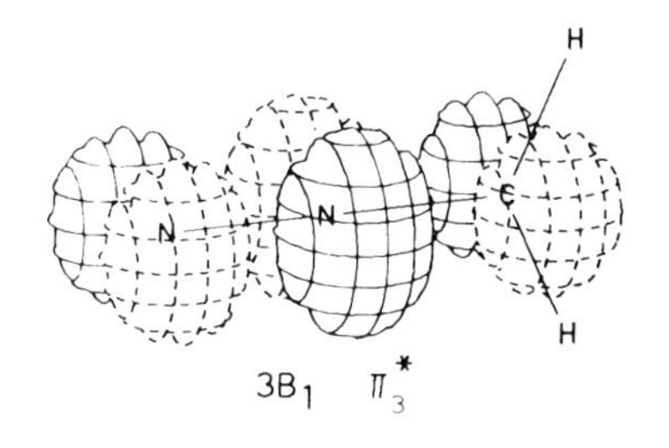

$3B_1$ π_3^*

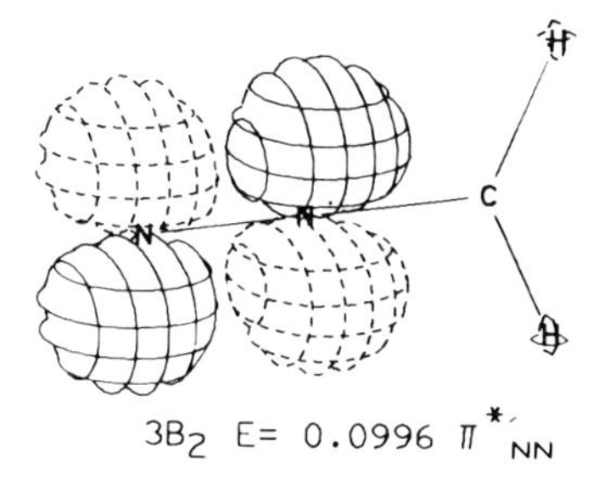

$3B_2$ E= 0.0996 π^*_{NN}

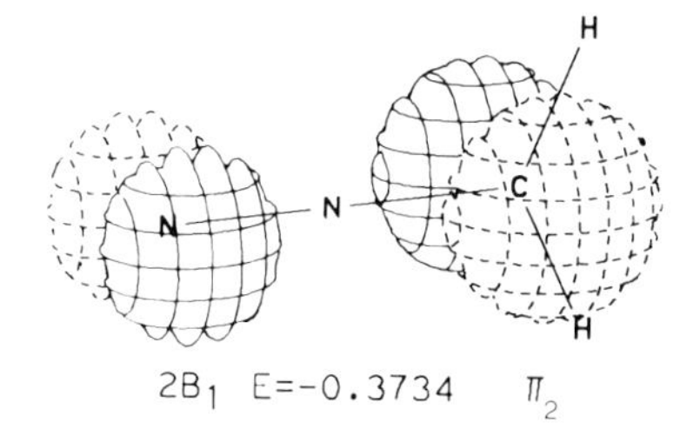

$2B_1$ E=-0.3734 π_2

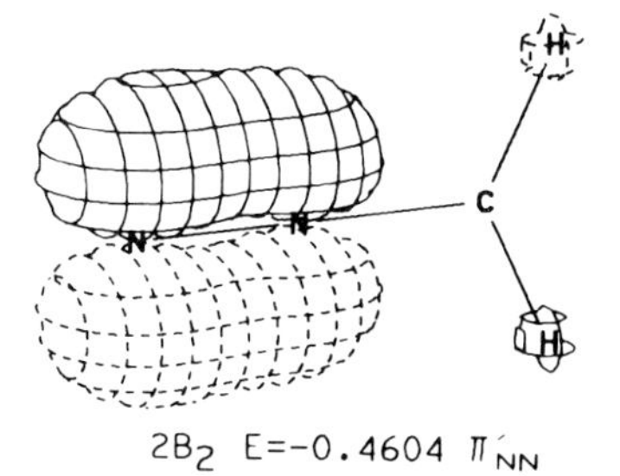

$2B_2$ E=-0.4604 π_{NN}

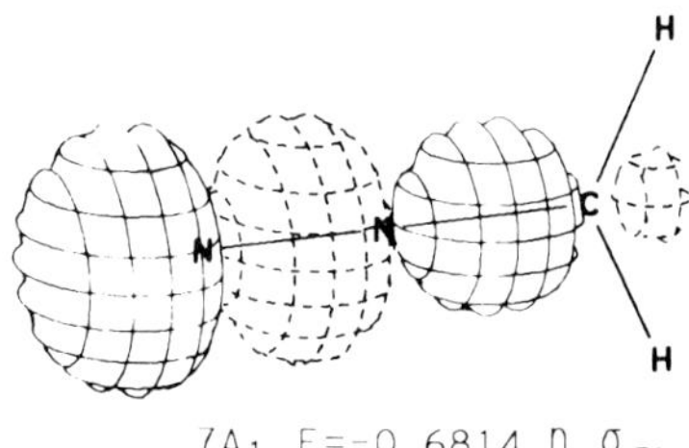

$7A_1$ E=-0.6814 n,σ_{CN}

44. Carbodiimide

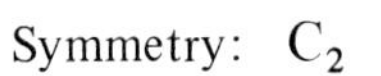

Symmetry: C_2

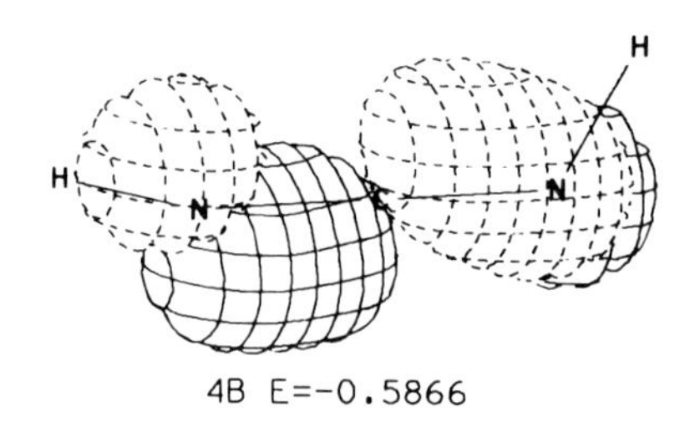

4B E=-0.5866

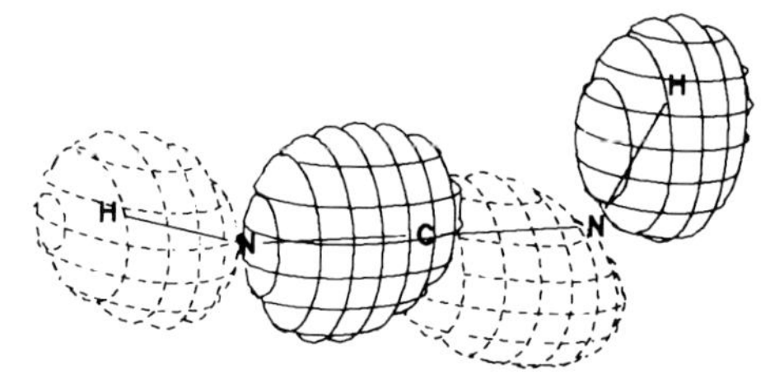

3B E=-0.7489 σ_{NC}, σ_{NH}

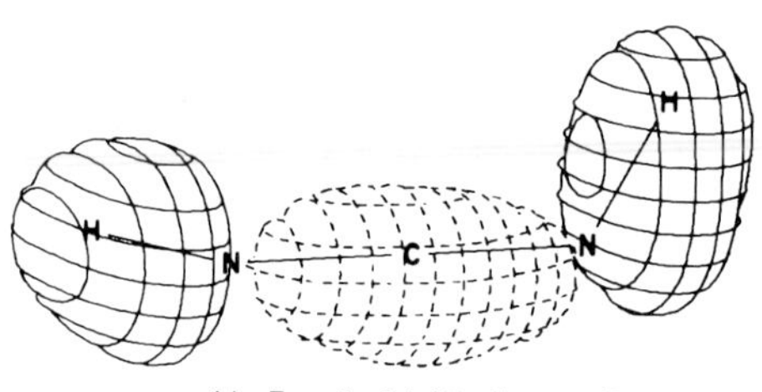

4A E=-0.8142 σ_{NH}, σ_{NC}

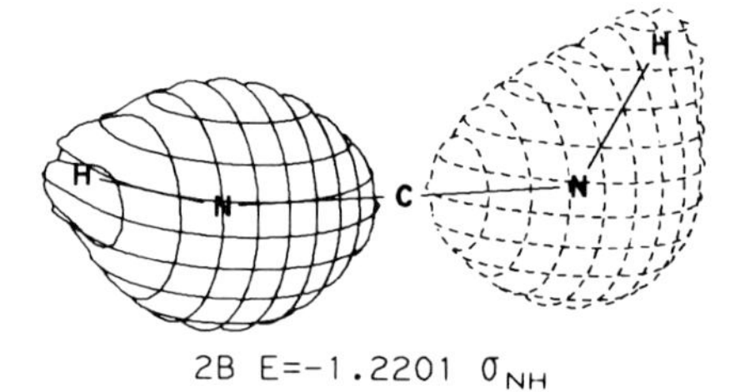

2B E=-1.2201 σ_{NH}

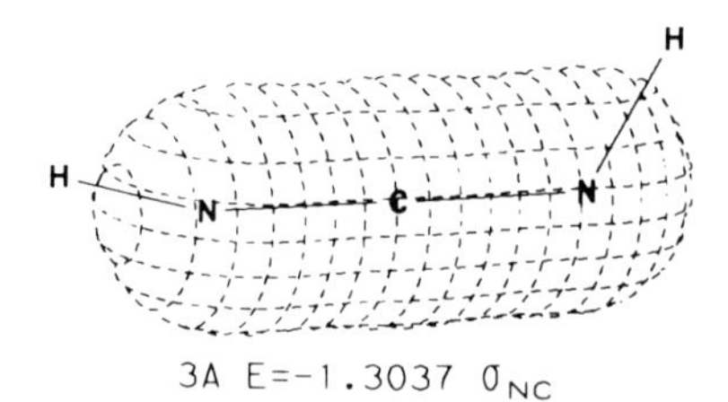

3A E=-1.3037 σ_{NC}

Carbodiimide (Continued)

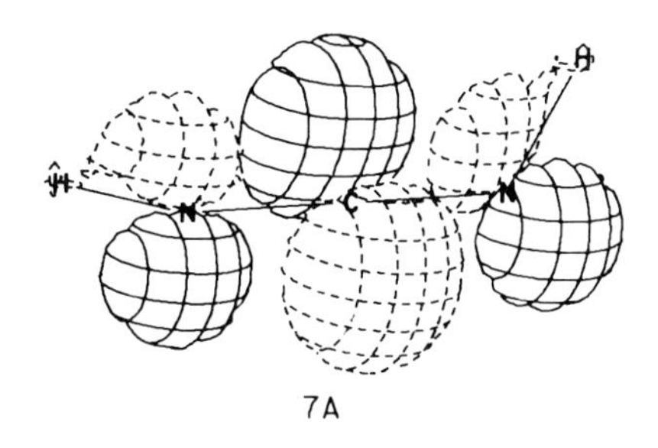

7A

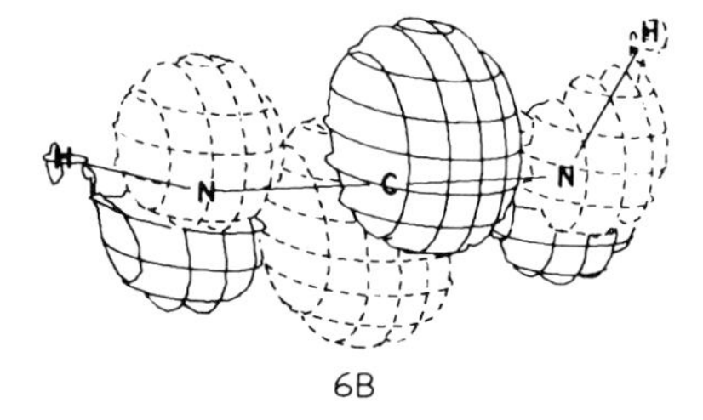

6B

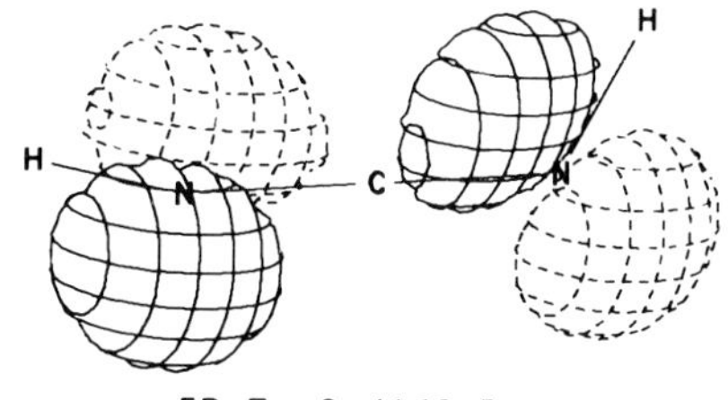

5B E=-0.4168 Π

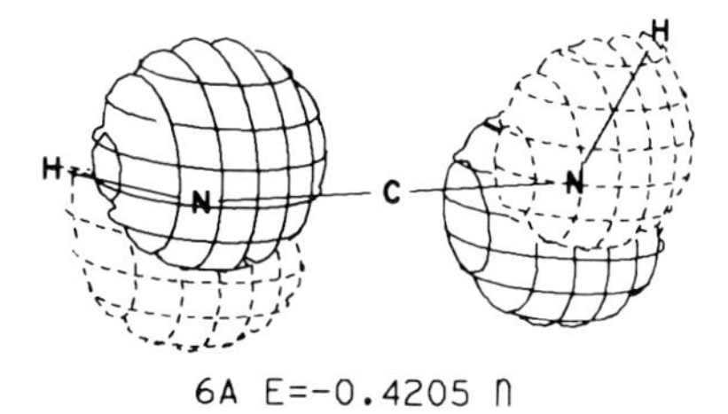

6A E=-0.4205 Π

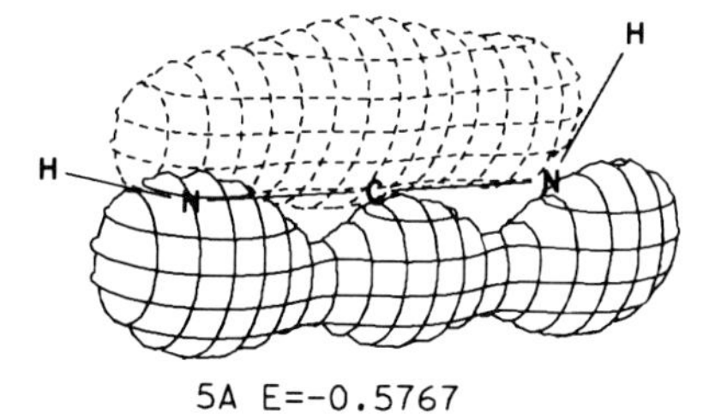

5A E=-0.5767

45. Carbon Dioxide

Symmetry: $D_{\infty h}$

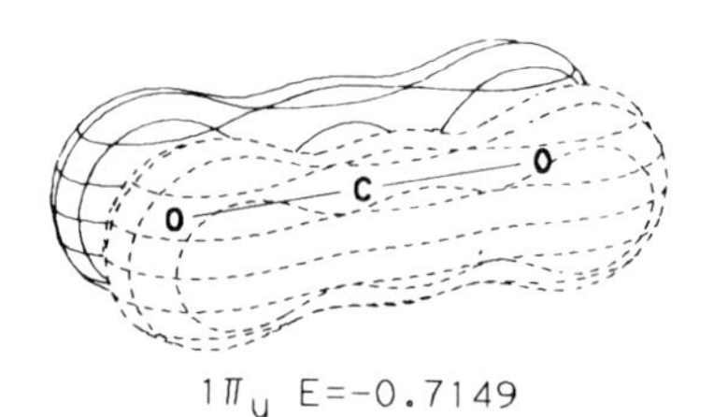

$1\pi_u$ E=-0.7149

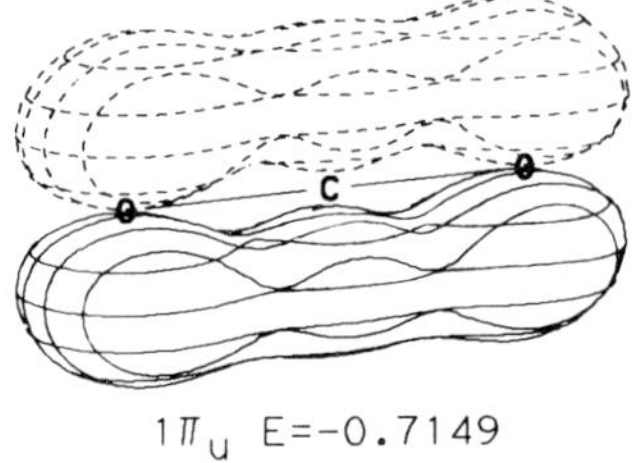

$1\pi_u$ E=-0.7149

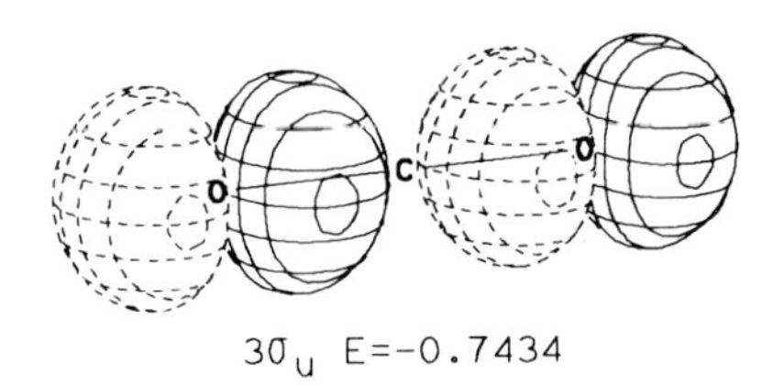

$3\sigma_u$ E=-0.7434

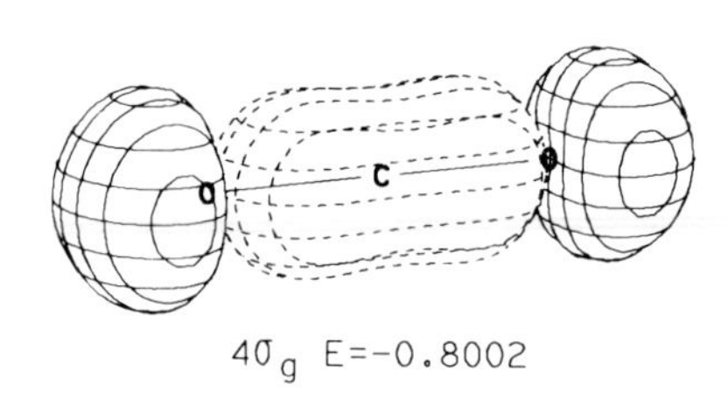

$4\sigma_g$ E=-0.8002

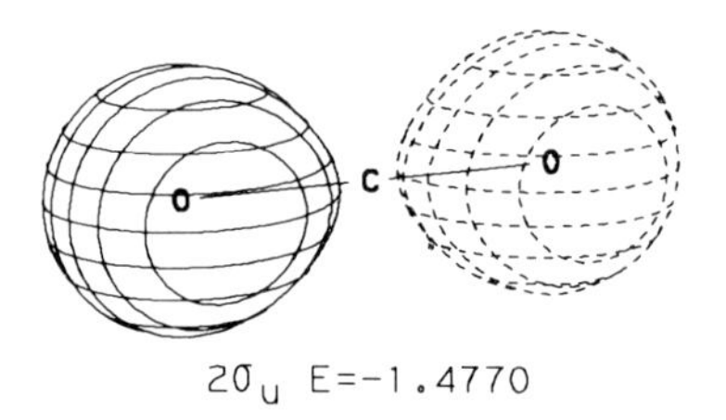

$2\sigma_u$ E=-1.4770

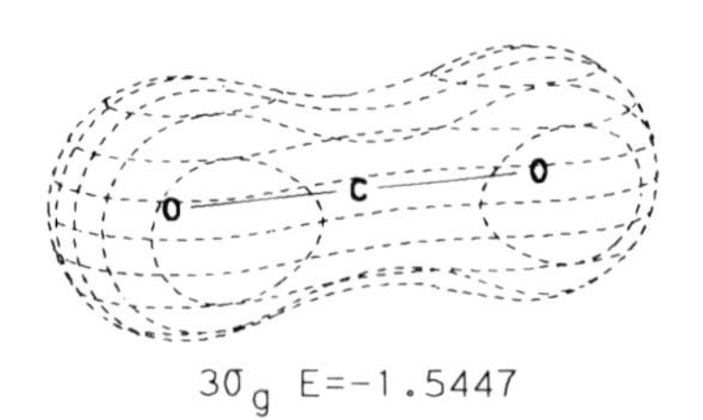

$3\sigma_g$ E=-1.5447

Carbon Dioxide (Continued)

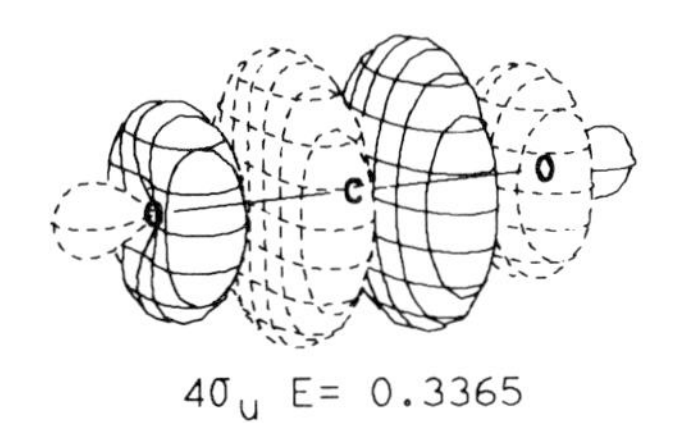

$4\sigma_u$ E= 0.3365

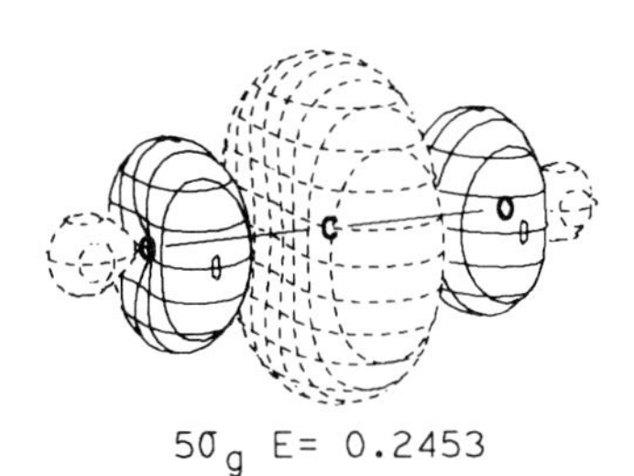

$5\sigma_g$ E= 0.2453

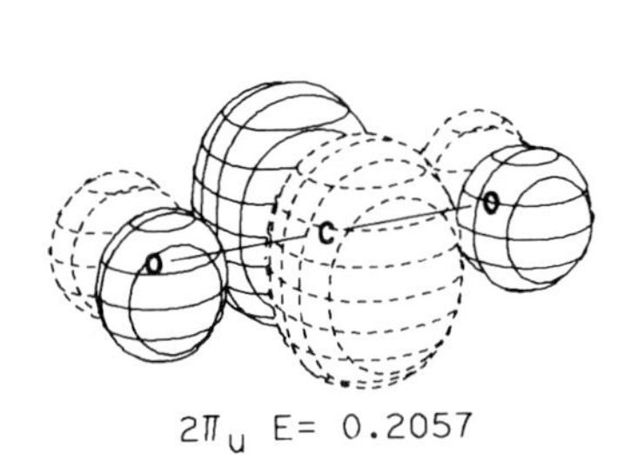

$2\pi_u$ E= 0.2057

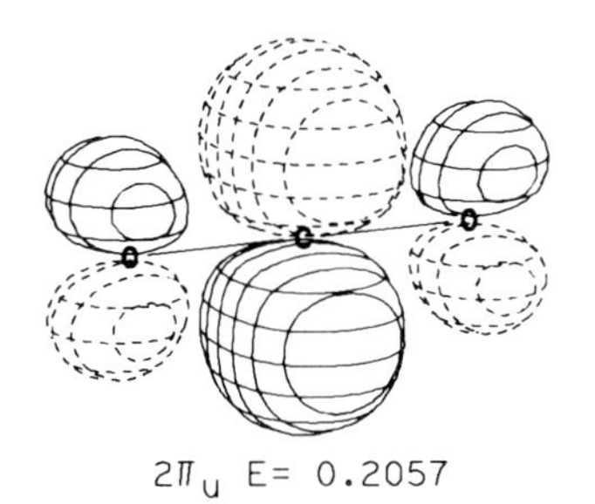

$2\pi_u$ E= 0.2057

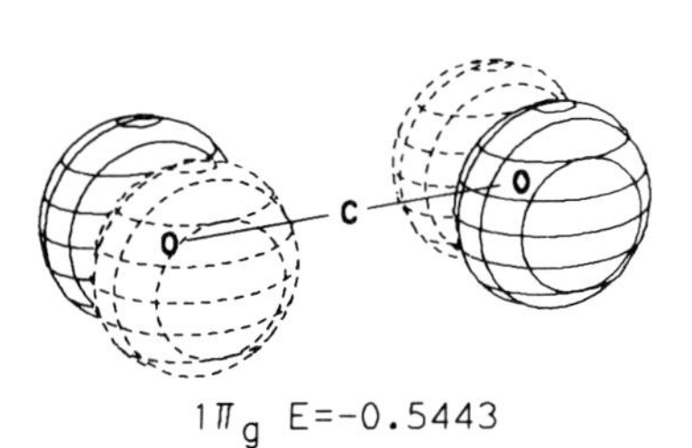

$1\pi_g$ E=-0.5443

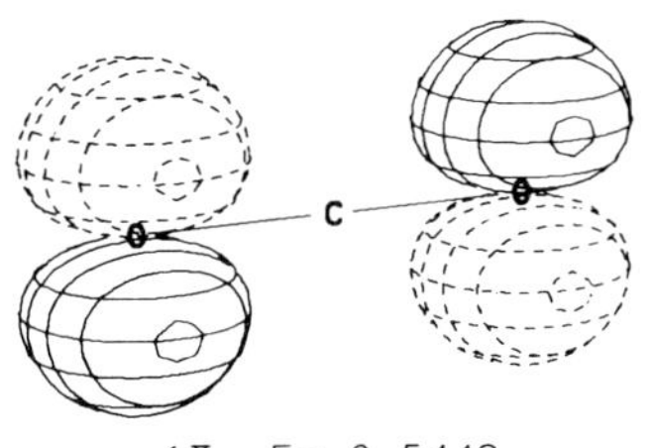

$1\pi_g$ E=-0.5443

46. Cyclopropene

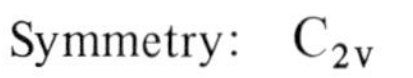

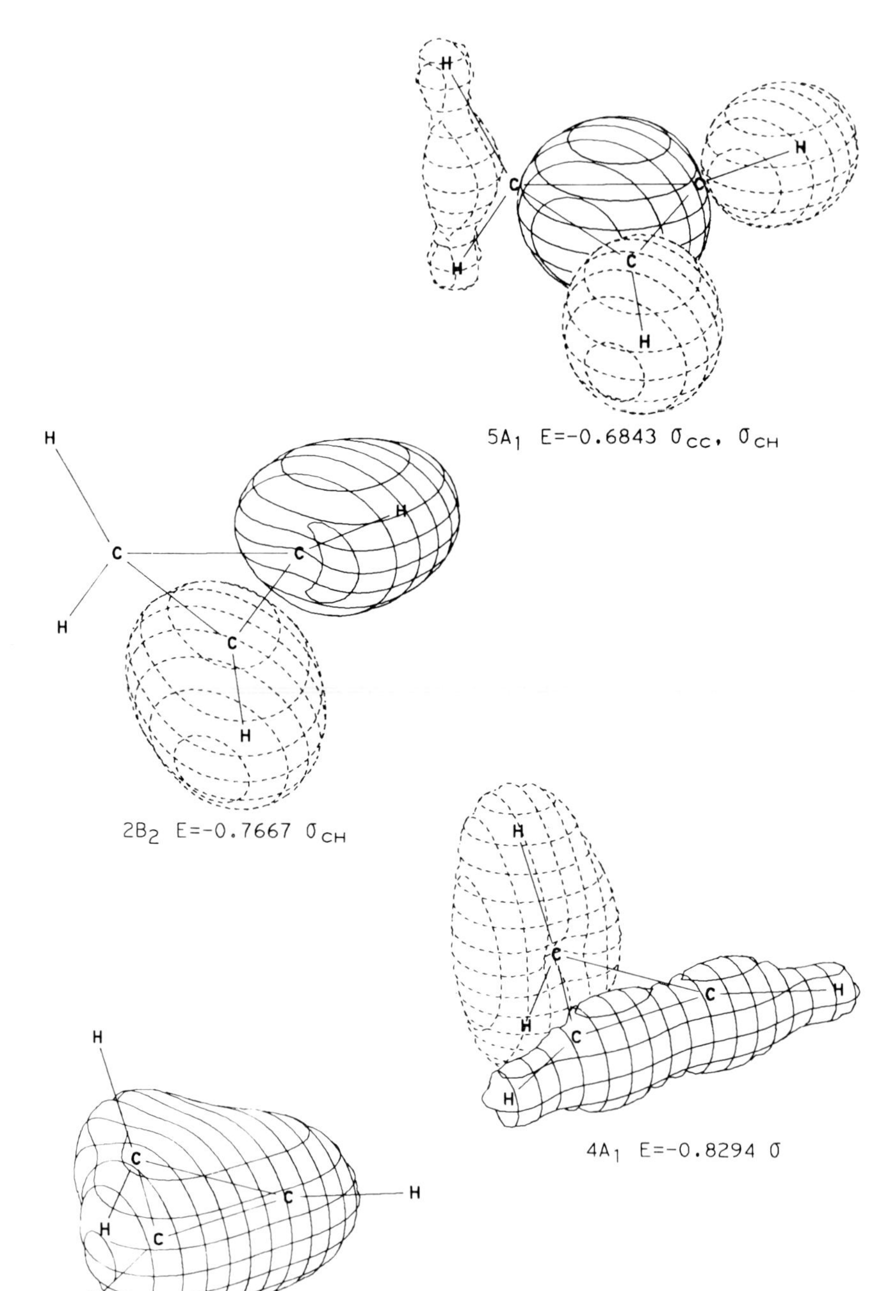

$5A_1$ E=-0.6843 σ_{CC}, σ_{CH}

$2B_2$ E=-0.7667 σ_{CH}

$4A_1$ E=-0.8294 σ

$3A_1$ E=-1.1743 σ_{CC}

Cyclopropene (Continued)

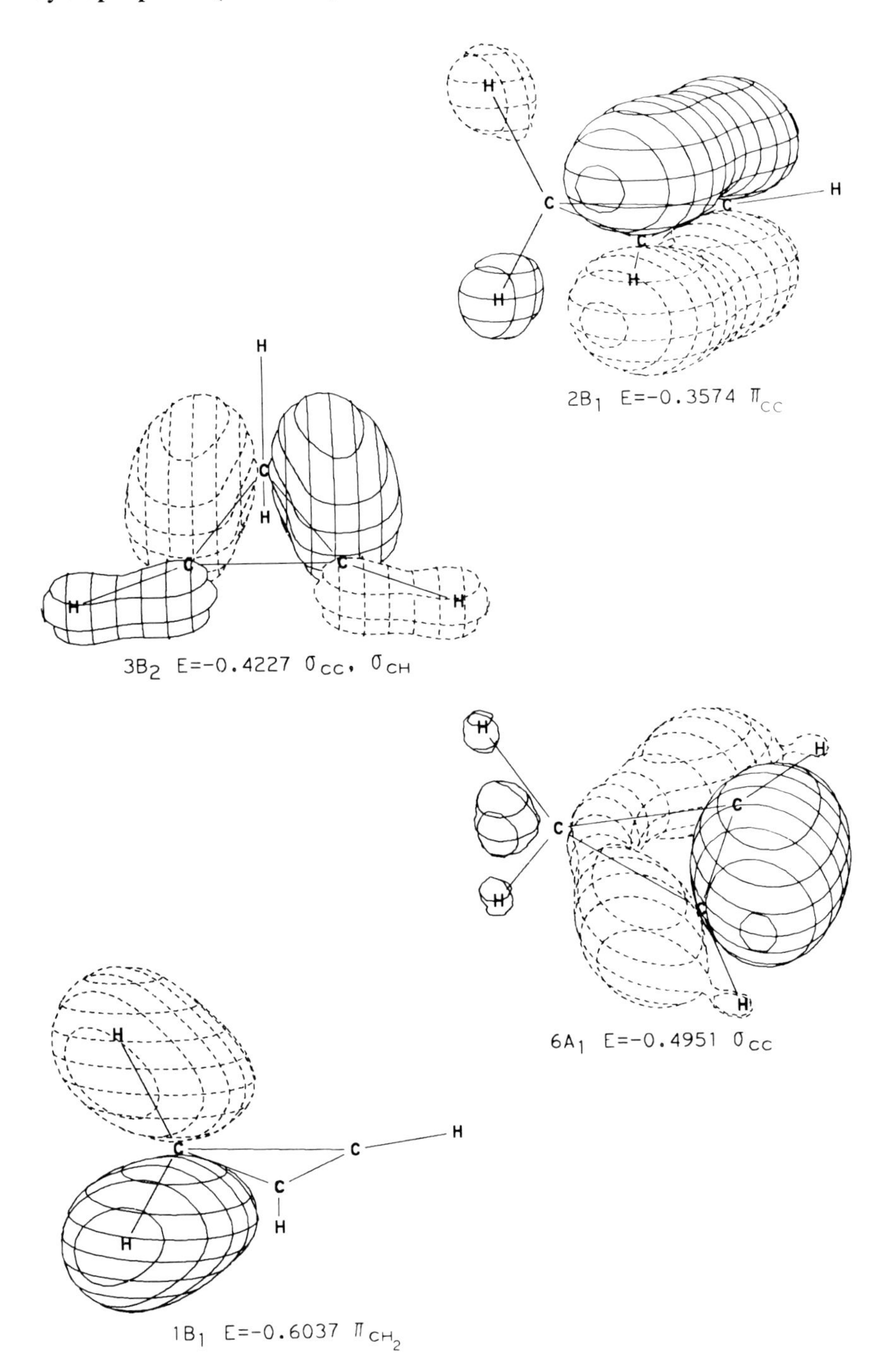

Cyclopropene (Continued)

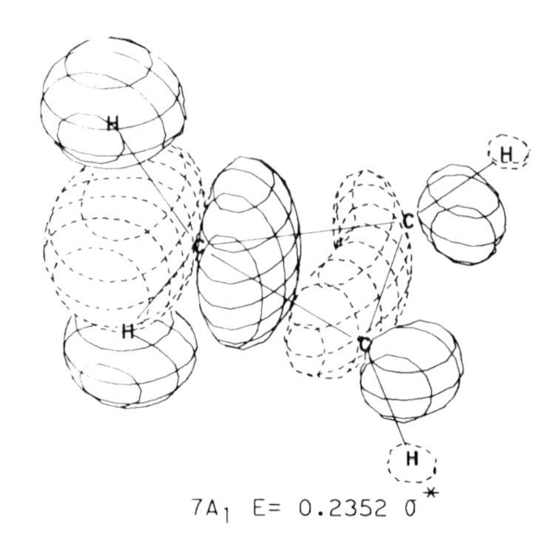

$7A_1$ E= 0.2352 σ^*

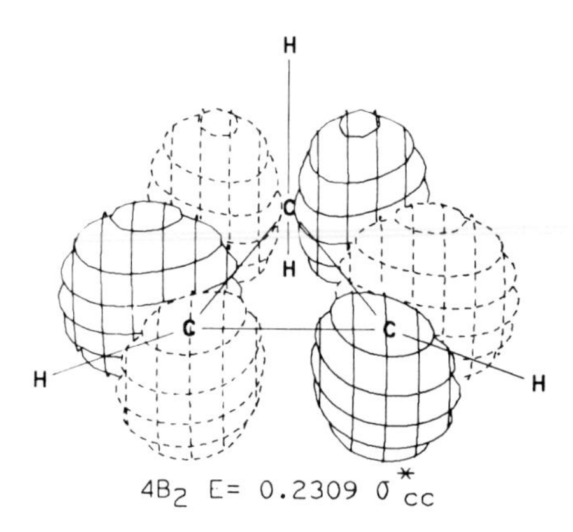

$4B_2$ E= 0.2309 σ^*_{cc}

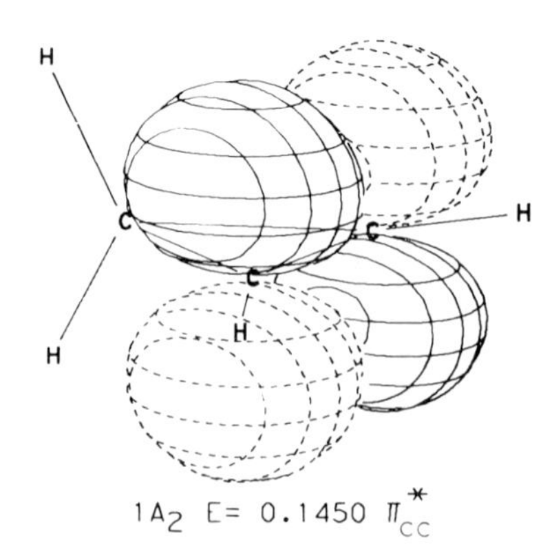

$1A_2$ E= 0.1450 π^*_{cc}

47. Diazirine

Symmetry: C_{2v}

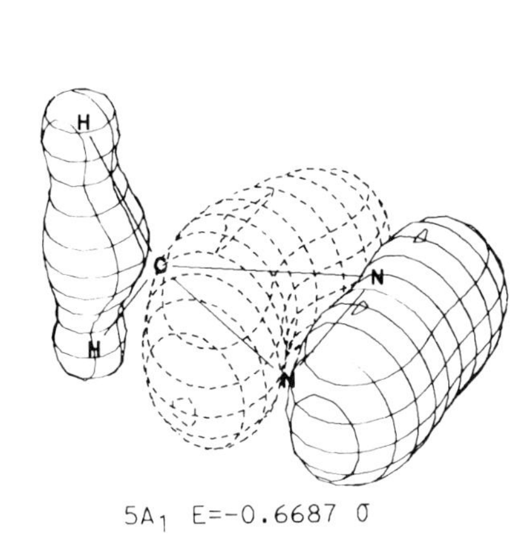

$5A_1$ E=-0.6687 σ

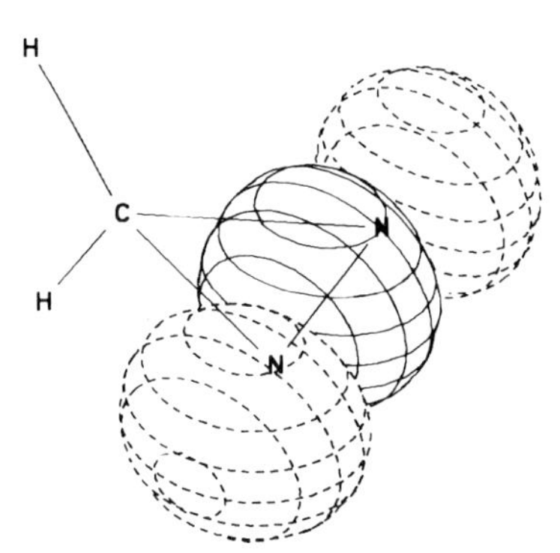

$6A_1$ E=-0.5777 n, σ_{NN}

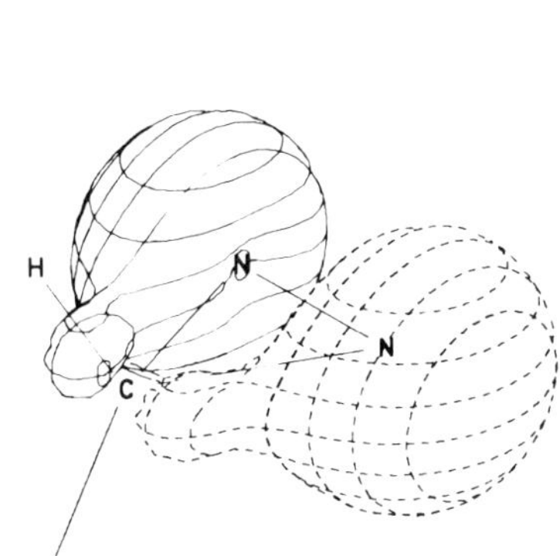

$2B_2$ E=-0.8540 n, (σ_{CN})

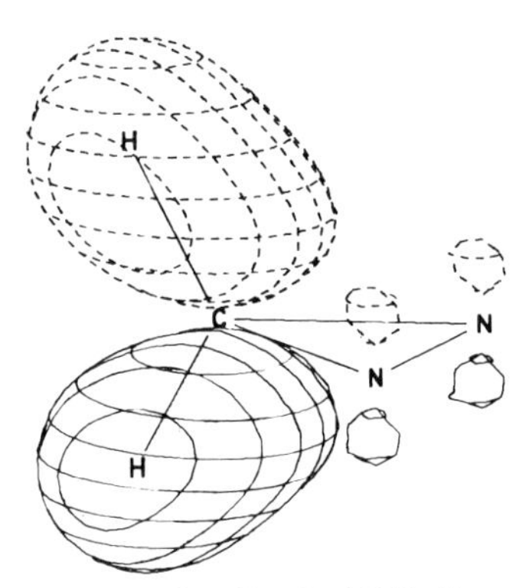

$1B_1$ E=-0.6970 π_{CH_2}

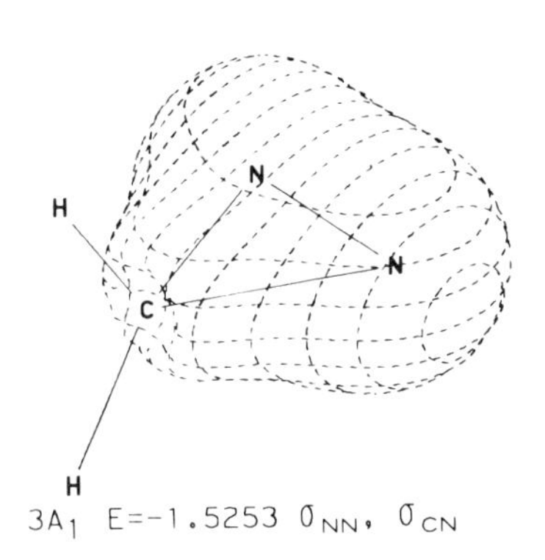

$3A_1$ E=-1.5253 σ_{NN}, σ_{CN}

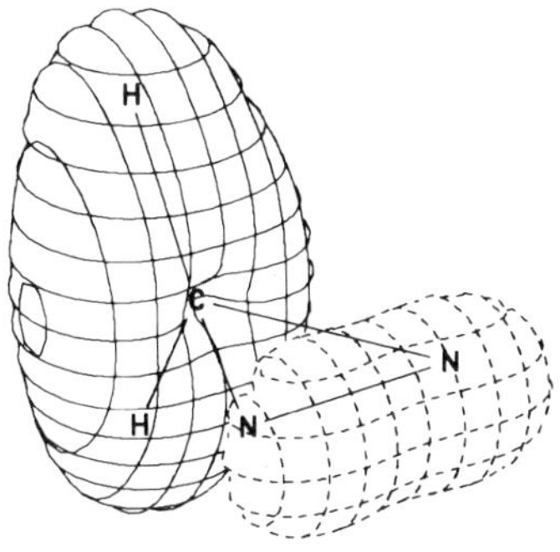

$4A_1$ E=-0.9385 σ_{CH_2}, σ_{NN}

Diazirine (Continued)

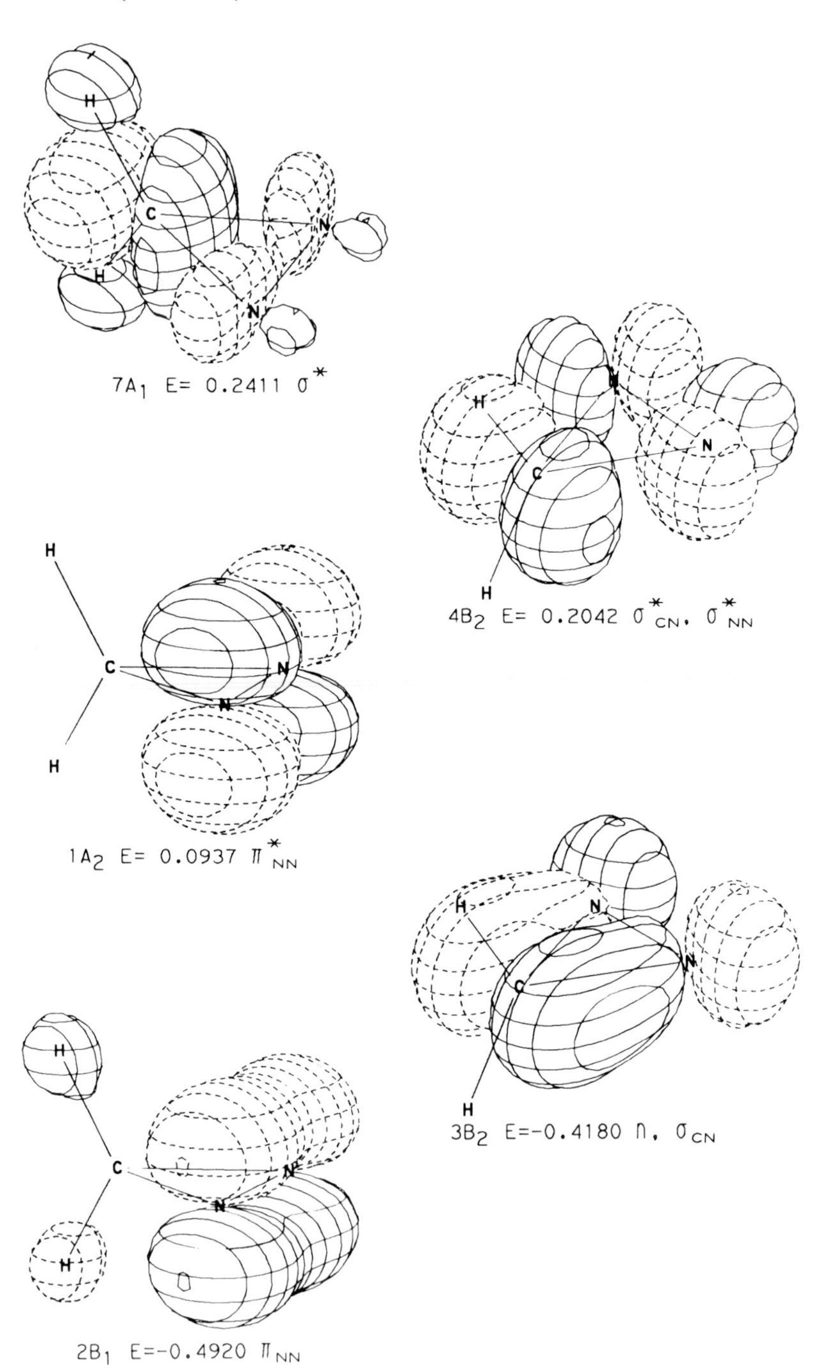

$7A_1$ E= 0.2411 σ^*

$4B_2$ E= 0.2042 σ^*_{CN}, σ^*_{NN}

$1A_2$ E= 0.0937 π^*_{NN}

$3B_2$ E=−0.4180 n, σ_{CN}

$2B_1$ E=−0.4920 π_{NN}

48. Allyl Cation

Symmetry: C_{2v}

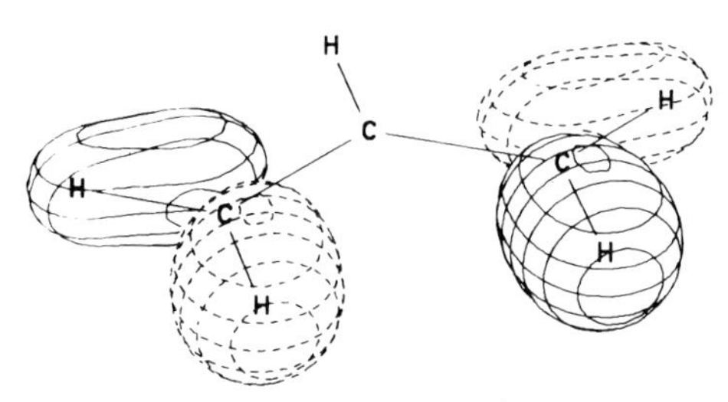

$3B_2$ E=-0.8937 π'_{CH_2}

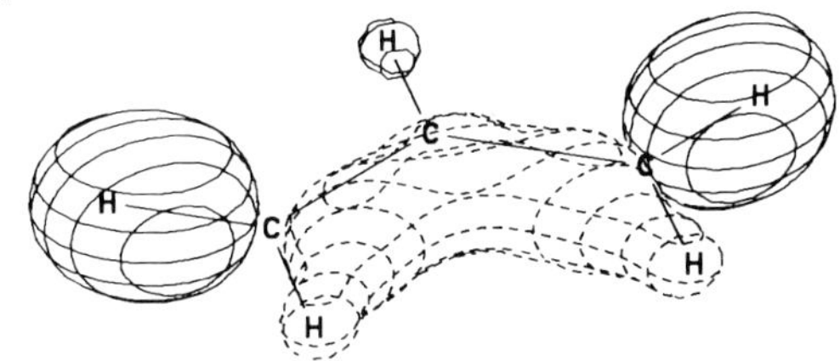

$5A_1$ E=-0.9502 σ_{CC}, π'_{CH_2}

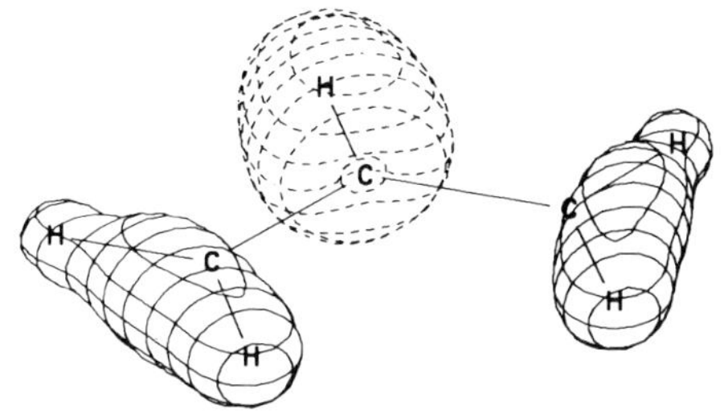

$4A_1$ E=-1.0243 σ_{CH}, σ_{CH_2}

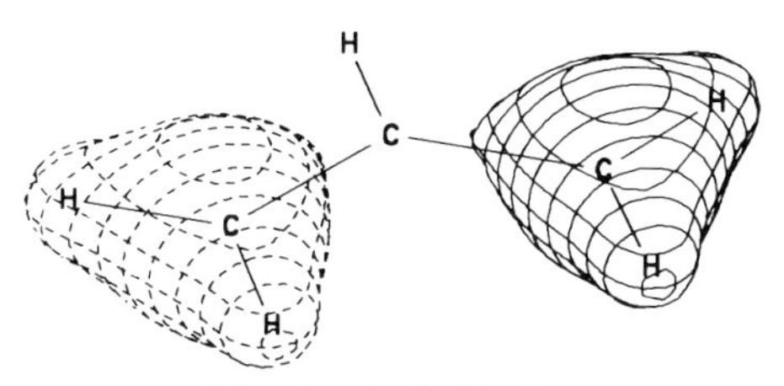

$2B_2$ E=-1.2009 σ_{CH_2}

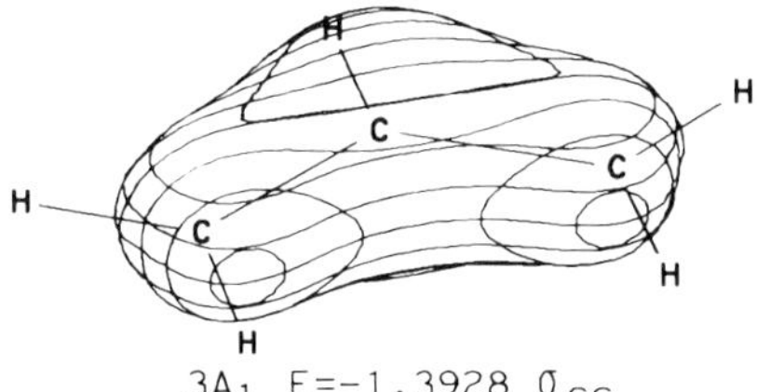

$3A_1$ E=-1.3928 σ_{CC}

Allyl Cation (Continued)

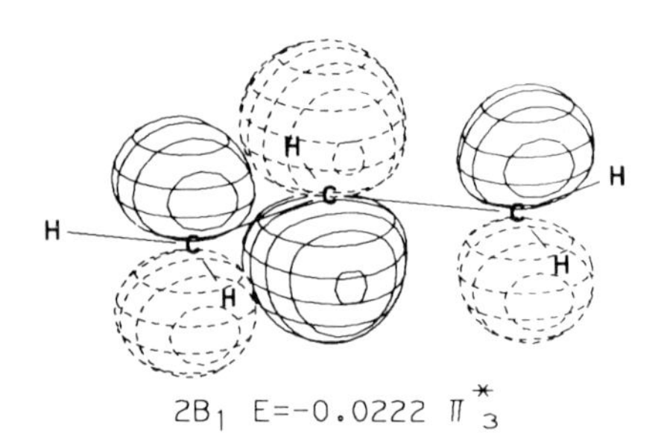

$2B_1$ E=-0.0222 π^*_3

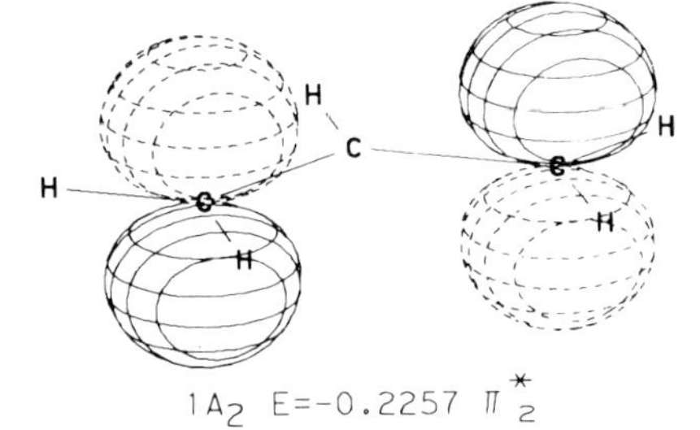

$1A_2$ E=-0.2257 π^*_2

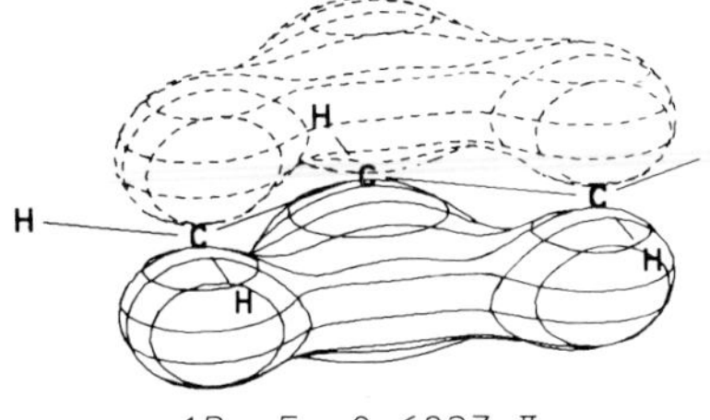

$1B_1$ E=-0.6827 π_1

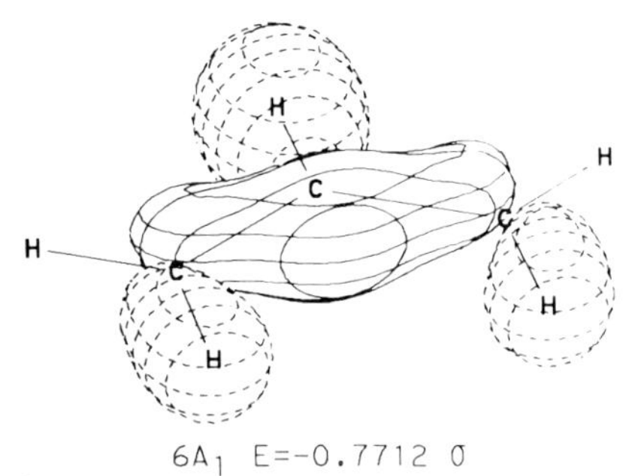

$6A_1$ E=-0.7712 σ

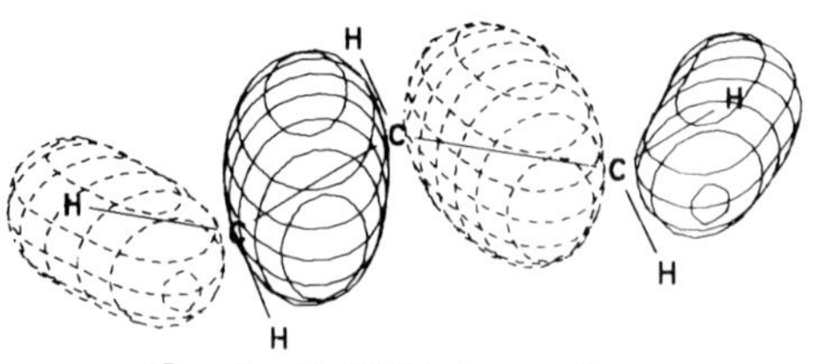

$4B_2$ E=-0.8124 σ_{CC}, σ_{CH_2}

49. Propylene

Symmetry: C_s

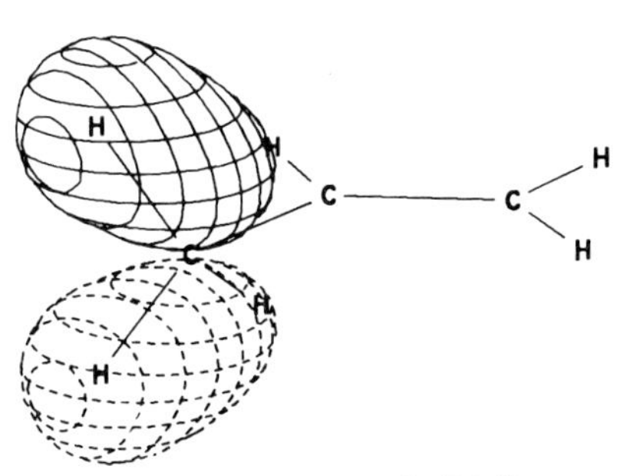

1A'' E=−0.5176 π_{CH_3}

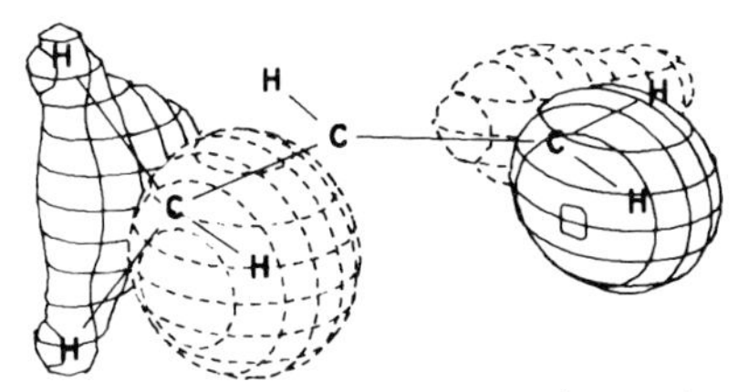

8A' E=−0.5312 π'_{CH_3}, π'_{CH_2}

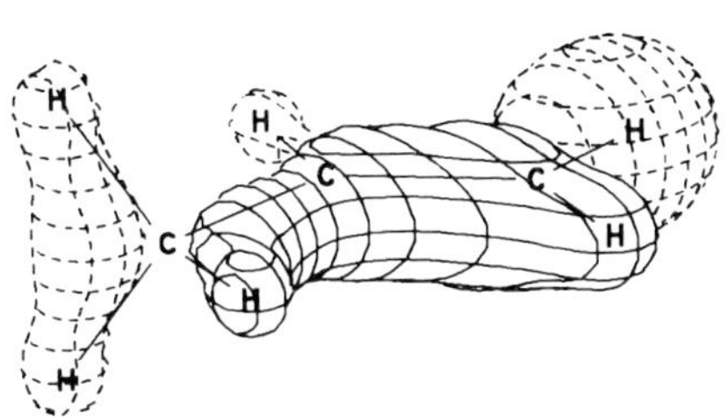

7A' E=−0.5951 π'_{CH_2}, π'_{CH_3}

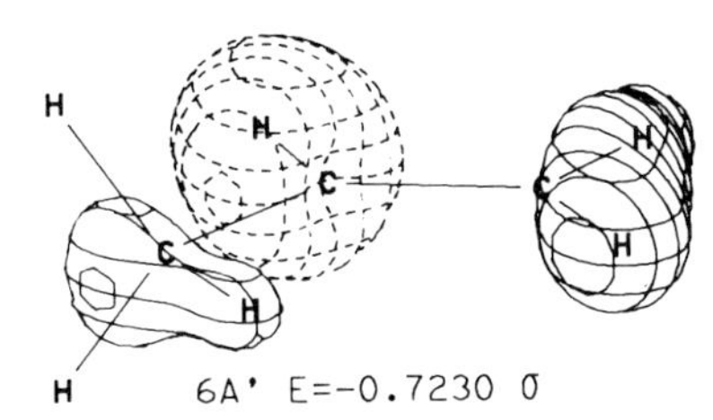

6A' E=−0.7230 σ

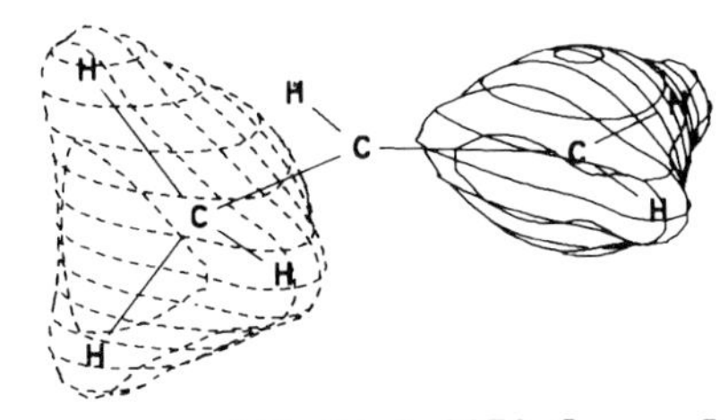

5A' E=−0.9070 σ_{CH_3}, σ_{CH_2}

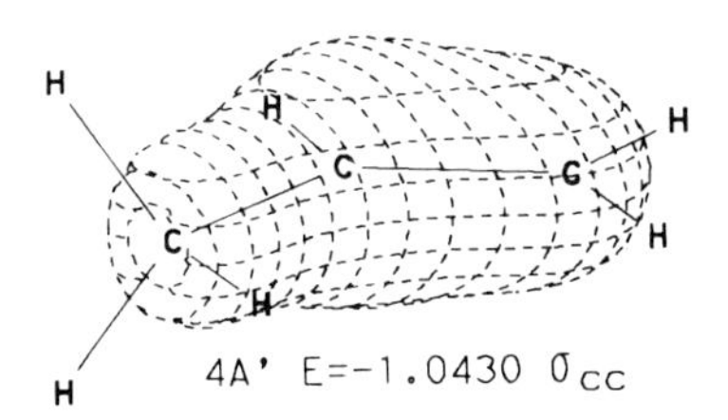

4A' E=−1.0430 σ_{CC}

Propylene (Continued)

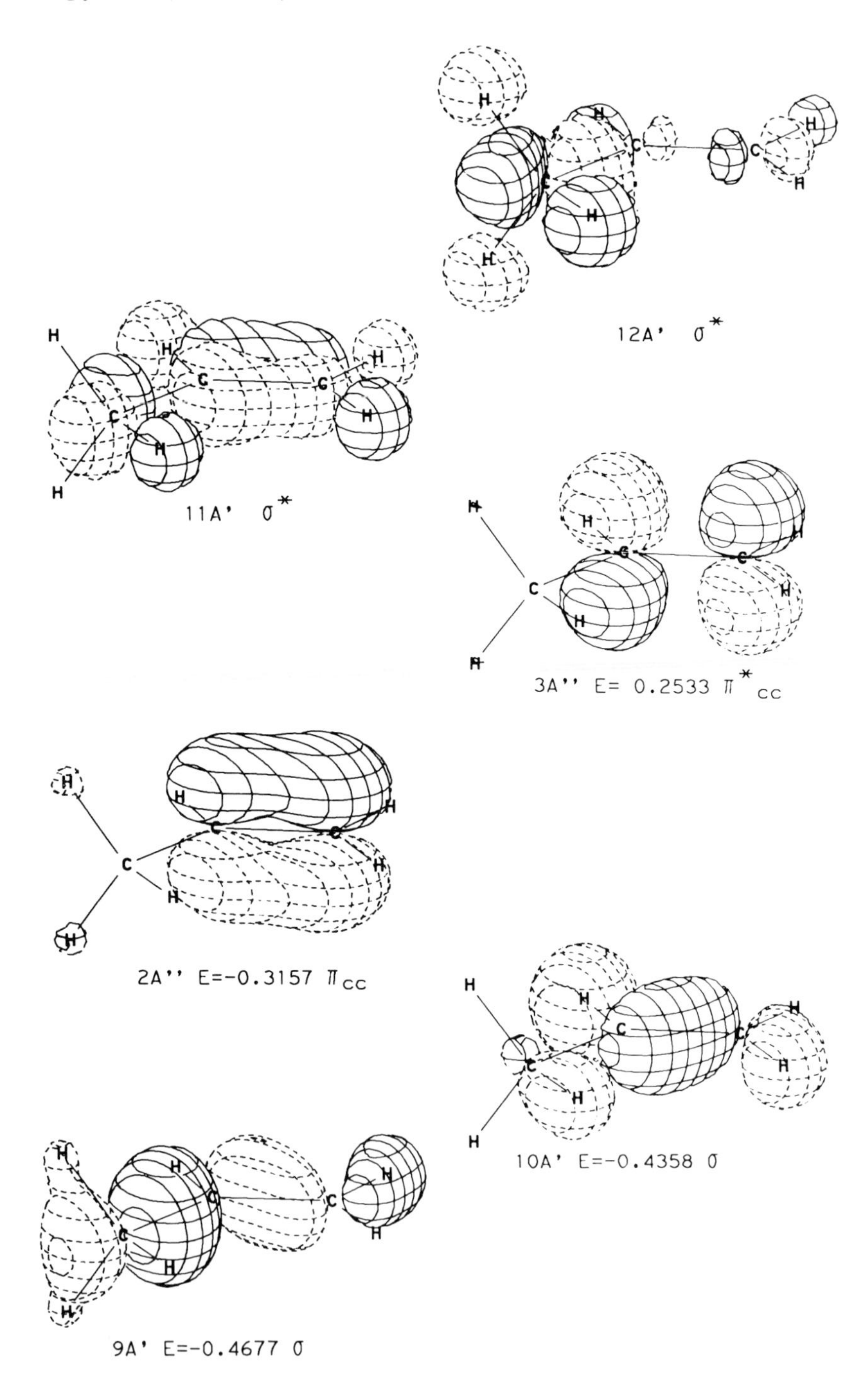

50. Acetaldehyde

Symmetry: C_s

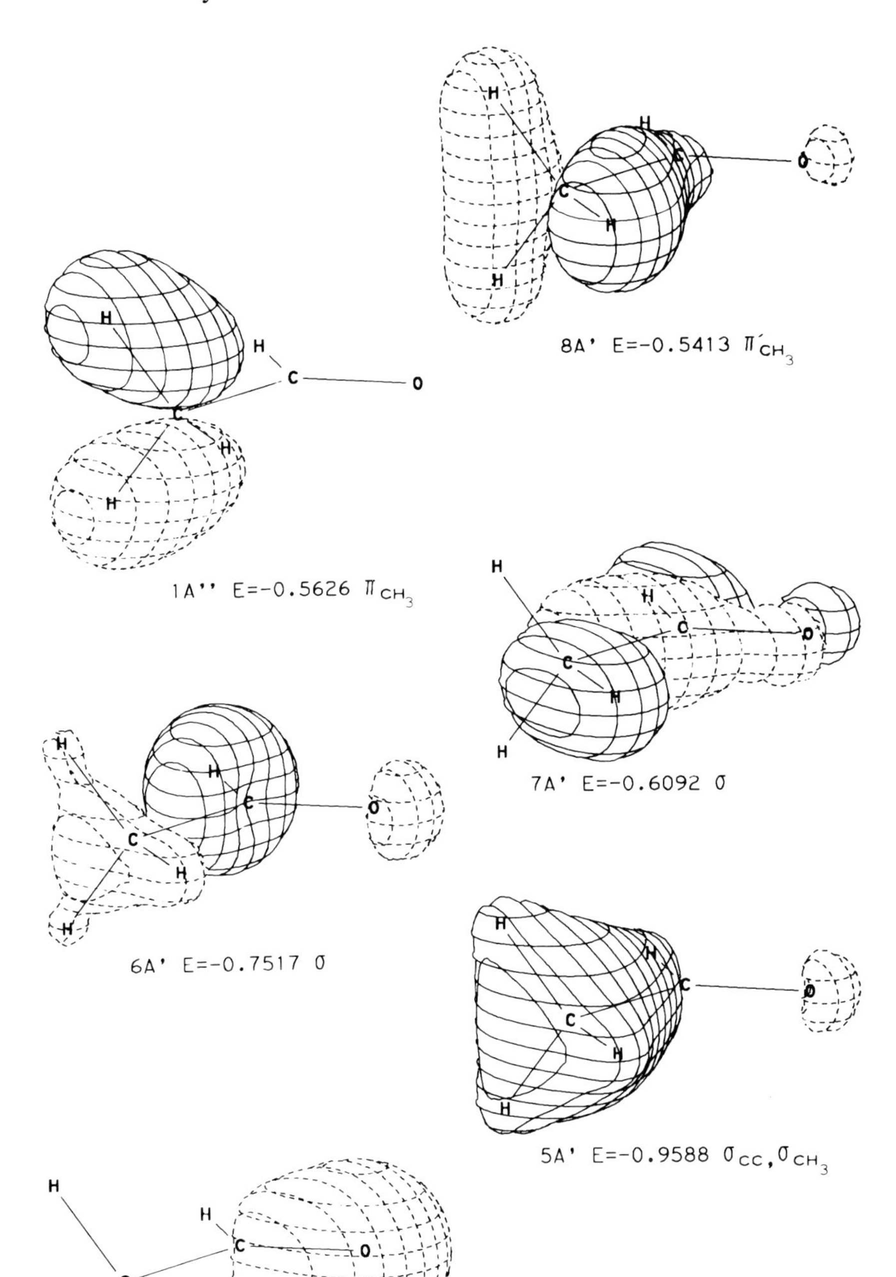

8A' E=-0.5413 π'_{CH_3}

1A'' E=-0.5626 π_{CH_3}

7A' E=-0.6092 σ

6A' E=-0.7517 σ

5A' E=-0.9588 σ_{CC}, σ_{CH_3}

4A' E=-1.3253 σ_{CO}

Acetaldehyde (Continued)

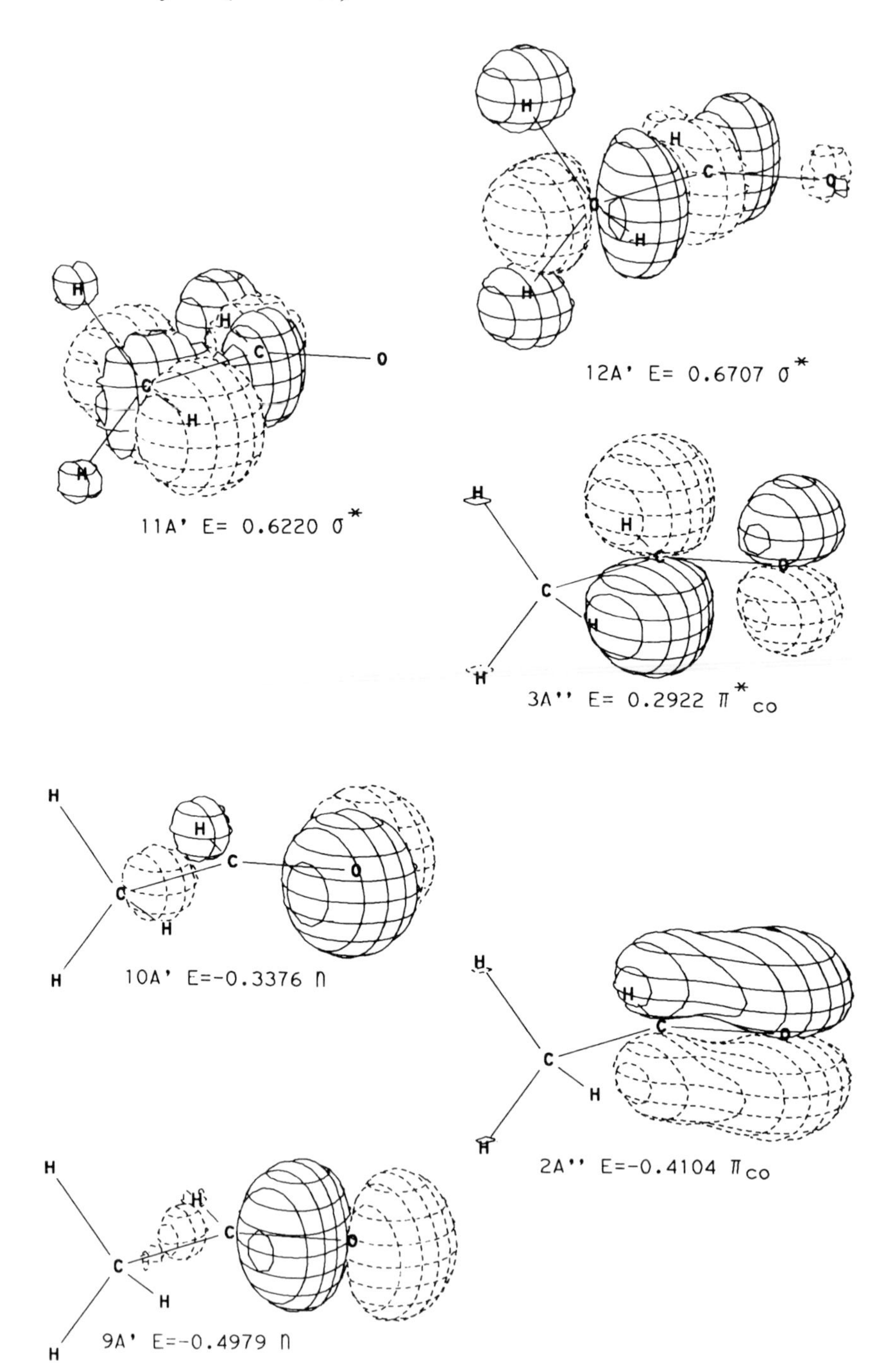

12A' E= 0.6707 σ^*

11A' E= 0.6220 σ^*

3A'' E= 0.2922 π^*_{CO}

10A' E=-0.3376 n

2A'' E=-0.4104 π_{CO}

9A' E=-0.4979 n

51. Formamide

Symmetry: C_s

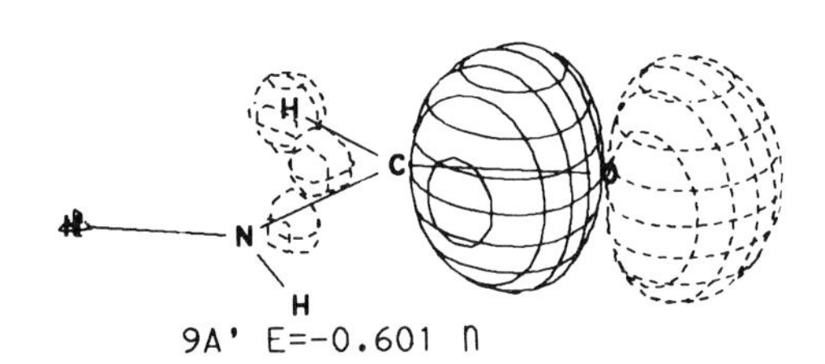

9A' E=-0.601 n

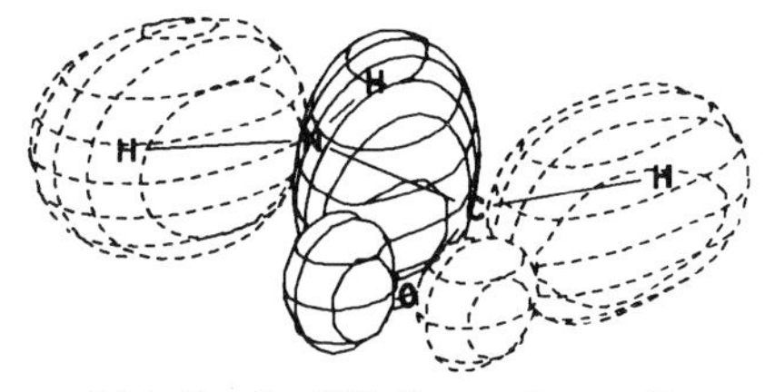

8A' E=-0.672 σ_{NC}, σ_{CH}, σ_{NH_2}

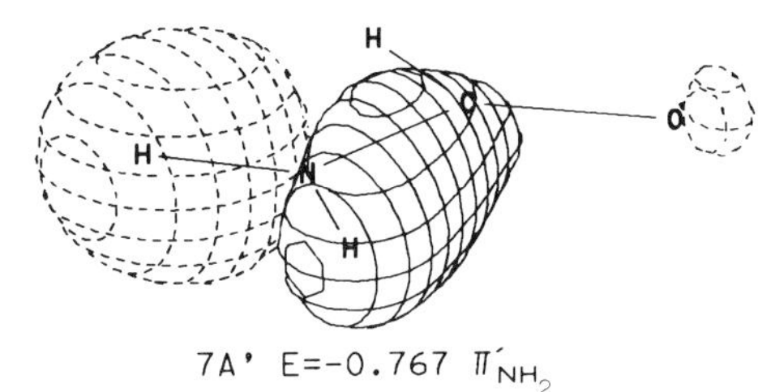

7A' E=-0.767 π'_{NH_2}

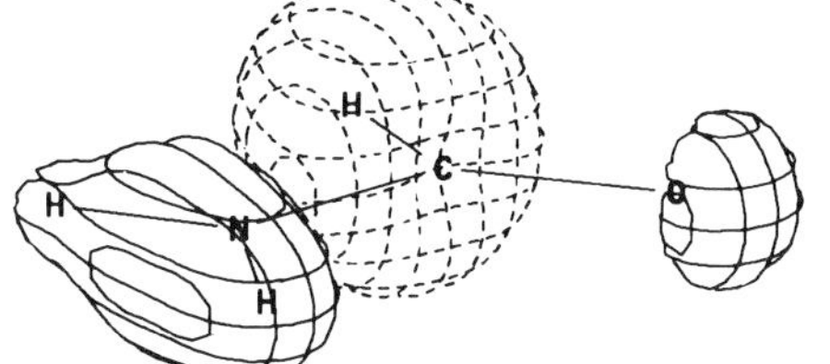

6A' E=-0.877 σ_{CH}, σ_{NH_2}

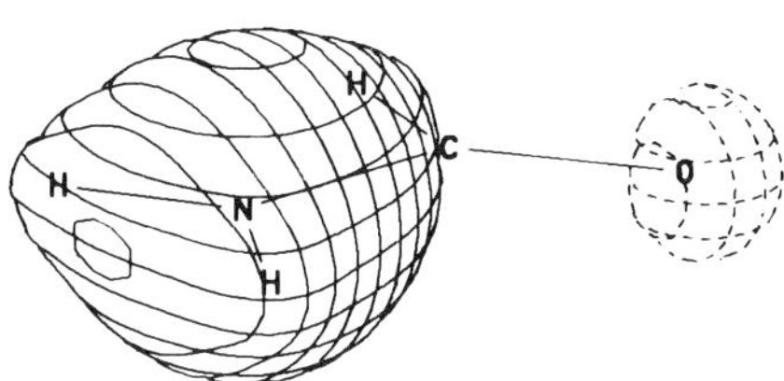

5A' E=-1.239 σ_{NC}, σ_{NH_2}

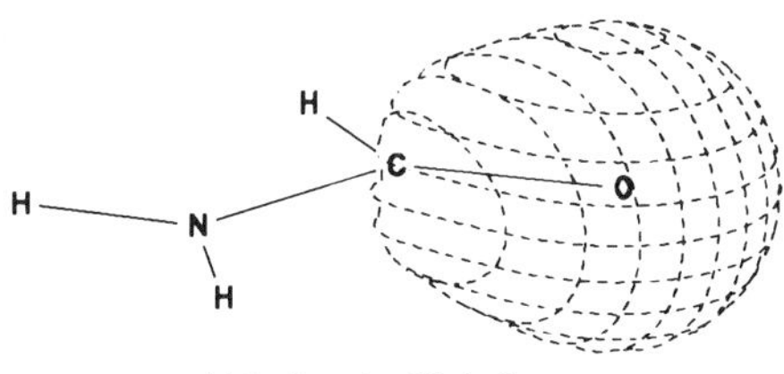

4A' E=-1.394 σ_{CO}

Formamide (Continued)

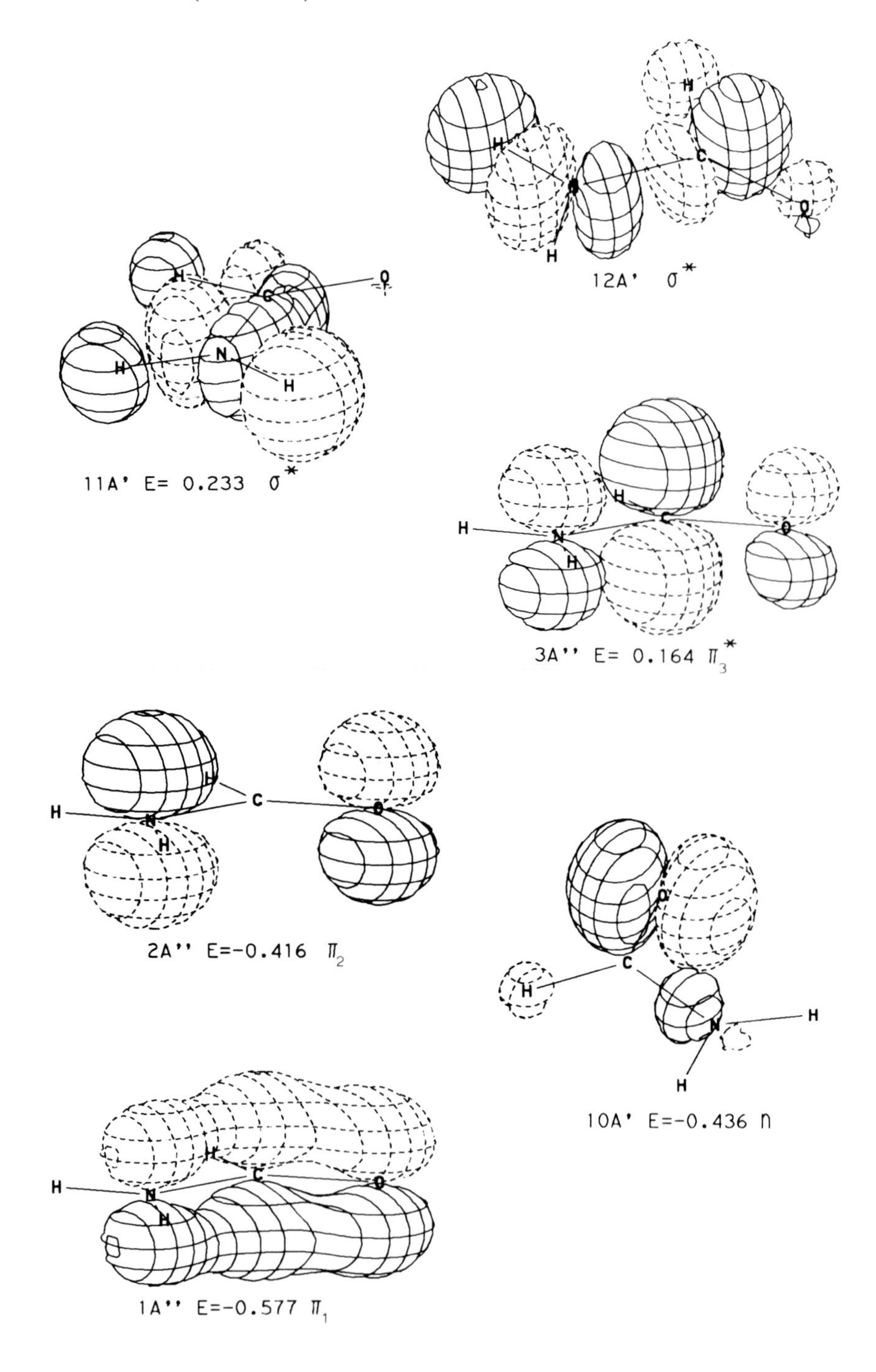

52. Formic Acid

Symmetry: C_s

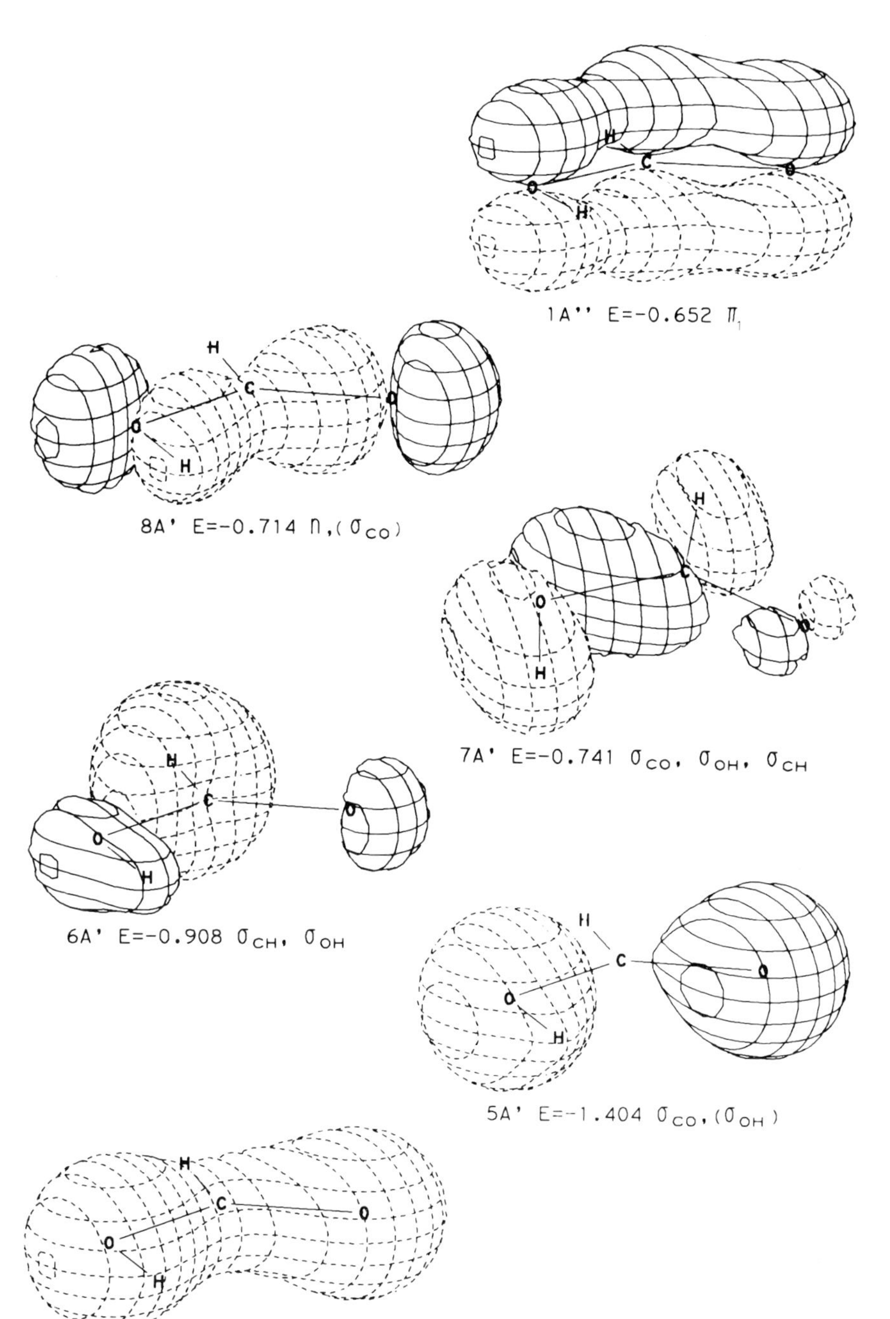

1A'' E=-0.652 π_1

8A' E=-0.714 n,(σ_{CO})

7A' E=-0.741 σ_{CO}, σ_{OH}, σ_{CH}

6A' E=-0.908 σ_{CH}, σ_{OH}

5A' E=-1.404 σ_{CO},(σ_{OH})

4A' E=-1.511 σ_{CO}

Formic Acid (Continued)

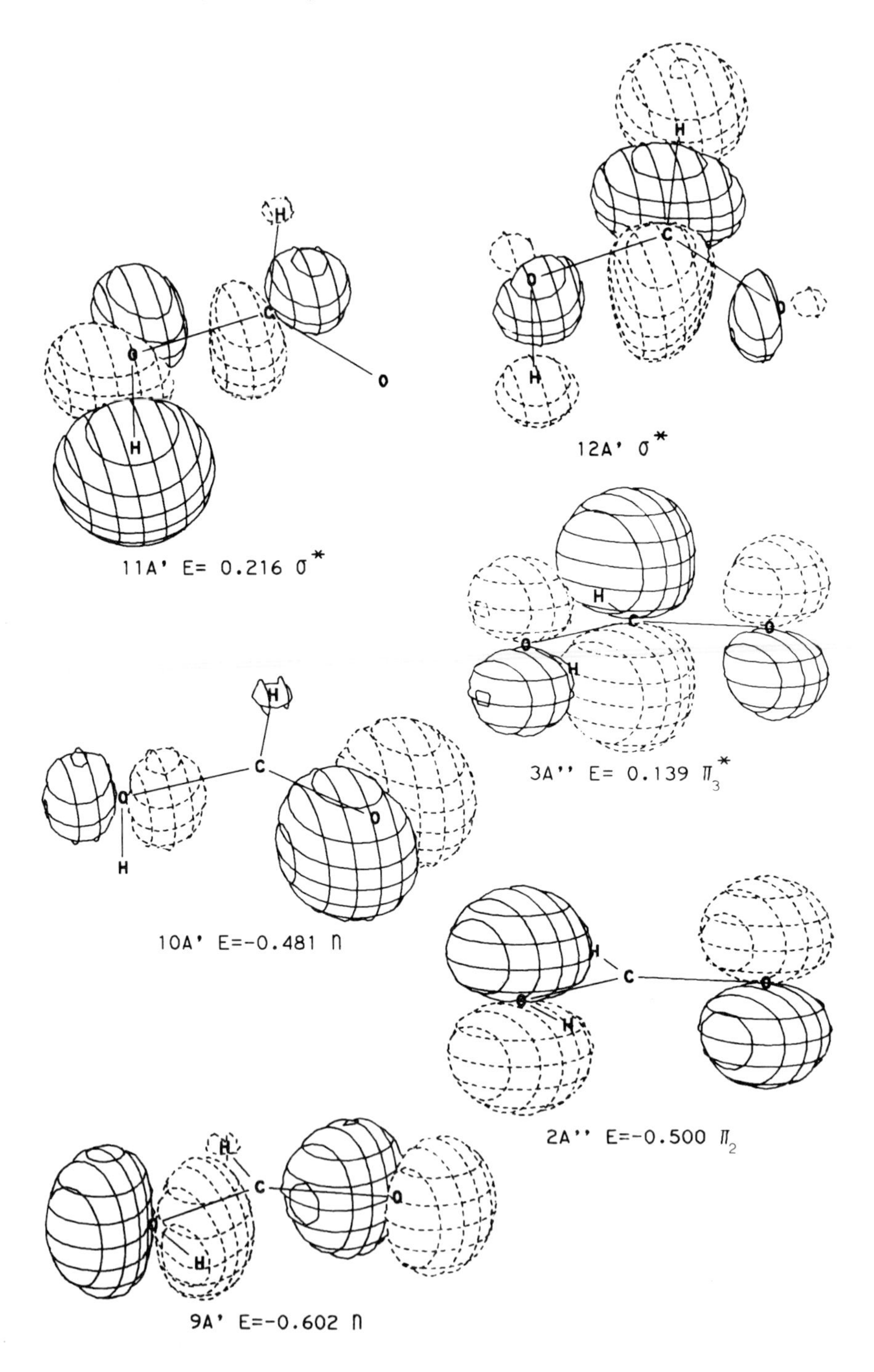

53. Formyl Fluoride

Symmetry: C_s

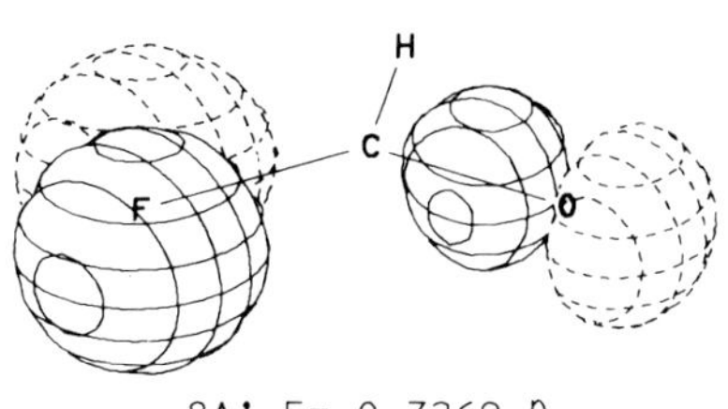

8A' E=-0.7269 n

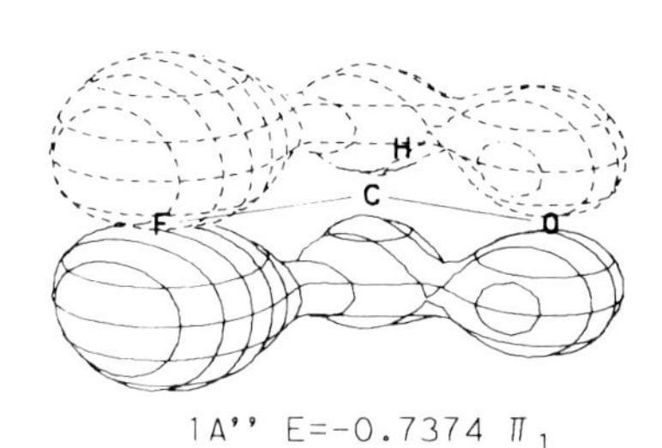

1A'' E=-0.7374 π_1

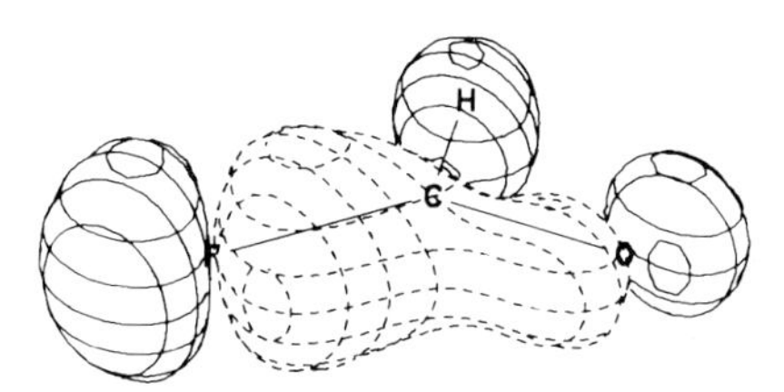

7A' E=-0.8028 σ_{CF}, n, σ_{CH}

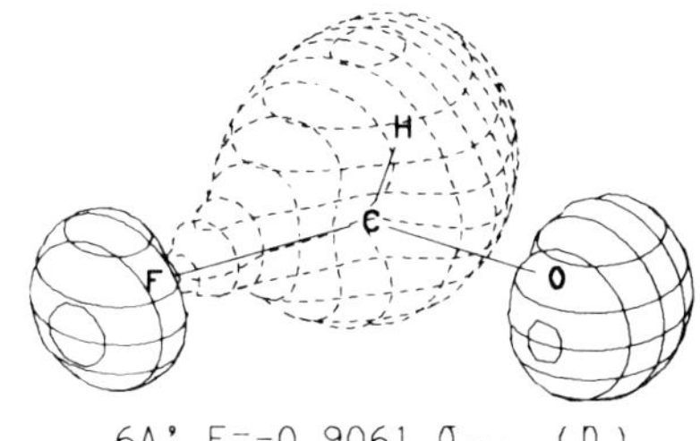

6A' E=-0.9061 σ_{CH}, (n)

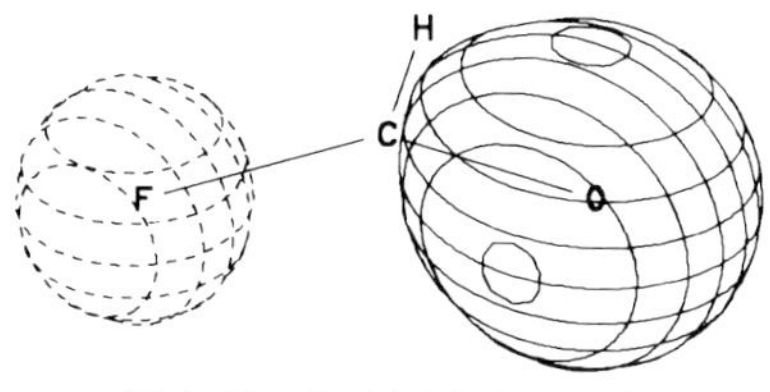

5A' E=-1.4915 σ_{CO}, n

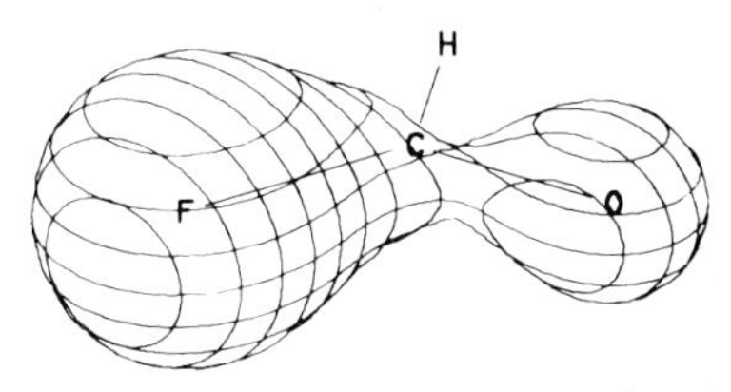

4A' E=-1.7123 σ_{CF}, (σ_{CO})

Formyl Fluoride (Continued)

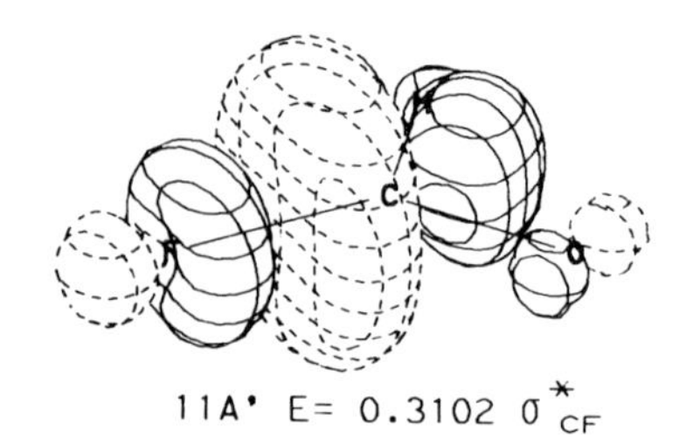

11A' E= 0.3102 σ^{*}_{CF}

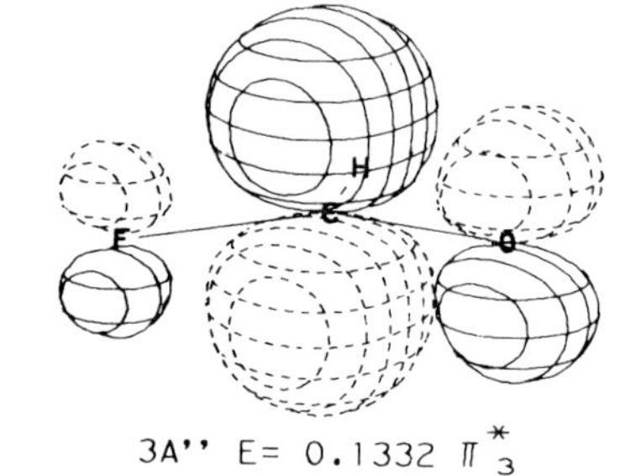

3A'' E= 0.1332 π^{*}_{3}

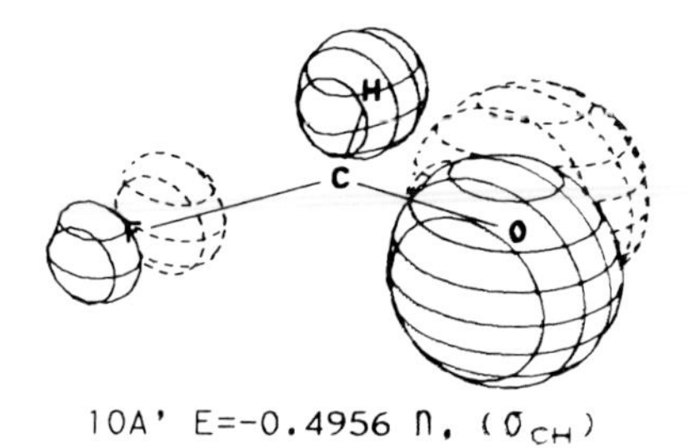

10A' E=-0.4956 n, (σ_{CH})

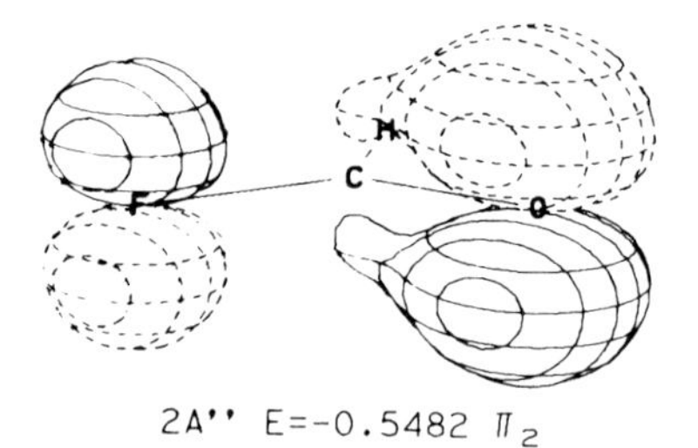

2A'' E=-0.5482 π_{2}

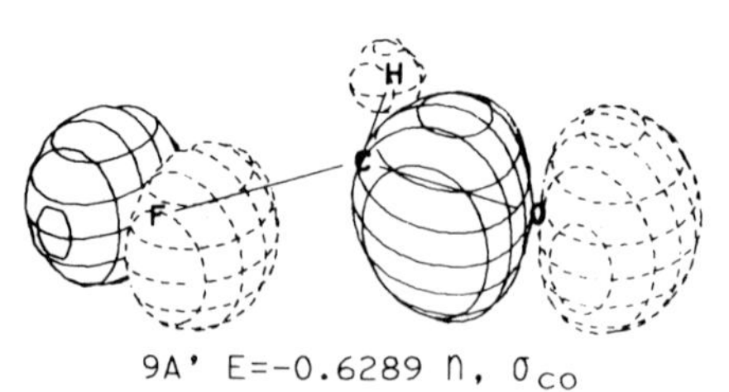

9A' E=-0.6289 n, σ_{CO}

54. Nitrosomethane

Symmetry: C_s

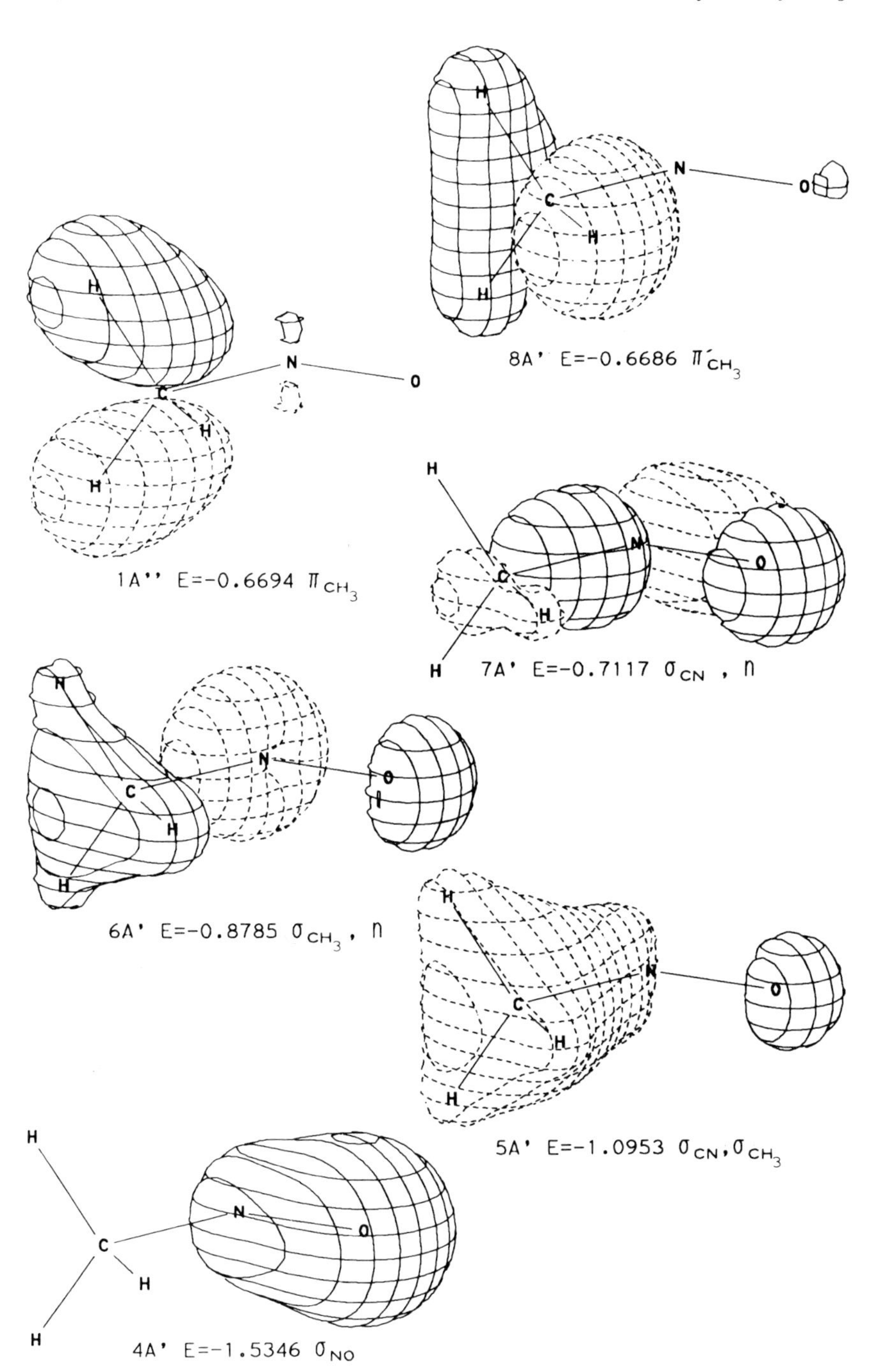

Nitrosomethane (Continued)

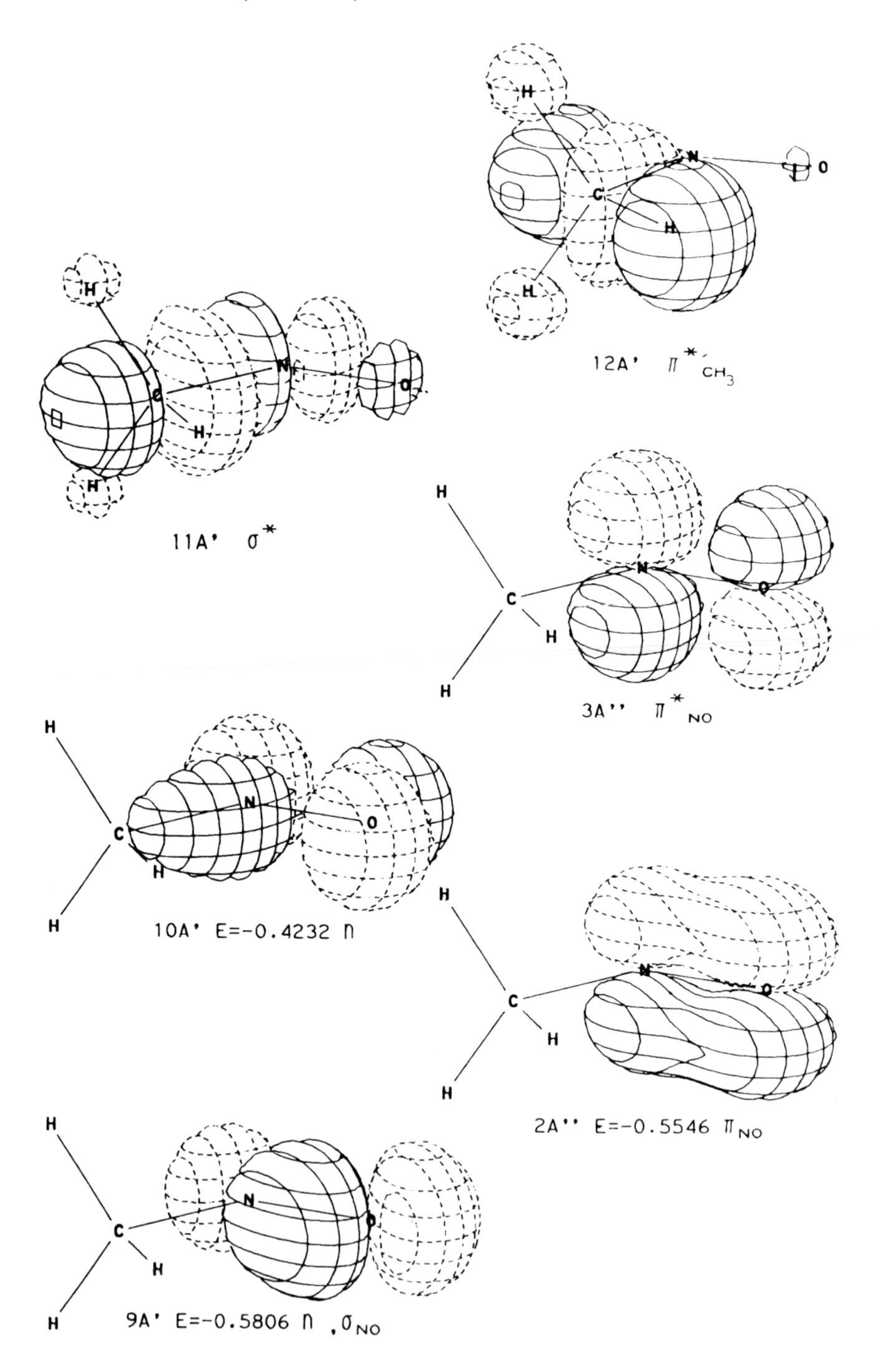

55. Ozone

Symmetry: C_{2v}

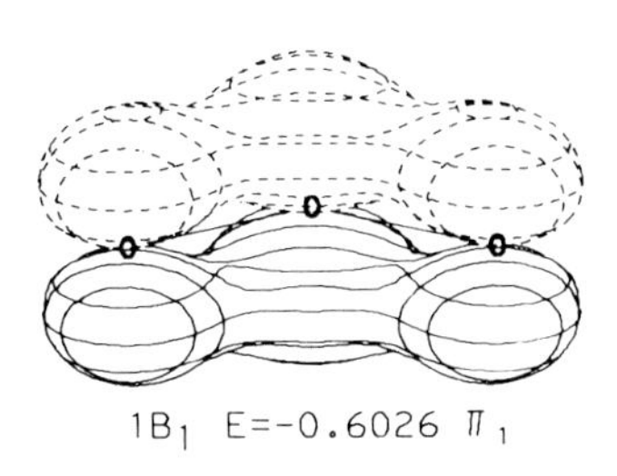

$1B_1$ E=−0.6026 π_1

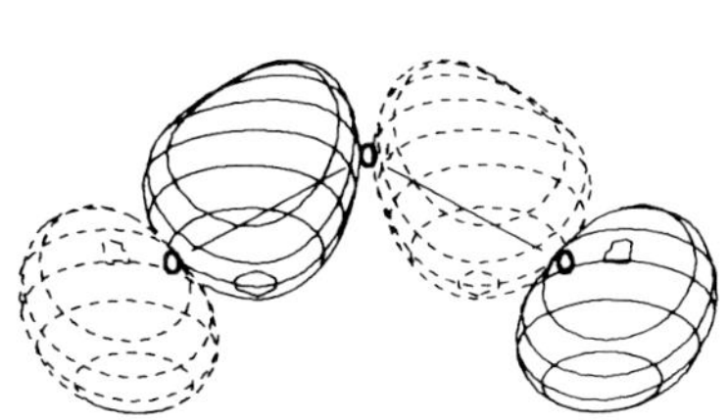

$3B_2$ E=−0.6219 σ_{OO}, n

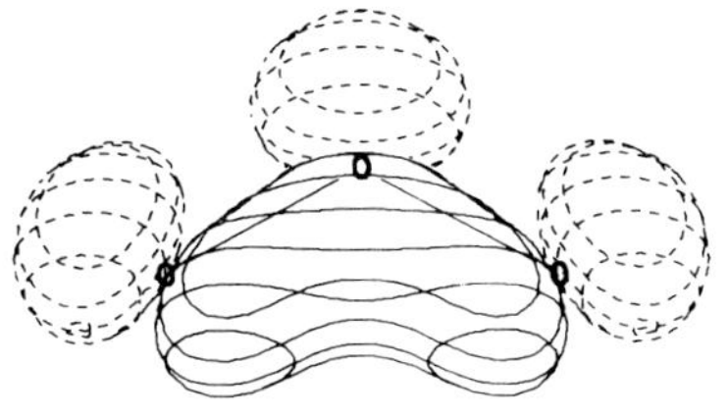

$5A_1$ E=−0.6452 σ_{OO}, n

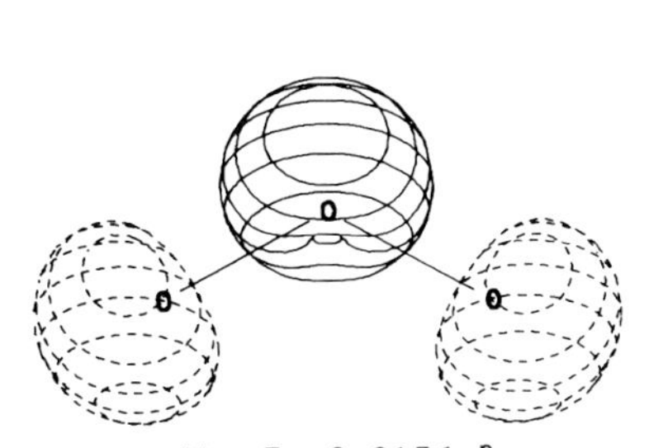

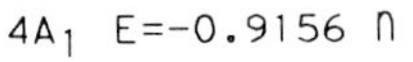

$4A_1$ E=−0.9156 n

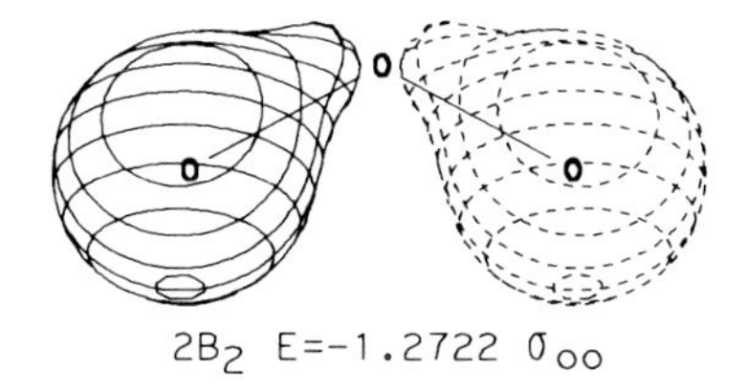

$2B_2$ E=−1.2722 σ_{OO}

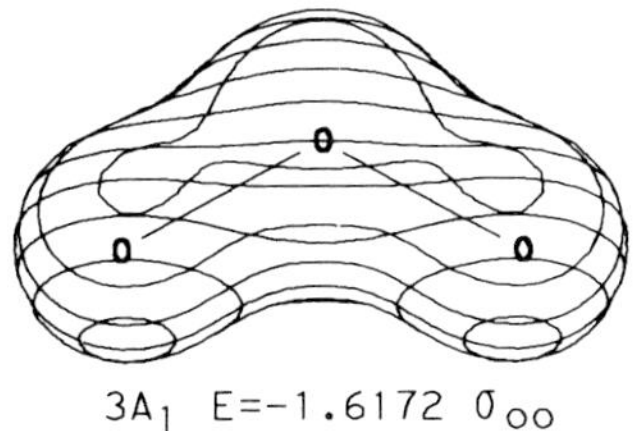

$3A_1$ E=−1.6172 σ_{OO}

Ozone (Continued)

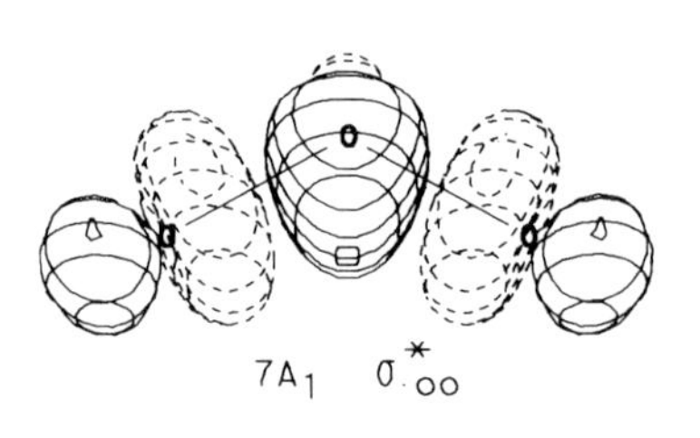

$7A_1$ σ^*_{OO}

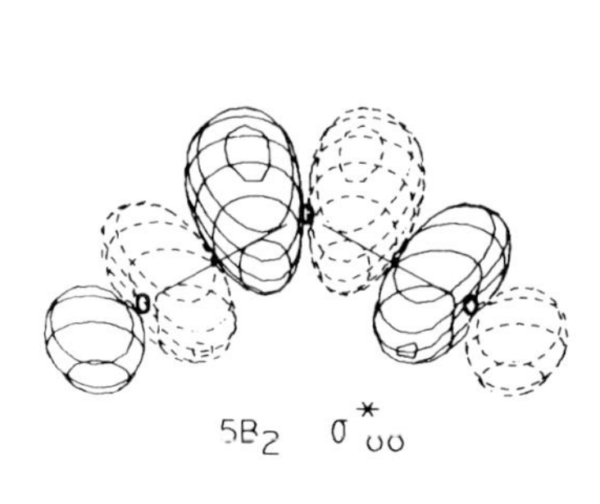

$5B_2$ σ^*_{OO}

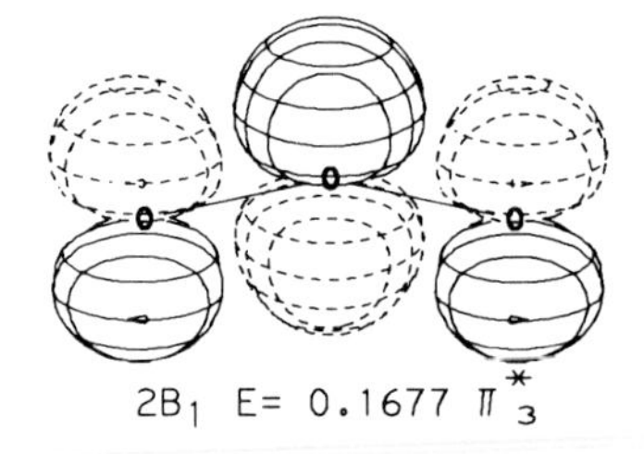

$2B_1$ E= 0.1677 π^*_3

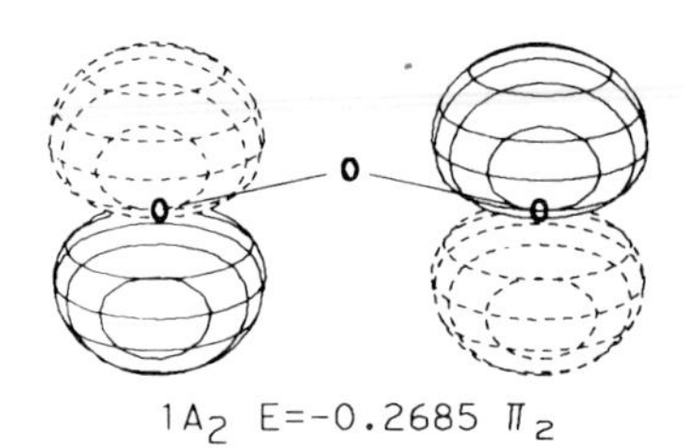

$1A_2$ E=-0.2685 π_2

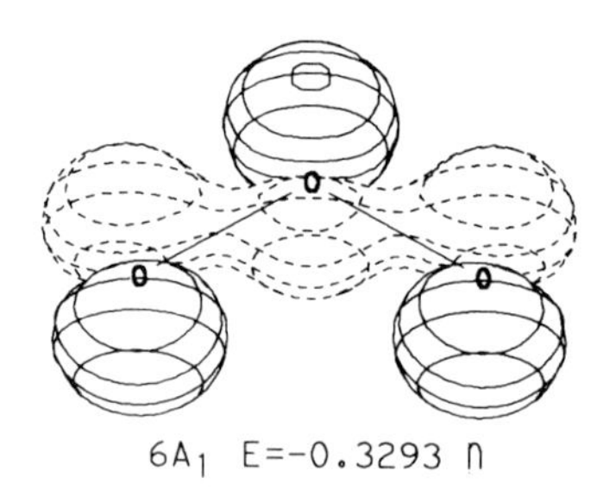

$6A_1$ E=-0.3293 n

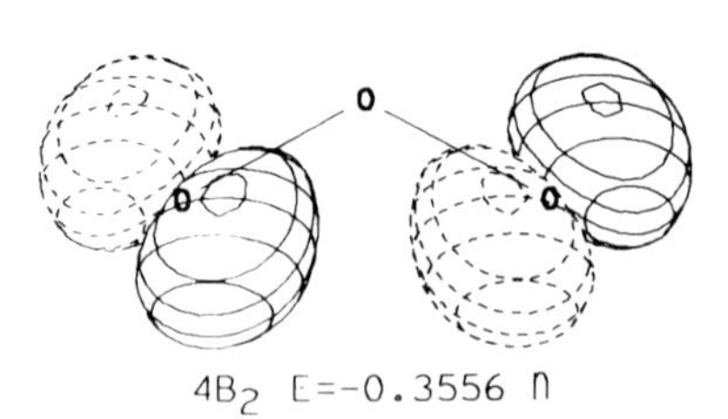

$4B_2$ E=-0.3556 n

56. Cyclopropane

Symmetry: D_{3h}

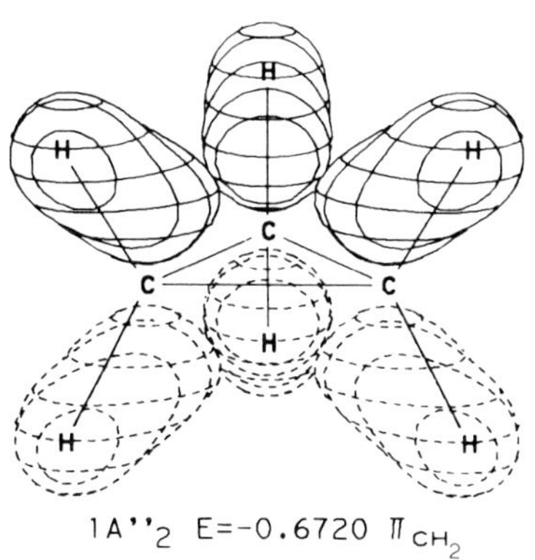

$1A''_2$ E=-0.6720 π_{CH_2}

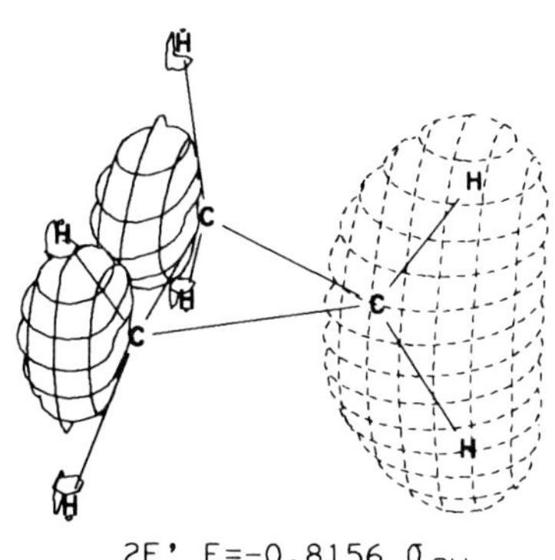

2E' E=-0.8156 σ_{CH_2}

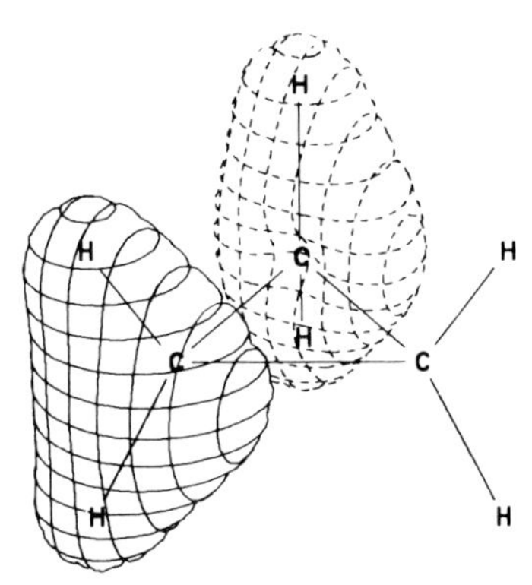

2E' E=-0.8156 σ_{CH_2}

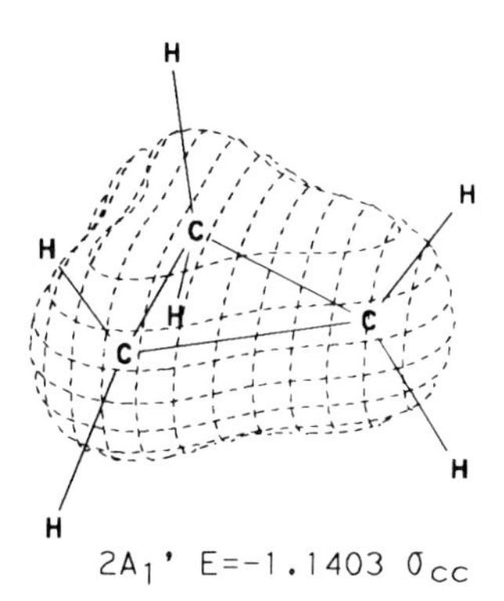

$2A_1'$ E=-1.1403 σ_{CC}

Cyclopropane (Continued)

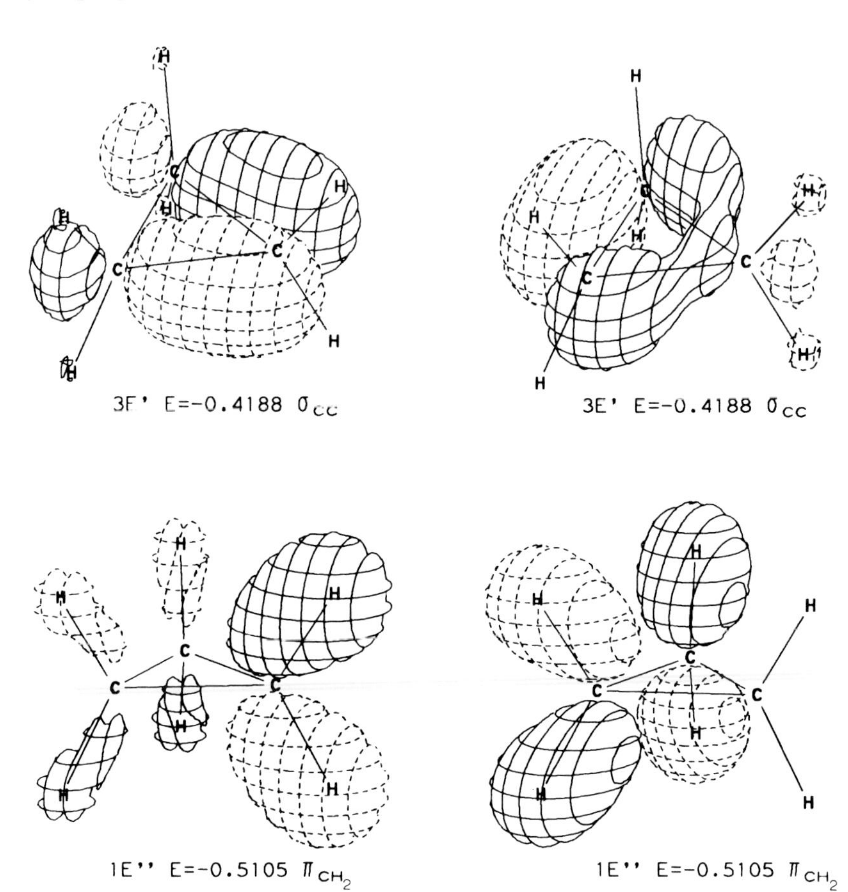

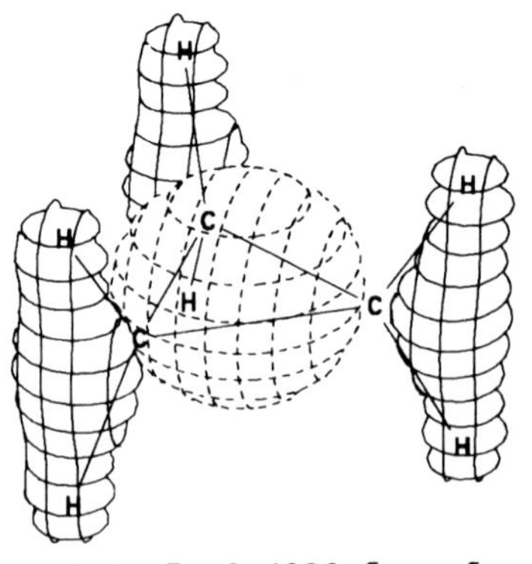

3A'1 E=-0.6228 σCC, σCH2

Cyclopropane (Continued)

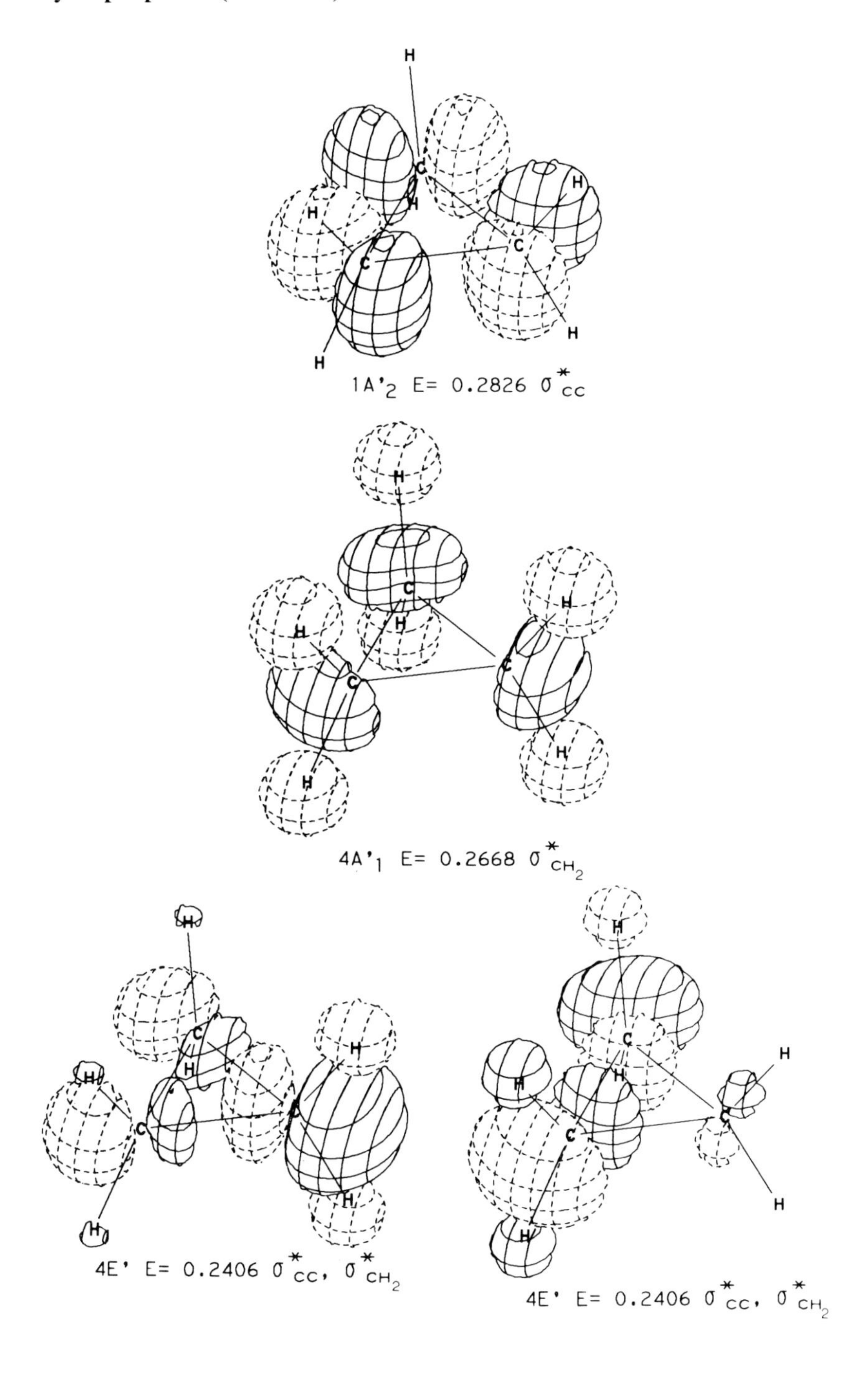

1A'$_2$ E= 0.2826 σ^*_{CC}

4A'$_1$ E= 0.2668 $\sigma^*_{CH_2}$

4E' E= 0.2406 σ^*_{CC}, $\sigma^*_{CH_2}$

4E' E= 0.2406 σ^*_{CC}, $\sigma^*_{CH_2}$

57. Aziridine

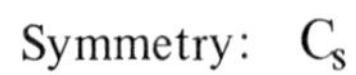
Symmetry: C_s

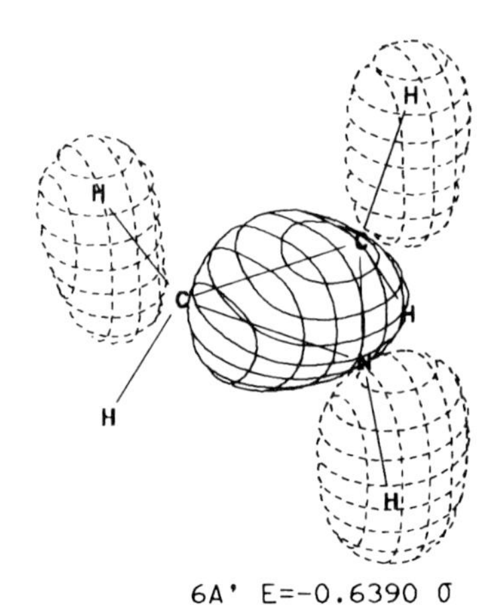

6A' E=-0.6390 σ

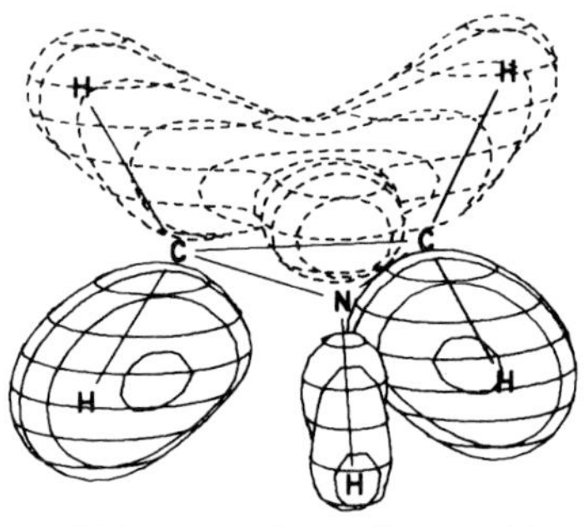

5A' E=-0.7040 π_{CH_2}, π_{NH}

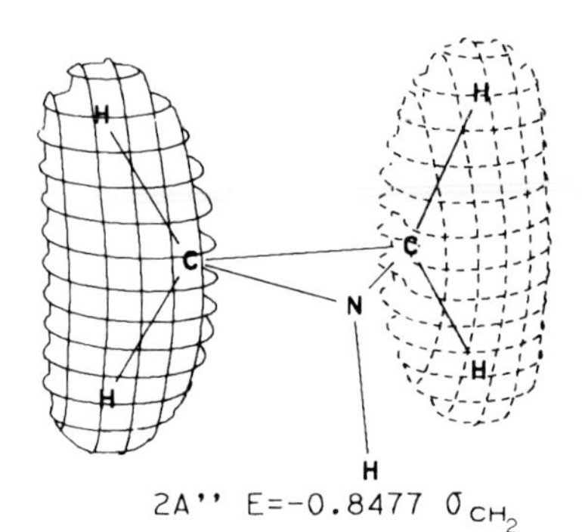

2A'' E=-0.8477 σ_{CH_2}

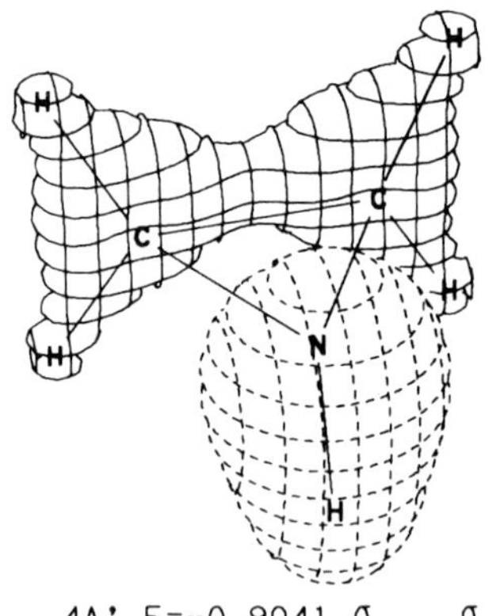

4A' E=-0.9041 σ_{NH}, σ_{CH_2}

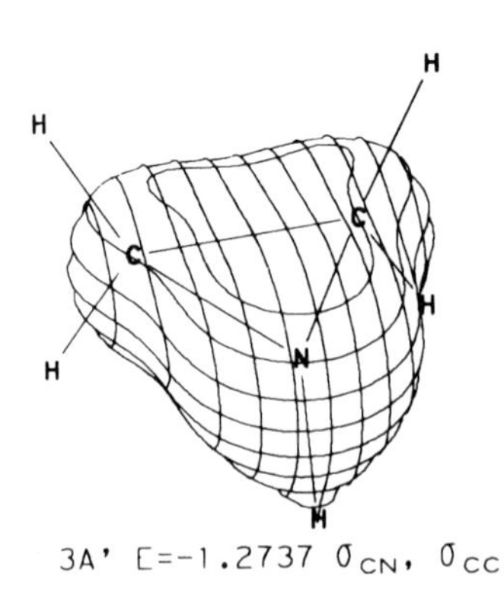

3A' E=-1.2737 σ_{CN}, σ_{CC}

Aziridine (Continued)

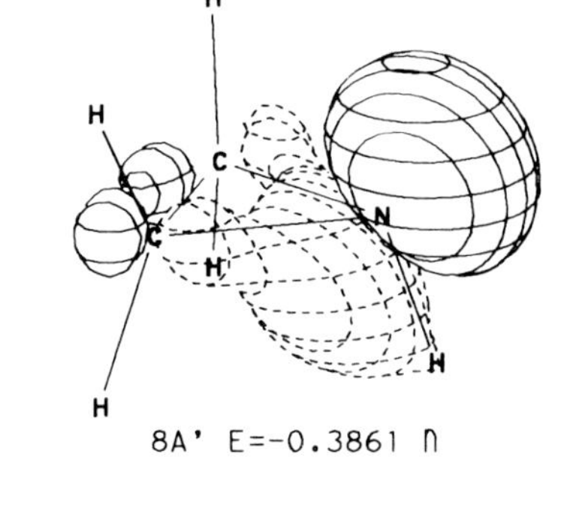

8A' E=-0.3861 n

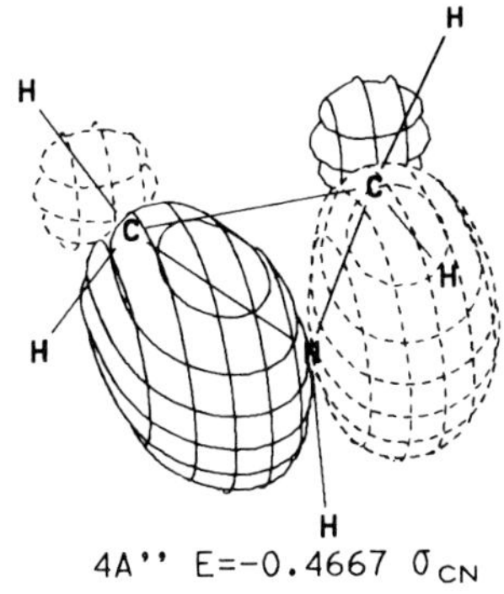

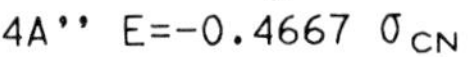

4A'' E=-0.4667 σ_{CN}

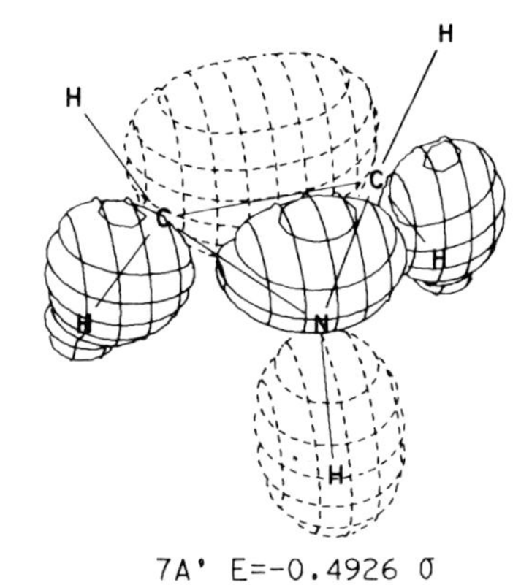

7A' E=-0.4926 σ

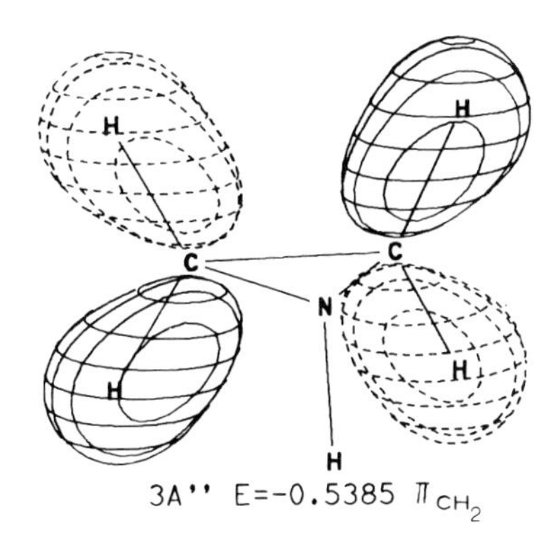

3A'' E=-0.5385 π_{CH_2}

Aziridine (Continued)

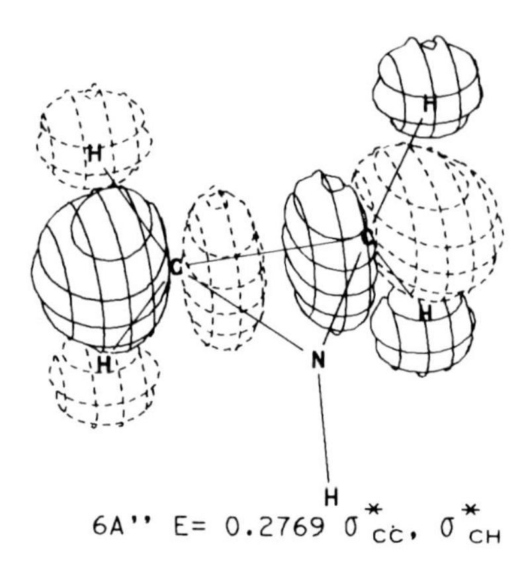

6A'' E= 0.2769 σ^*_{CC}, σ^*_{CH}

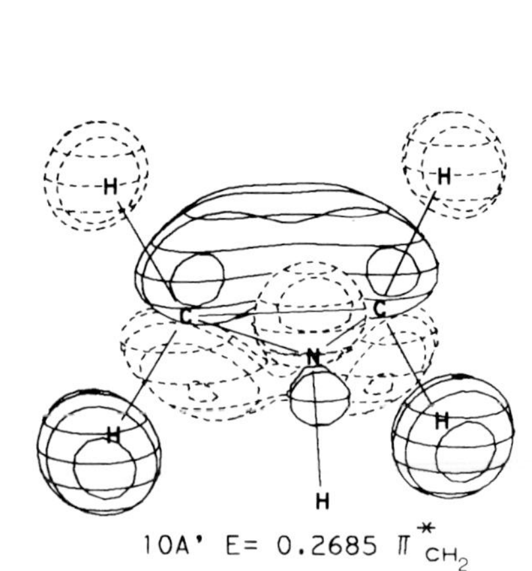

10A' E= 0.2685 $\pi^*_{CH_2}$

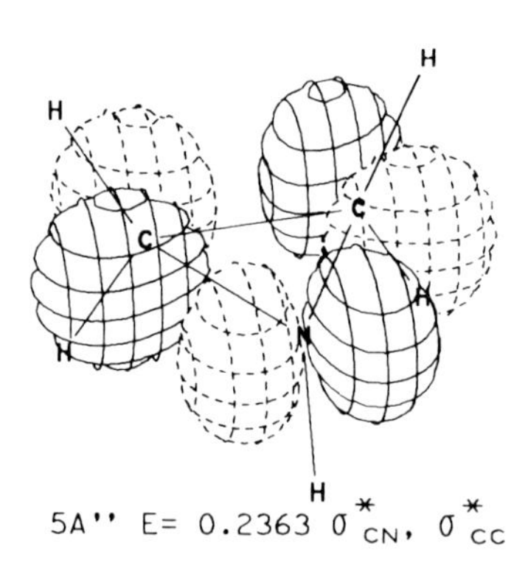

5A'' E= 0.2363 σ^*_{CN}, σ^*_{CC}

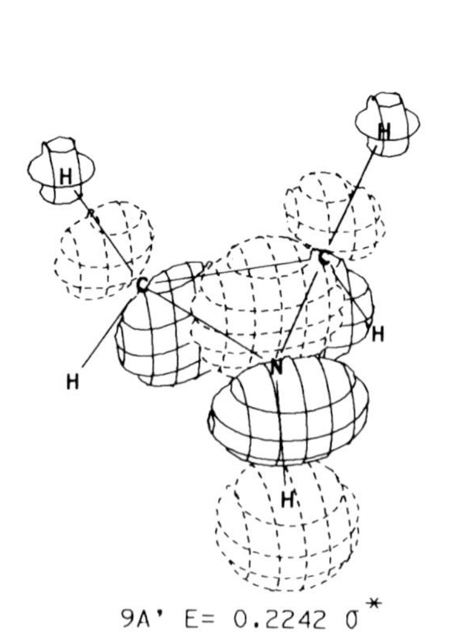

9A' E= 0.2242 σ^*

58. Ethylene Oxide

Symmetry: C_{2v}

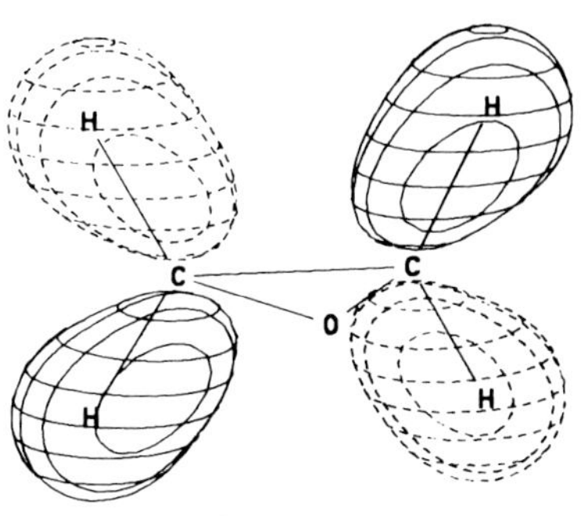

$1A_2$ E=-0.5531 π_{CH_2}

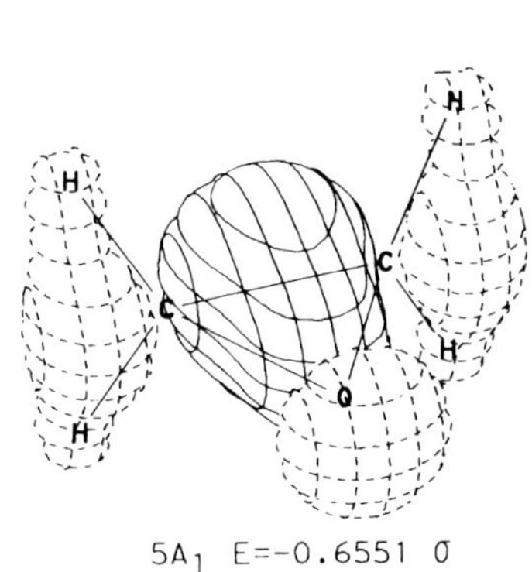
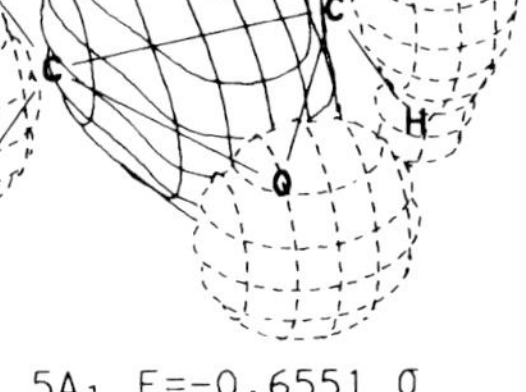

$5A_1$ E=-0.6551 σ

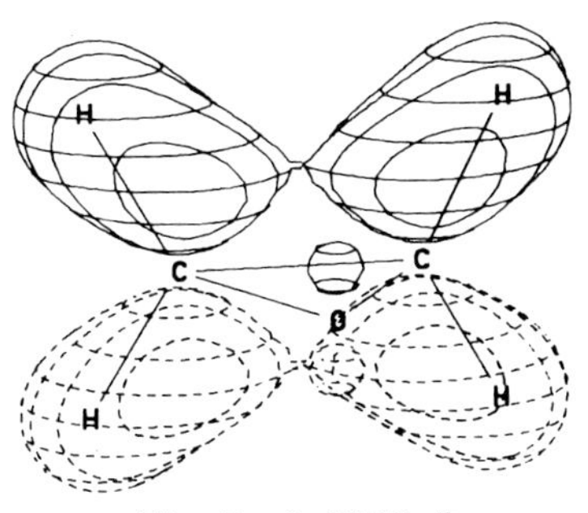

$1B_1$ E=-0.7182 π_{CH_2}

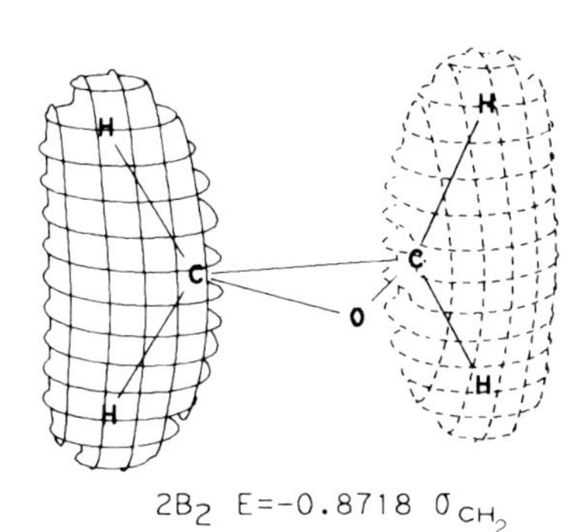

$2B_2$ E=-0.8718 σ_{CH_2}

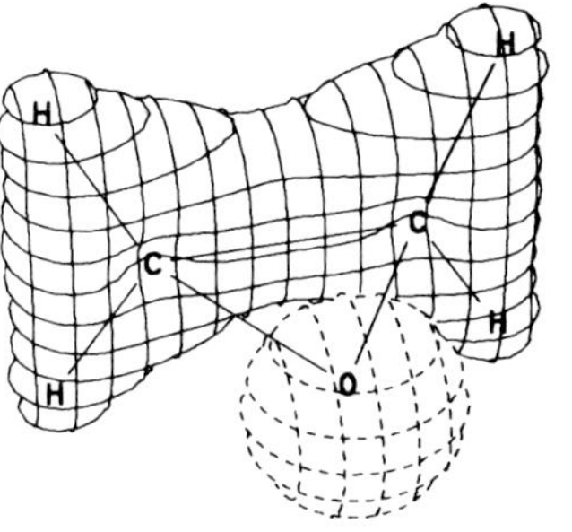

$4A_1$ E=-0.9389 σ_{CC}, σ_{CH_2}, n

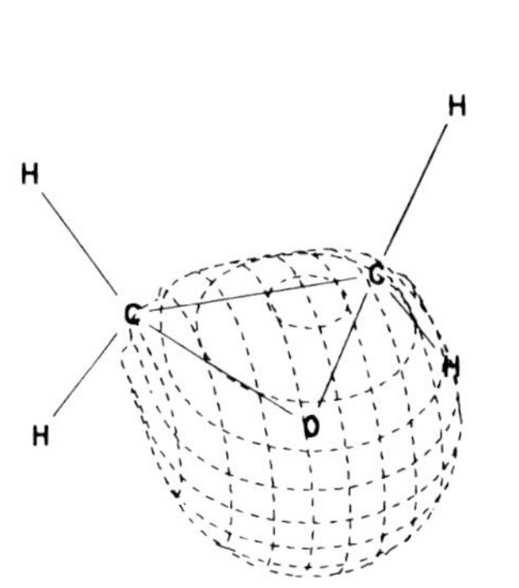

$3A_1$ E=-1.4284 σ_{CO}, σ_{CC}

Ethylene Oxide (Continued)

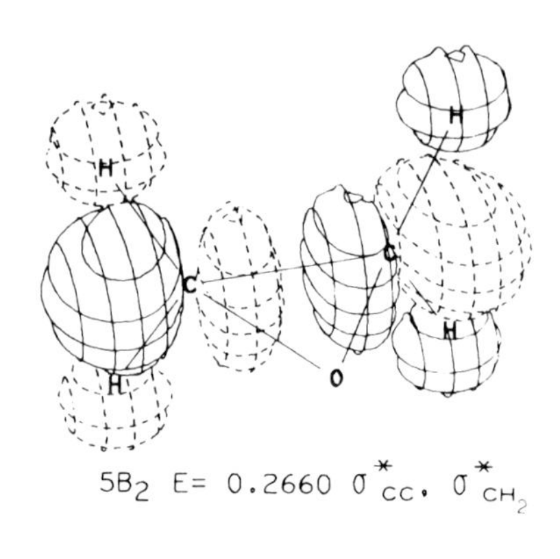

$5B_2$ E= 0.2660 σ^*_{CC}, $\sigma^*_{CH_2}$

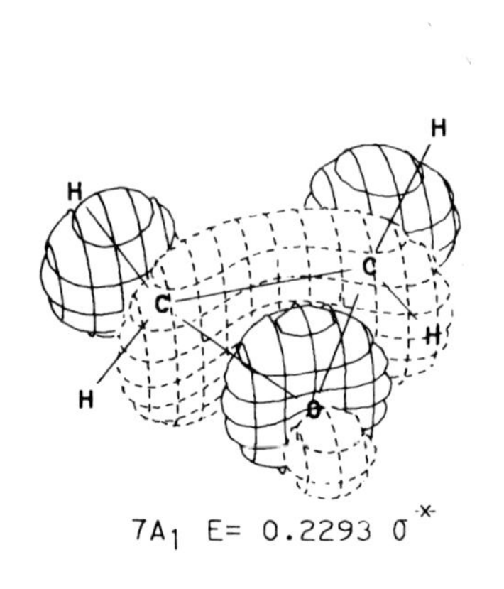

$7A_1$ E= 0.2293 σ^*

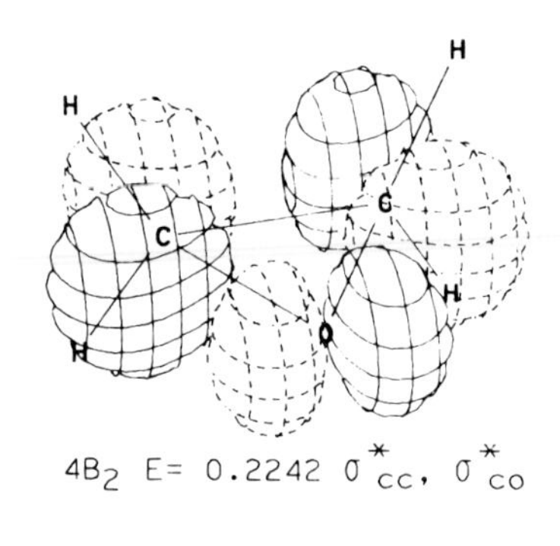

$4B_2$ E= 0.2242 σ^*_{CC}, σ^*_{CO}

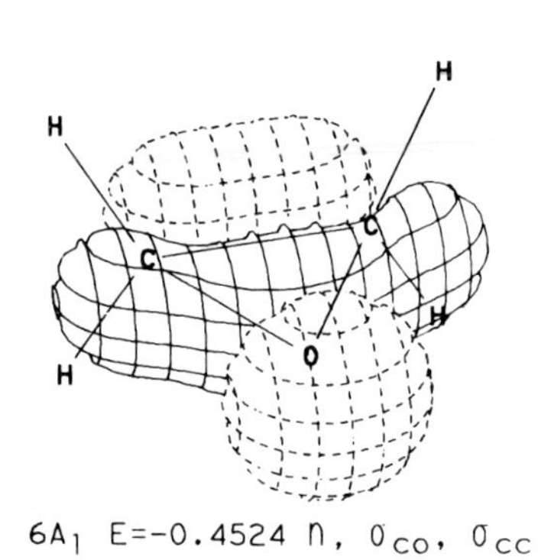

$6A_1$ E=-0.4524 n, σ_{CO}, σ_{CC}

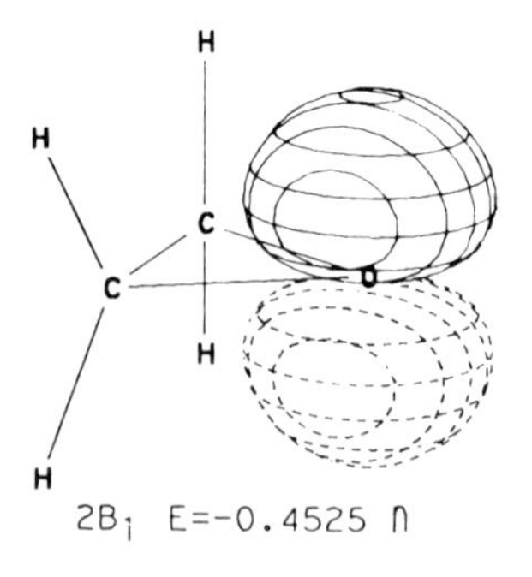

$2B_1$ E=-0.4525 n

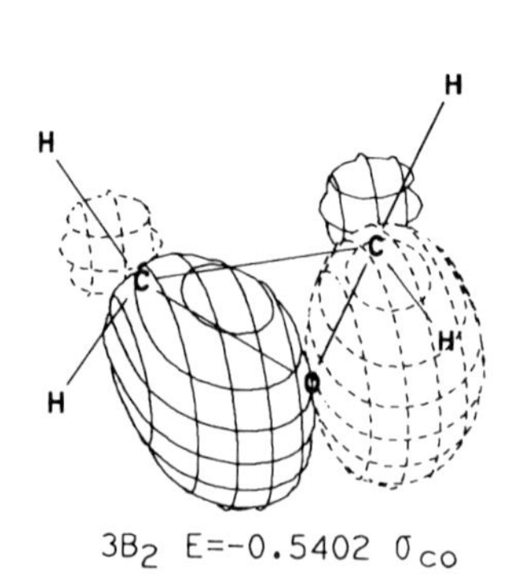

$3B_2$ E=-0.5402 σ_{CO}

59. Trimethylene, Edge-to-Edge

Symmetry: C_{2v}

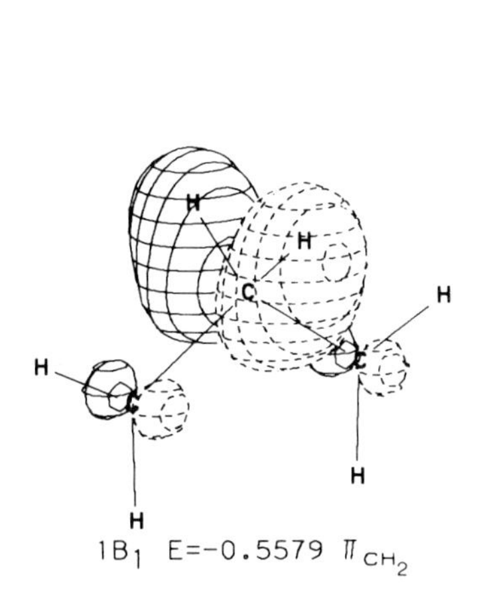

$1B_1$ E=-0.5579 π_{CH_2}

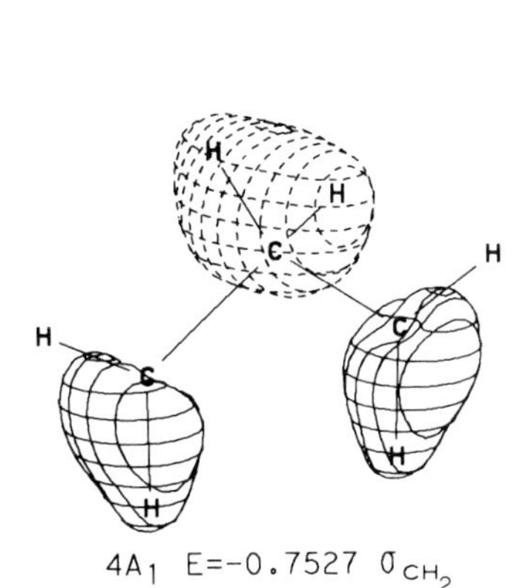

$4A_1$ E=-0.7527 σ_{CH_2}

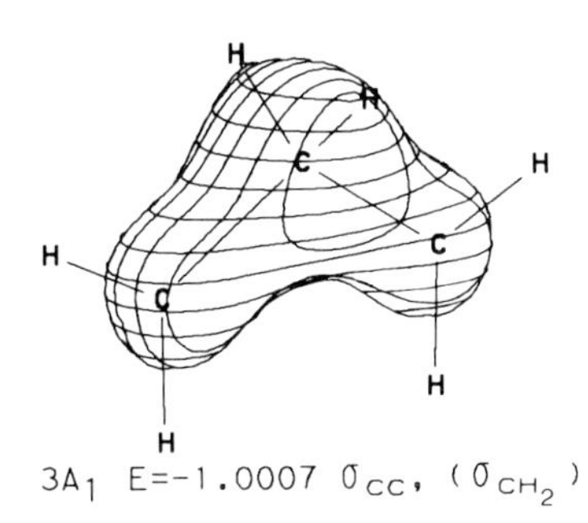

$3A_1$ E=-1.0007 σ_{CC}, (σ_{CH_2})

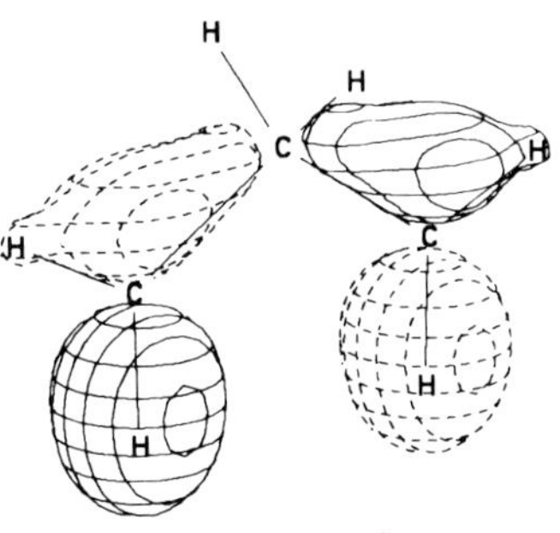

$3B_2$ E=-0.5379 π'_{CH_2}, σ_{CC}

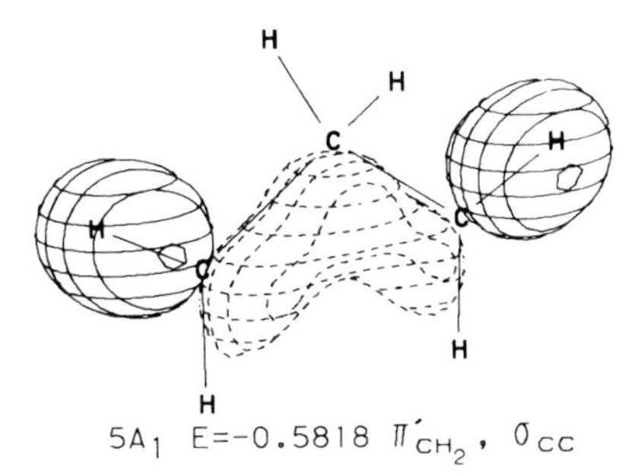

$5A_1$ E=-0.5818 π'_{CH_2}, σ_{CC}

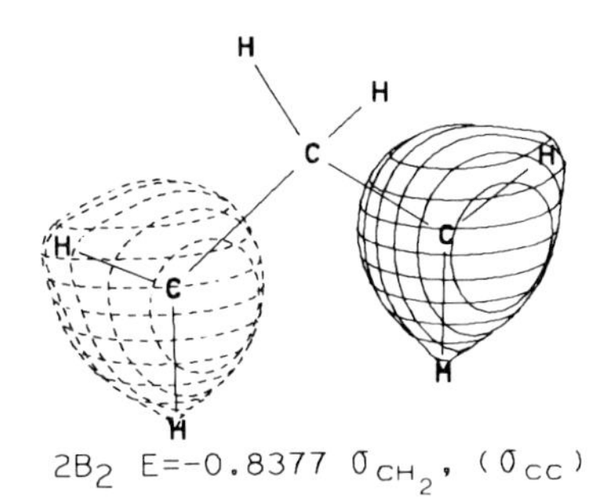

$2B_2$ E=-0.8377 σ_{CH_2}, (σ_{CC})

Trimethylene, Edge-to-Edge (Continued)

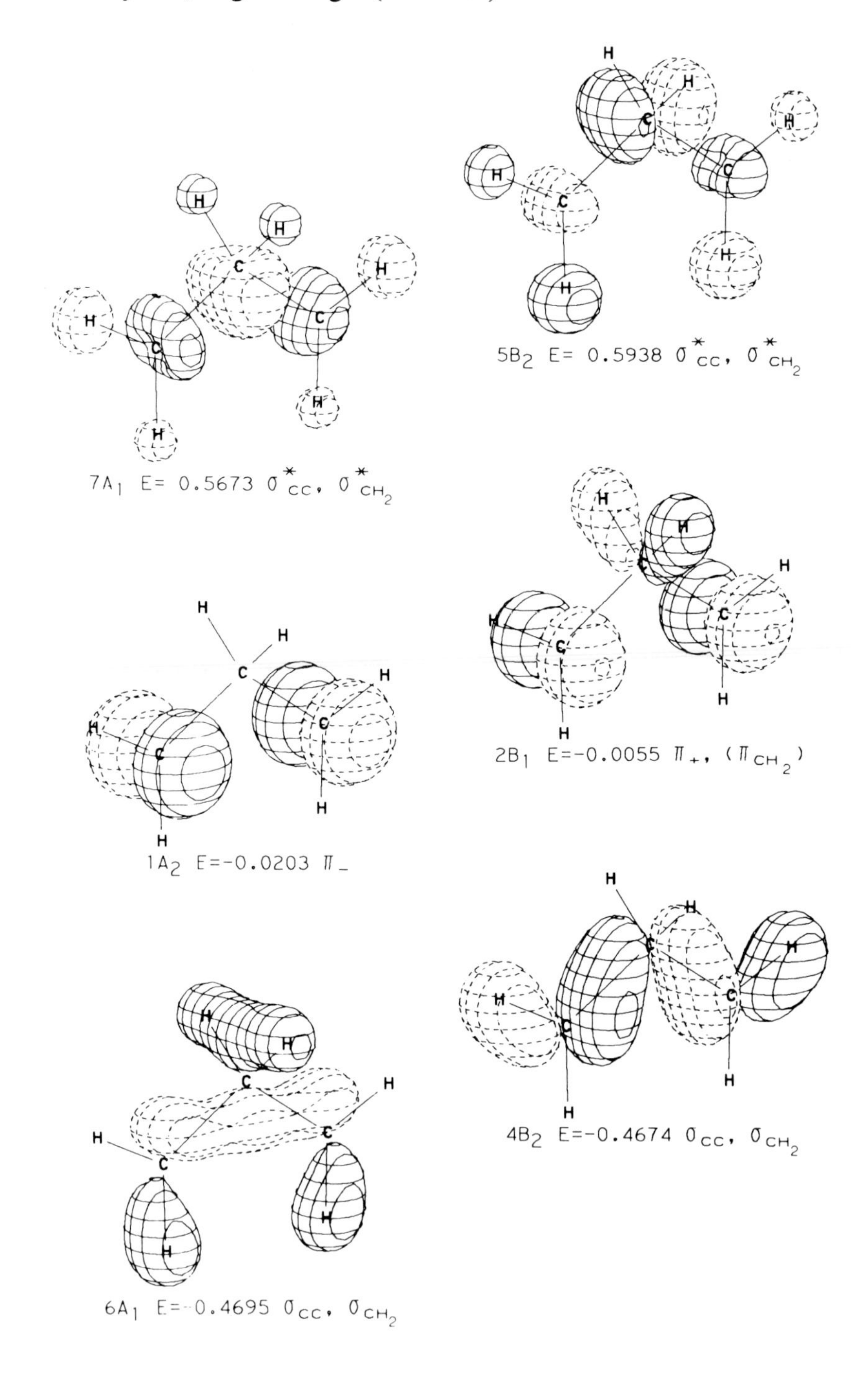

7A1 E= 0.5673 σ^*_{CC}, $\sigma^*_{CH_2}$

5B2 E= 0.5938 σ^*_{CC}, $\sigma^*_{CH_2}$

1A2 E=-0.0203 π_-

2B1 E=-0.0055 π_+, (π_{CH_2})

6A1 E=-0.4695 σ_{CC}, σ_{CH_2}

4B2 E=-0.4674 σ_{CC}, σ_{CH_2}

60. *n*-Propyl Cation, Bisected

Symmetry: C_s

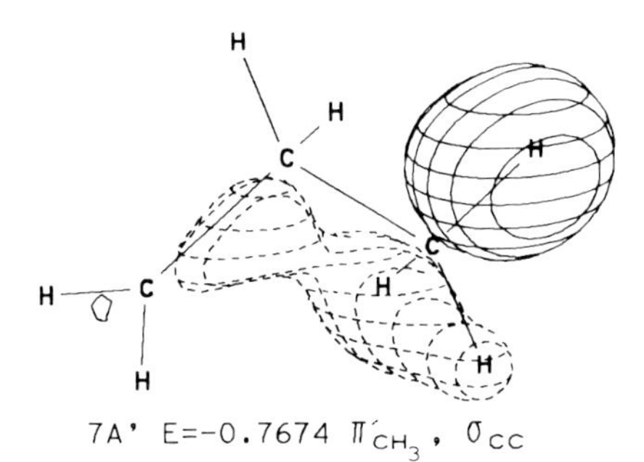

7A' E=-0.7674 π'_{CH_3}, σ_{CC}

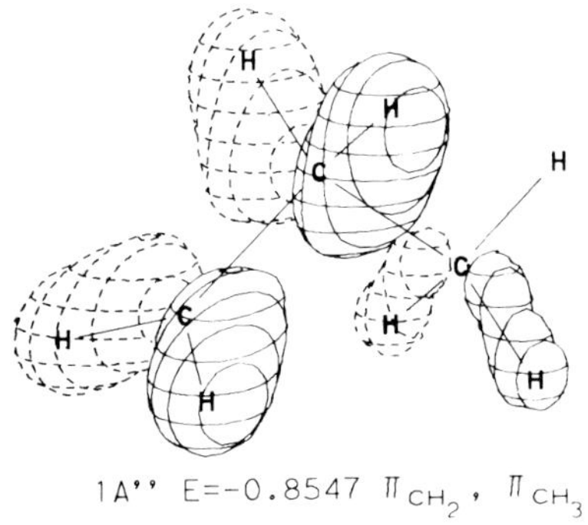

1A'' E=-0.8547 π_{CH_2}, π_{CH_3}

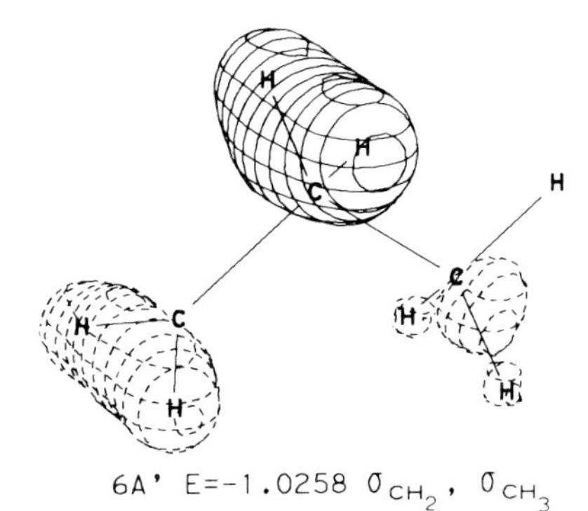

6A' E=-1.0258 σ_{CH_2}, σ_{CH_3}

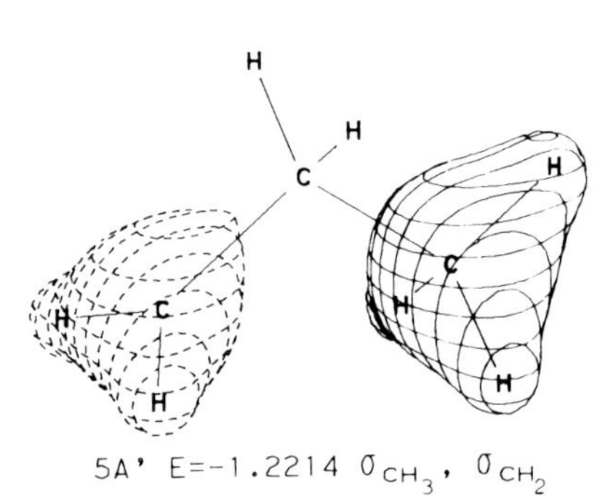

5A' E=-1.2214 σ_{CH_3}, σ_{CH_2}

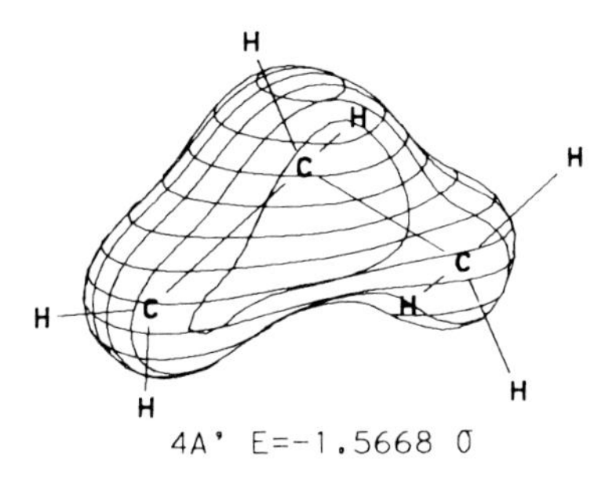

4A' E=-1.5668 σ

n-Propyl Cation, Bisected (Continued)

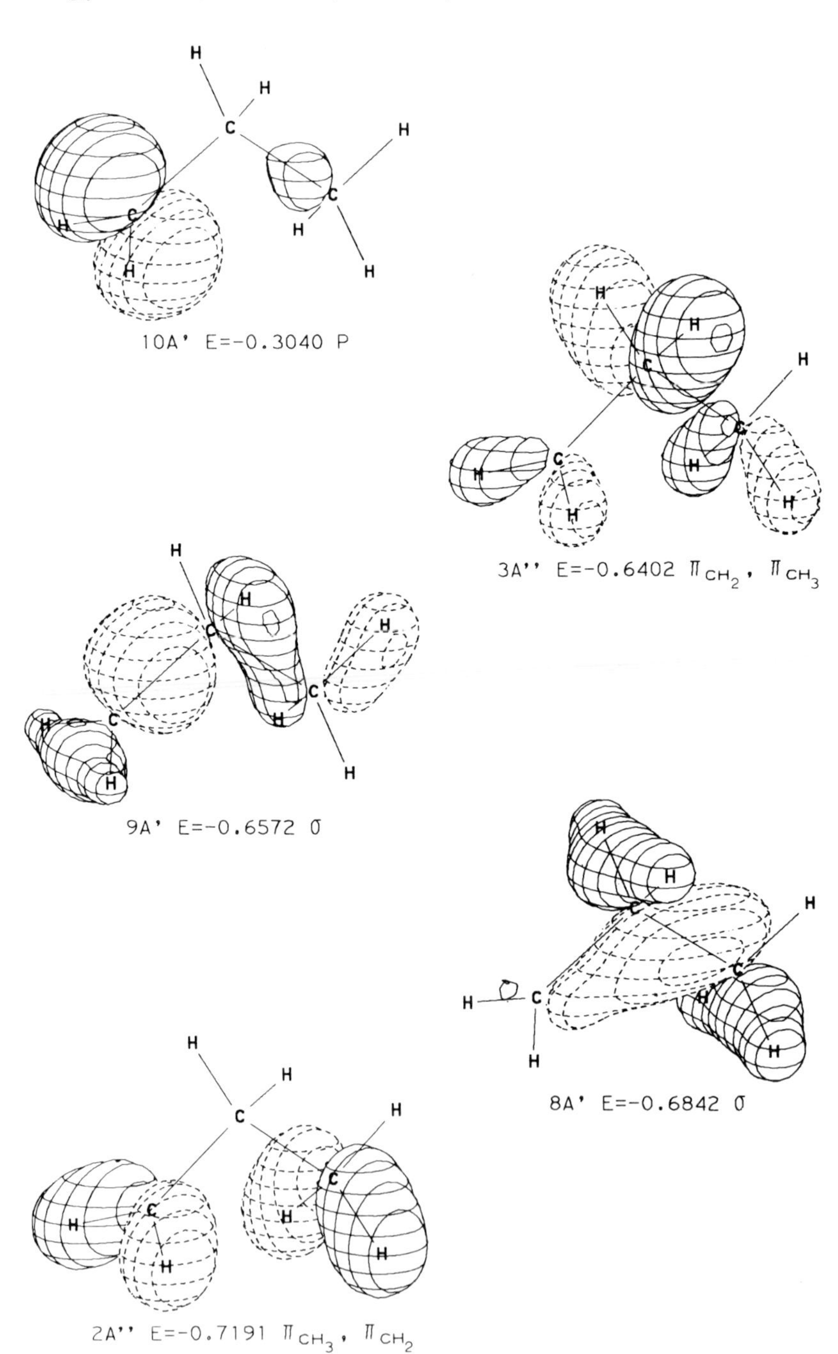

10A' E=-0.3040 P

3A'' E=-0.6402 π_{CH_2}, π_{CH_3}

9A' E=-0.6572 σ

8A' E=-0.6842 σ

2A'' E=-0.7191 π_{CH_3}, π_{CH_2}

61. Propane

Symmetry: C_{2v}

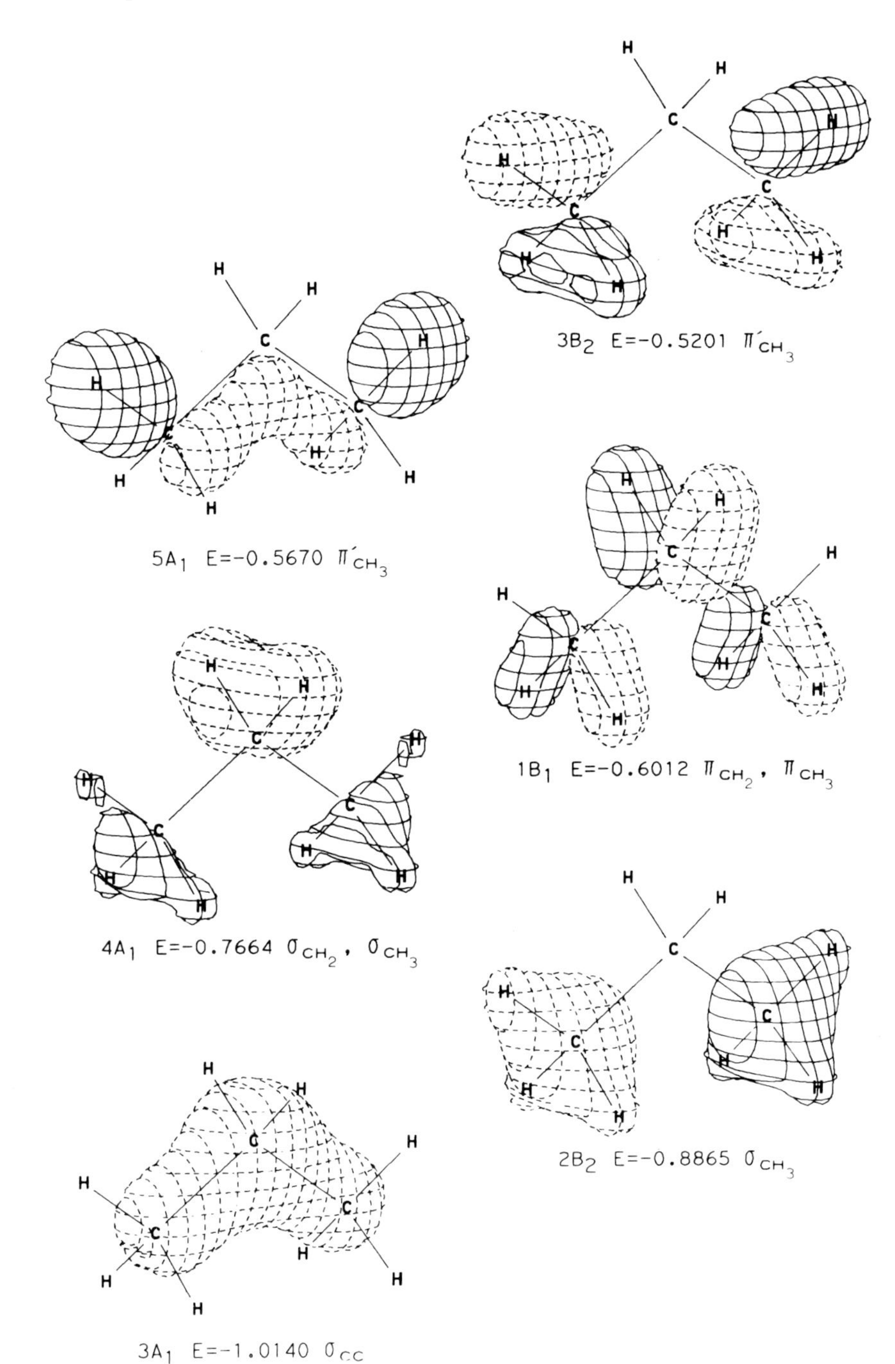

$3B_2$ $E=-0.5201$ π'_{CH_3}

$5A_1$ $E=-0.5670$ π'_{CH_3}

$1B_1$ $E=-0.6012$ π_{CH_2}, π_{CH_3}

$4A_1$ $E=-0.7664$ σ_{CH_2}, σ_{CH_3}

$2B_2$ $E=-0.8865$ σ_{CH_3}

$3A_1$ $E=-1.0140$ σ_{CC}

Propane (Continued)

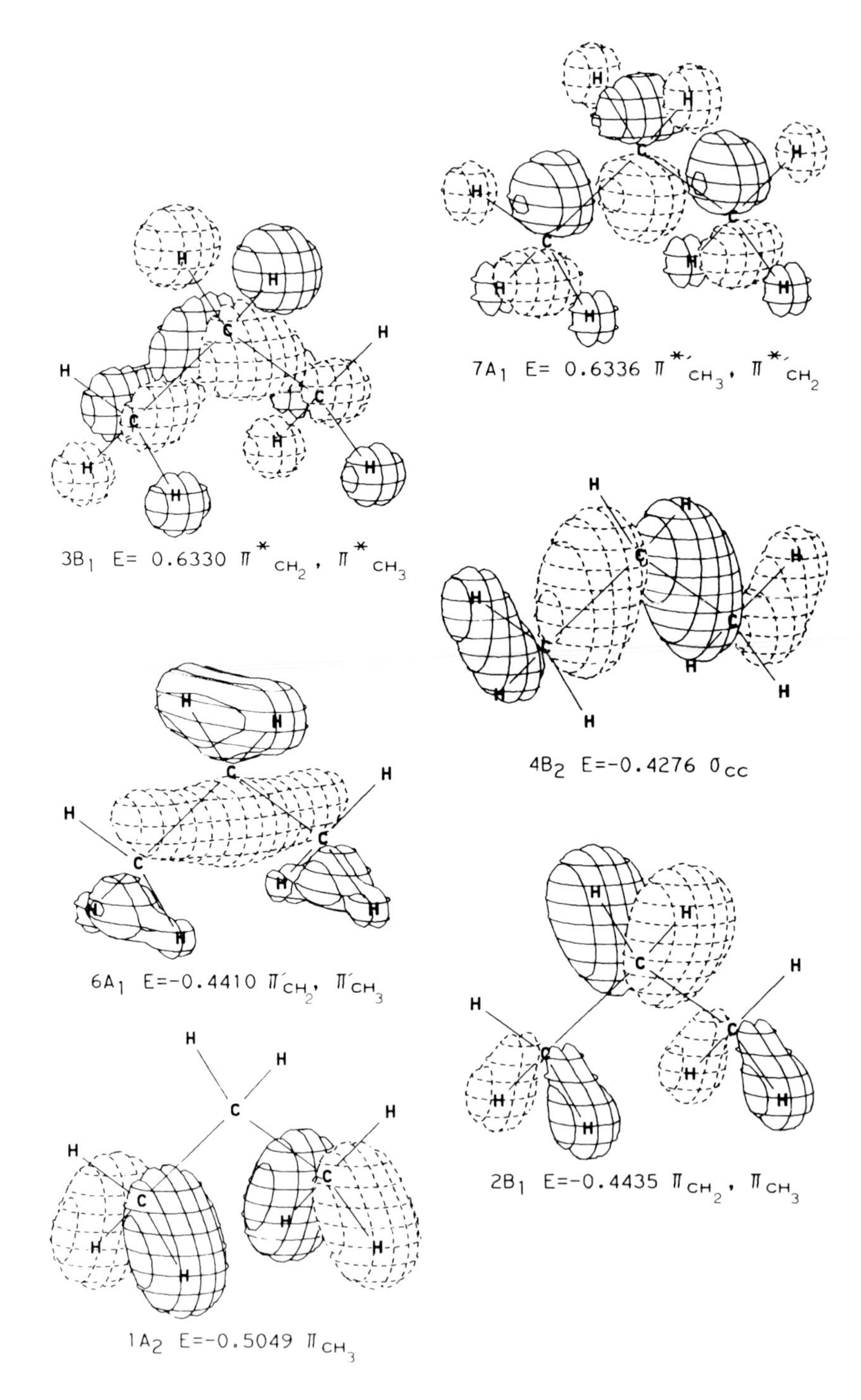

$7A_1$ E= 0.6336 $\pi^{*\prime}_{CH_3}$, $\pi^{*\prime}_{CH_2}$

$3B_1$ E= 0.6330 $\pi^{*}_{CH_2}$, $\pi^{*}_{CH_3}$

$4B_2$ E=−0.4276 σ_{CC}

$6A_1$ E=−0.4410 π'_{CH_2}, π'_{CH_3}

$2B_1$ E=−0.4435 π_{CH_2}, π_{CH_3}

$1A_2$ E=−0.5049 π_{CH_3}

62. Dimethylether

Symmetry: C_{2v}

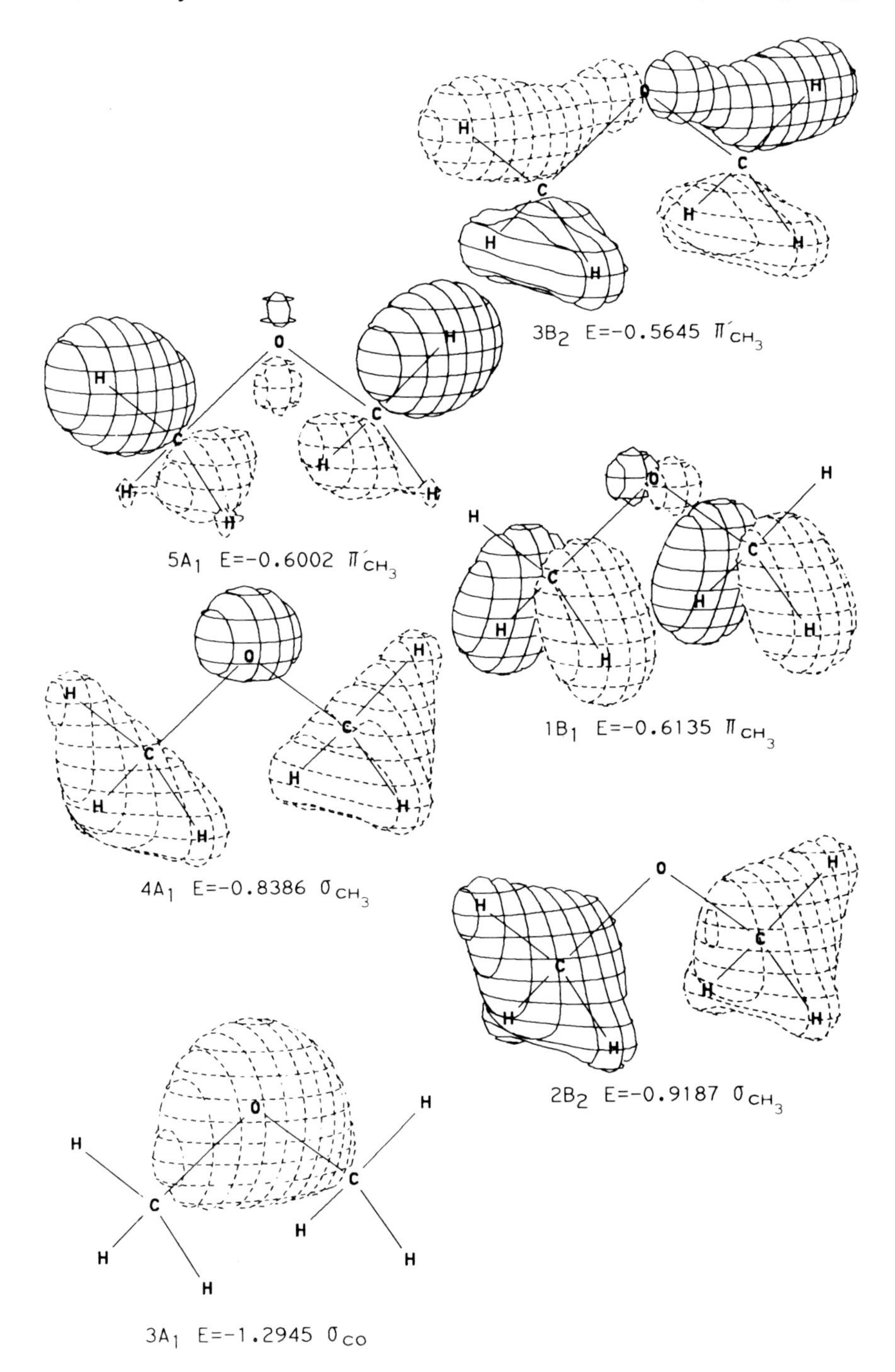

Dimethylether (Continued)

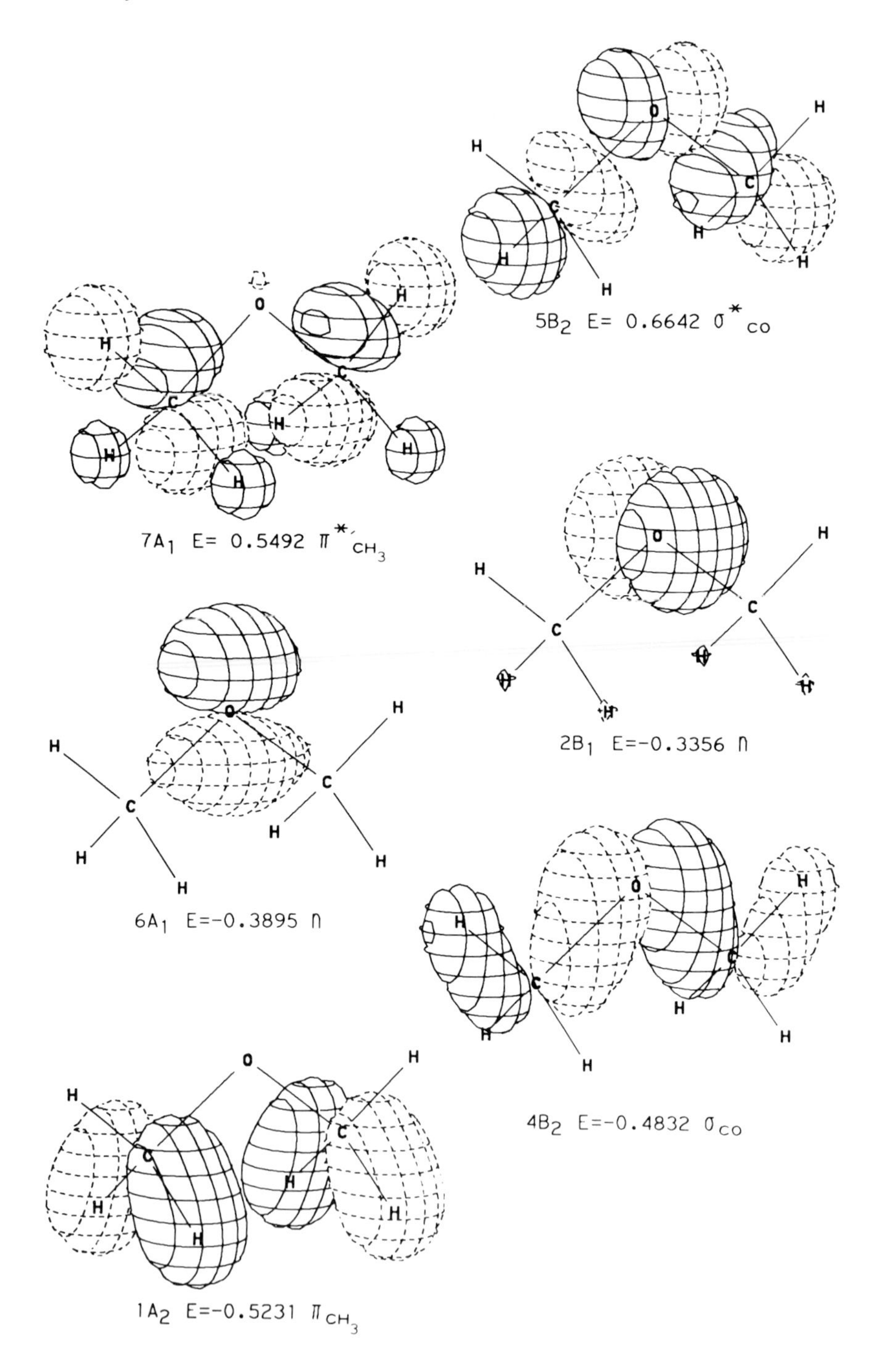

$5B_2$ E= 0.6642 σ^*_{CO}

$7A_1$ E= 0.5492 $\pi^{*\prime}_{CH_3}$

$2B_1$ E=-0.3356 n

$6A_1$ E=-0.3895 n

$4B_2$ E=-0.4832 σ_{CO}

$1A_2$ E=-0.5231 π_{CH_3}

63. Ethyl Fluoride

Symmetry: C_s

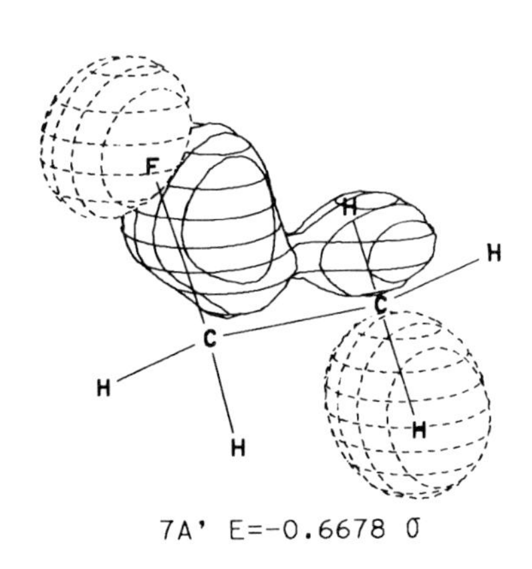

7A' E=-0.6678 σ

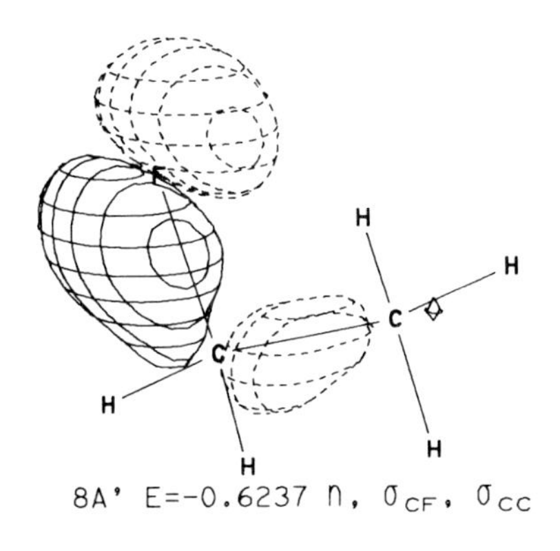

8A' E=-0.6237 n, σ_{CF}, σ_{CC}

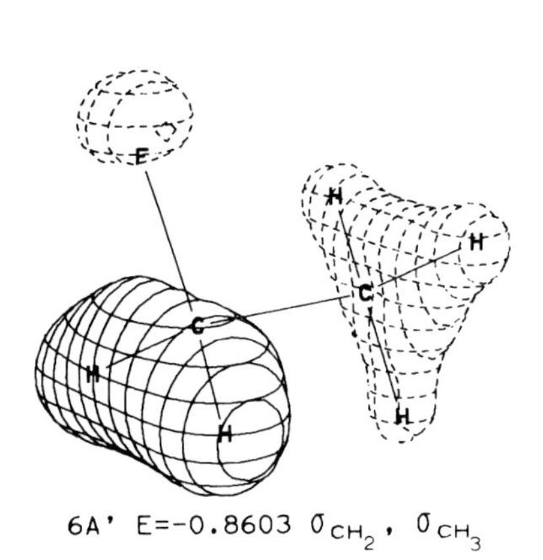

6A' E=-0.8603 σ_{CH_2}, σ_{CH_3}

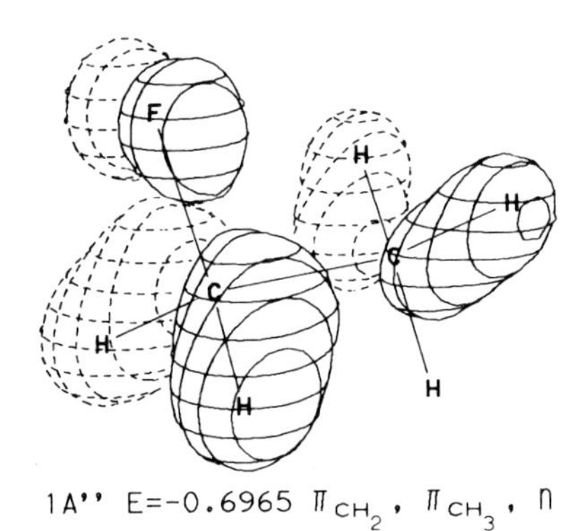

1A'' E=-0.6965 π_{CH_2}, π_{CH_3}, n

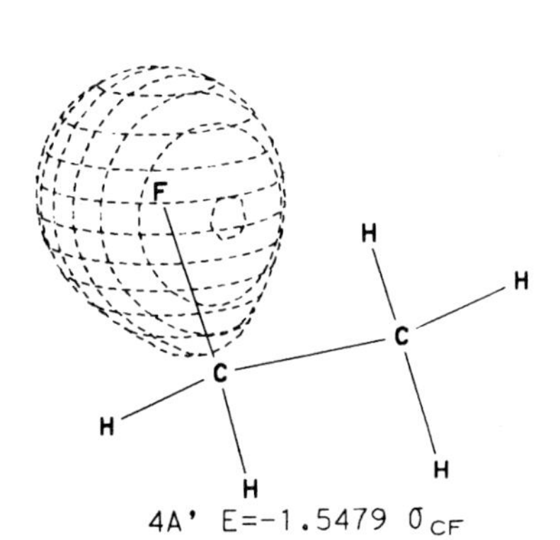

4A' E=-1.5479 σ_{CF}

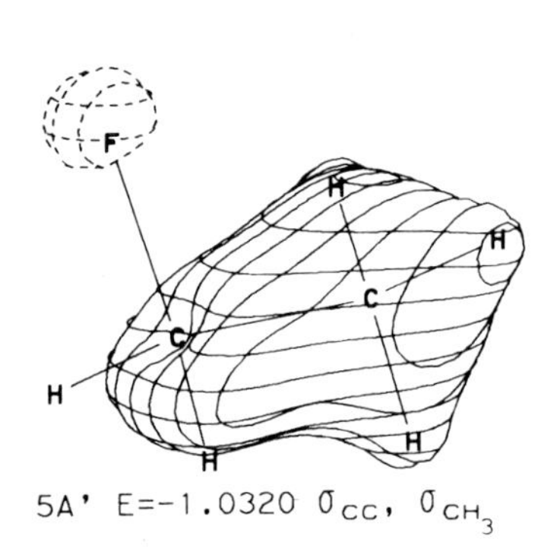

5A' E=-1.0320 σ_{CC}, σ_{CH_3}

Ethyl Fluoride (Continued)

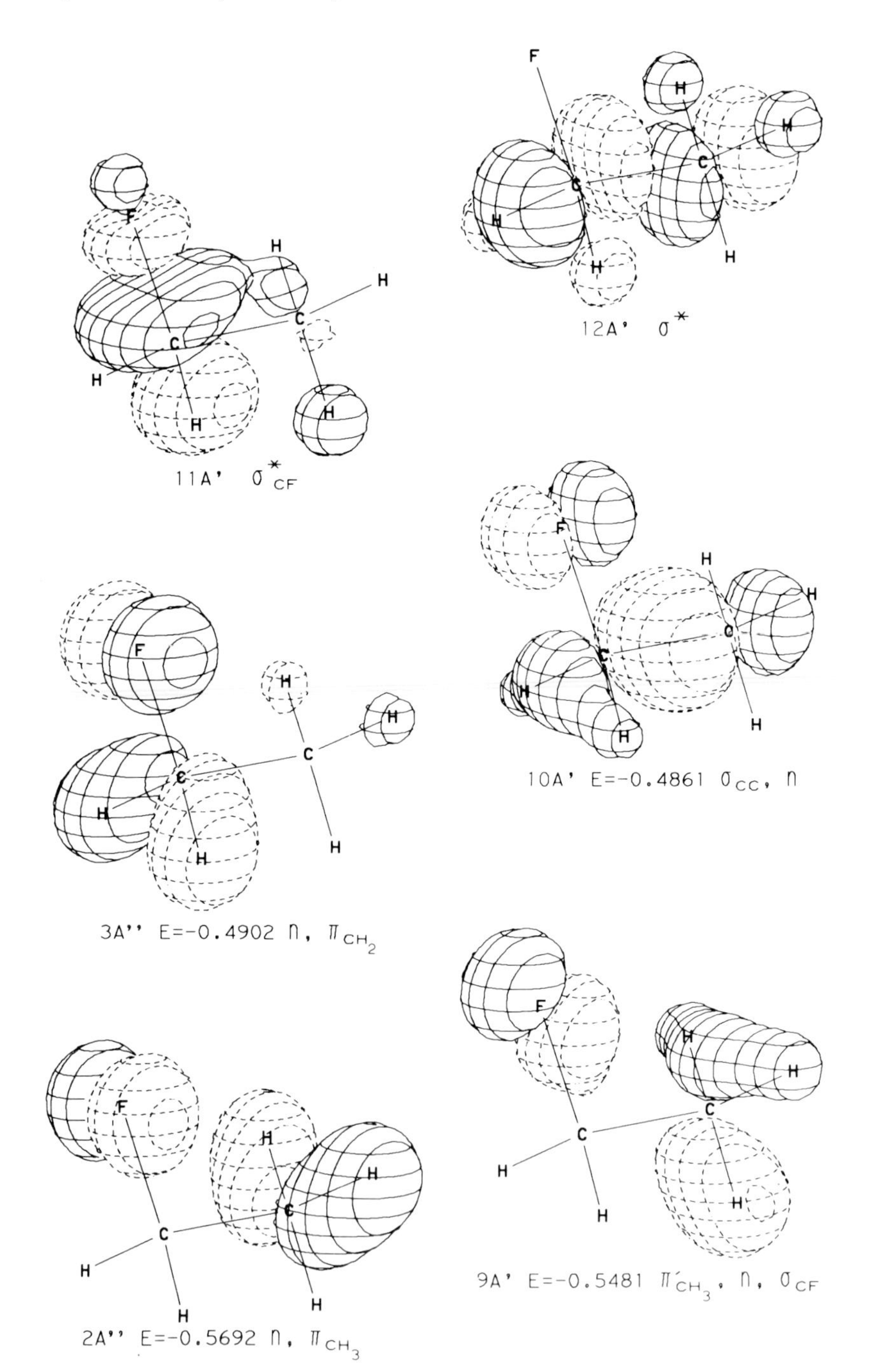

64. Cyclobutadiene (Rectangular Singlet) Symmetry: D_{2h}

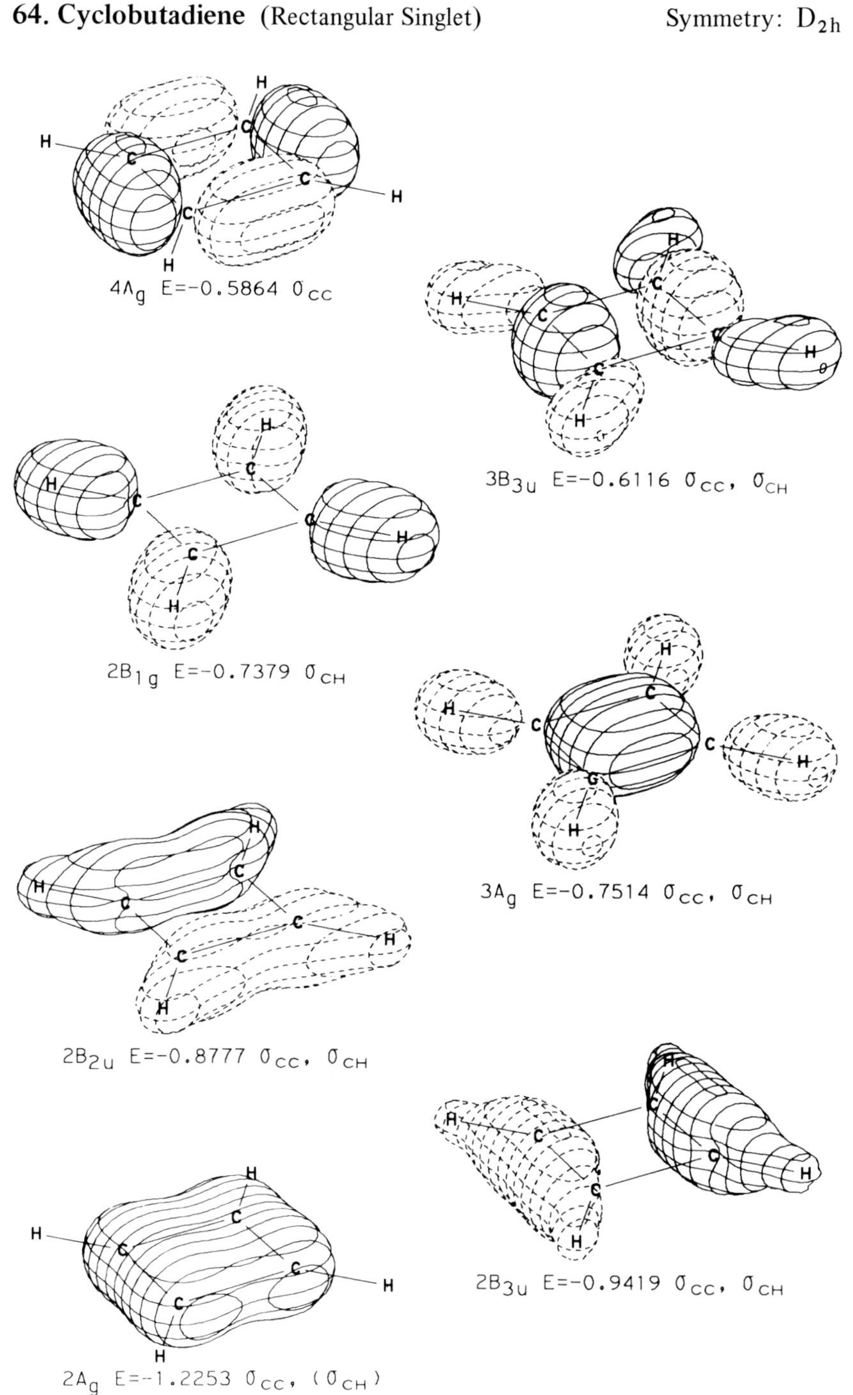

$4A_g$ E=-0.5864 σ_{CC}

$3B_{3u}$ E=-0.6116 σ_{CC}, σ_{CH}

$2B_{1g}$ E=-0.7379 σ_{CH}

$3A_g$ E=-0.7514 σ_{CC}, σ_{CH}

$2B_{2u}$ E=-0.8777 σ_{CC}, σ_{CH}

$2B_{3u}$ E=-0.9419 σ_{CC}, σ_{CH}

$2A_g$ E=-1.2253 σ_{CC}, (σ_{CH})

Cyclobutadiene (Rectangular Singlet) (Continued)

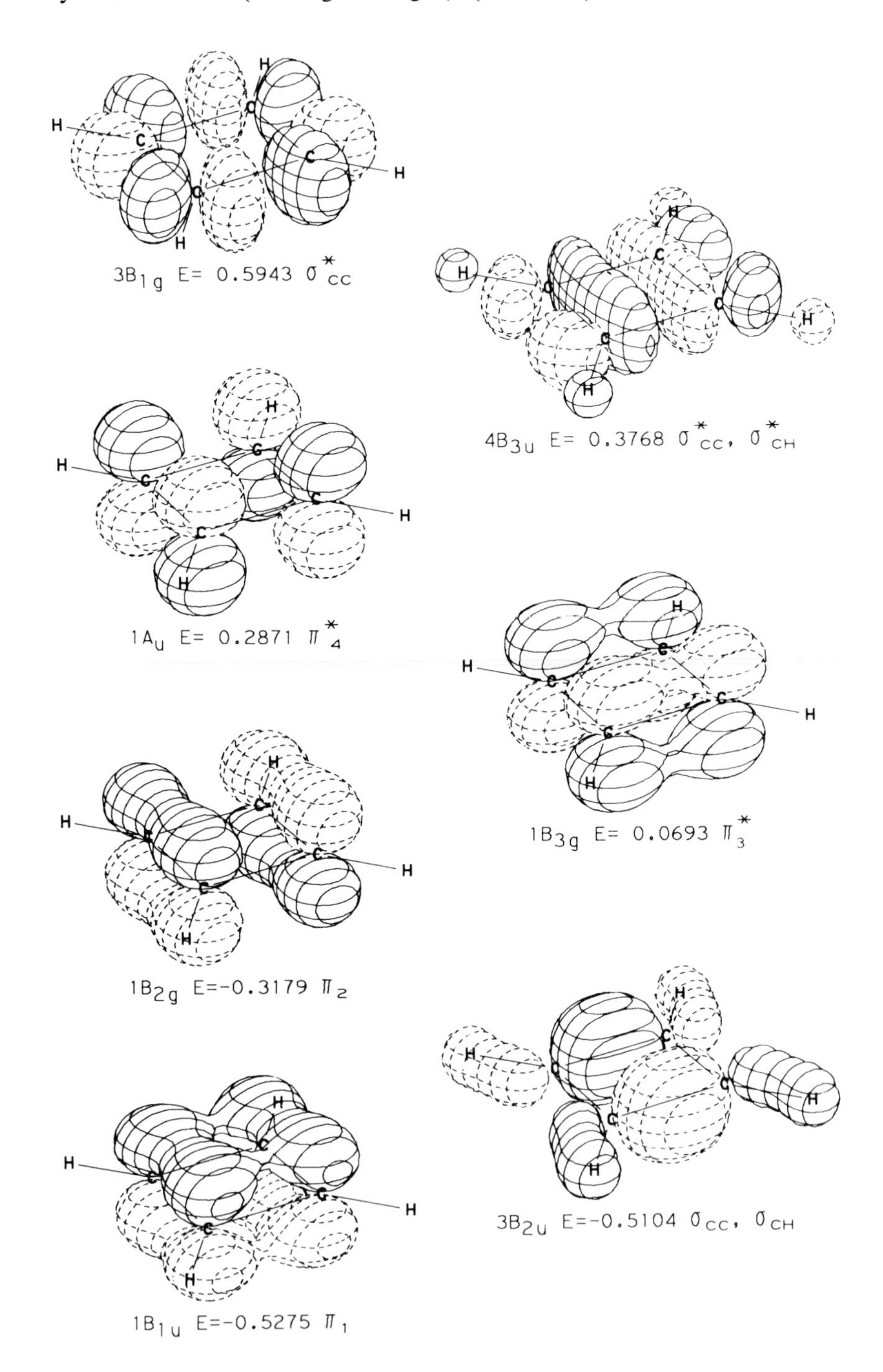

$3B_{1g}$ E= 0.5943 σ^*_{CC}

$4B_{3u}$ E= 0.3768 σ^*_{CC}, σ^*_{CH}

$1A_u$ E= 0.2871 π^*_4

$1B_{3g}$ E= 0.0693 π^*_3

$1B_{2g}$ E=−0.3179 π_2

$3B_{2u}$ E=−0.5104 σ_{CC}, σ_{CH}

$1B_{1u}$ E=−0.5275 π_1

65. 1,3-Butadiene, Transoid

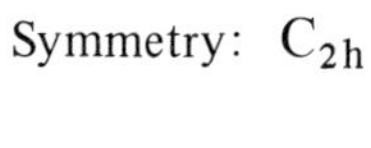

Symmetry: C_{2h}

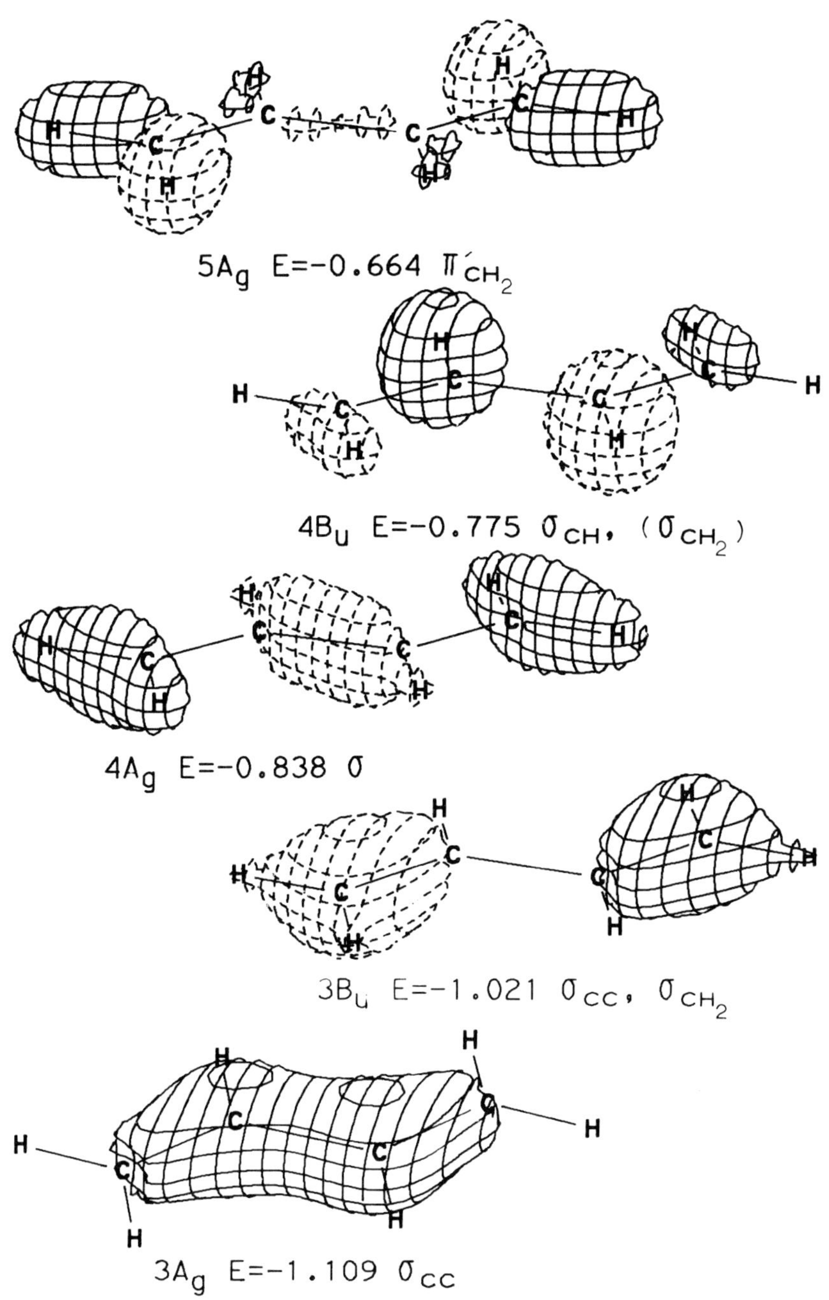

$5A_g$ E=−0.664 π'_{CH_2}

$4B_u$ E=−0.775 σ_{CH}, (σ_{CH_2})

$4A_g$ E=−0.838 σ

$3B_u$ E=−1.021 σ_{CC}, σ_{CH_2}

$3A_g$ E=−1.109 σ_{CC}

1,3-Butadiene, Transoid (Continued)

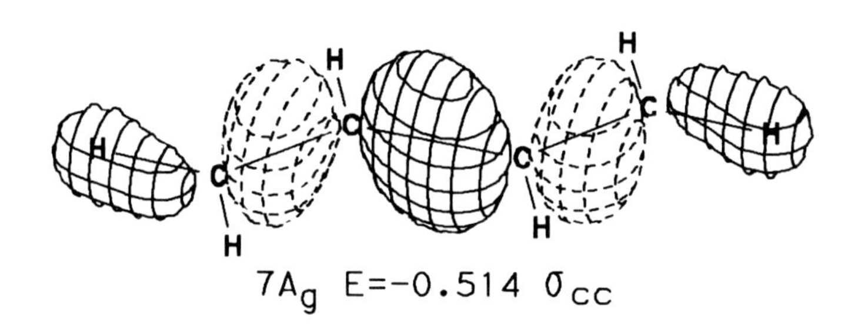

$7A_g$ E=−0.514 σ_{CC}

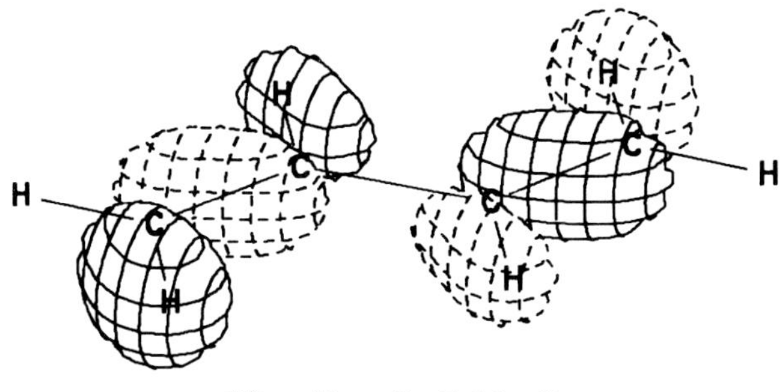

$6B_u$ E=−0.569 σ

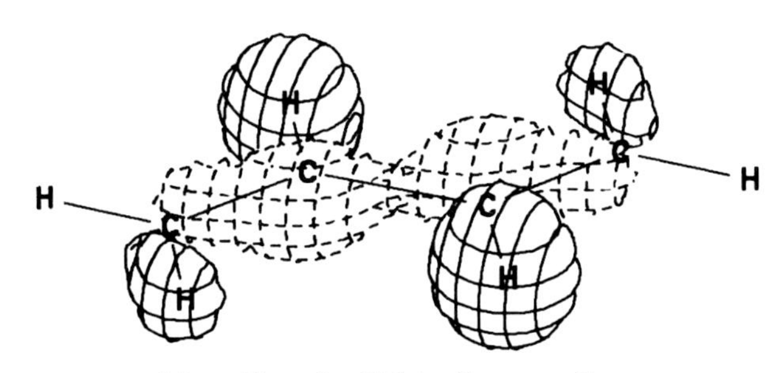

$6A_g$ E=−0.574 σ_{CH}, σ_{CC}

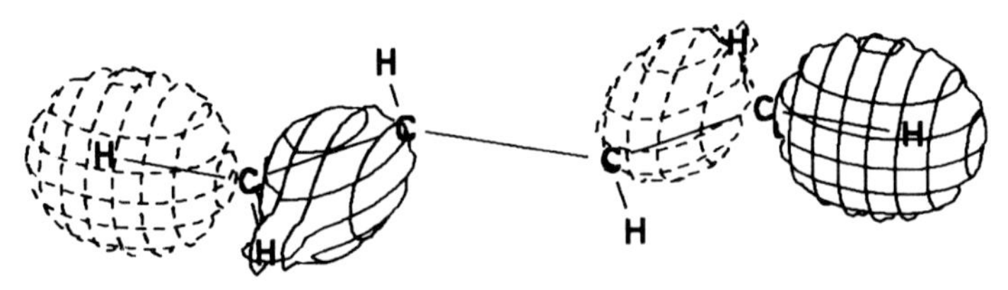

$5B_u$ E=−0.660 π'_{CH_2}, (σ_{CC})

1,3-Butadiene, Transoid (Continued)

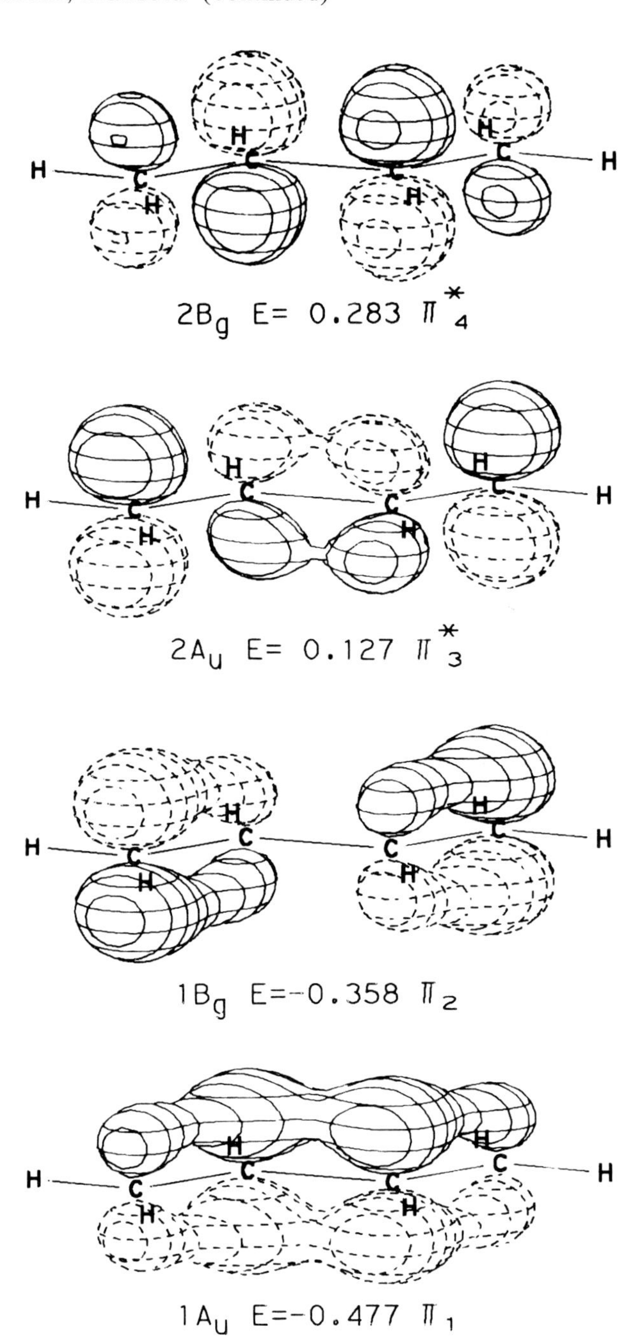

$2B_g$ E= 0.283 π^*_4

$2A_u$ E= 0.127 π^*_3

$1B_g$ E=−0.358 π_2

$1A_u$ E=−0.477 π_1

66. 1,3-Butadiene, Cisoid

Symmetry: C_{2v}

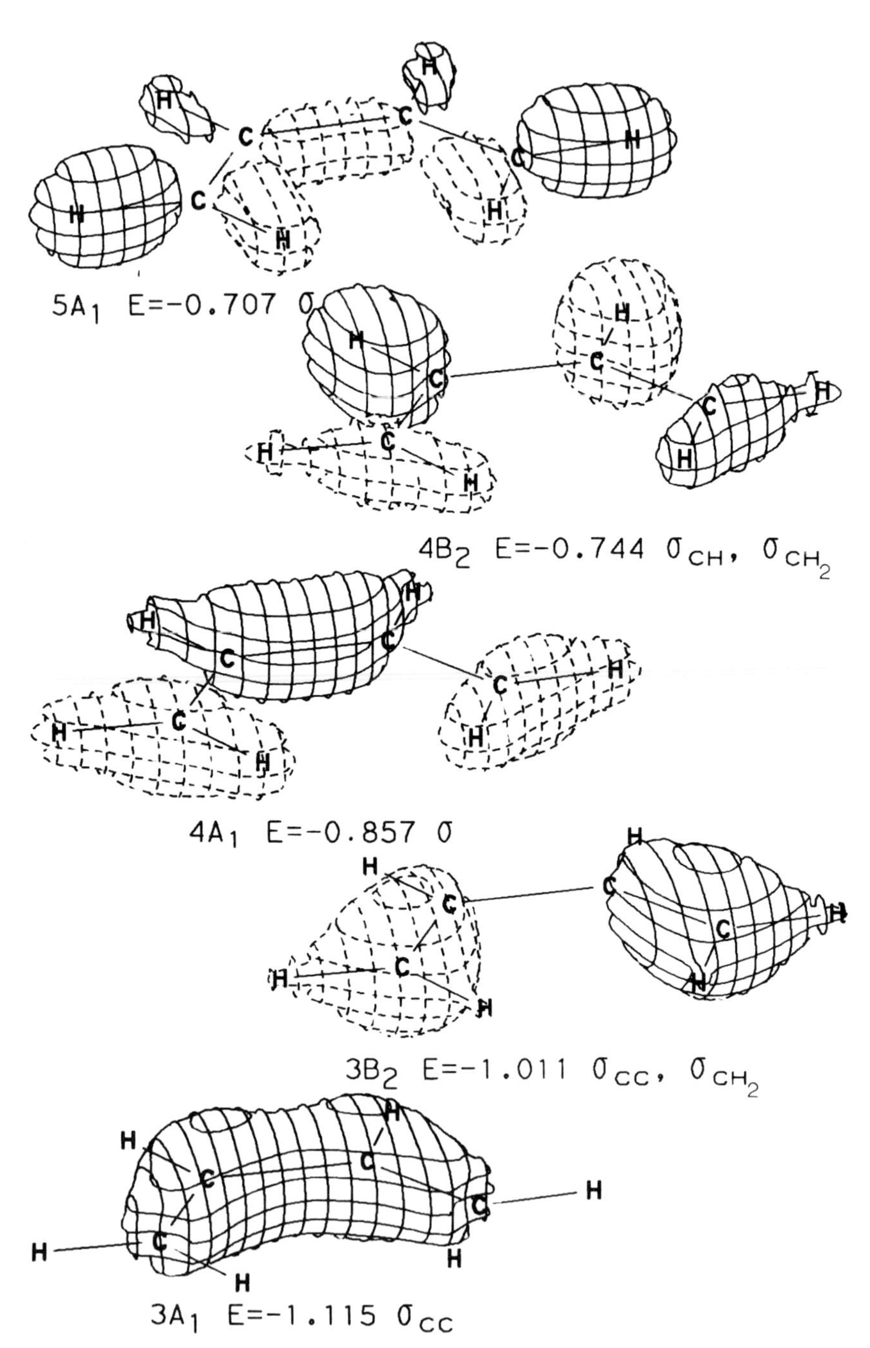

1,3-Butadiene, Cisoid (Continued)

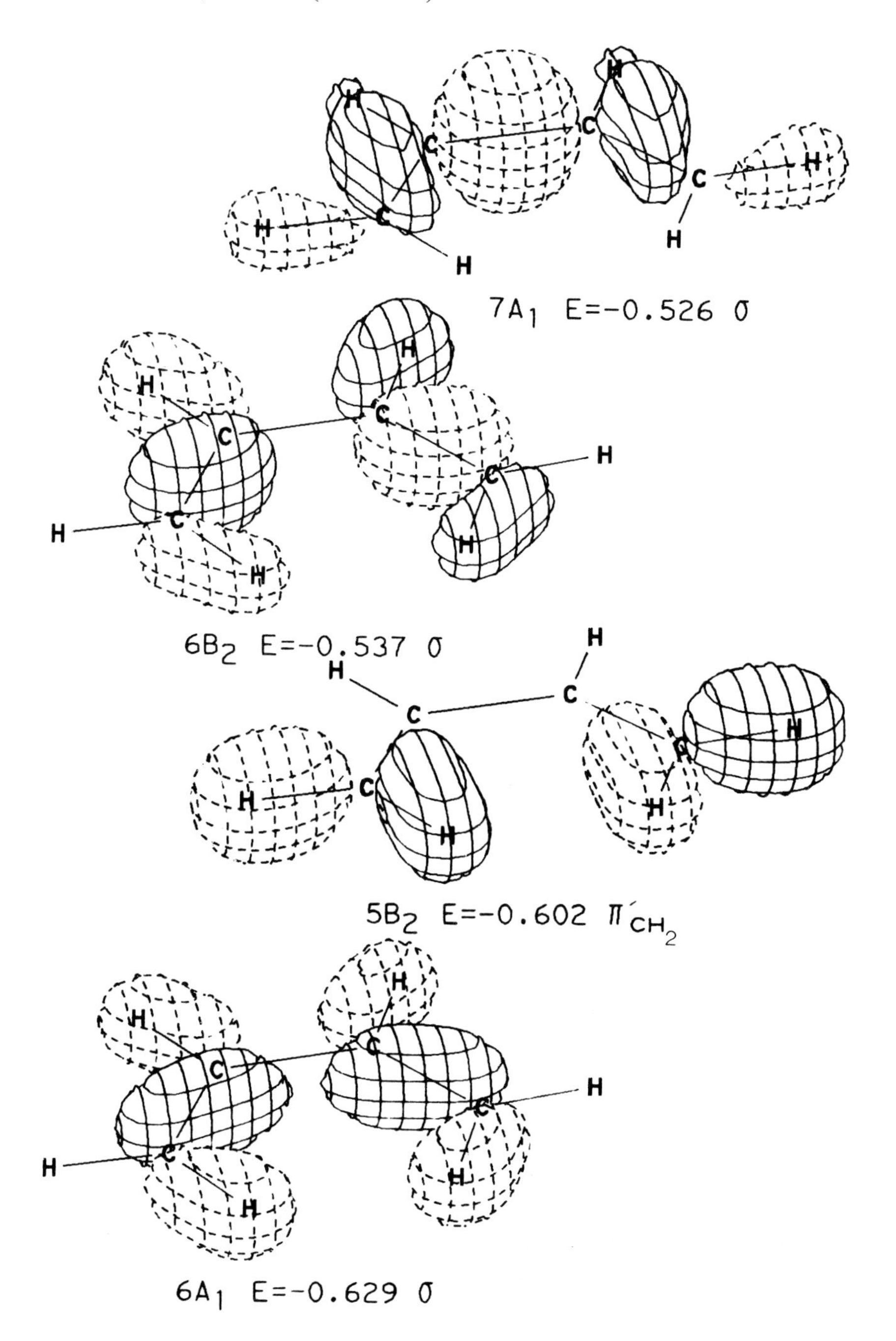

1,3-Butadiene, Cisoid (Continued)

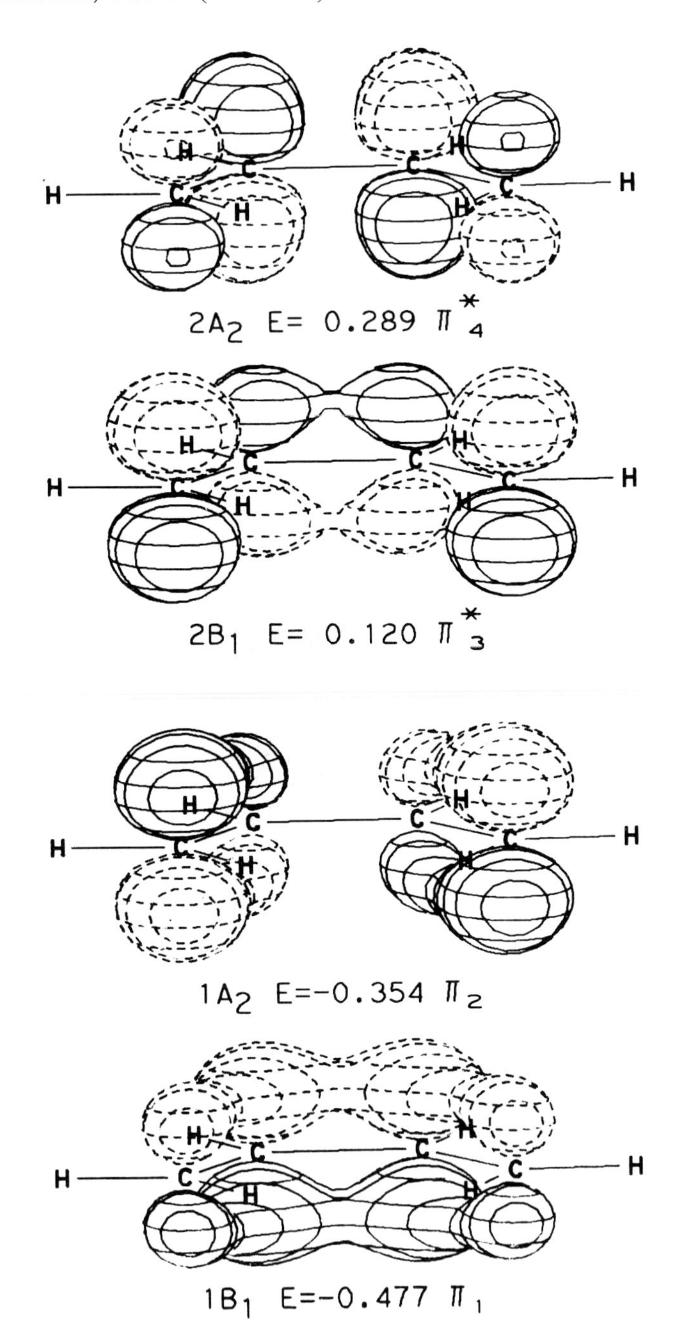

67. Acrolein, Transoid

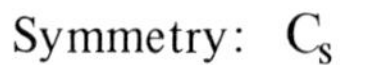
Symmetry: C_s

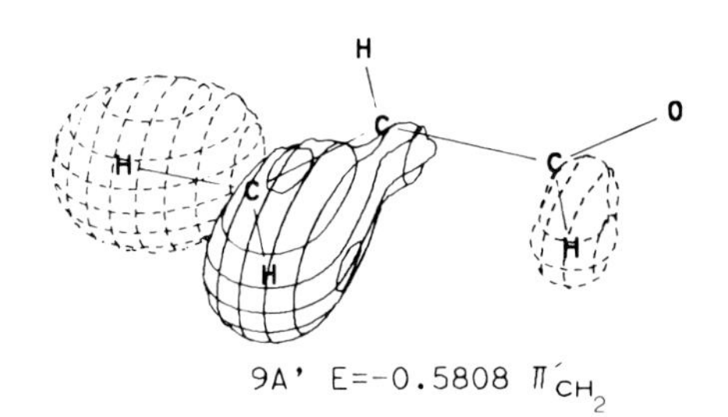

9A' E=-0.5808 π'_{CH_2}

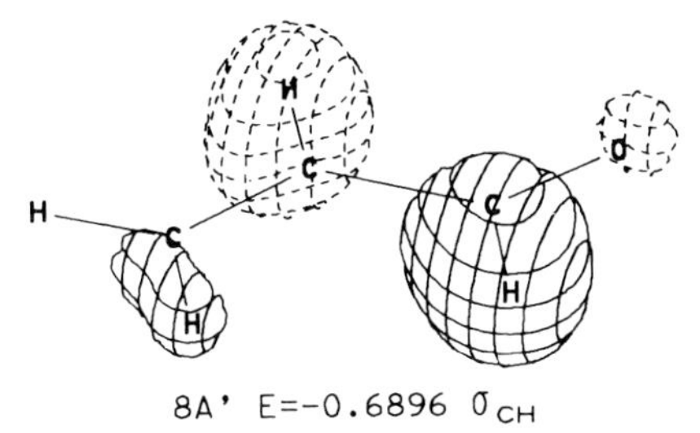

8A' E=-0.6896 σ_{CH}

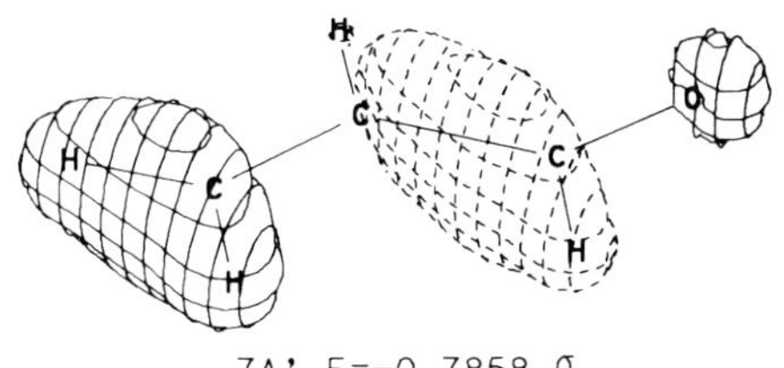

7A' E=-0.7858 σ

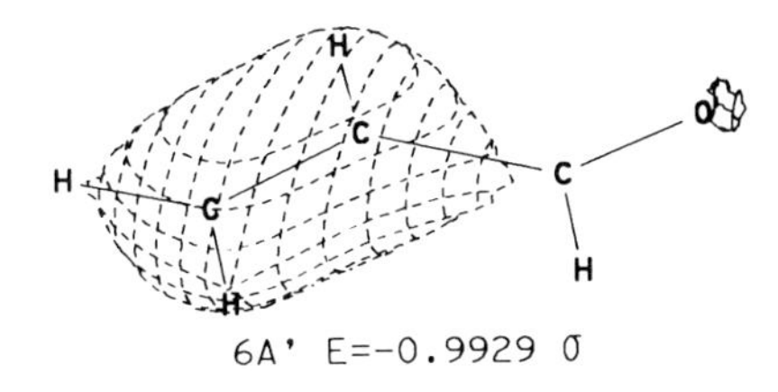

6A' E=-0.9929 σ

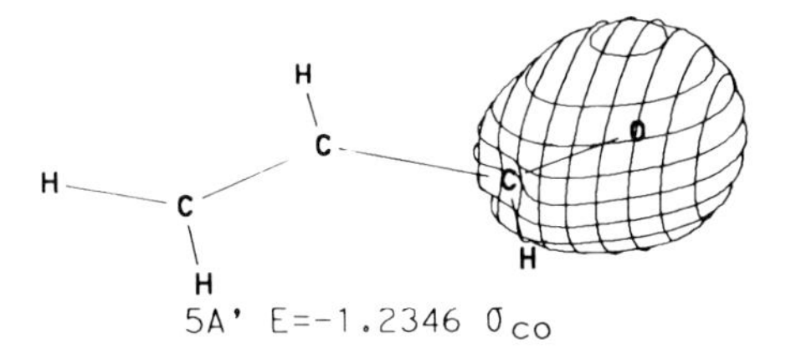

5A' E=-1.2346 σ_{CO}

Acrolein, Transoid (Continued)

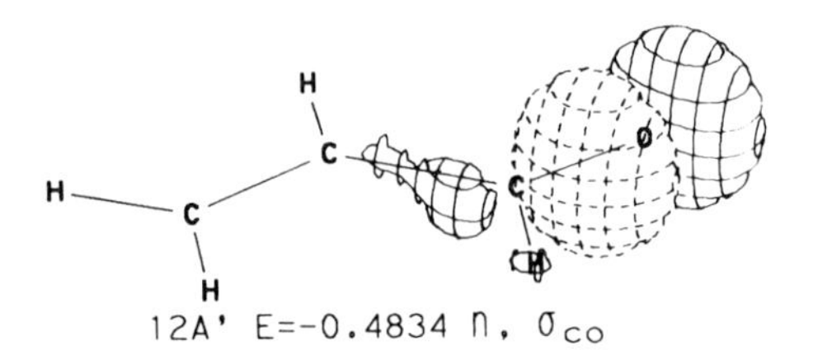

12A' E=-0.4834 n, σ_{CO}

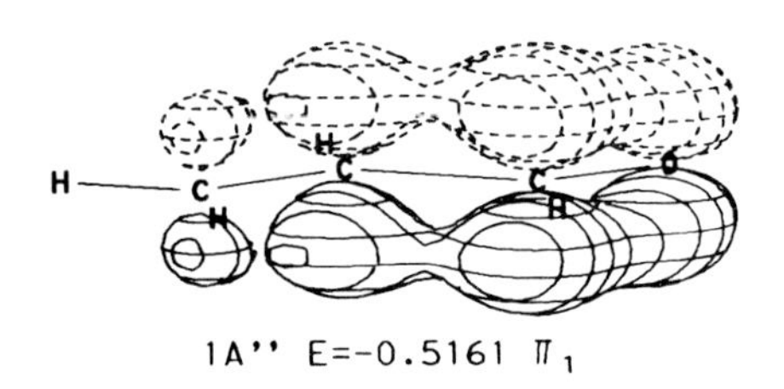

1A'' E=-0.5161 π_1

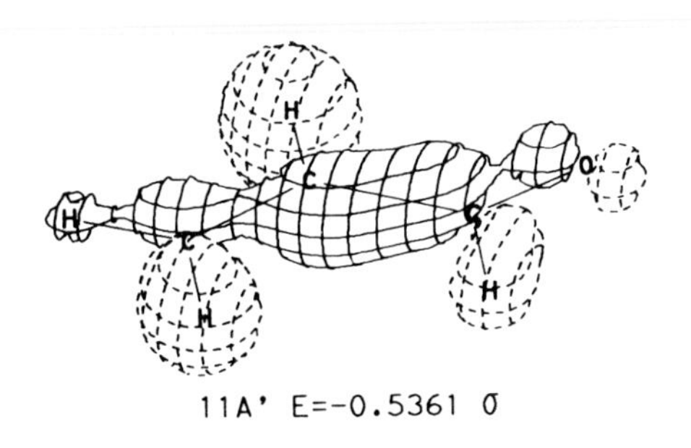

11A' E=-0.5361 σ

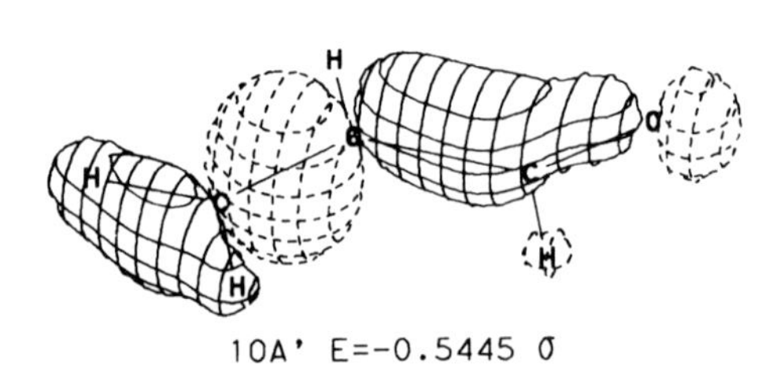

10A' E=-0.5445 σ

Acrolein, Transoid (Continued)

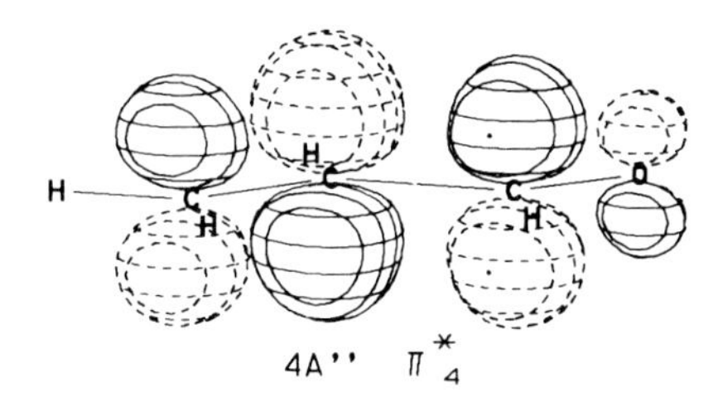

4A'' π^*_4

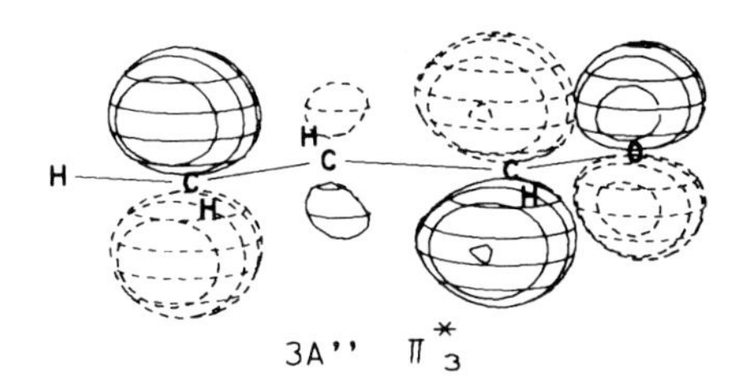

3A'' π^*_3

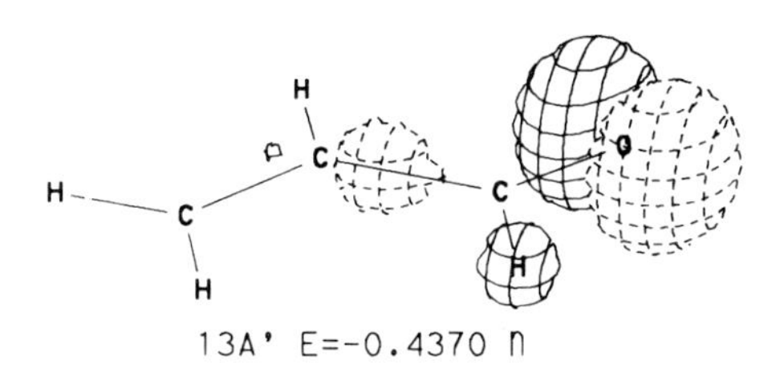

13A' E=-0.4370 n

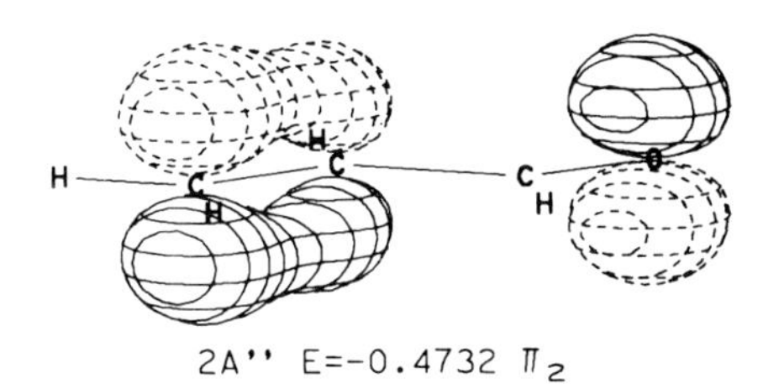

2A'' E=-0.4732 π_2

68. Acrolein, Cisoid

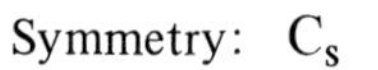
Symmetry: C_s

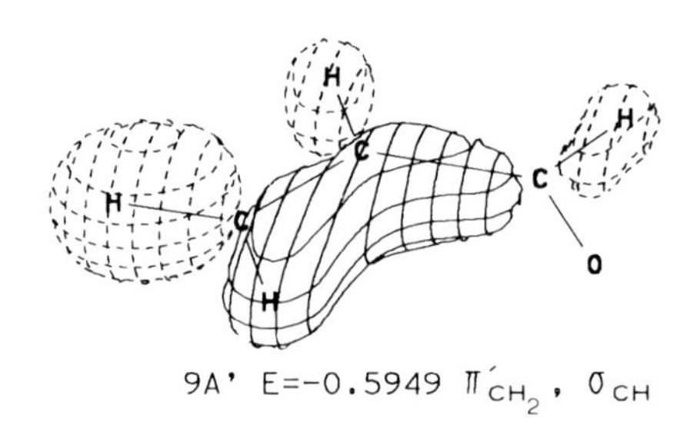

9A' E=−0.5949 π'_{CH_2}, σ_{CH}

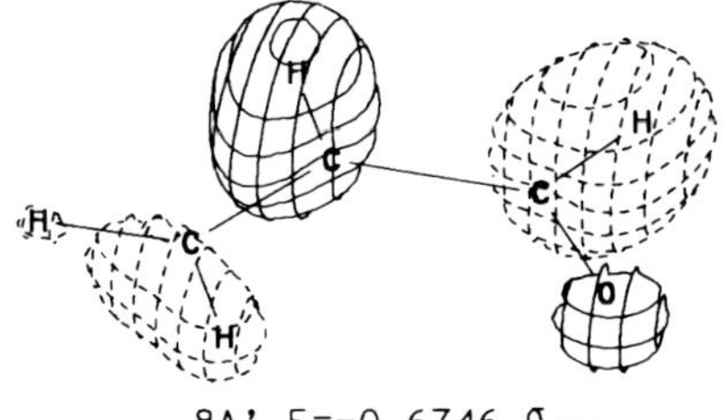

8A' E=−0.6746 σ_{CH}

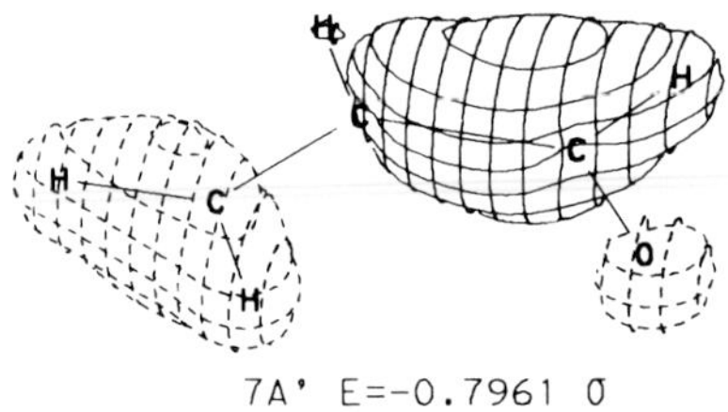

7A' E=−0.7961 σ

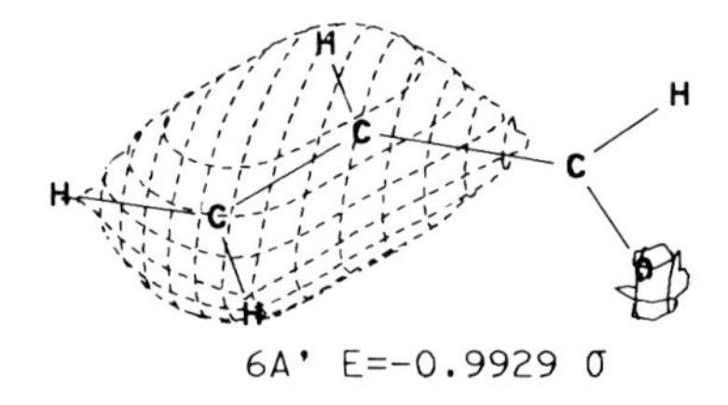

6A' E=−0.9929 σ

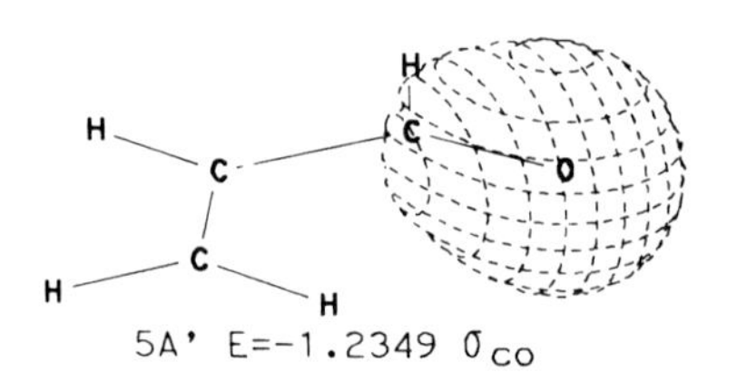

5A' E=−1.2349 σ_{CO}

Acrolein, Cisoid (Continued)

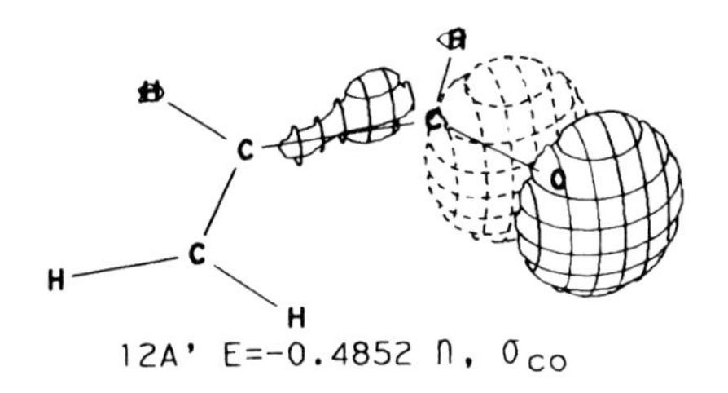

12A' E=-0.4852 n, σ_{CO}

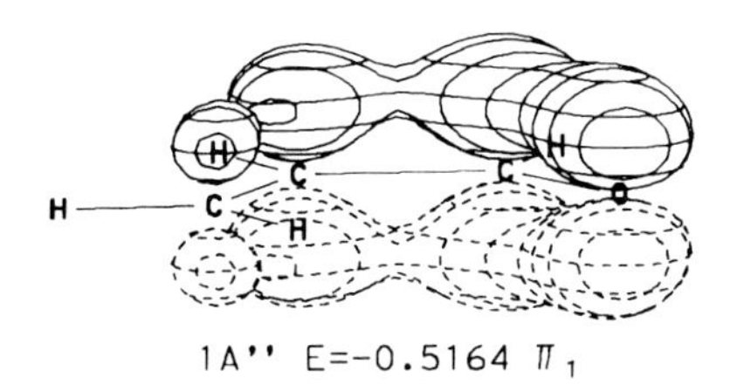

1A'' E=-0.5164 π_1

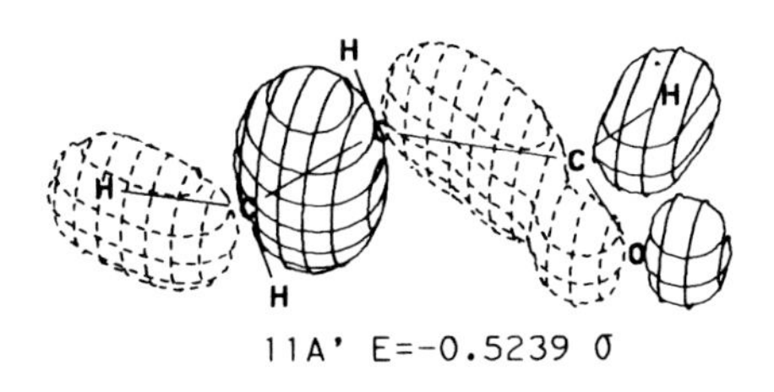

11A' E=-0.5239 σ

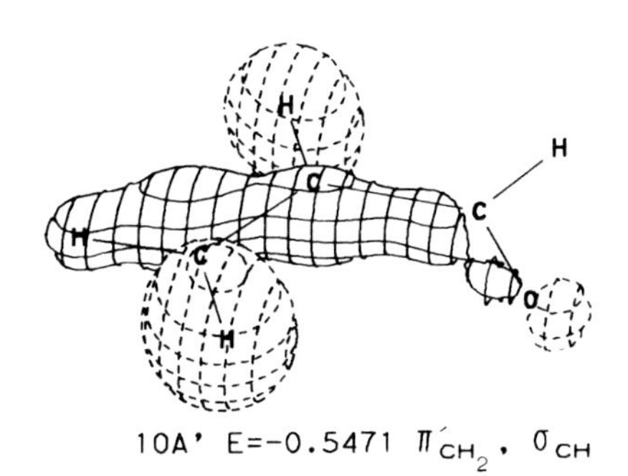

10A' E=-0.5471 π'_{CH_2}, σ_{CH}

Acrolein, Cisoid (Continued)

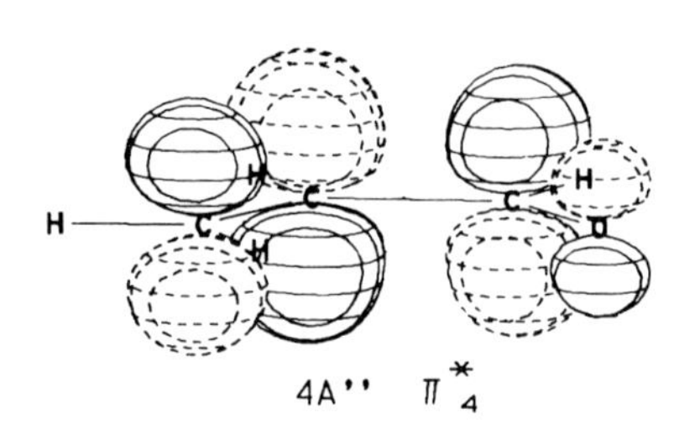

4A'' π^*_4

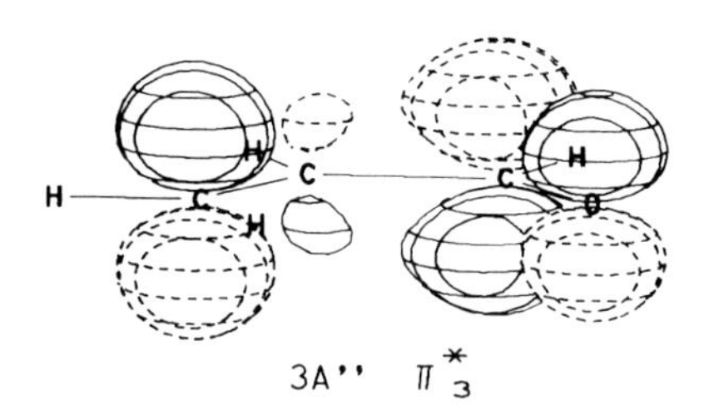

3A'' π^*_3

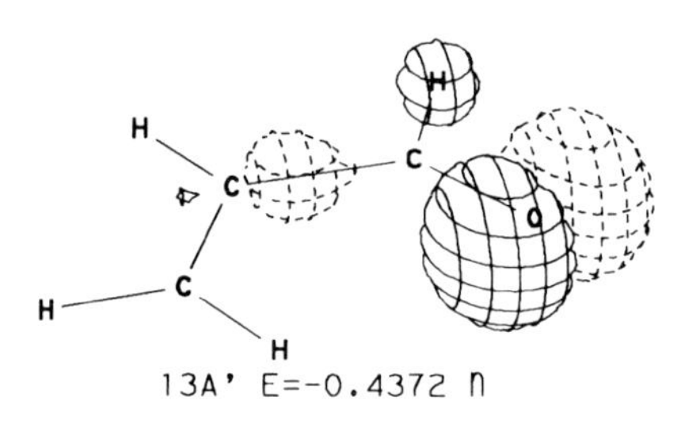

13A' E=-0.4372 n

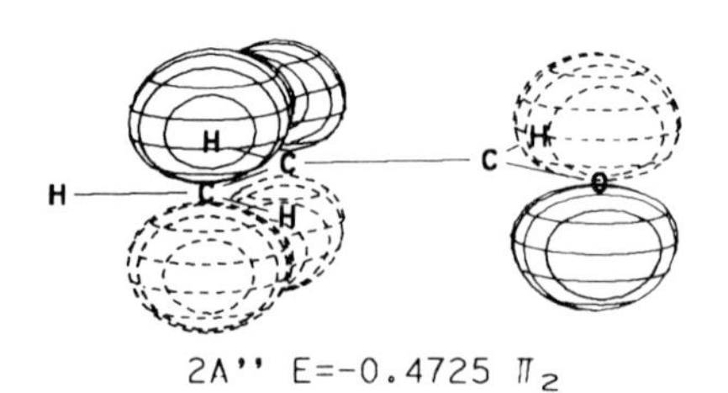

2A'' E=-0.4725 π_2

69. Glyoxal, Transoid

Symmetry: C_{2h}

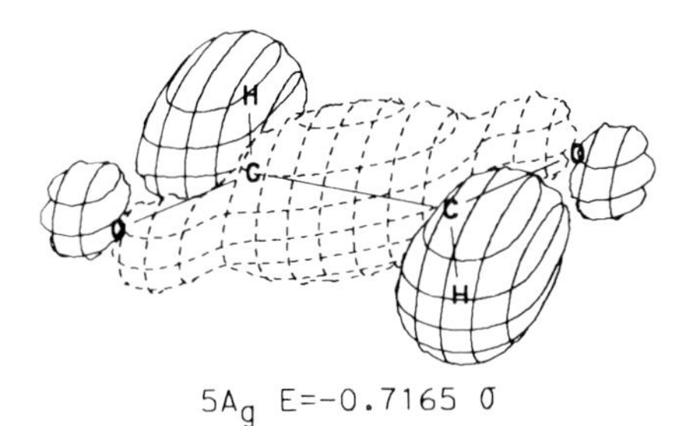

$5A_g$ E=−0.7165 σ

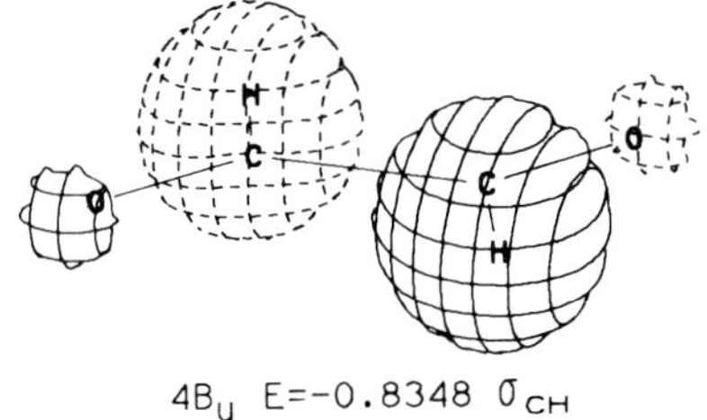

$4B_u$ E=−0.8348 σ_{CH}

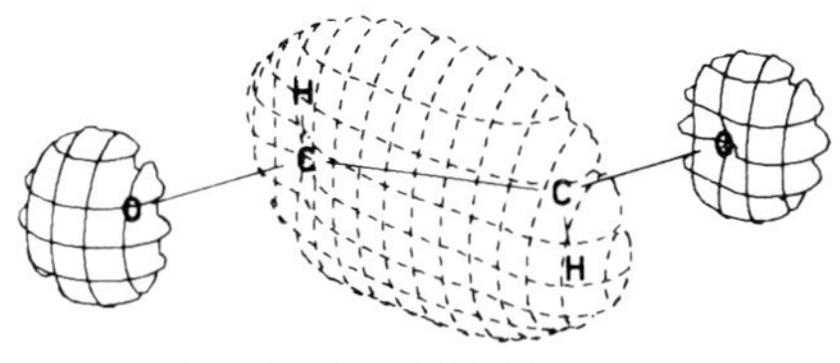

$4A_g$ E=−0.9882 σ_{CC}, σ_{CH}

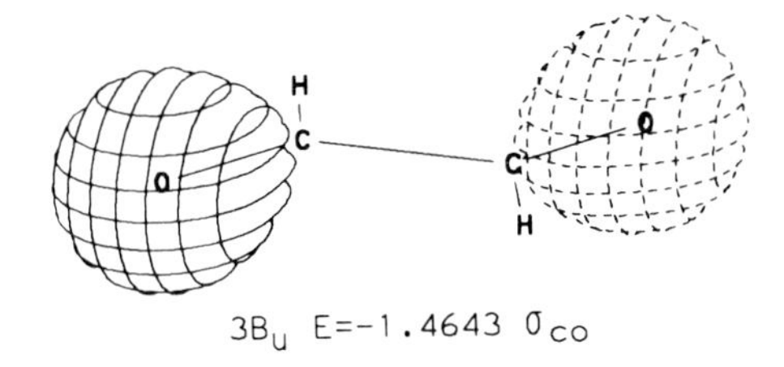

$3B_u$ E=−1.4643 σ_{CO}

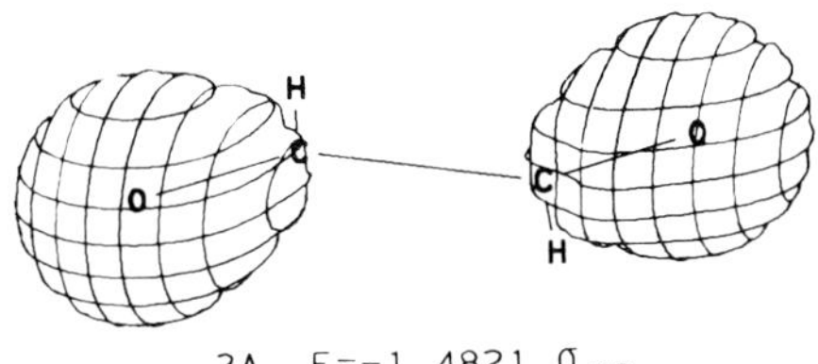

$3A_g$ E=−1.4821 σ_{CO}

Glyoxal, Transoid (Continued)

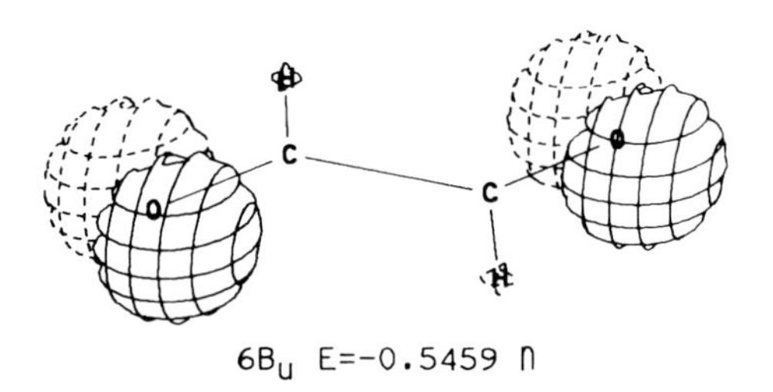

$6B_u$ E=−0.5459 n

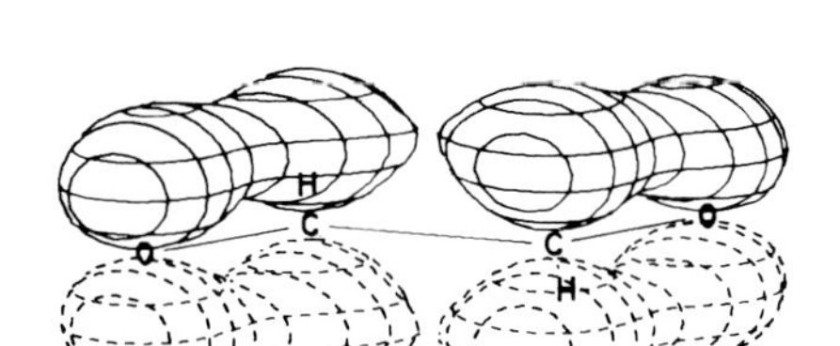

$1A_u$ E=−0.6098 π_1

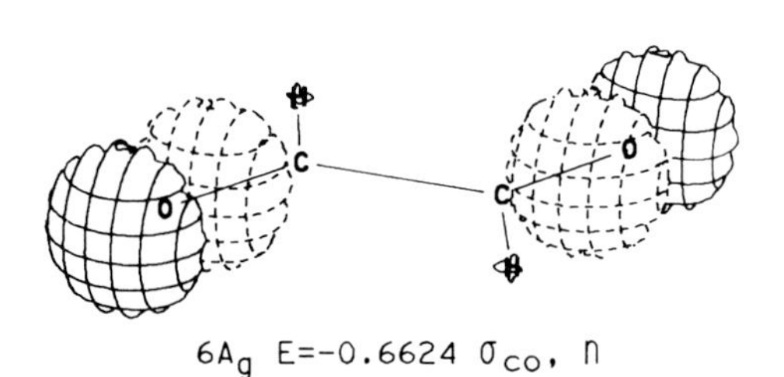

$6A_g$ E=−0.6624 σ_{CO}, n

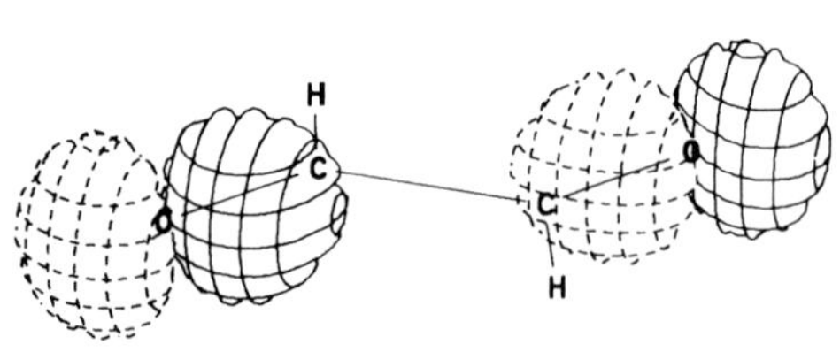

$5B_u$ E=−0.7121 σ_{CO}, n

Glyoxal, Transoid (Continued)

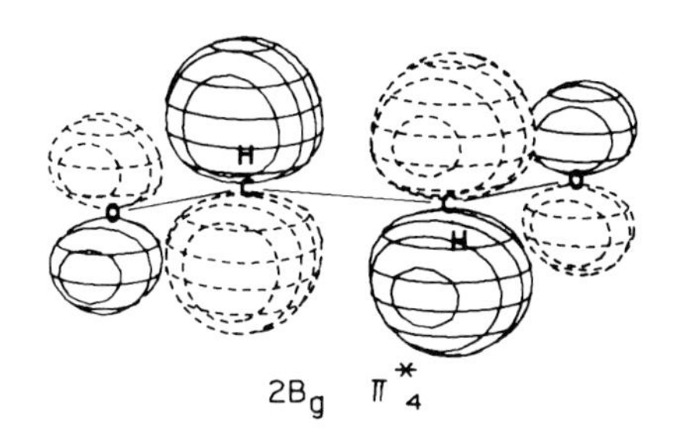

$2B_g$ π_4^*

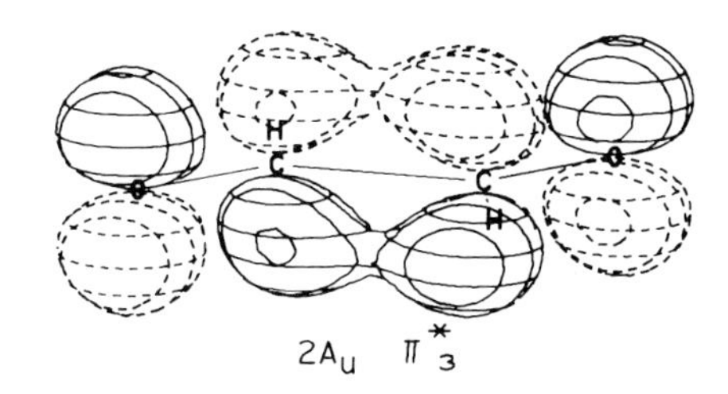

$2A_u$ π_3^*

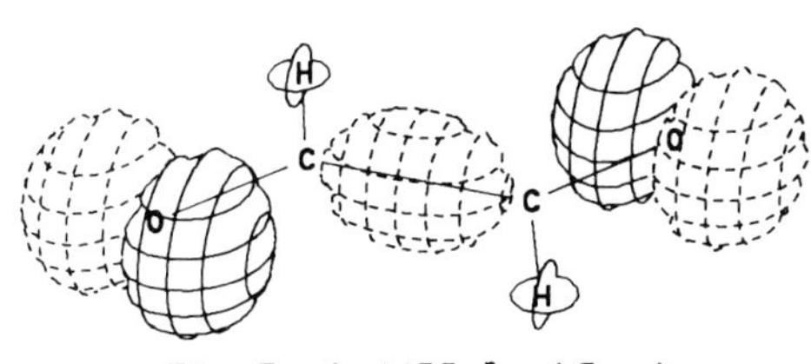

$7A_g$ E=-0.4455 n, (σ_{CC})

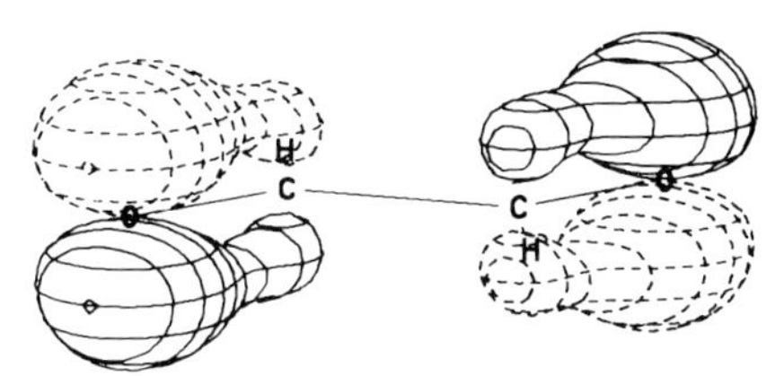

$1B_g$ E=-0.5387 π_2

70. Glyoxal, Cisoid

Symmetry: C_{2v}

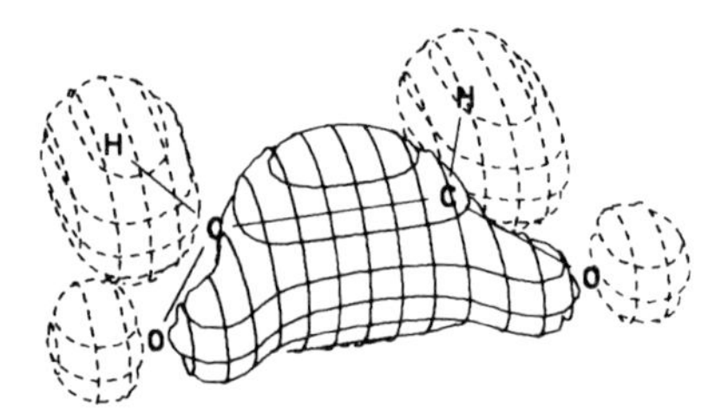

$5A_1'$ E=-0.7583 σ

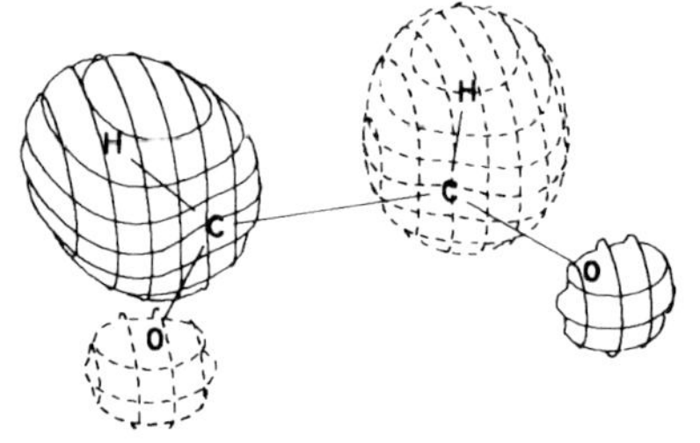

$4B_2$ E=-0.8022 $σ_{CH}$

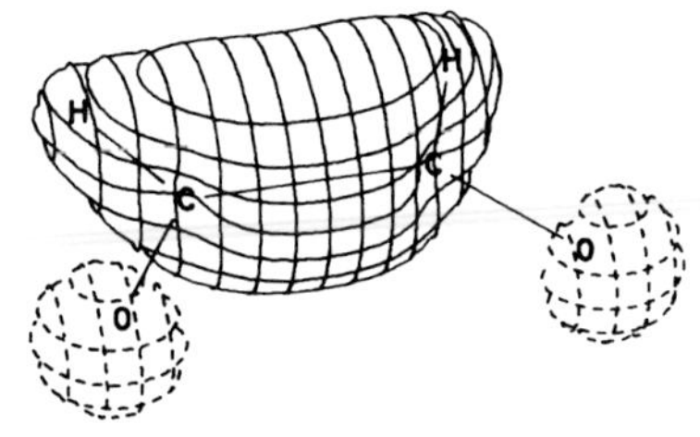

$4A_1$ E=-0.9981 $σ_{CC}$, $σ_{CH}$

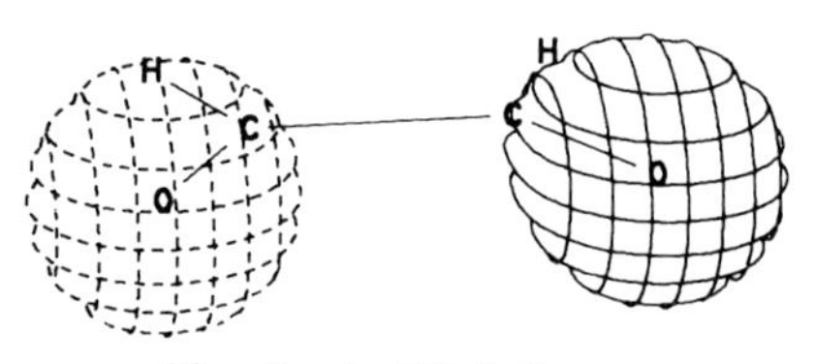

$3B_2$ E=-1.4514 $σ_{CO}$

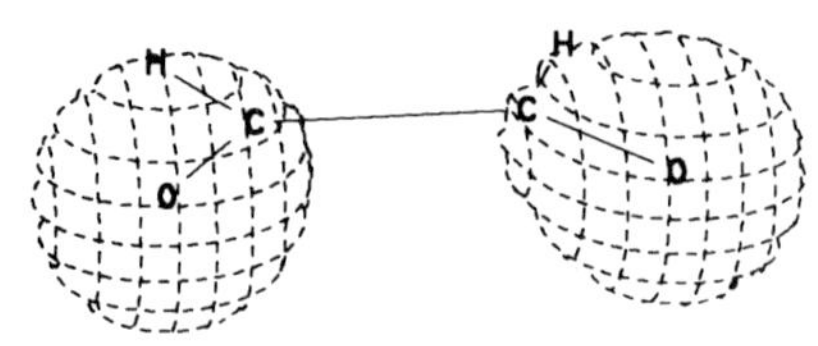

$3A_1$ E=-1.4827 $σ_{CO}$

Glyoxal, Cisoid (Continued)

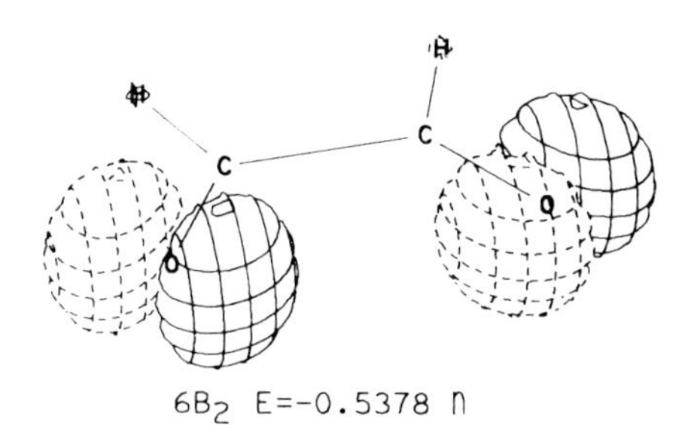

$6B_2$ E=-0.5378 n

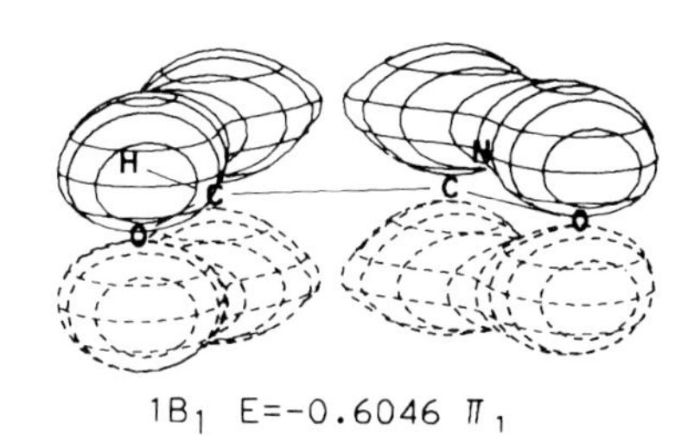

$1B_1$ E=-0.6046 π_1

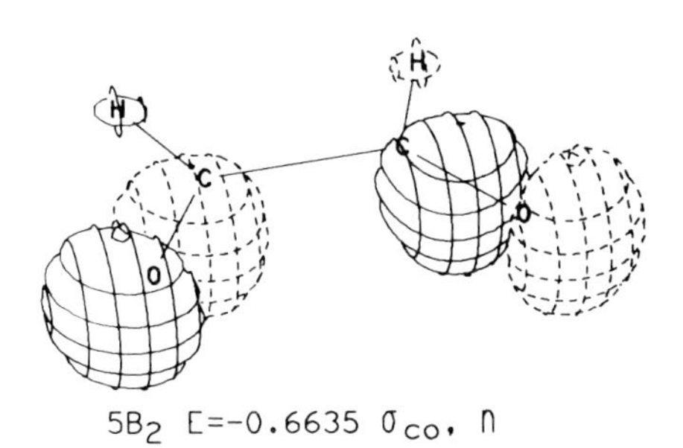

$5B_2$ E=-0.6635 σ_{CO}, n

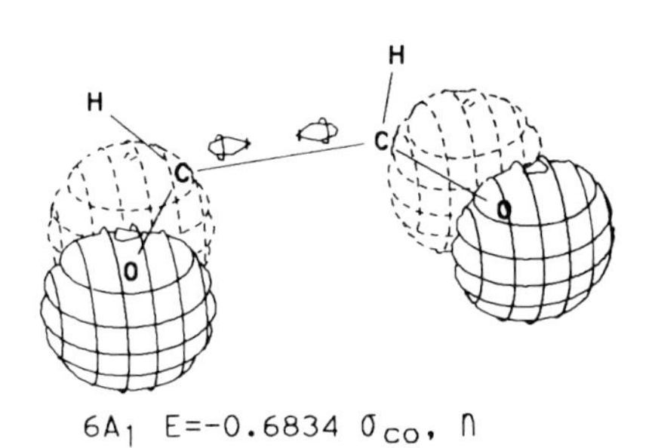

$6A_1$ E=-0.6834 σ_{CO}, n

Glyoxal, Cisoid (Continued)

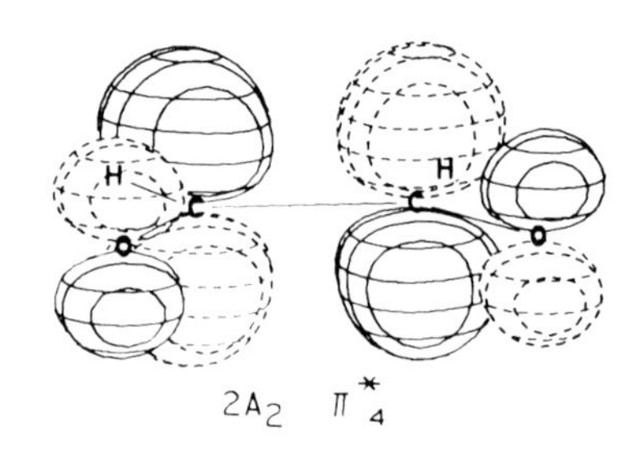

$2A_2$ π^*_4

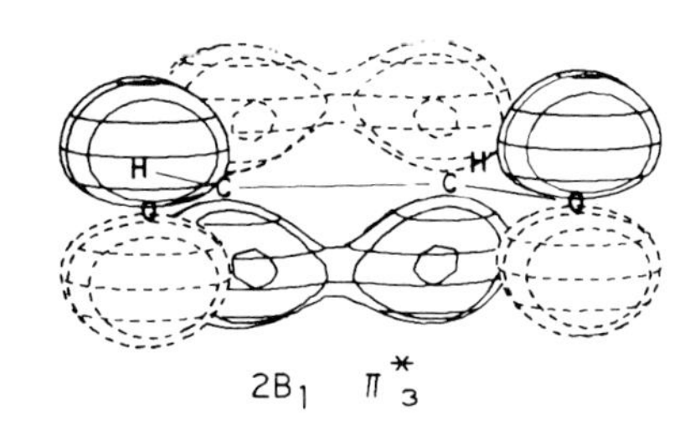

$2B_1$ π^*_3

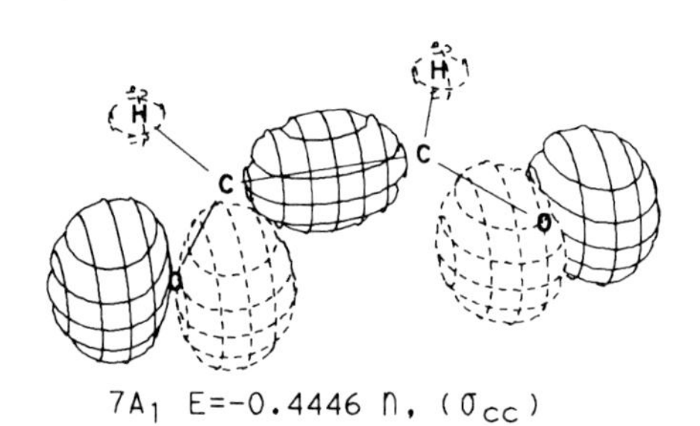

$7A_1$ E=-0.4446 n, (σ_{CC})

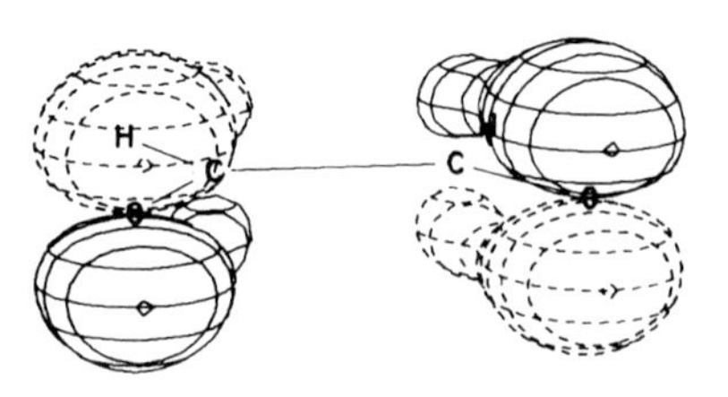

$1A_2$ E=-0.5318 π_2

71. Methylazide

Symmetry: C_s

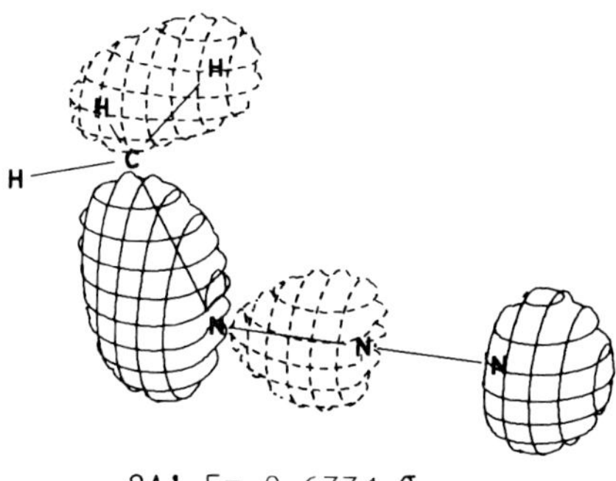

8A' E=-0.6774 σ

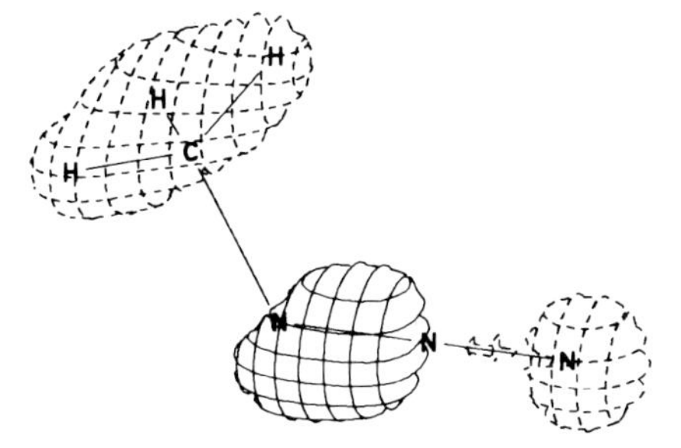

7A' E=-0.9502 σ_{CH_3}, σ_{NN}

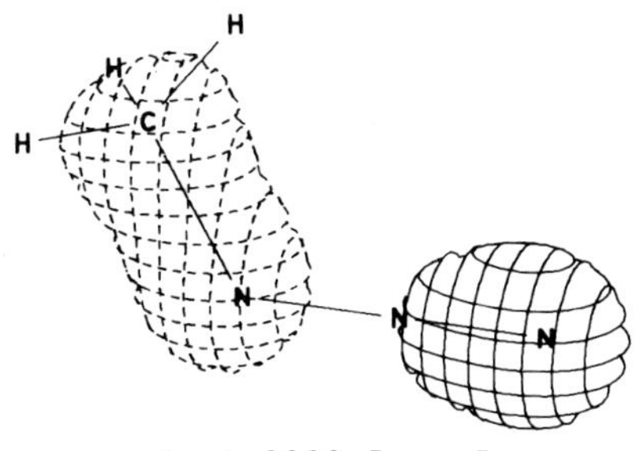

6A' E=-1.2238 σ_{CN}, σ_{NN}

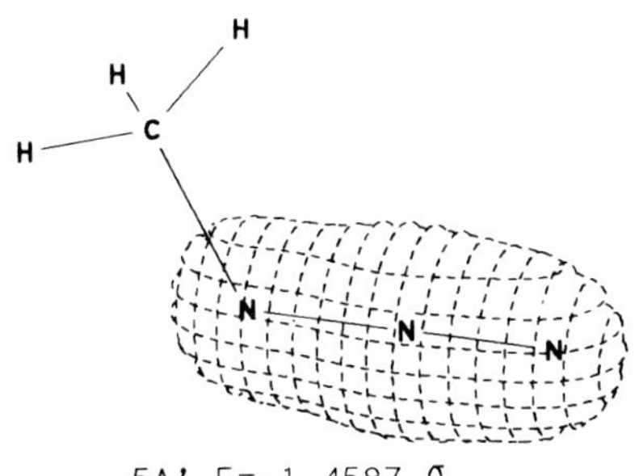

5A' E=-1.4587 σ_{NN}

Methylazide (Continued)

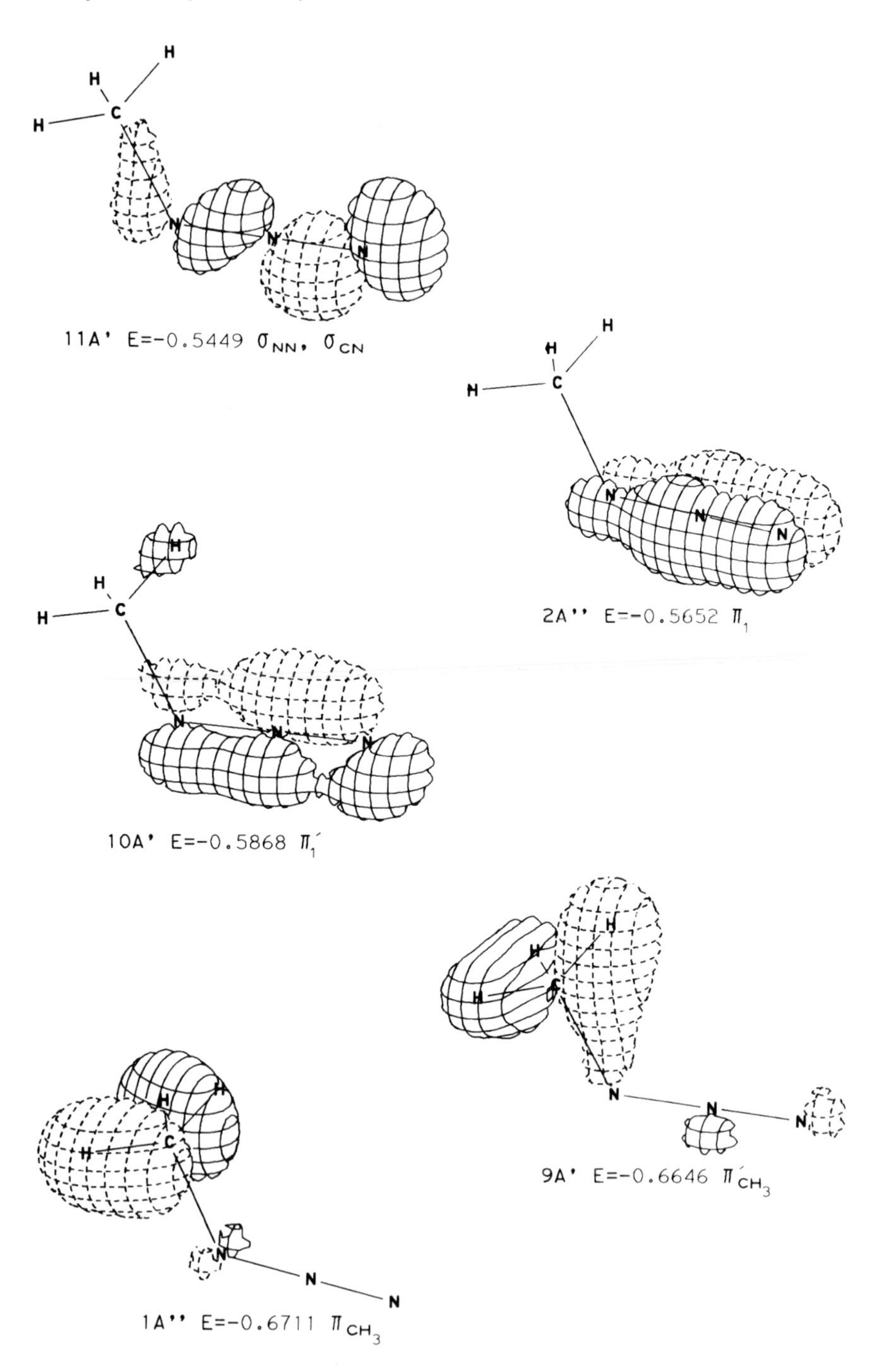

Methylazide (Continued)

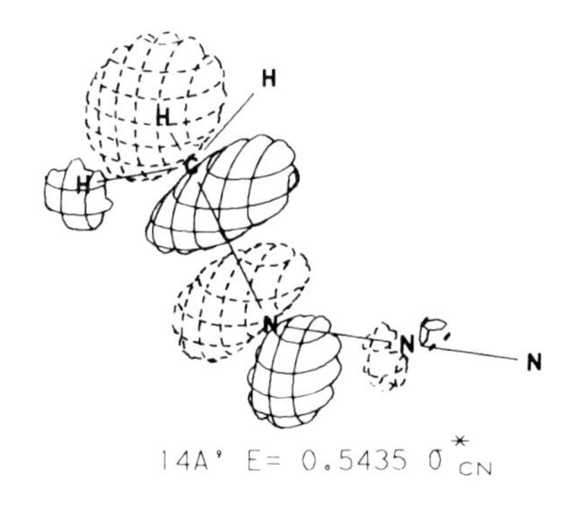

14A' E= 0.5435 σ^*_{CN}

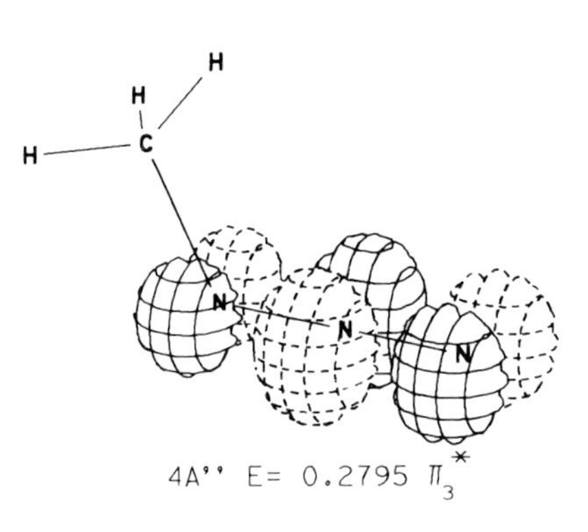

4A'' E= 0.2795 π_3^*

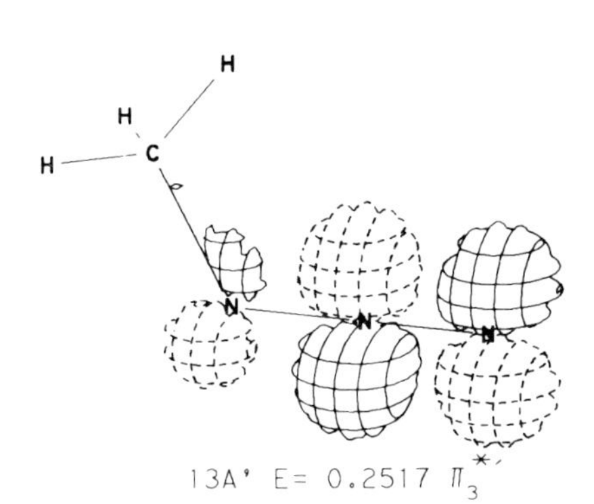

13A' E= 0.2517 π_3^*

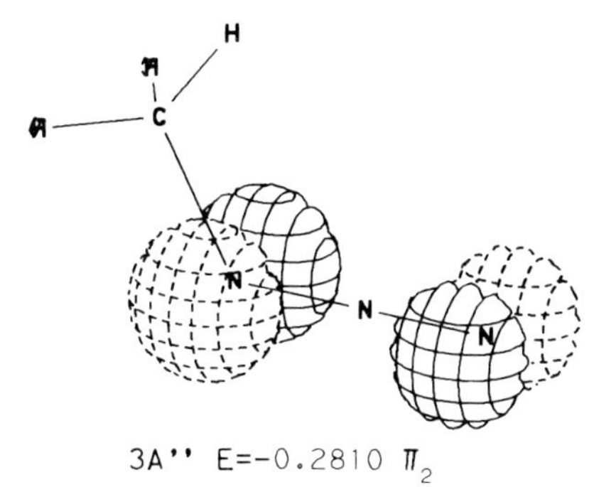

3A'' E=-0.2810 π_2

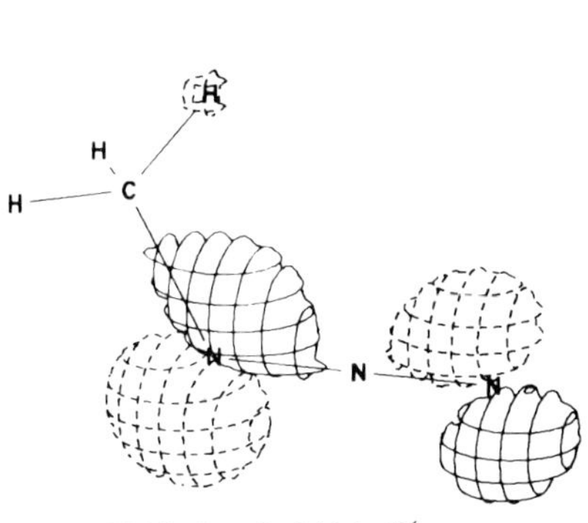

12A' E=-0.3414 π_2'

72. Methylenecyclopropane

Symmetry: C_{2v}

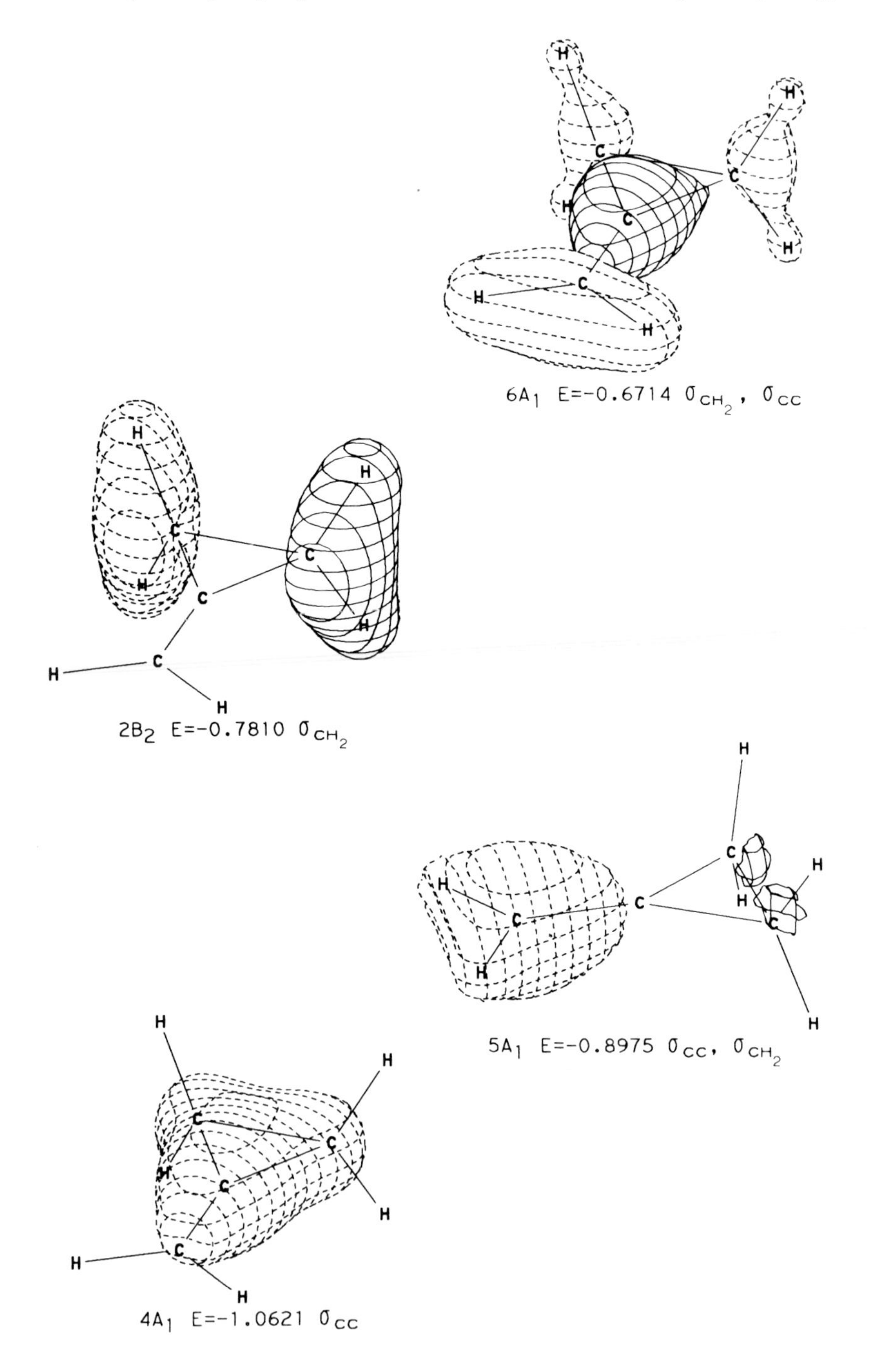

$6A_1$ E=−0.6714 σ_{CH_2}, σ_{CC}

$2B_2$ E=−0.7810 σ_{CH_2}

$5A_1$ E=−0.8975 σ_{CC}, σ_{CH_2}

$4A_1$ E=−1.0621 σ_{CC}

Methylenecyclopropane (Continued)

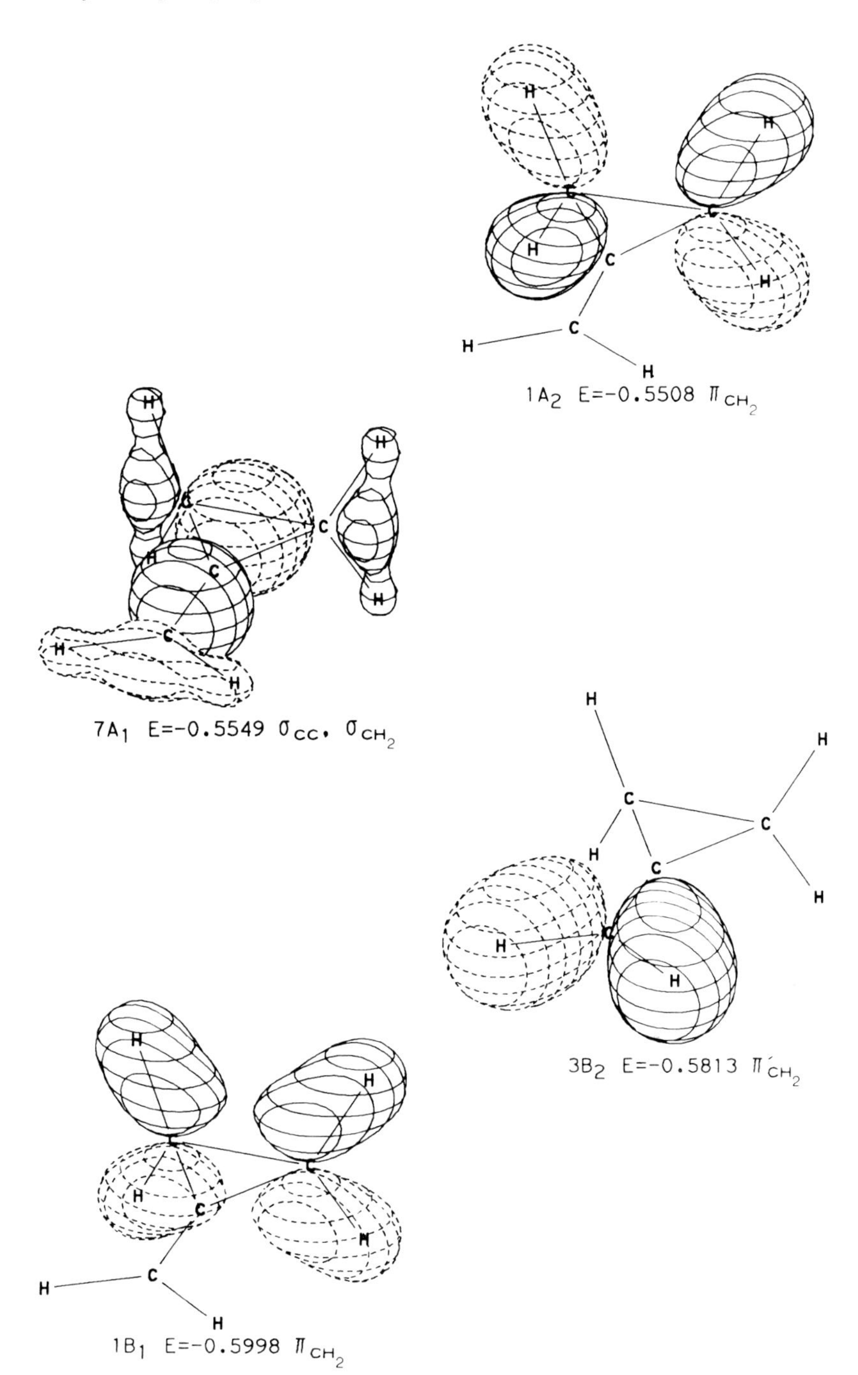

Methylenecyclopropane (Continued)

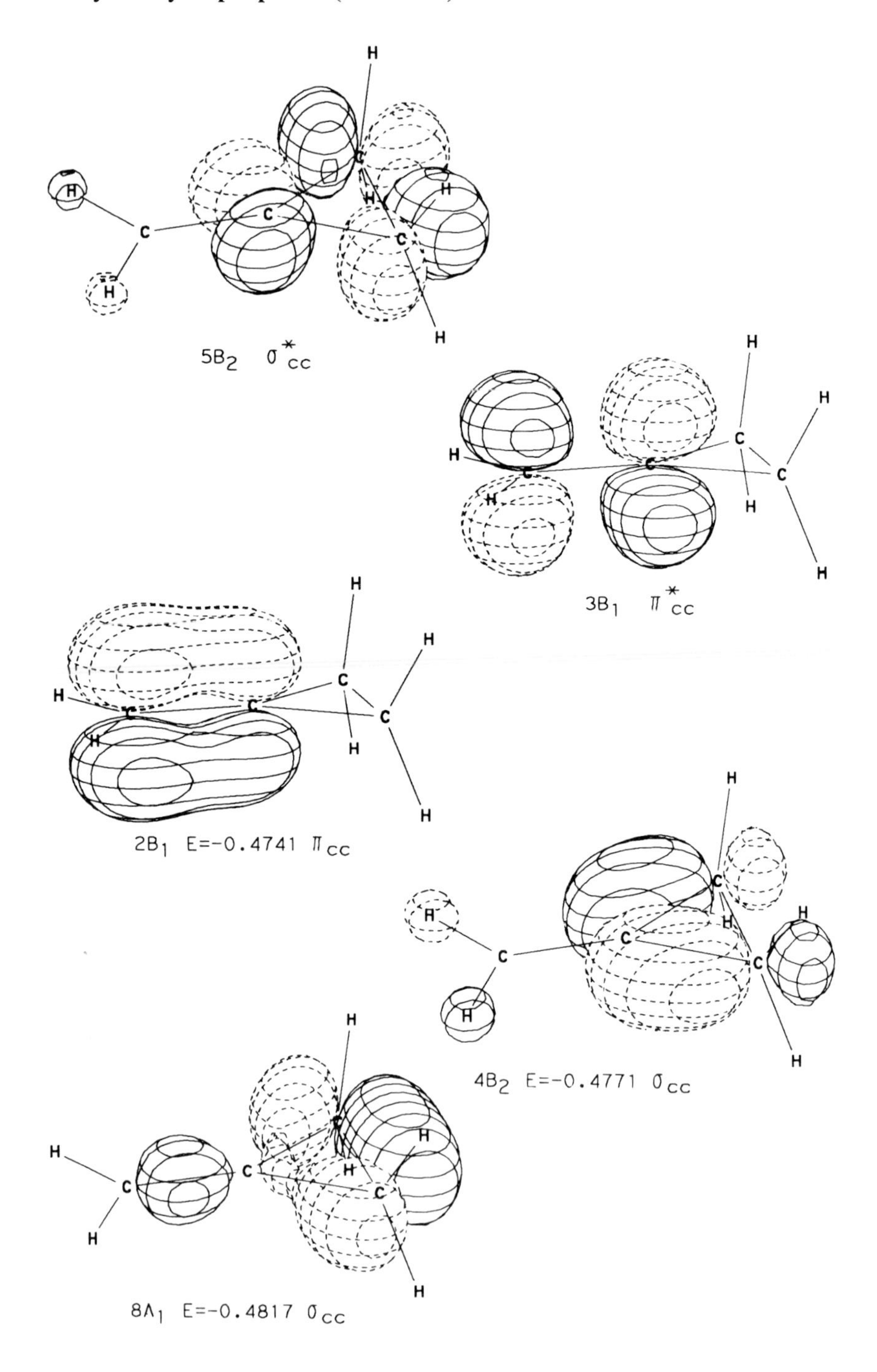

5B$_2$ σ^*_{cc}

3B$_1$ π^*_{cc}

2B$_1$ E=-0.4741 π_{cc}

4B$_2$ E=-0.4771 σ_{cc}

8A$_1$ E=-0.4817 σ_{cc}

73. Cyclopropanone

Symmetry: C_{2v}

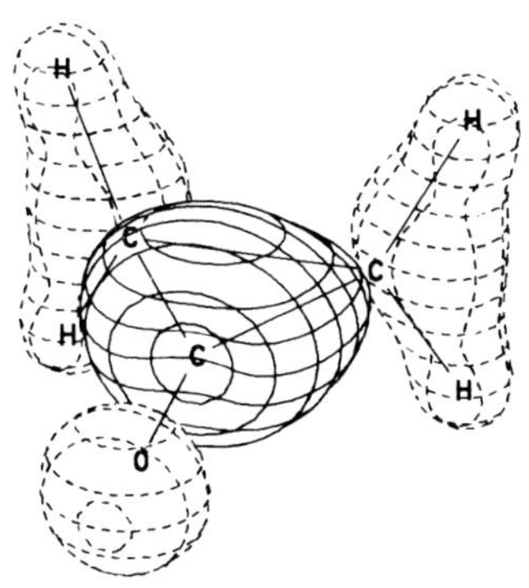

$6A_1$ E=-0.6484 σ_{CH_2}, σ_{CC}

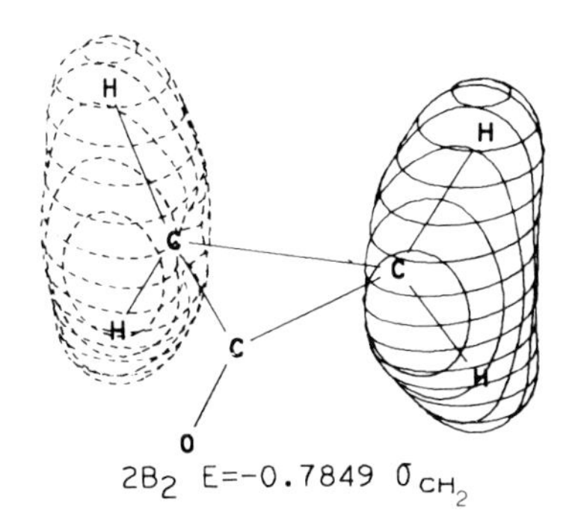

$2B_2$ E=-0.7849 σ_{CH_2}

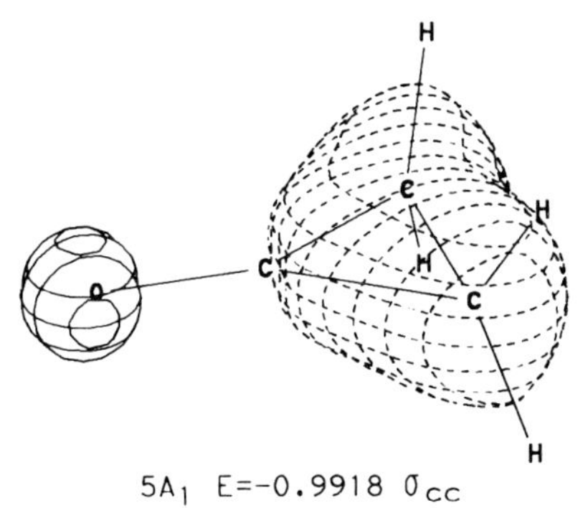

$5A_1$ E=-0.9918 σ_{CC}

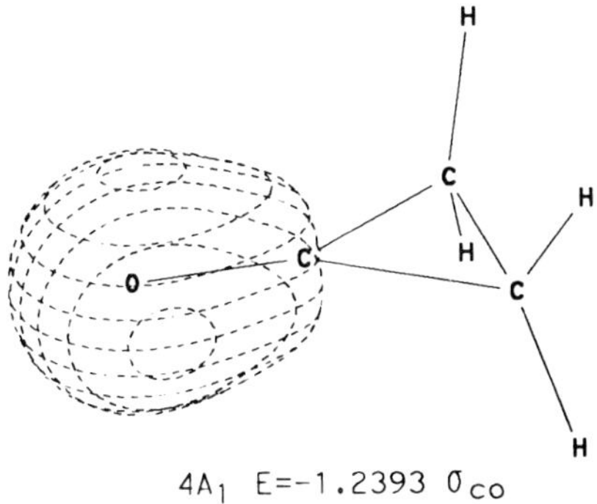

$4A_1$ E=-1.2393 σ_{CO}

Cyclopropanone (Continued)

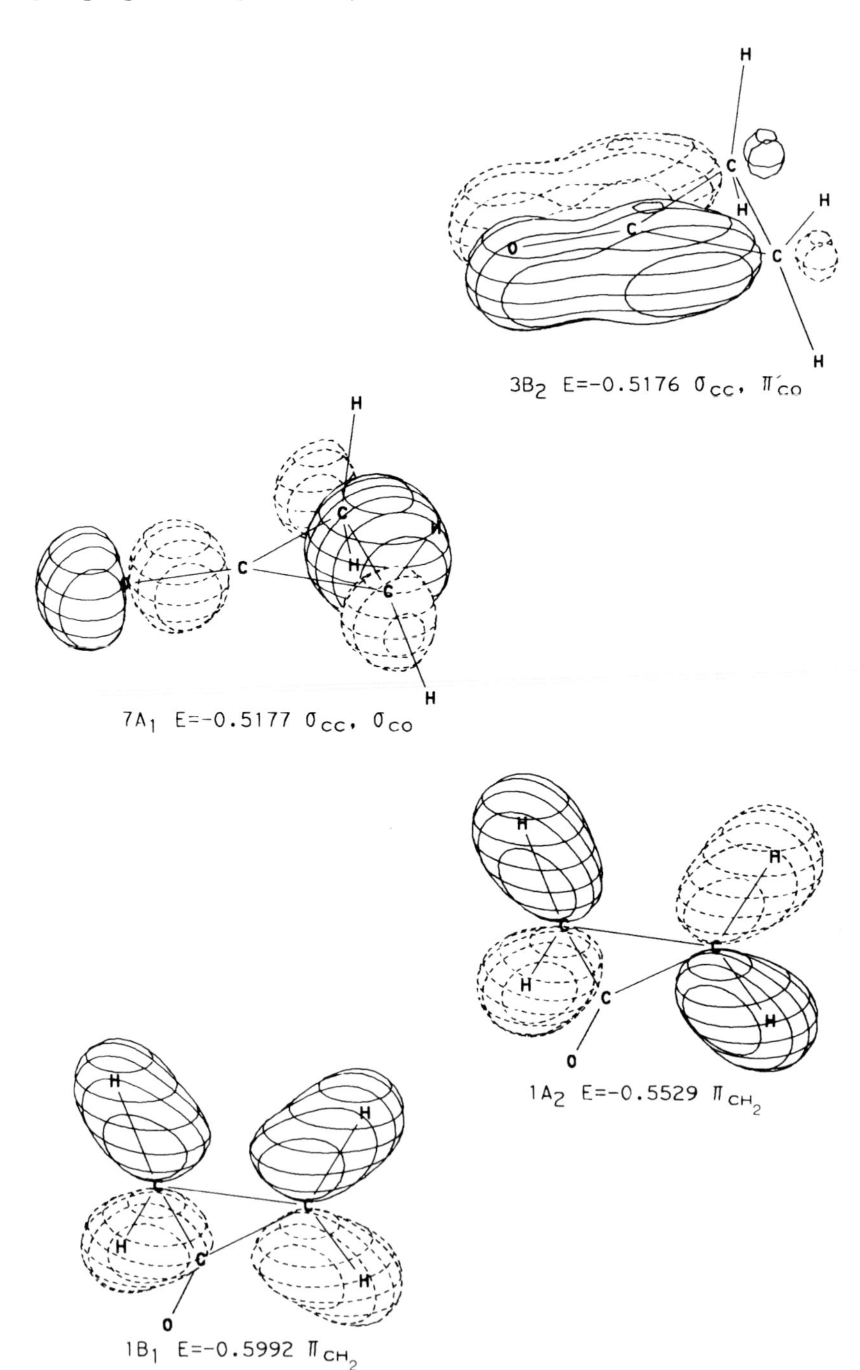

3B$_2$ E=−0.5176 σ_{CC}, π'_{CO}

7A$_1$ E=−0.5177 σ_{CC}, σ_{CO}

1A$_2$ E=−0.5529 π_{CH_2}

1B$_1$ E=−0.5992 π_{CH_2}

Cyclopropanone (Continued)

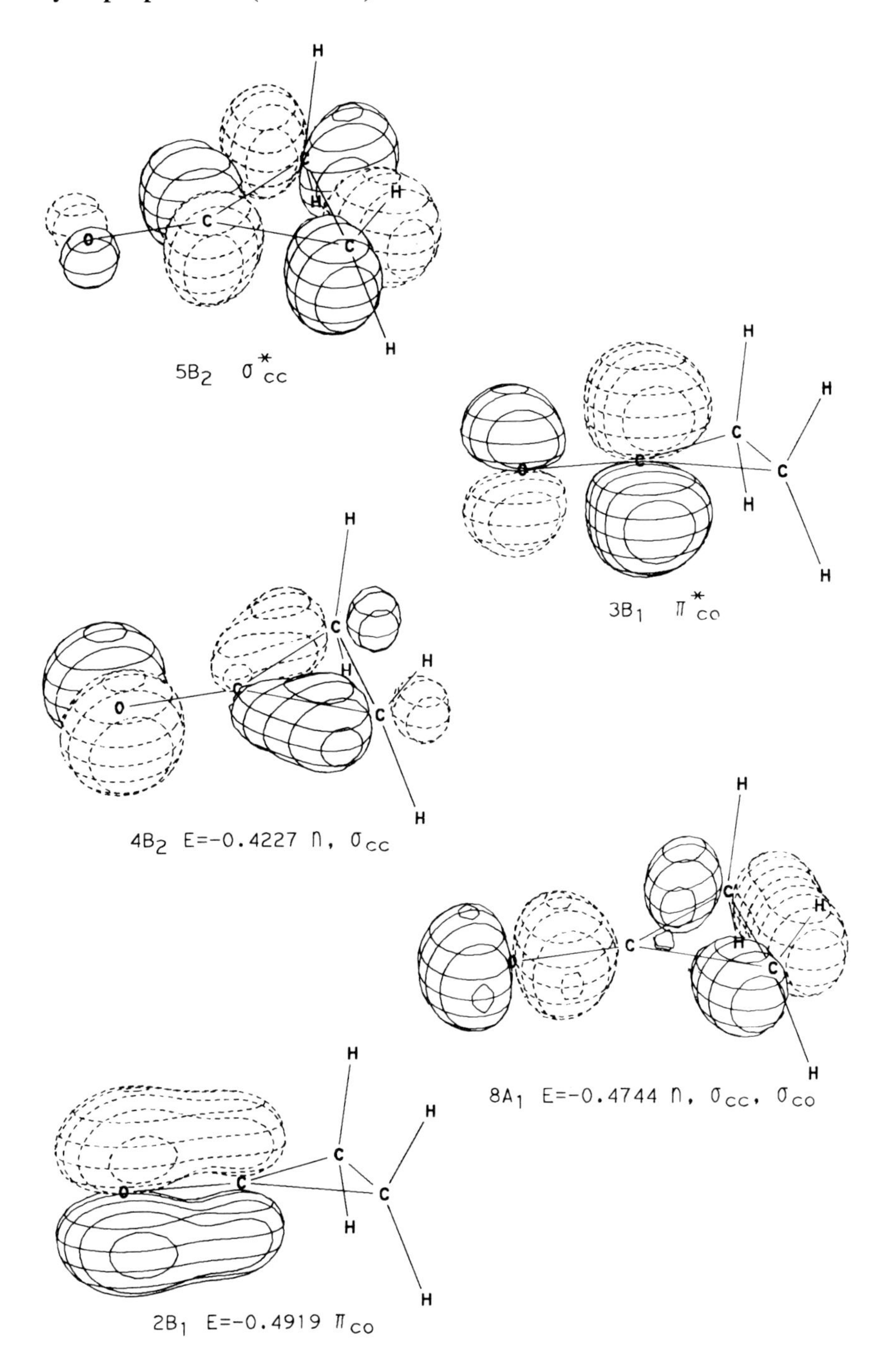

$5B_2$ σ^*_{cc}

$3B_1$ π^*_{co}

$4B_2$ E=−0.4227 n, σ_{cc}

$8A_1$ E=−0.4744 n, σ_{cc}, σ_{co}

$2B_1$ E=−0.4919 π_{co}

74. Cyclobutene

Symmetry: C_{2v}

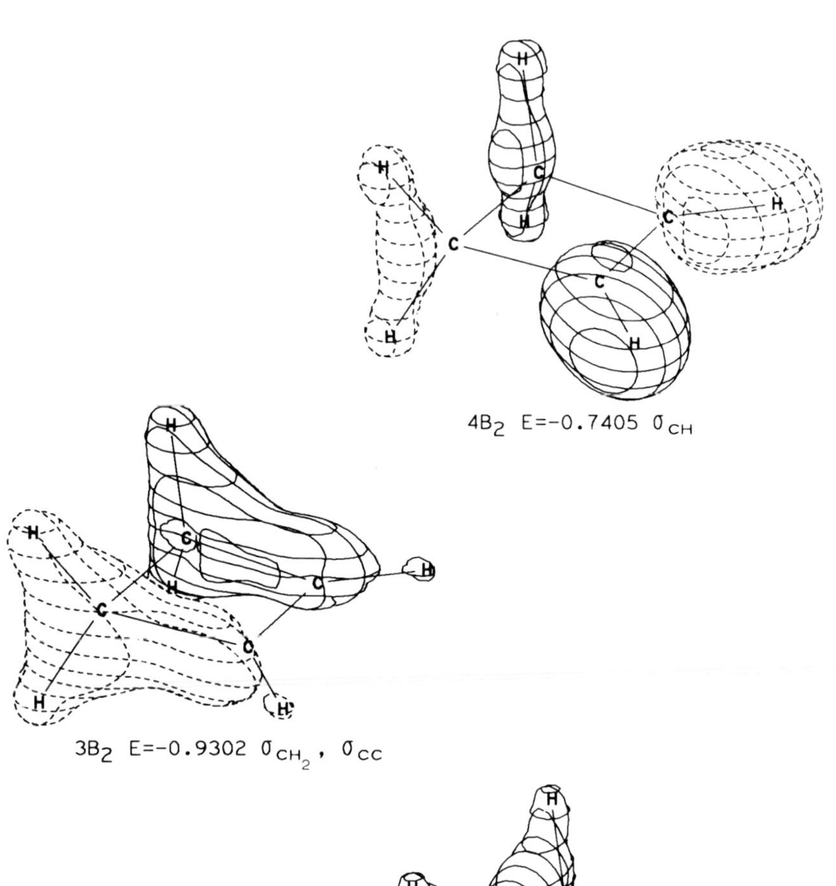

$4B_2$ E=−0.7405 σ_{CH}

$3B_2$ E=−0.9302 σ_{CH_2}, σ_{CC}

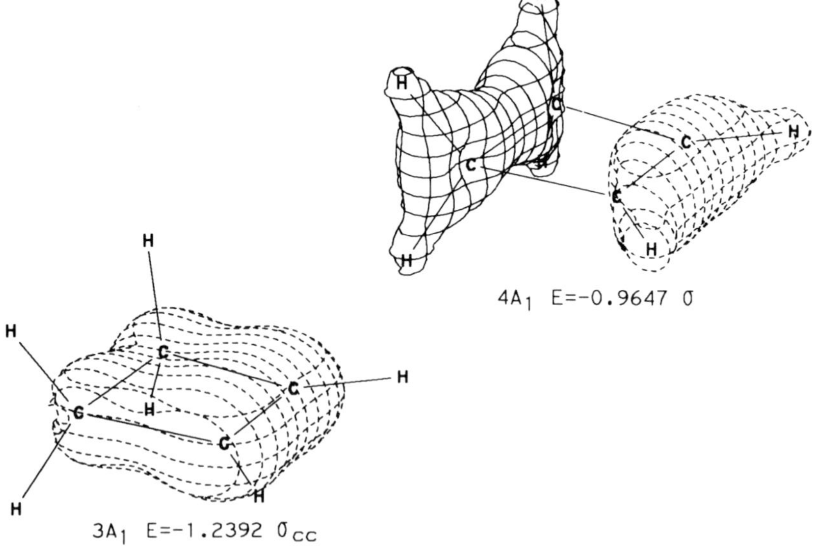

$4A_1$ E=−0.9647 σ

$3A_1$ E=−1.2392 σ_{CC}

Cyclobutene (Continued)

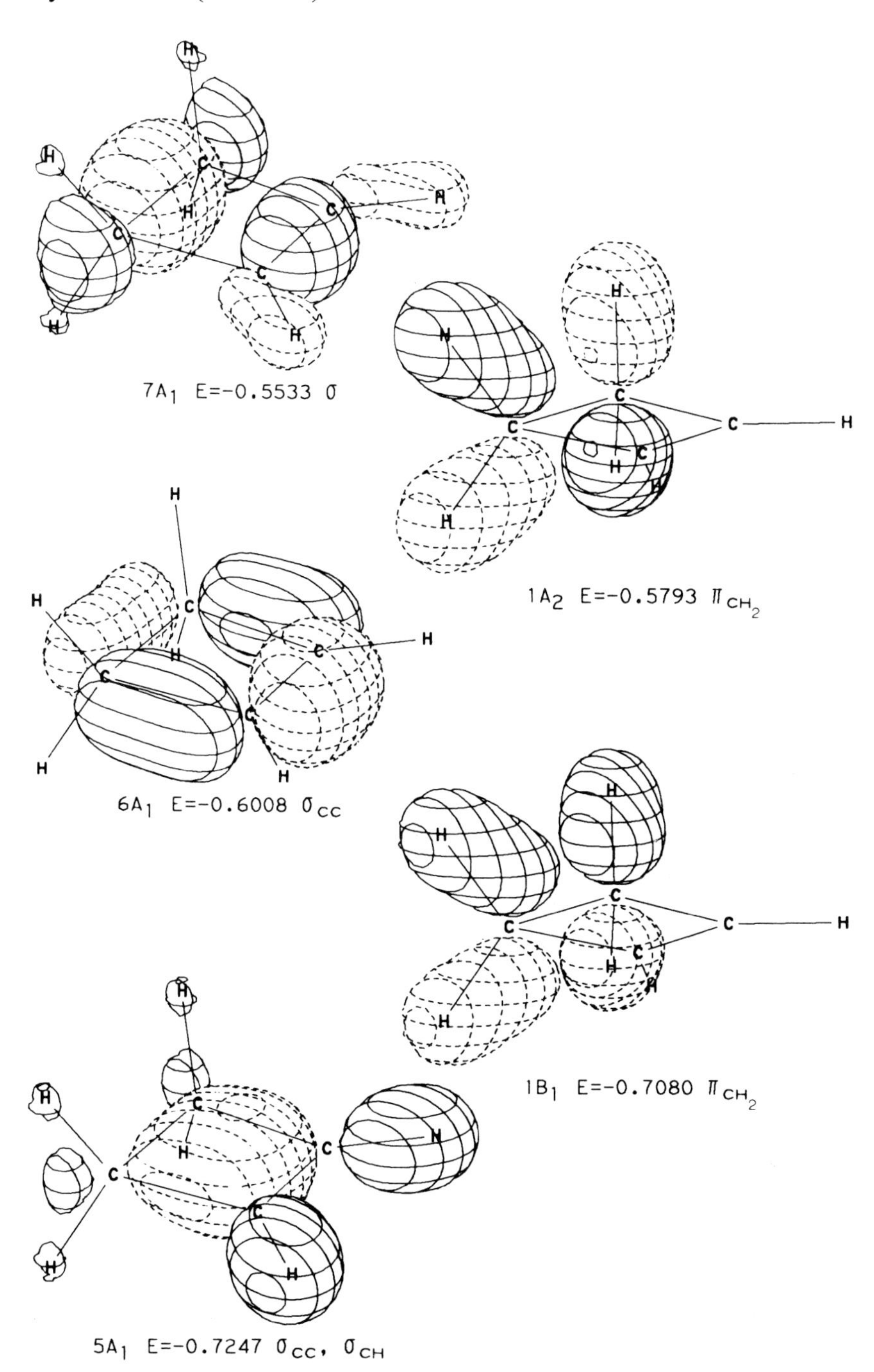

$7A_1$ E=−0.5533 σ

$1A_2$ E=−0.5793 π_{CH_2}

$6A_1$ E=−0.6008 σ_{CC}

$1B_1$ E=−0.7080 π_{CH_2}

$5A_1$ E=−0.7247 σ_{CC}, σ_{CH}

Cyclobutene (Continued)

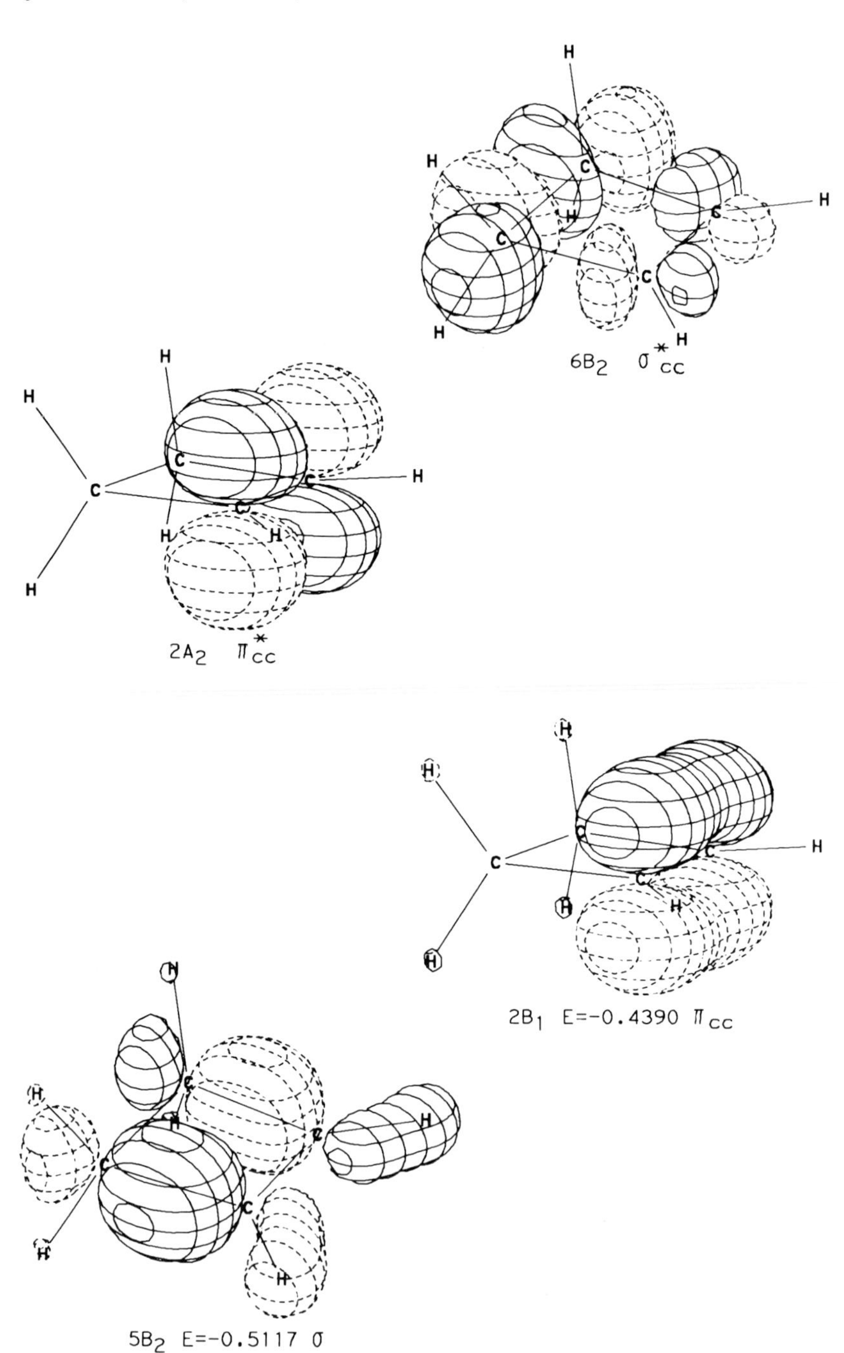

75. Bicyclobutane

Symmetry: C_{2v}

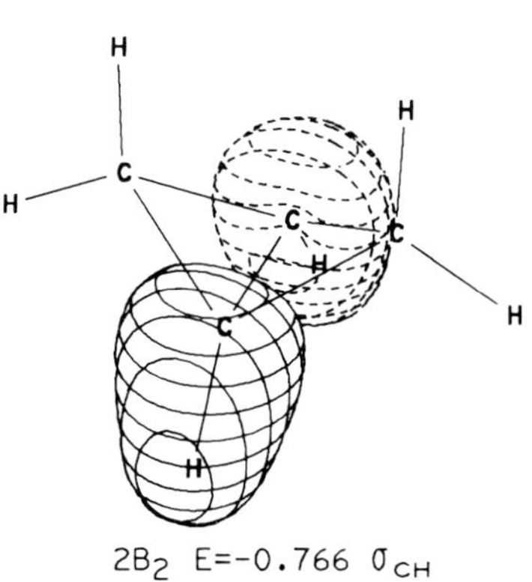

$2B_2$ E=-0.766 σ_{CH}

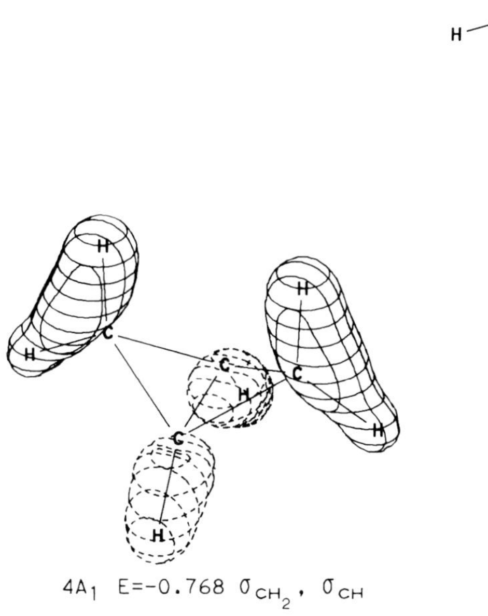

$4A_1$ E=-0.768 σ_{CH_2}, σ_{CH}

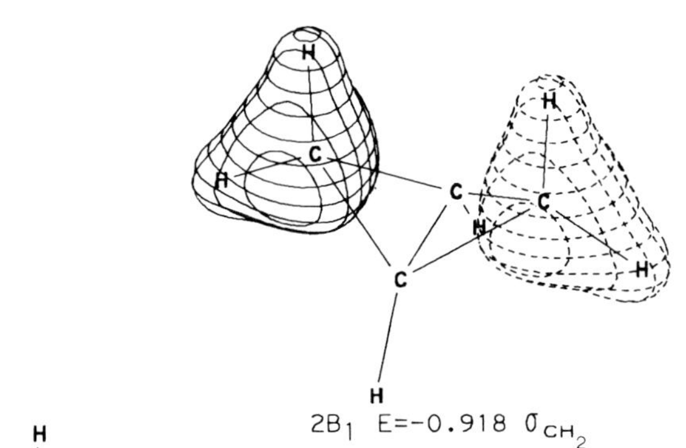

$2B_1$ E=-0.918 σ_{CH_2}

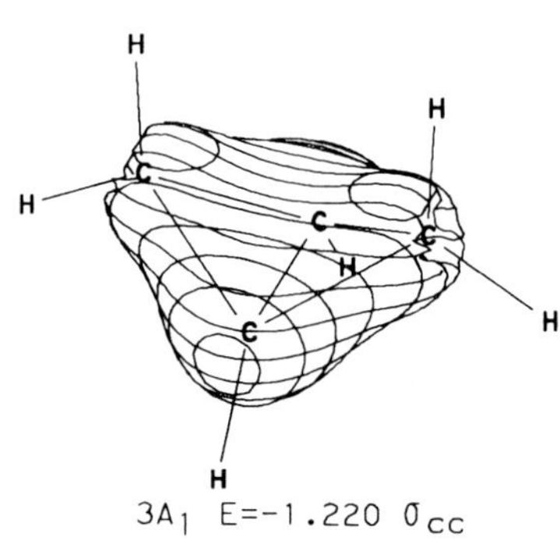

$3A_1$ E=-1.220 σ_{CC}

Bicyclobutane (Continued)

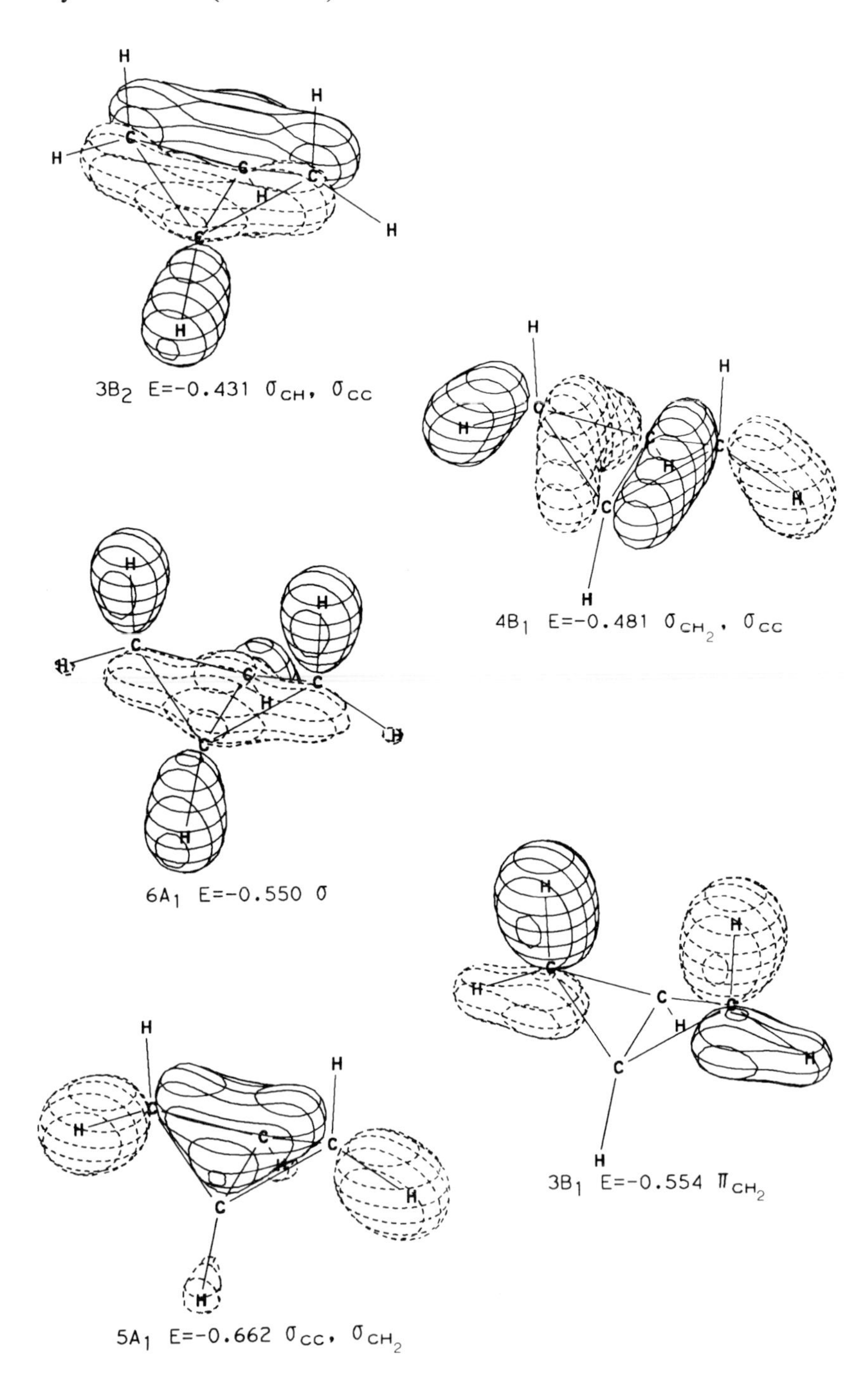

$3B_2$ E=-0.431 σ_{CH}, σ_{CC}

$4B_1$ E=-0.481 σ_{CH_2}, σ_{CC}

$6A_1$ E=-0.550 σ

$3B_1$ E=-0.554 π_{CH_2}

$5A_1$ E=-0.662 σ_{CC}, σ_{CH_2}

Bicyclobutane (Continued)

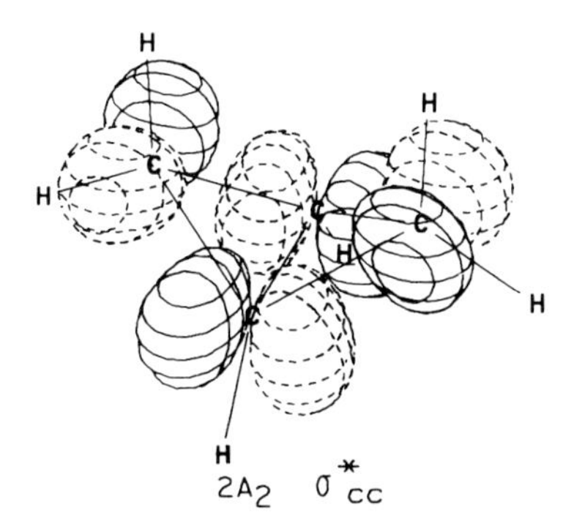

$2A_2$ σ^*_{cc}

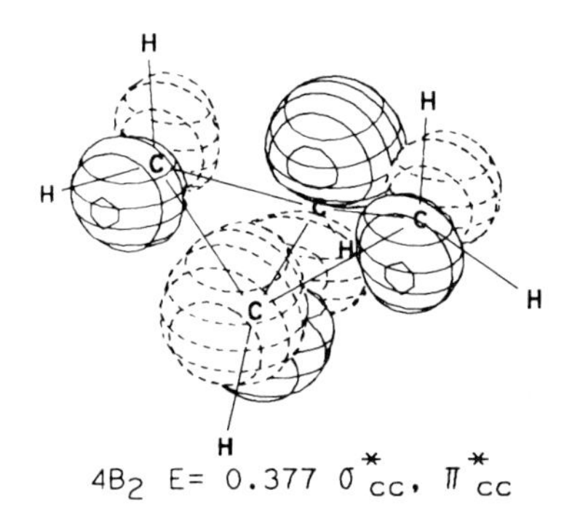

$4B_2$ E= 0.377 σ^*_{cc}, π^*_{cc}

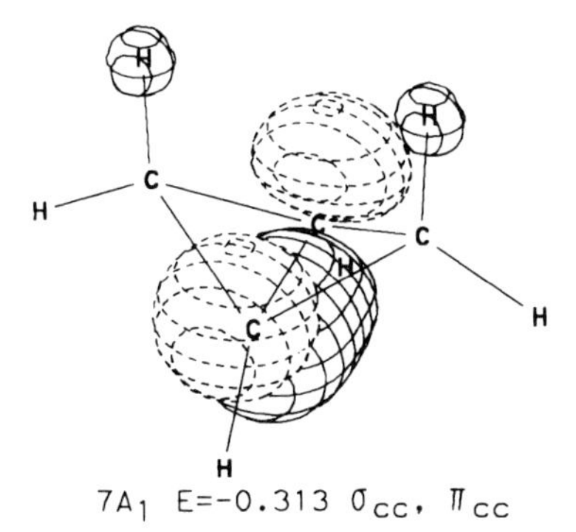

$7A_1$ E=−0.313 σ_{cc}, π_{cc}

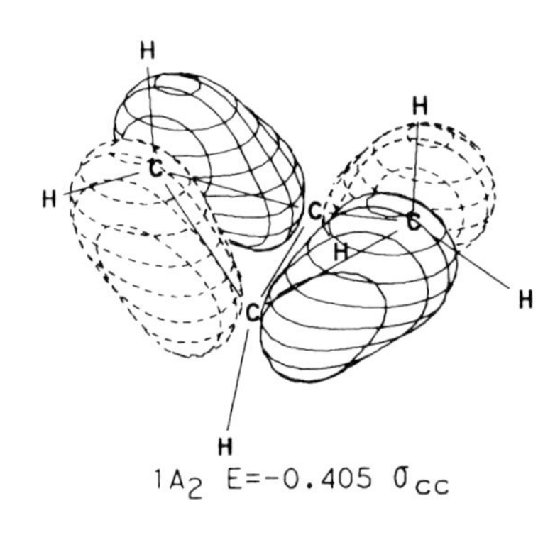

$1A_2$ E=−0.405 σ_{cc}

76. Cyclopropylcarbinyl Cation, Bisected

Symmetry: C_s

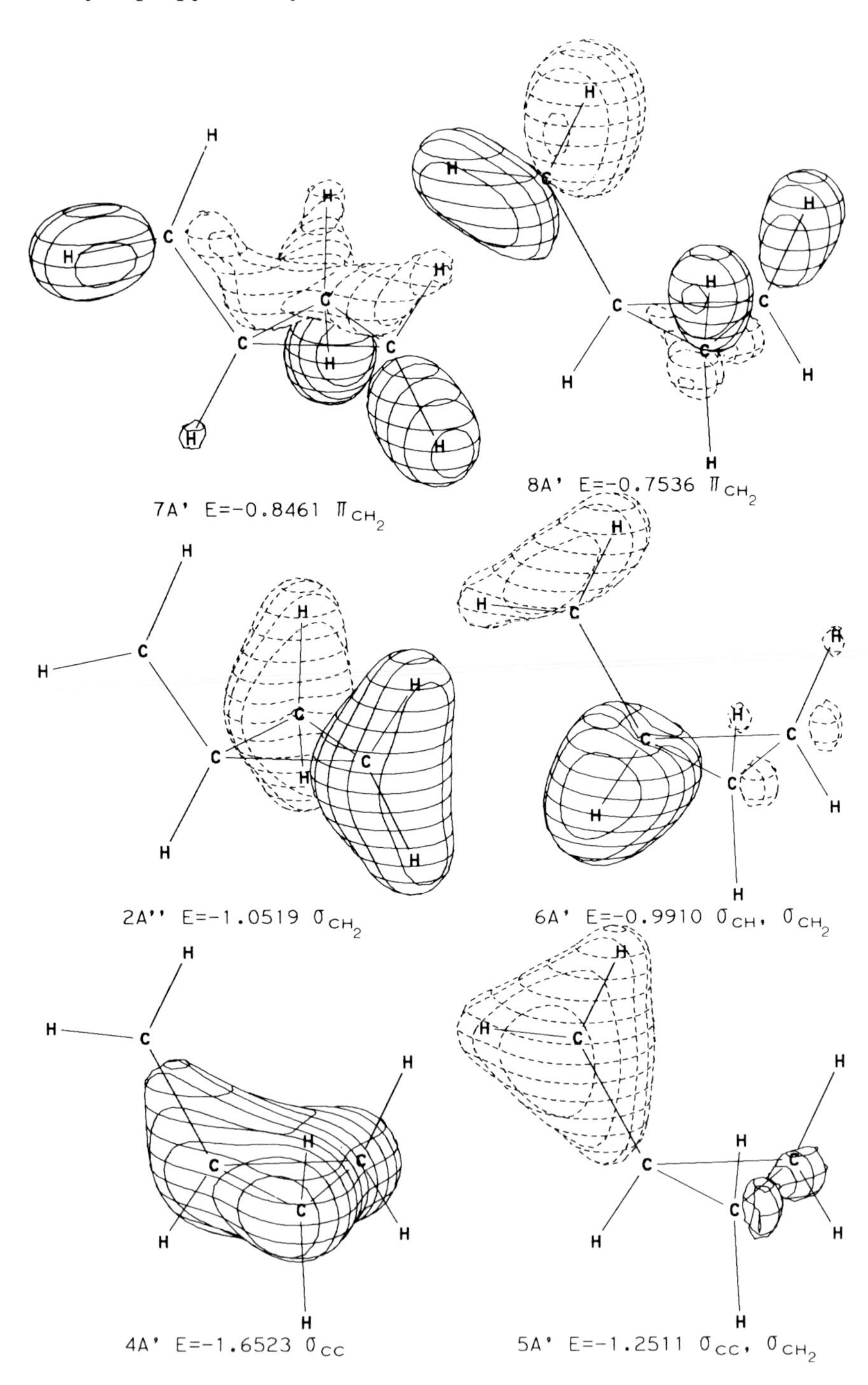

7A' E=-0.8461 π_{CH_2}

8A' E=-0.7536 π_{CH_2}

2A'' E=-1.0519 σ_{CH_2}

6A' E=-0.9910 σ_{CH}, σ_{CH_2}

4A' E=-1.6523 σ_{CC}

5A' E=-1.2511 σ_{CC}, σ_{CH_2}

Cyclopropylcarbinyl Cation, Bisected (Continued)

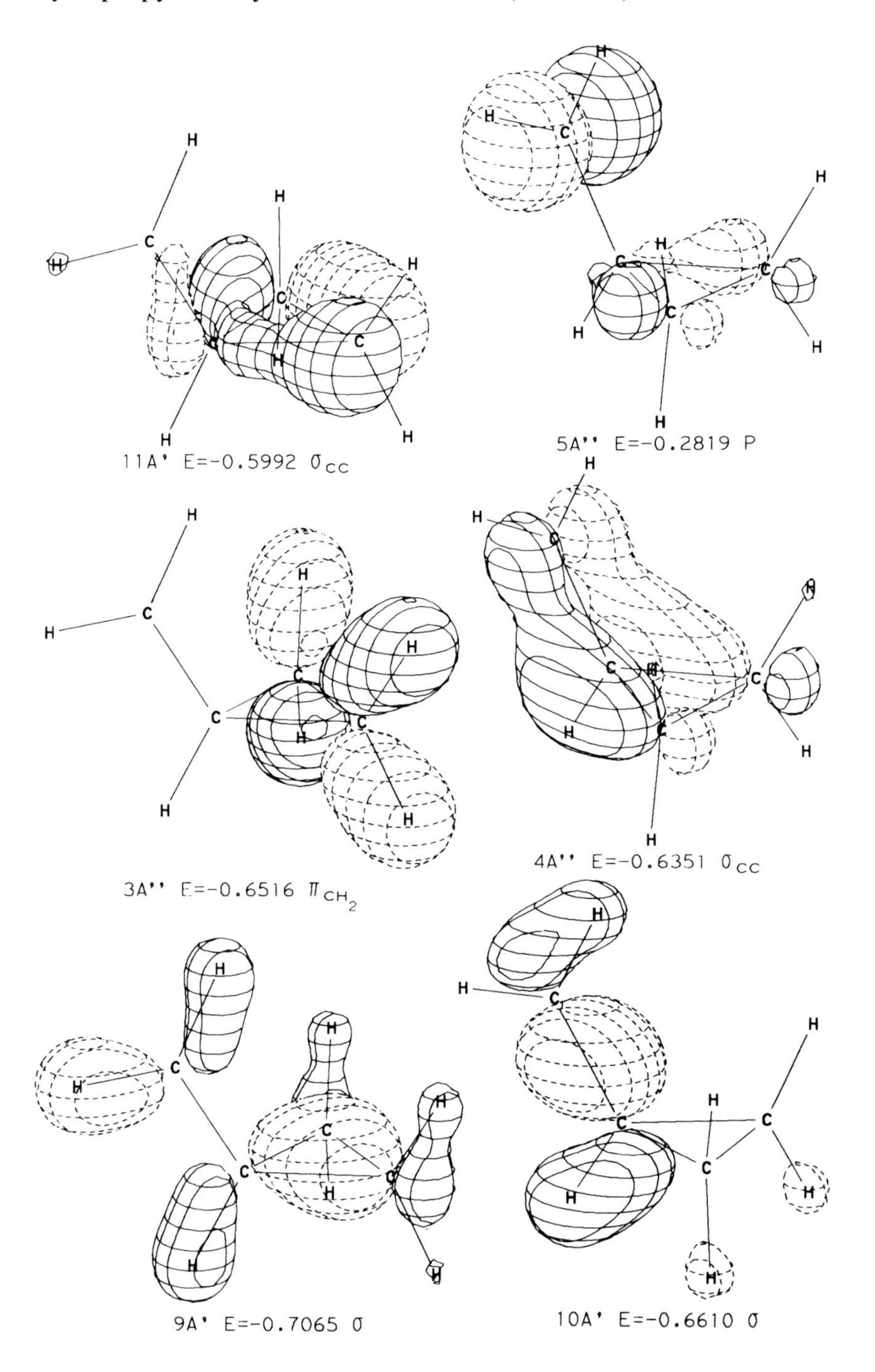

11A' E=-0.5992 σ_{CC}

5A'' E=-0.2819 P

3A'' E=-0.6516 π_{CH_2}

4A'' E=-0.6351 σ_{CC}

9A' E=-0.7065 σ

10A' E=-0.6610 σ

77. Cyclopropylcarbinyl Cation, Perpendicular

Symmetry: C_s

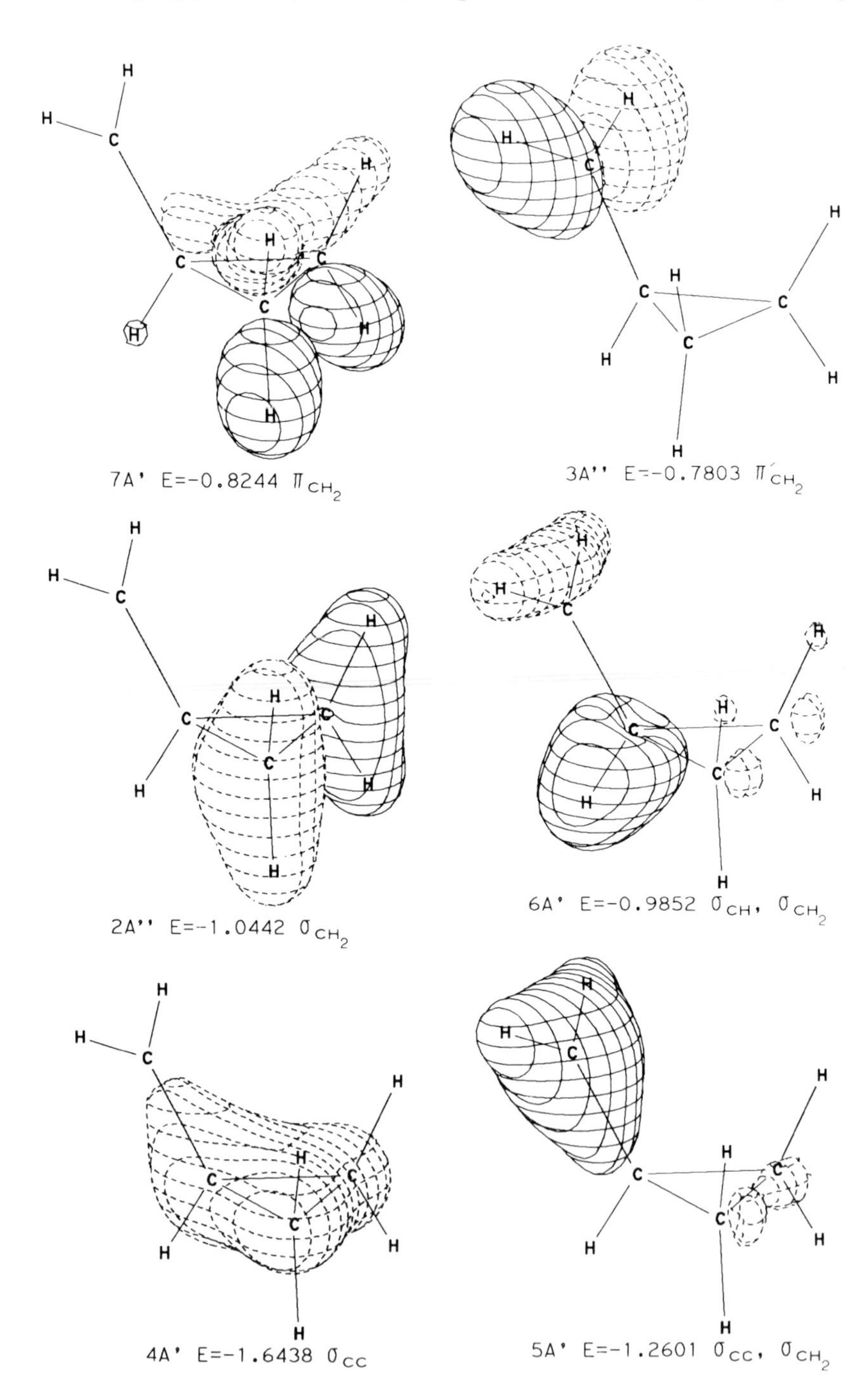

Cyclopropylcarbinyl Cation, Perpendicular (Continued)

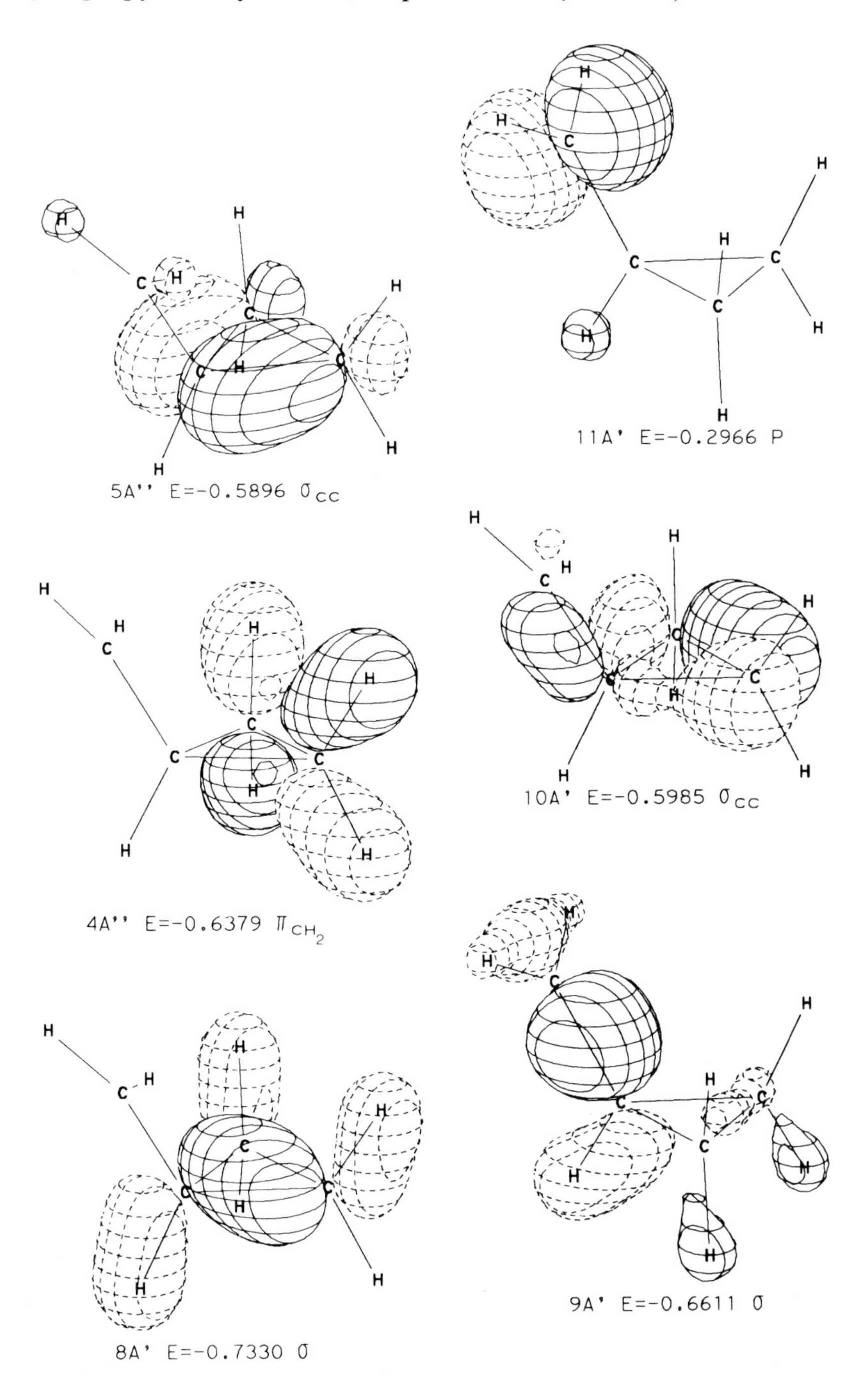

11A' E=-0.2966 P

5A'' E=-0.5896 σ_{CC}

10A' E=-0.5985 σ_{CC}

4A'' E=-0.6379 π_{CH_2}

9A' E=-0.6611 σ

8A' E=-0.7330 σ

78. *Trans*-2-Butene

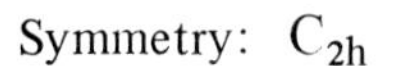

Symmetry: C_{2h}

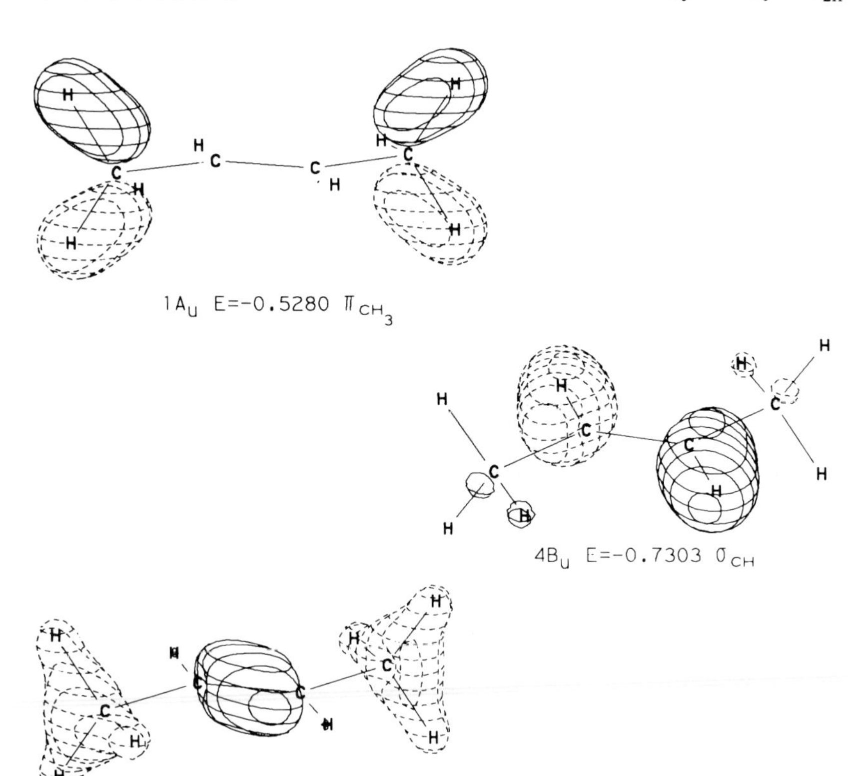

$1A_u$ E=-0.5280 π_{CH_3}

$4B_u$ E=-0.7303 σ_{CH}

$4A_g$ E=-0.8719 σ_{CH_3}, σ_{CC}

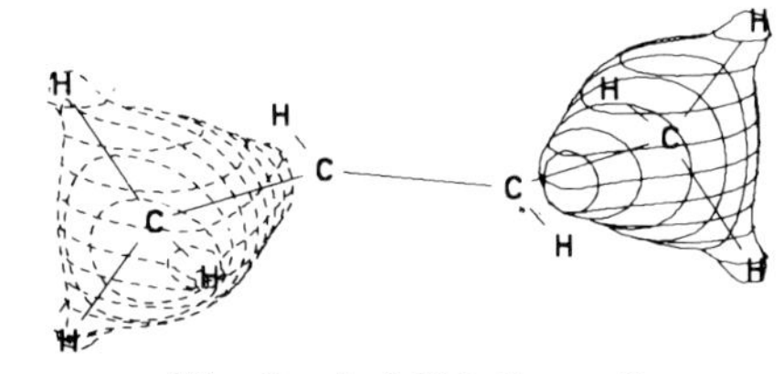

$3B_u$ E=-1.1666 σ_{CC}, σ_{CH_3}

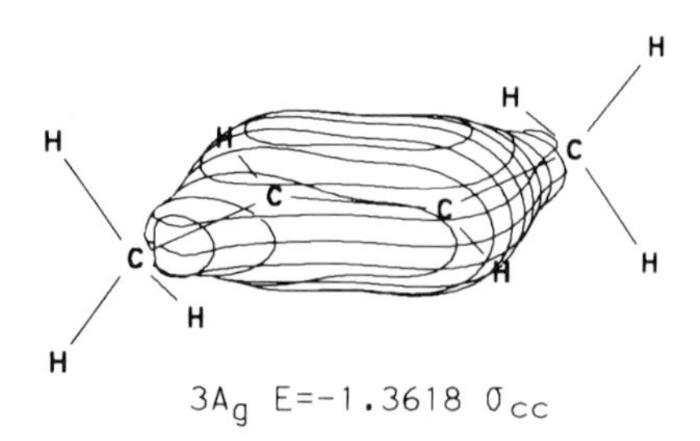

$3A_g$ E=-1.3618 σ_{CC}

***Trans*-2-Butene** (Continued)

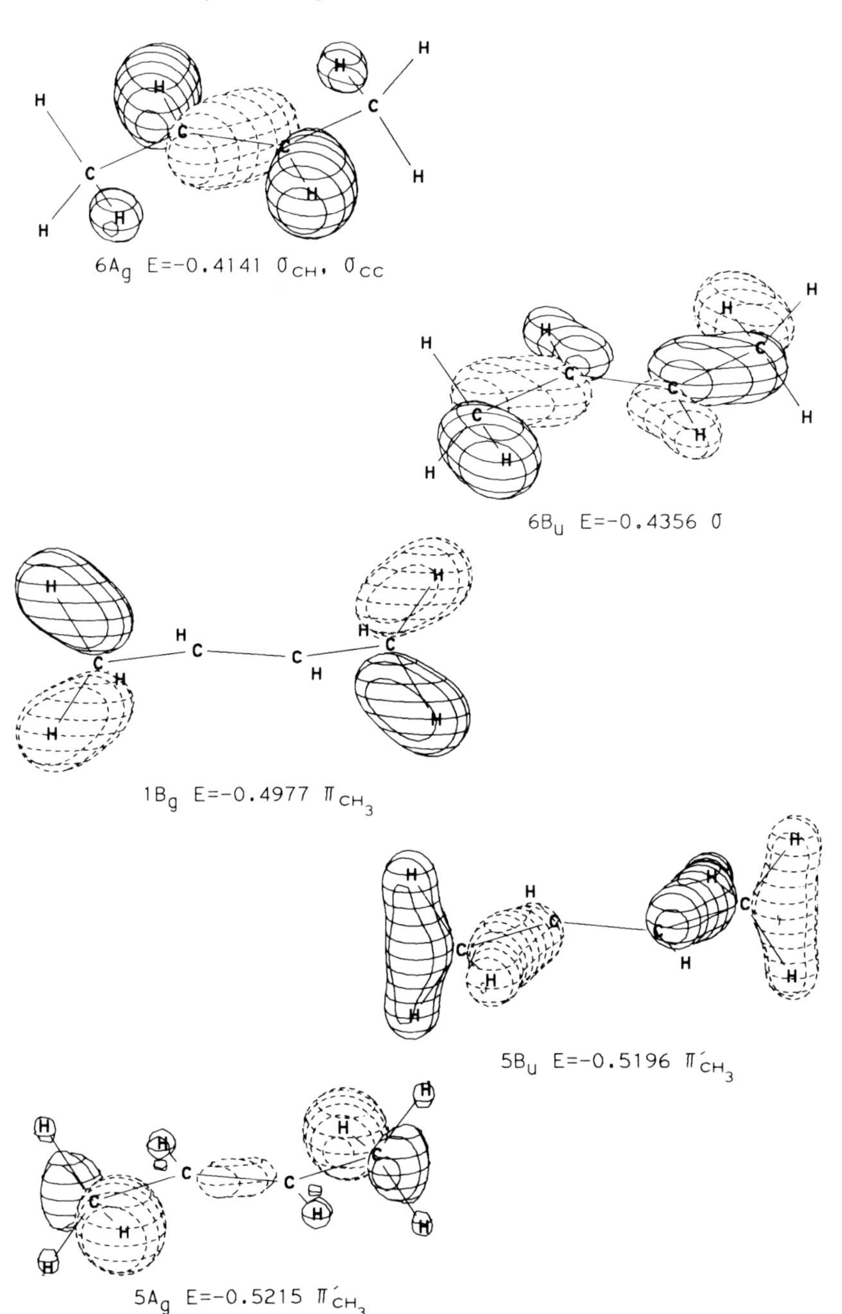

$6A_g$ E=-0.4141 σ_{CH}, σ_{CC}

$6B_u$ E=-0.4356 σ

$1B_g$ E=-0.4977 π_{CH_3}

$5B_u$ E=-0.5196 π'_{CH_3}

$5A_g$ E=-0.5215 π'_{CH_3}

Trans-2-Butene (Continued)

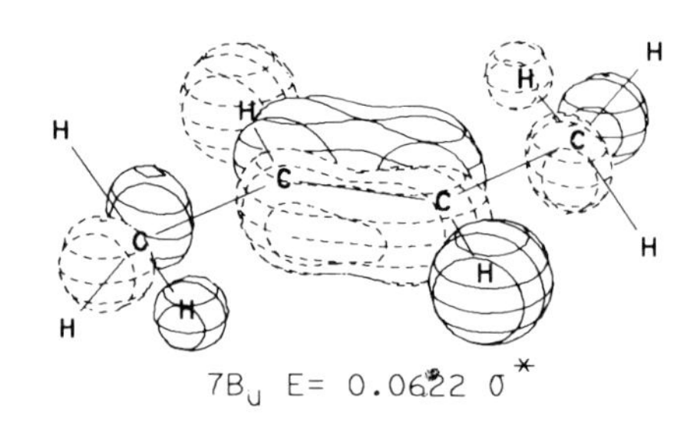

$7B_u$ E= 0.0622 σ^*

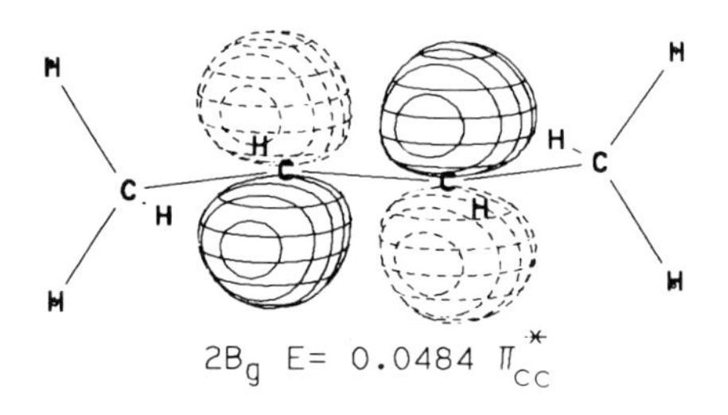

$2B_g$ E= 0.0484 π^*_{CC}

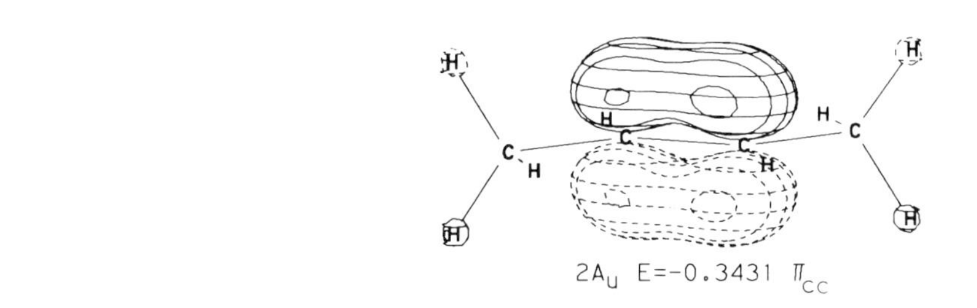

$2A_u$ E=-0.3431 π_{CC}

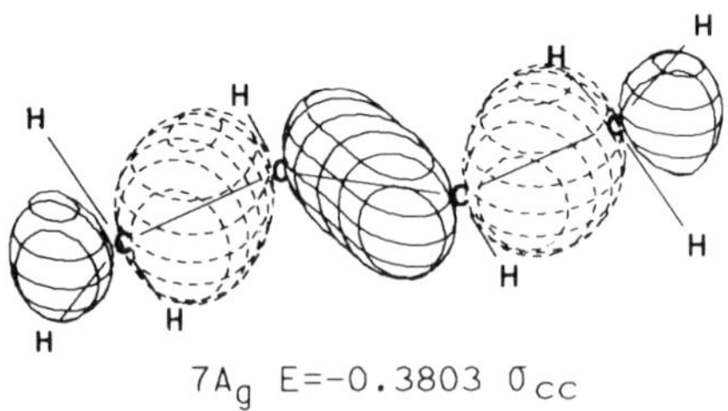

$7A_g$ E=-0.3803 σ_{CC}

79. Acetone

Symmetry: C_{2v}

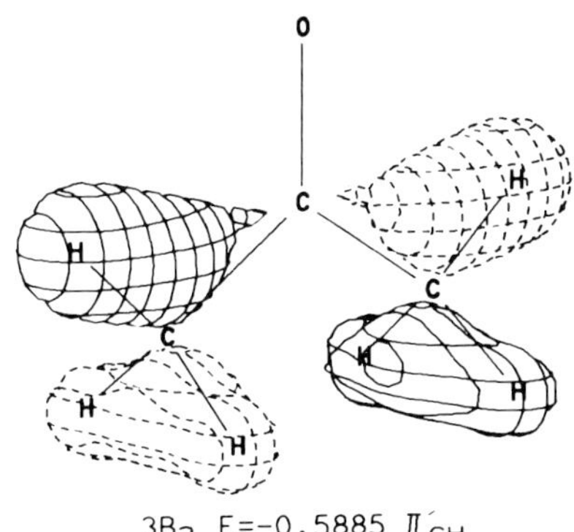

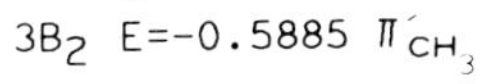

3B2 E=-0.5885 π'_{CH_3}

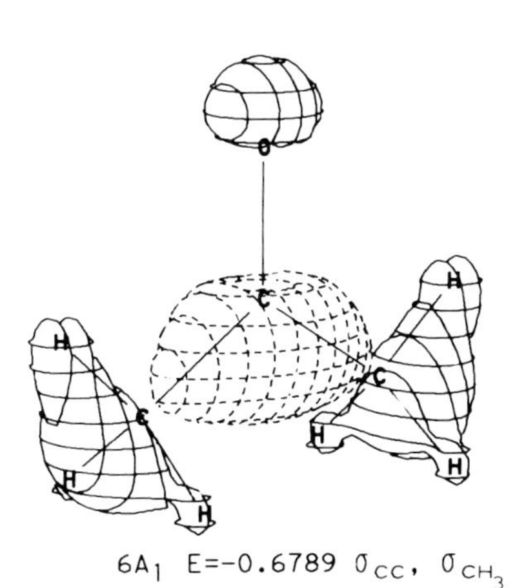

6A1 E=-0.6789 σ_{CC}, σ_{CH_3}

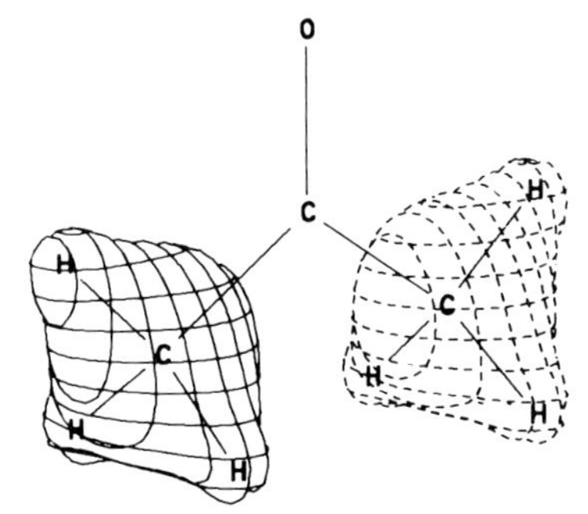

2B2 E=-0.9204 σ_{CH_3}

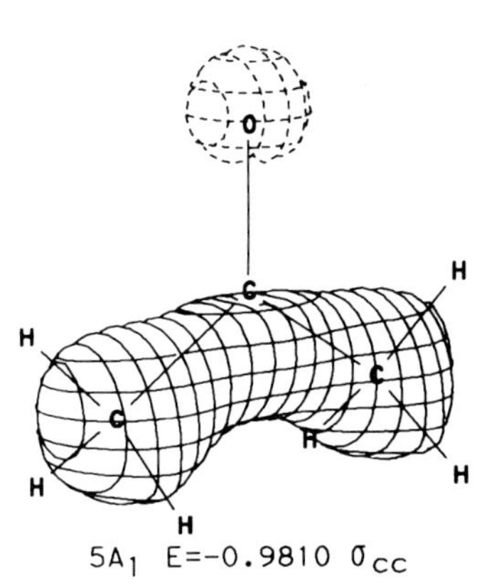

5A1 E=-0.9810 σ_{CC}

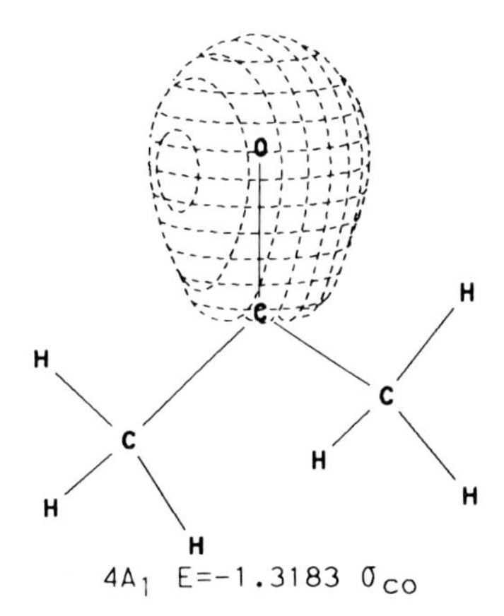

4A1 E=-1.3183 σ_{CO}

Acetone (Continued)

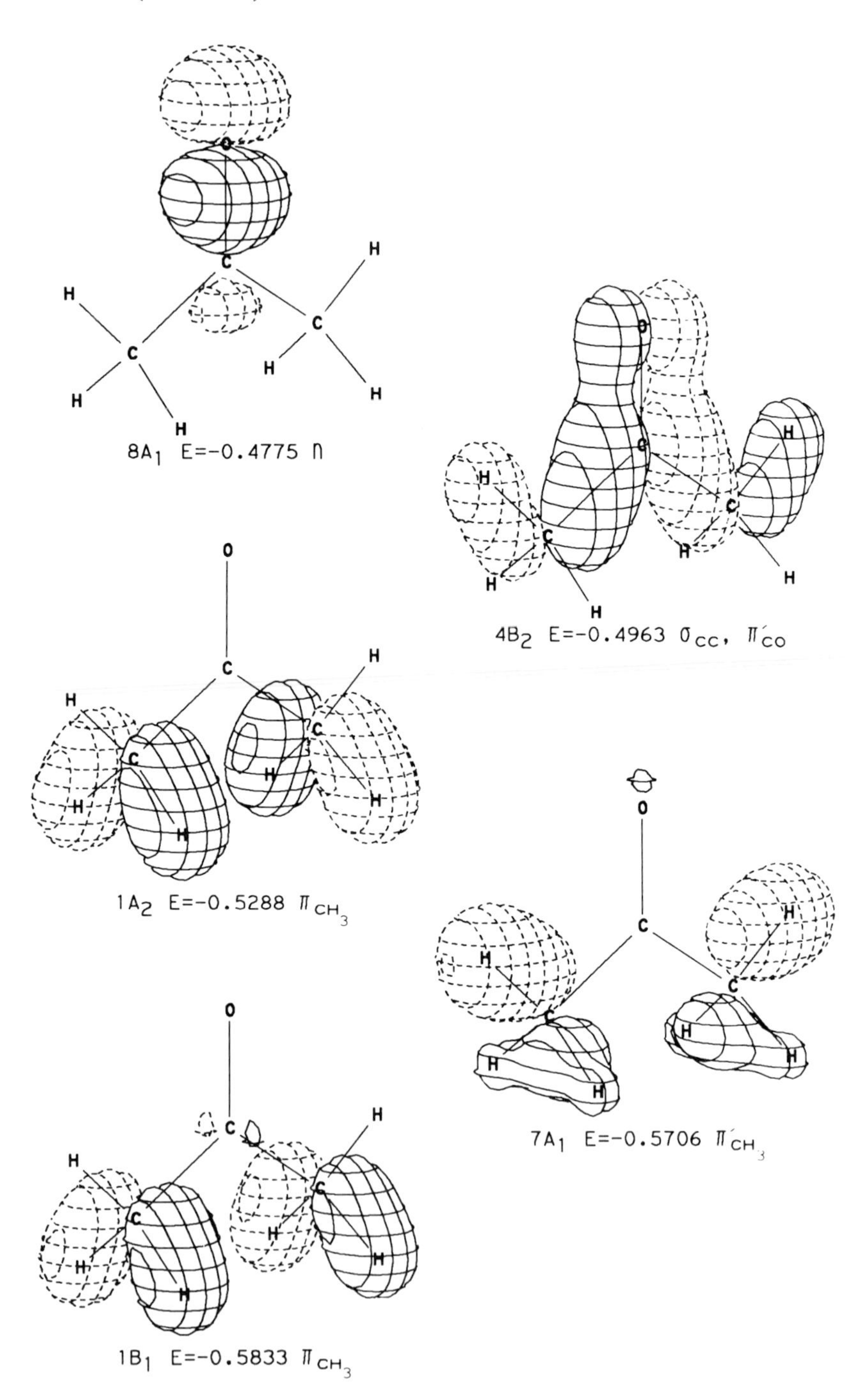

8A$_1$ E=-0.4775 n

4B$_2$ E=-0.4963 σ_{CC}, π'_{CO}

1A$_2$ E=-0.5288 π_{CH_3}

7A$_1$ E=-0.5706 π'_{CH_3}

1B$_1$ E=-0.5833 π_{CH_3}

Acetone (Continued)

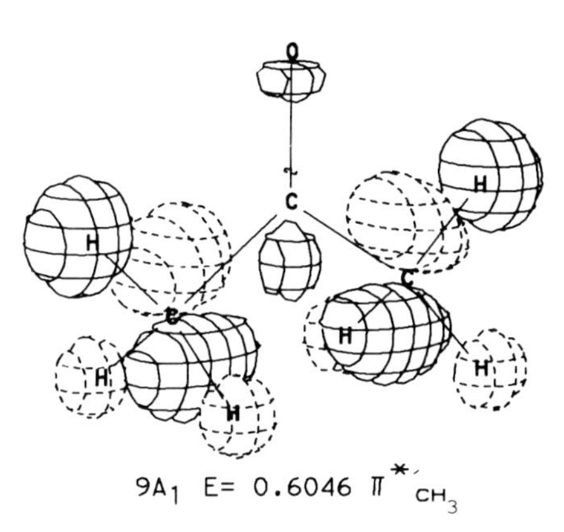

9A1 E= 0.6046 $\pi^{*}_{CH_3}$

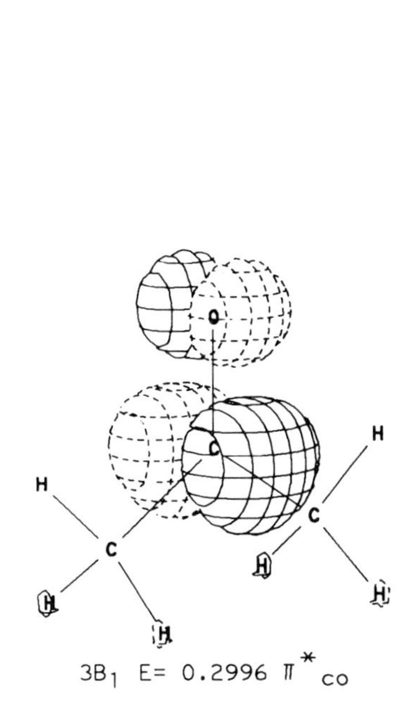

3B1 E= 0.2996 π^{*}_{CO}

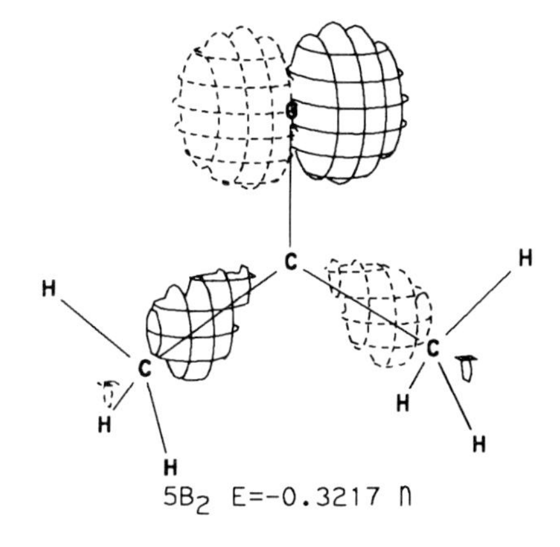

5B2 E=-0.3217 n

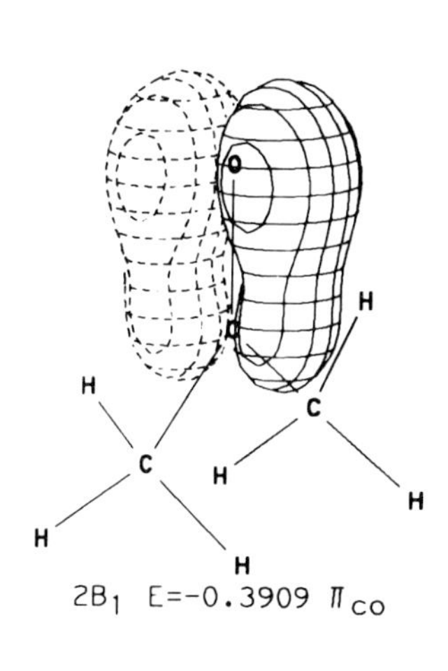

2B1 E=-0.3909 π_{CO}

80. Isopropenol

Symmetry: C_s

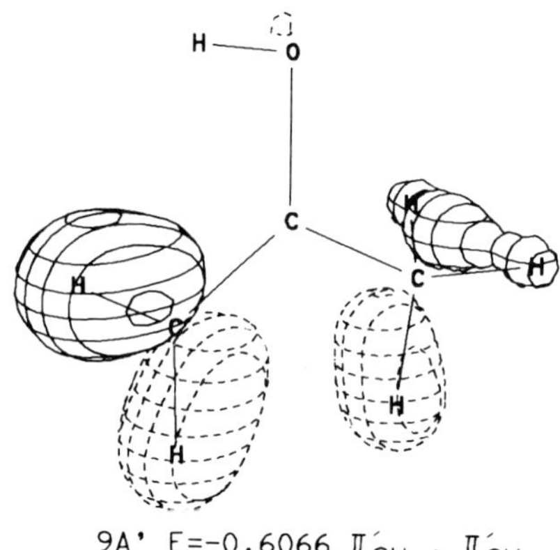

9A' E=-0.6066 π'_{CH_2}, π'_{CH_3}

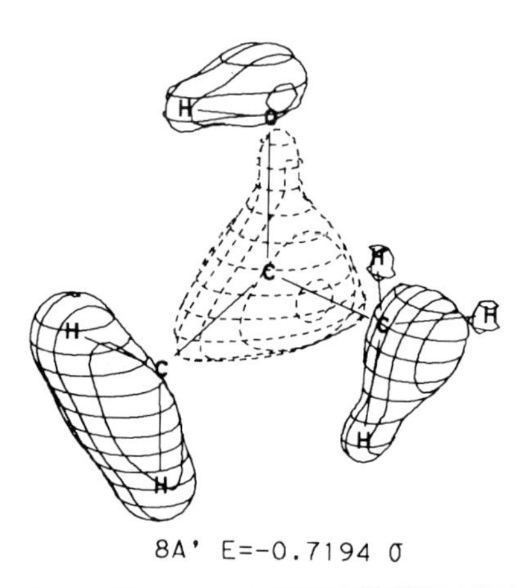

8A' E=-0.7194 σ

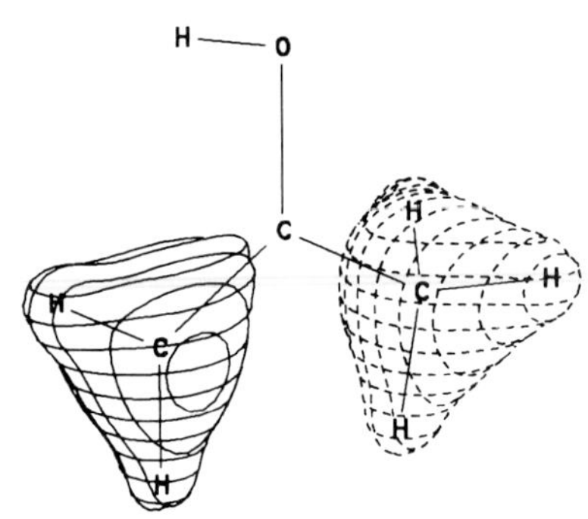

7A' E=-0.8908 σ_{CH_2}, σ_{CH_3}

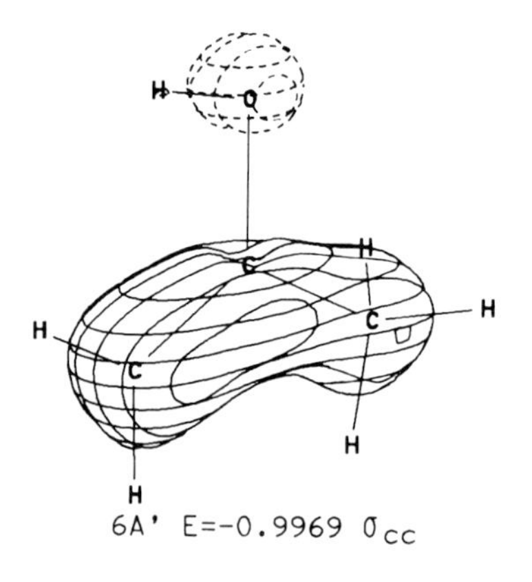

6A' E=-0.9969 σ_{CC}

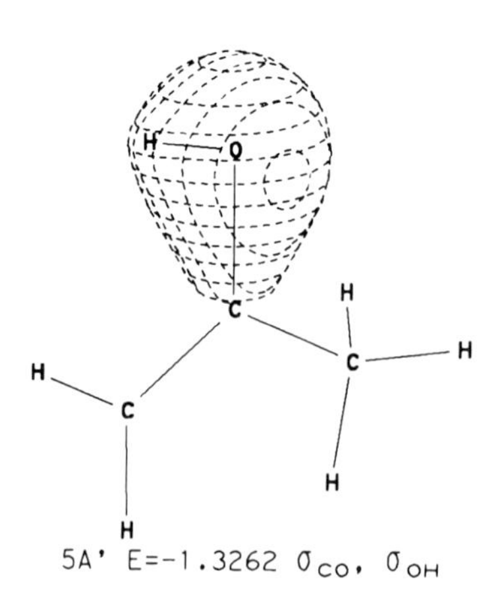

5A' E=-1.3262 σ_{CO}, σ_{OH}

Isopropenol (Continued)

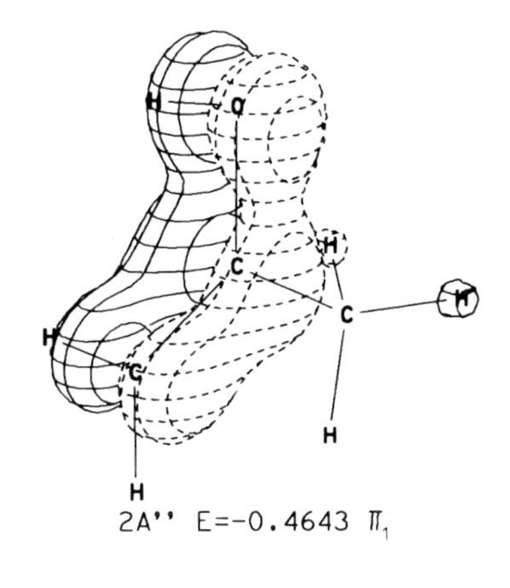

2A'' E=-0.4643 π_1

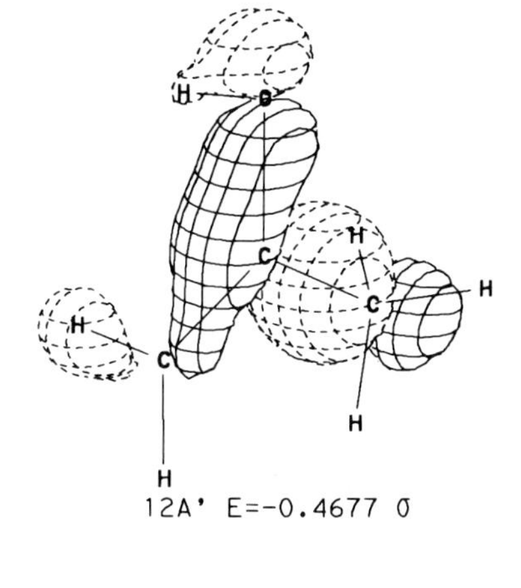

12A' E=-0.4677 σ

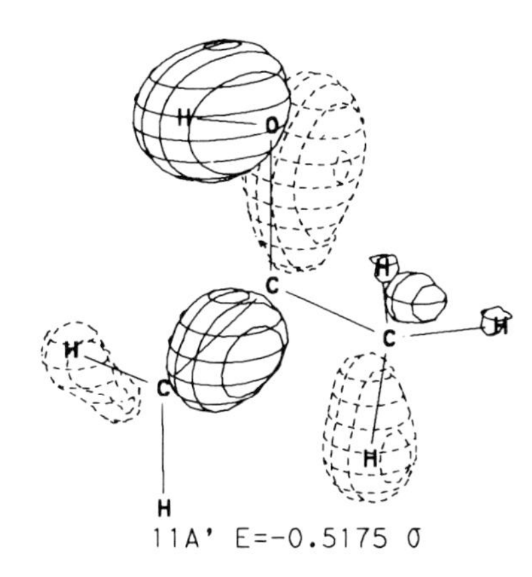

11A' E=-0.5175 σ

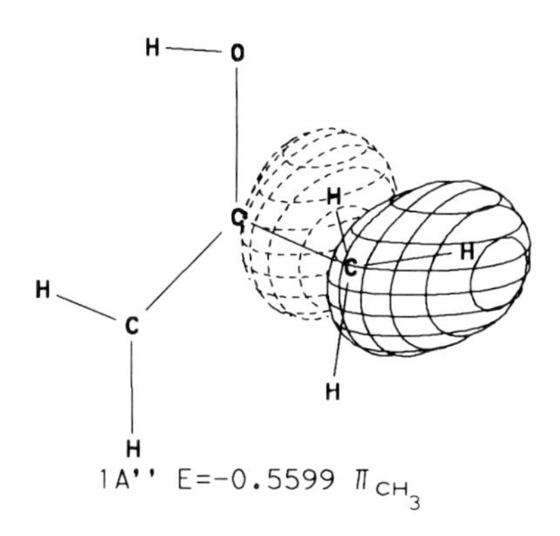

1A'' E=-0.5599 π_{CH_3}

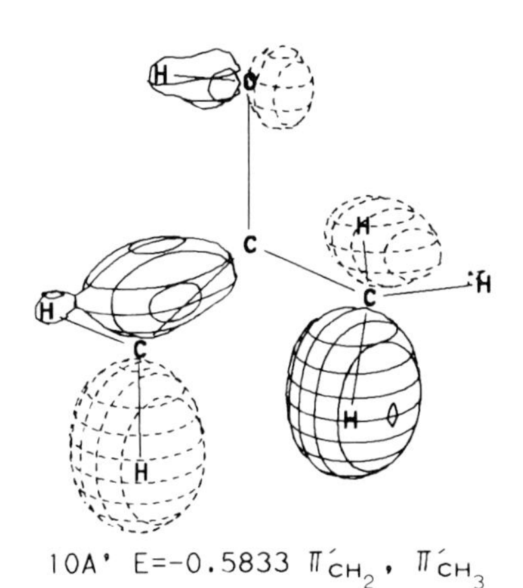

10A' E=-0.5833 π'_{CH_2}, π'_{CH_3}

Isopropenol (Continued)

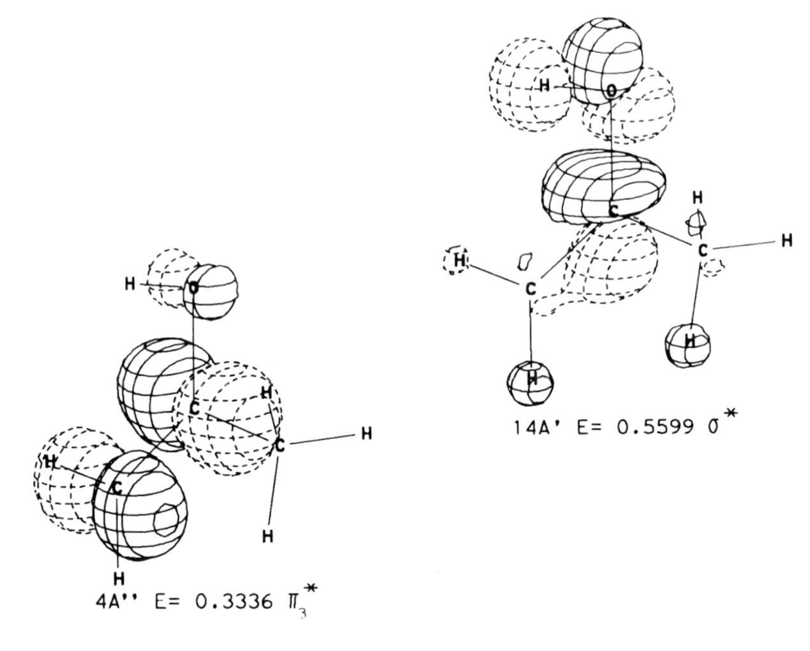

4A'' E= 0.3336 π_3^*

14A' E= 0.5599 σ^*

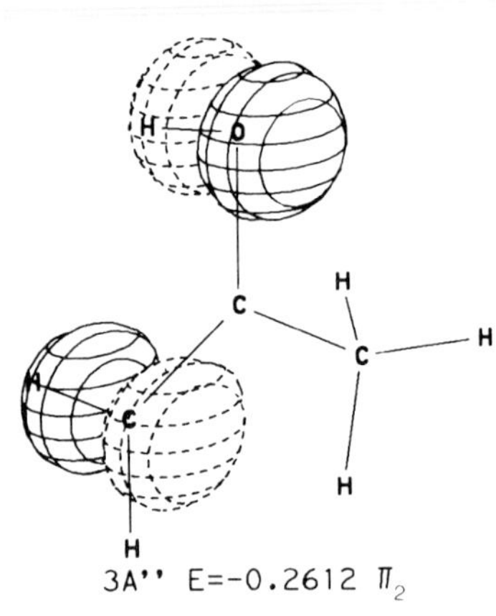

3A'' E=-0.2612 π_2

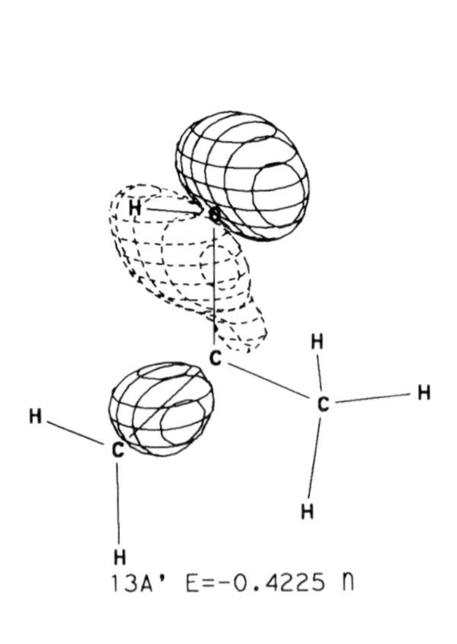

13A' E=-0.4225 n

81. Nitromethane

Symmetry: C_s

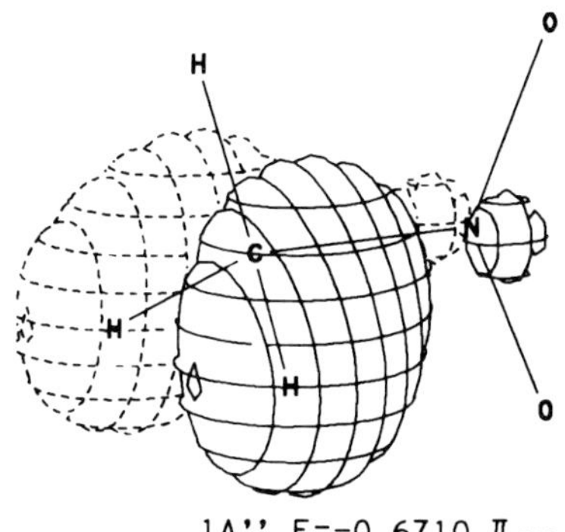

1A'' E=−0.6710 π_{CH_3}

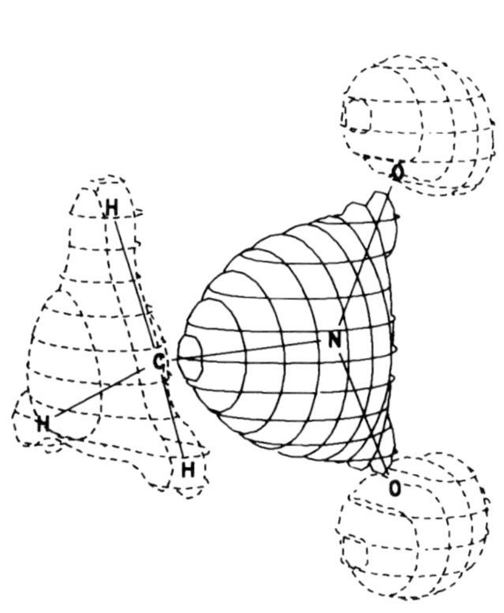

8A' E=−0.7968 σ

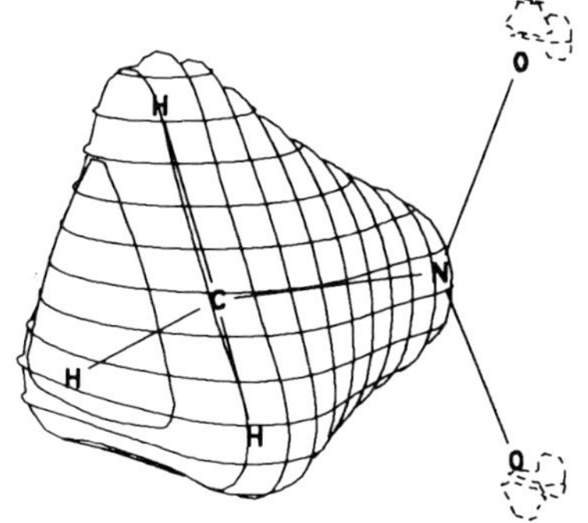

7A' E=−1.0500 σ_{CH_3}

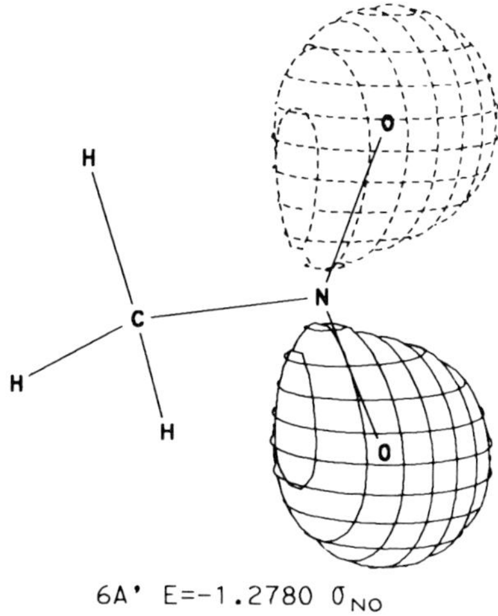

6A' E=−1.2780 σ_{NO}

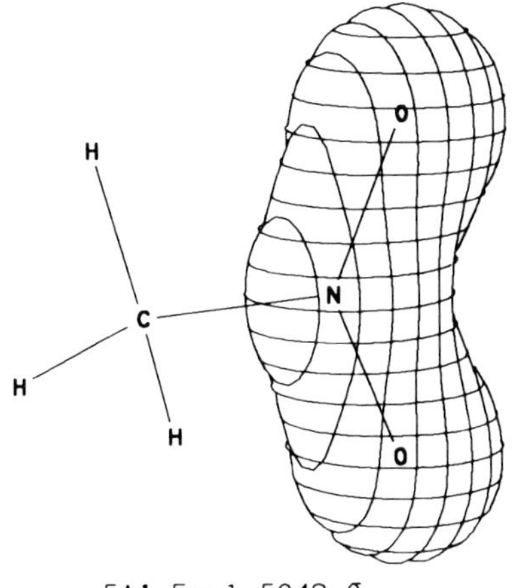

5A' E=−1.5049 σ_{NO}

Nitromethane (Continued)

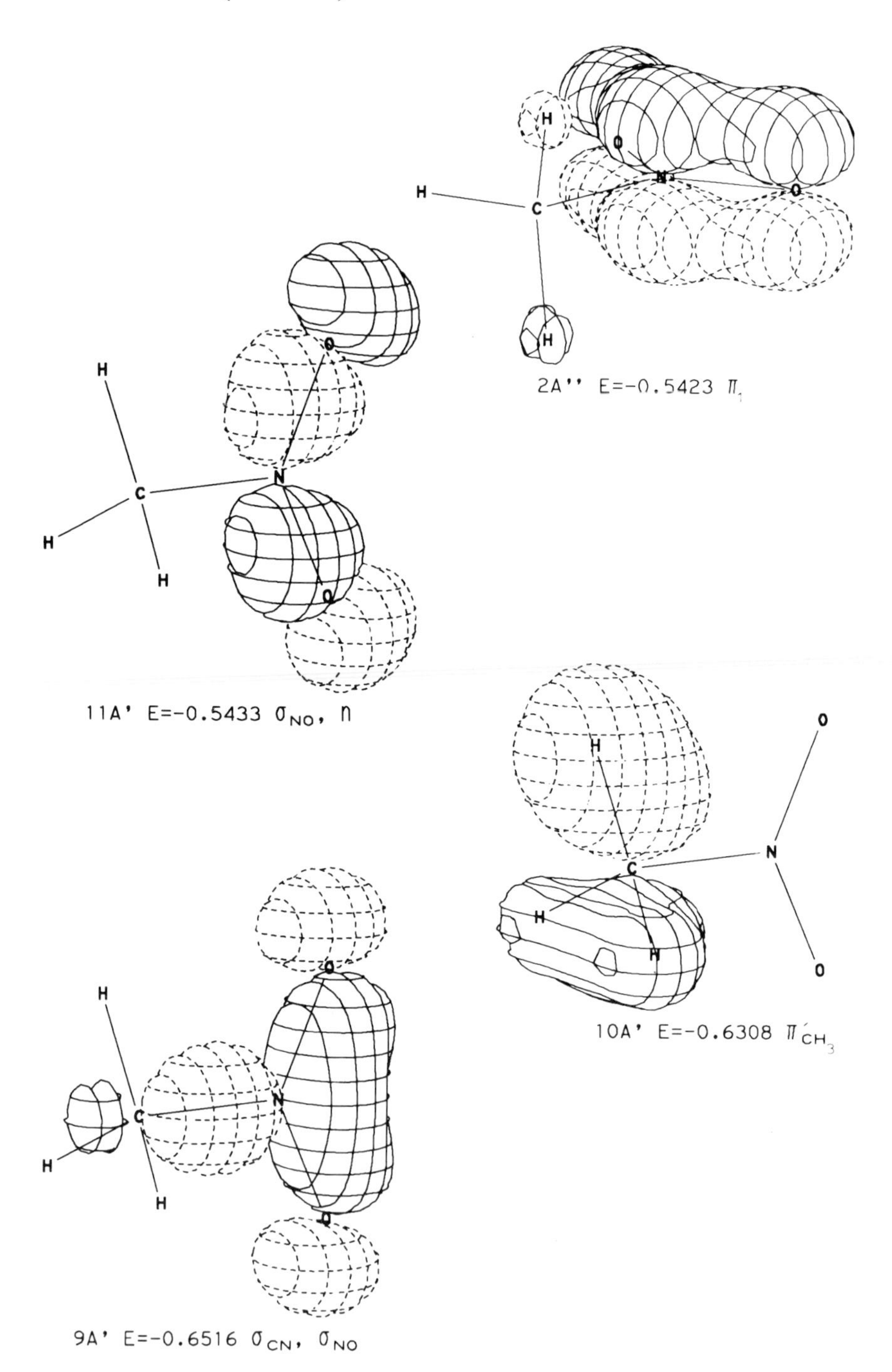

2A'' E=−0.5423 π_1

11A' E=−0.5433 σ_{NO}, n

10A' E=−0.6308 π'_{CH_3}

9A' E=−0.6516 σ_{CN}, σ_{NO}

Nitromethane (Continued)

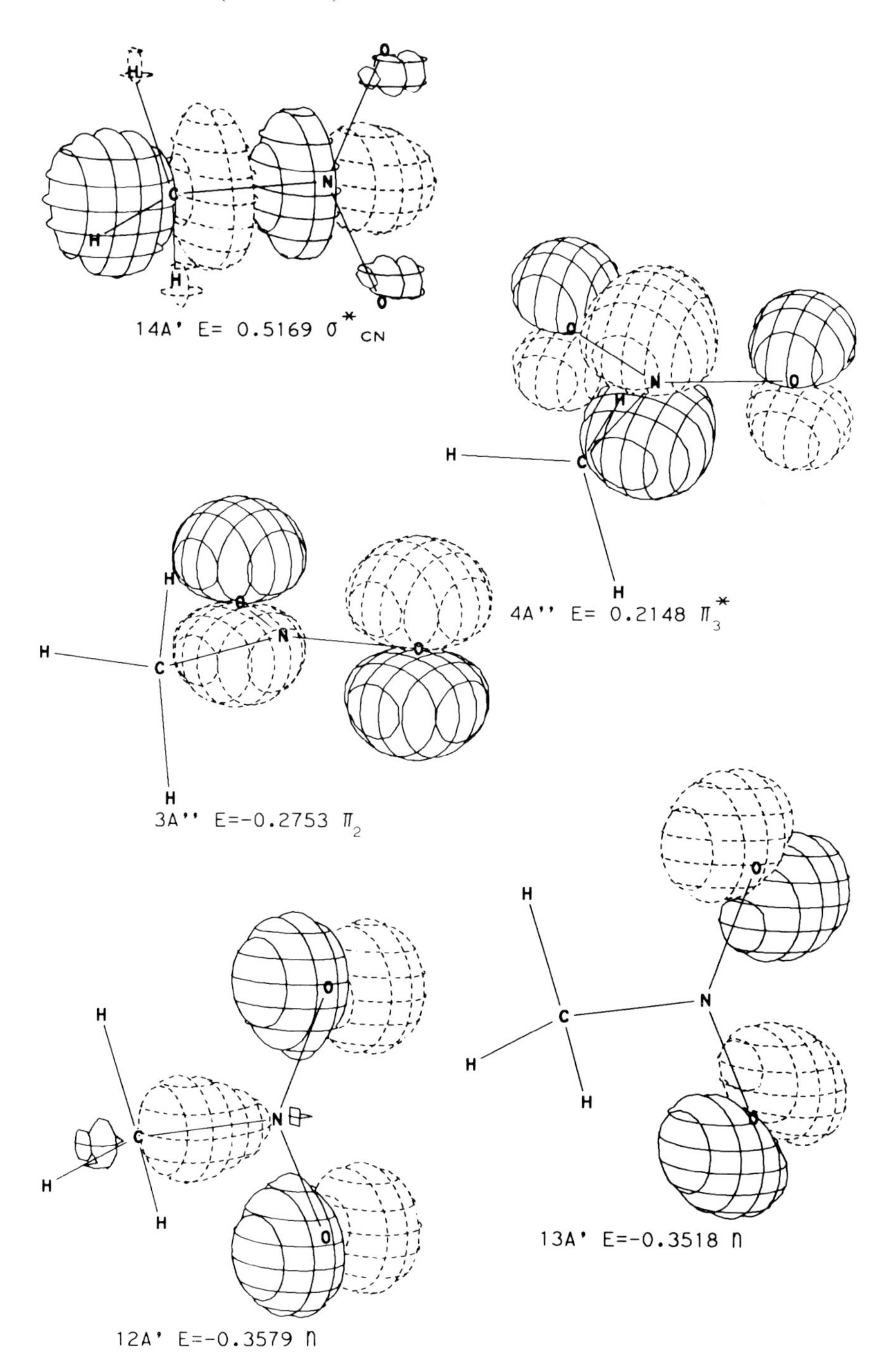

82. Cyclobutane, Planar

Symmetry: D_{4h}

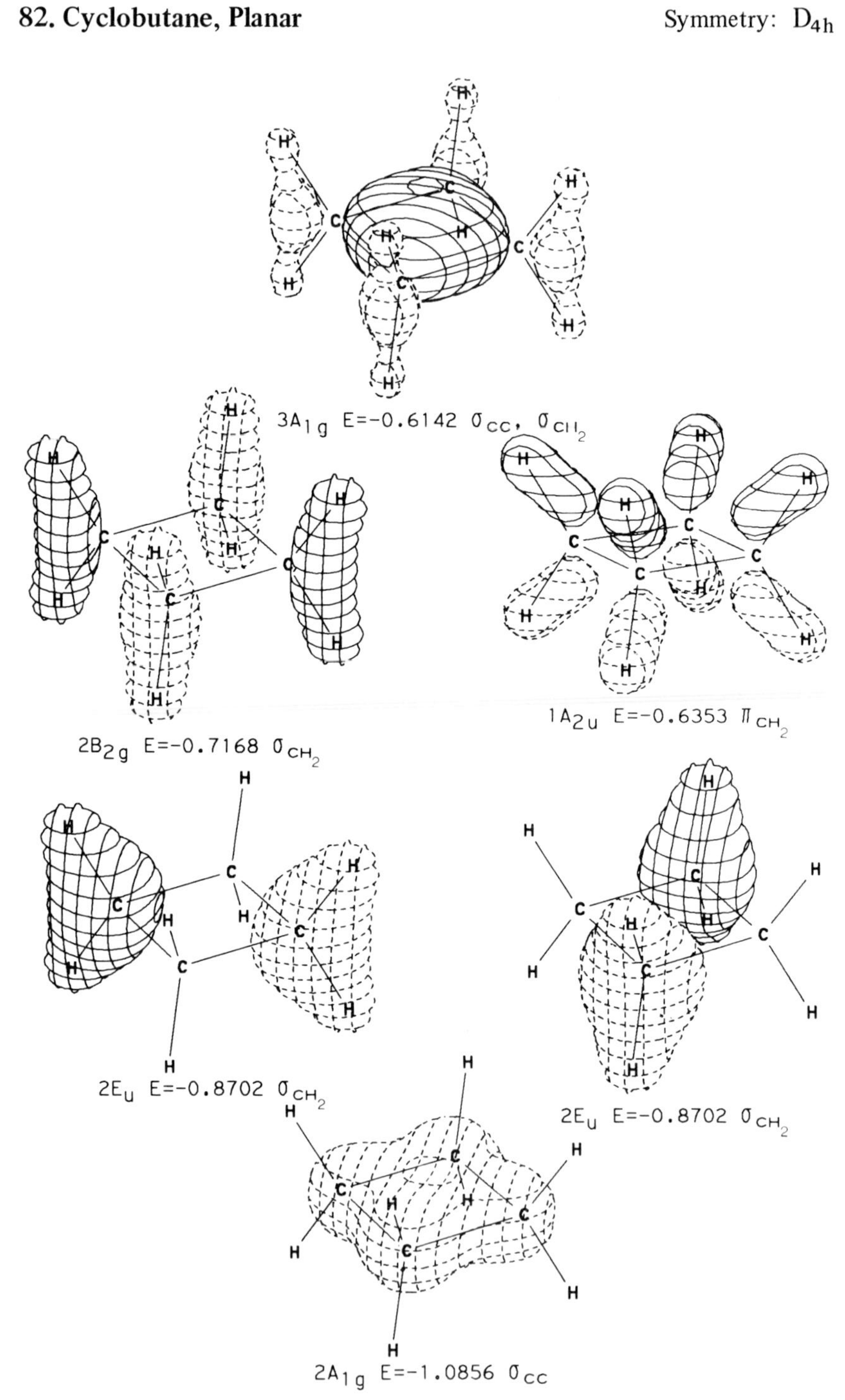

$3A_{1g}$ E=-0.6142 σ_{CC}, σ_{CH_2}

$2B_{2g}$ E=-0.7168 σ_{CH_2}

$1A_{2u}$ E=-0.6353 π_{CH_2}

$2E_u$ E=-0.8702 σ_{CH_2}

$2E_u$ E=-0.8702 σ_{CH_2}

$2A_{1g}$ E=-1.0856 σ_{CC}

Cyclobutane, Planar (Continued)

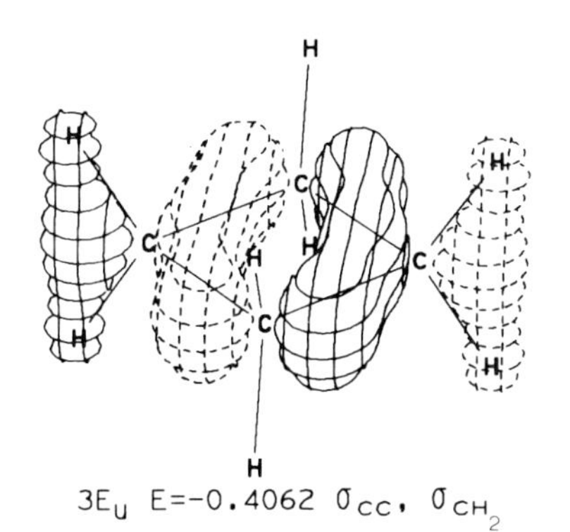

$3E_u$ E=-0.4062 σ_{CC}, σ_{CH_2}

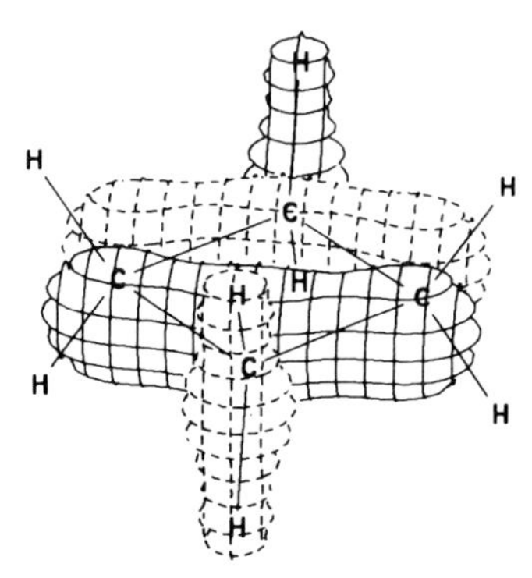

$3E_u$ E=-0.4062 σ_{CC}, σ_{CH_2}

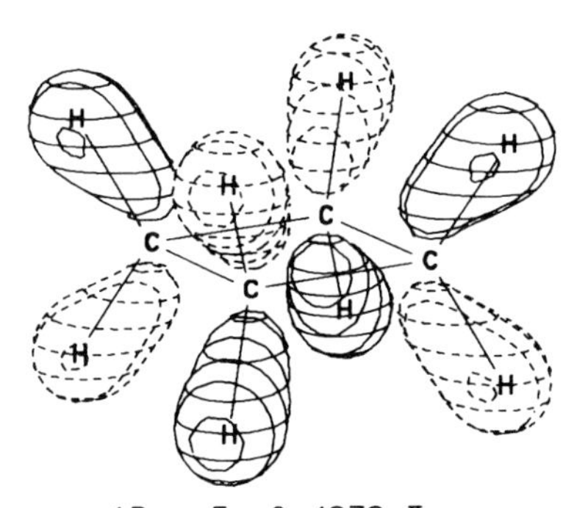

$1B_{1u}$ E=-0.4272 π_{CH_2}

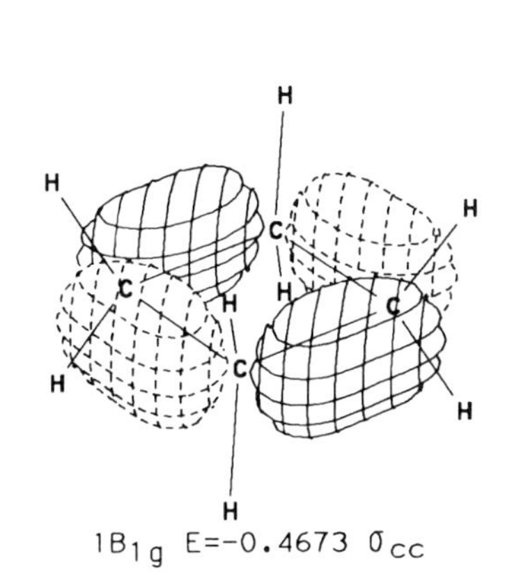

$1B_{1g}$ E=-0.4673 σ_{CC}

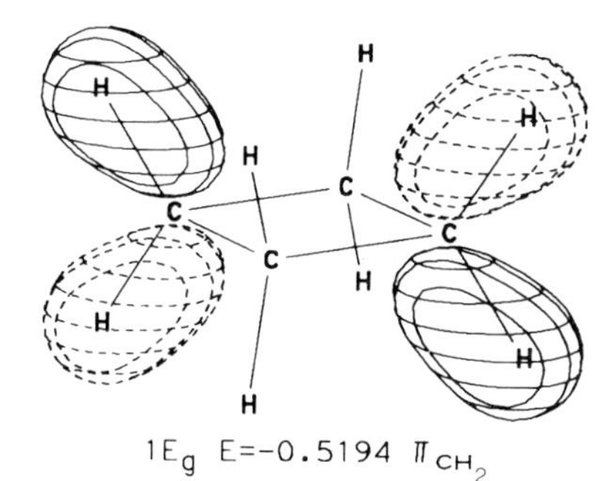

$1E_g$ E=-0.5194 π_{CH_2}

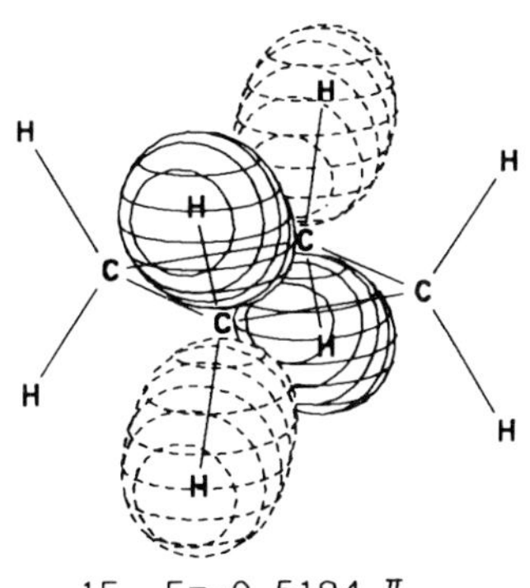

$1E_g$ E=-0.5194 π_{CH_2}

Cyclobutane, Planar (Continued)

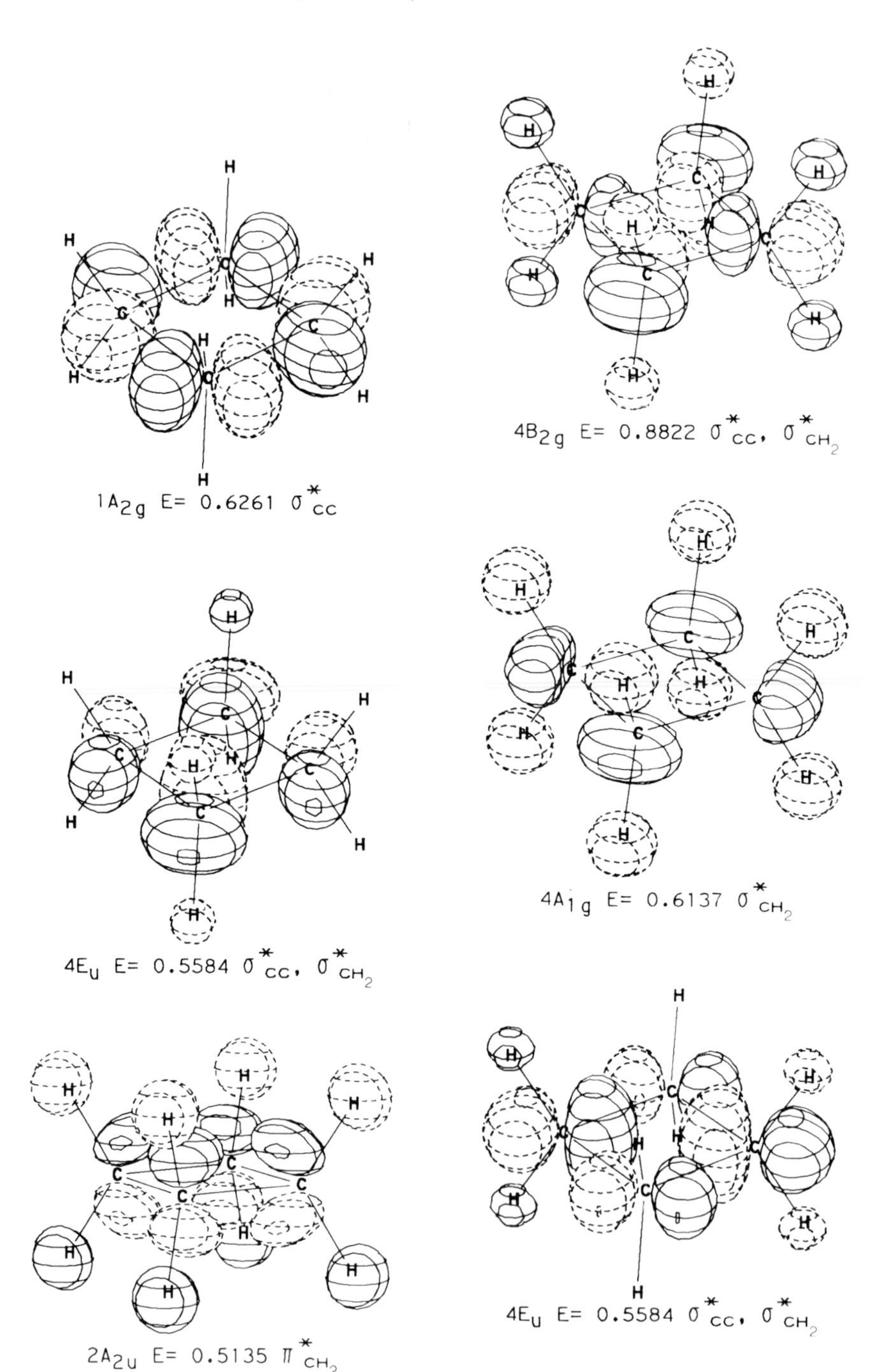

83. Cyclopentadiene

Symmetry: C_{2v}

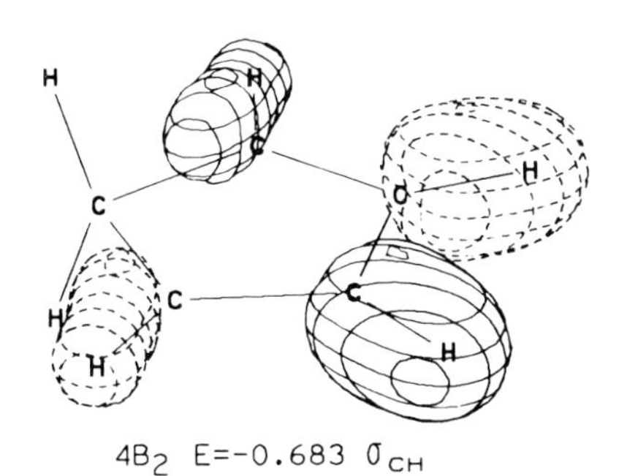

$4B_2$ E=-0.683 σ_{CH}

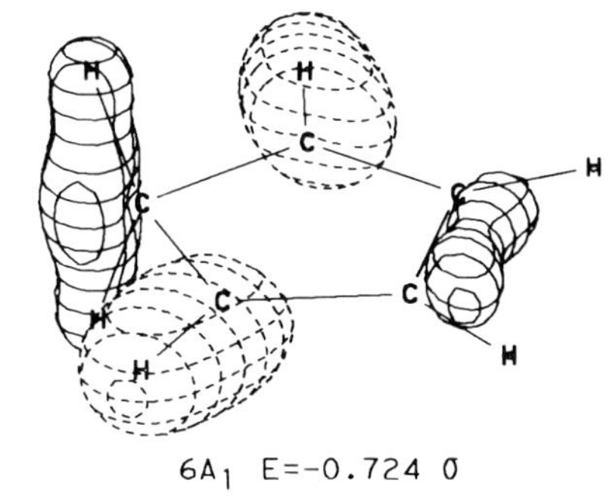

$6A_1$ E=-0.724 σ

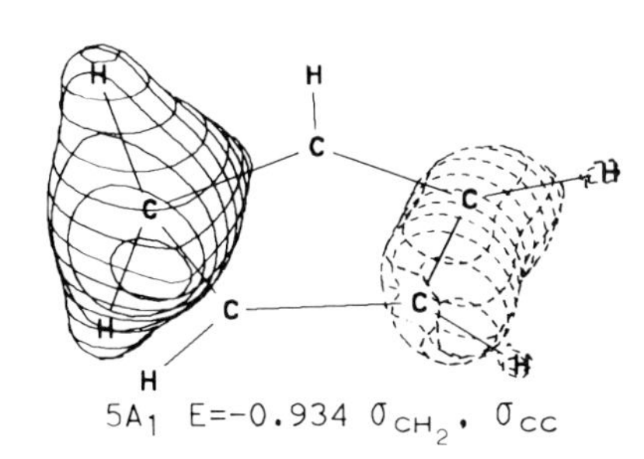

$5A_1$ E=-0.934 σ_{CH_2}, σ_{CC}

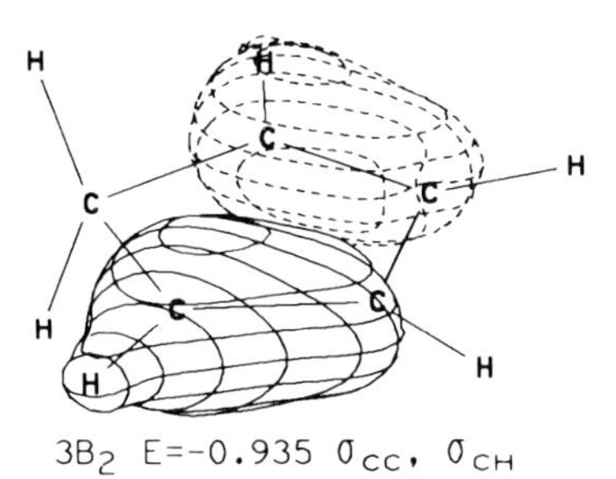

$3B_2$ E=-0.935 σ_{CC}, σ_{CH}

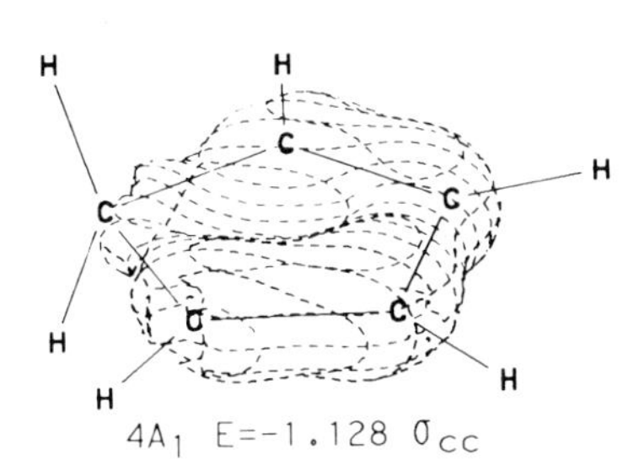

$4A_1$ E=-1.128 σ_{CC}

Cyclopentadiene (Continued)

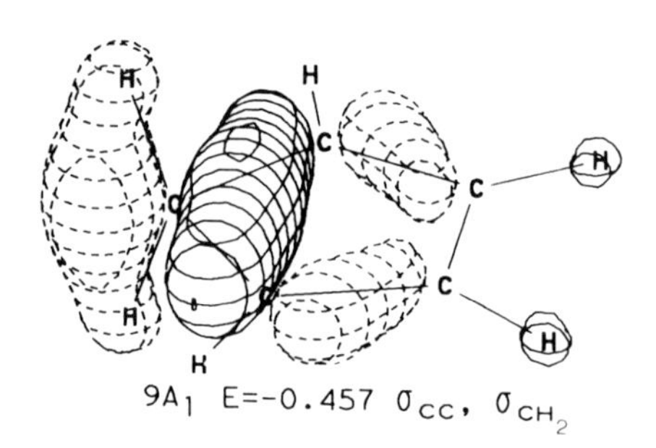

$9A_1$ E=−0.457 σ_{CC}, σ_{CH_2}

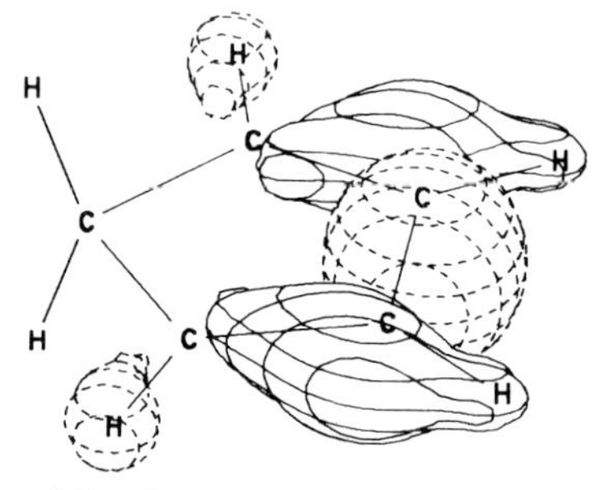

$8A_1$ E=−0.470 σ_{CC}, σ_{CH}

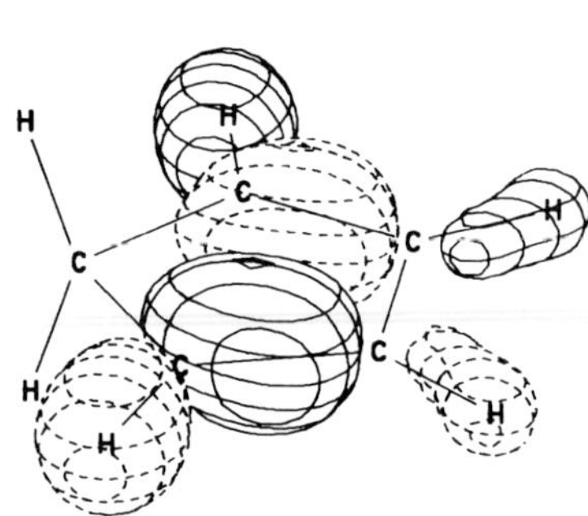

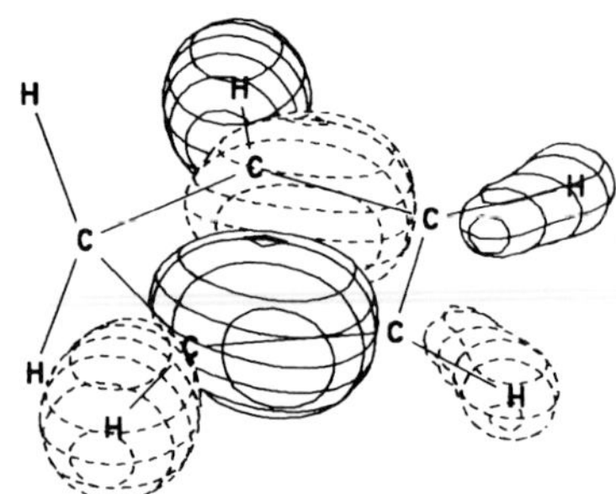

$5B_2$ E=−0.499 σ_{CC}, σ_{CH}

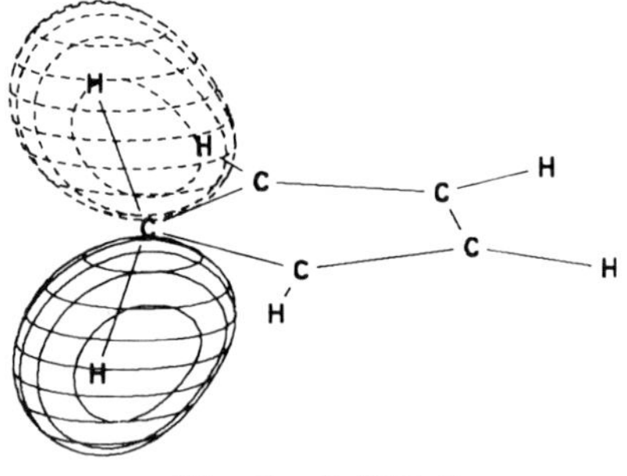

$1B_1$ E=−0.563 π_{CH_2}

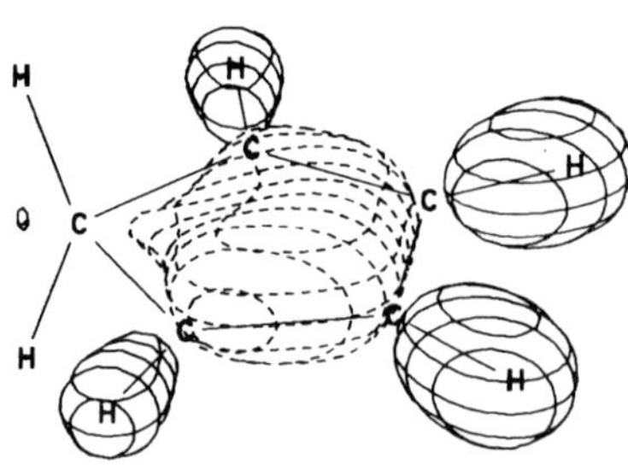

$7A_1$ E=−0.651 σ_{CC}, σ_{CH}

Cyclopentadiene (Continued)

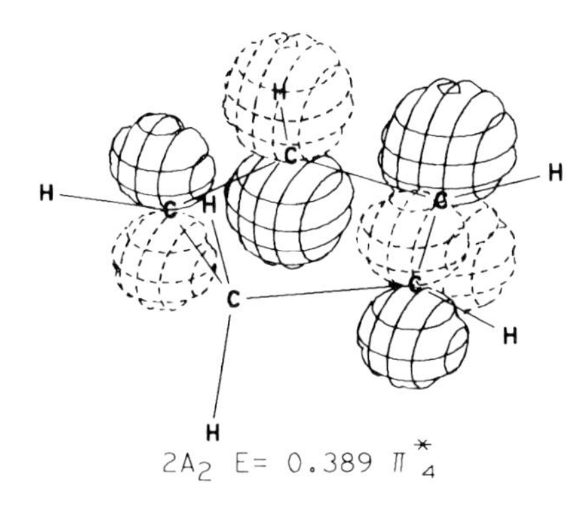

$2A_2$ E= 0.389 π^*_4

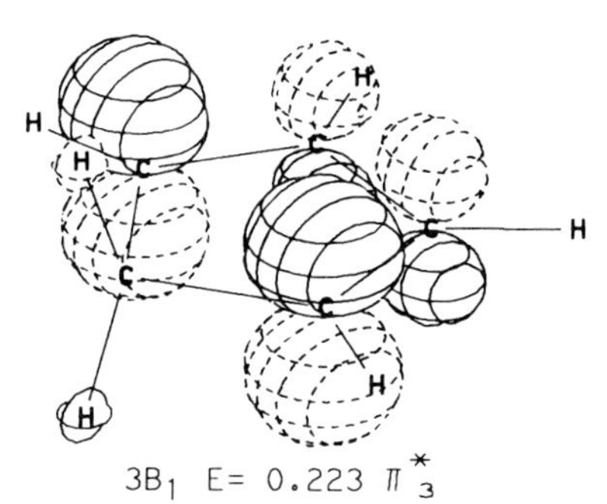

$3B_1$ E= 0.223 π^*_3

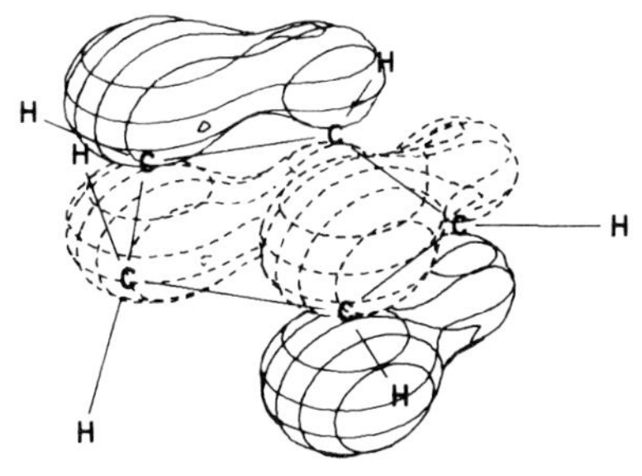

$1A_2$ E=-0.259 π_2

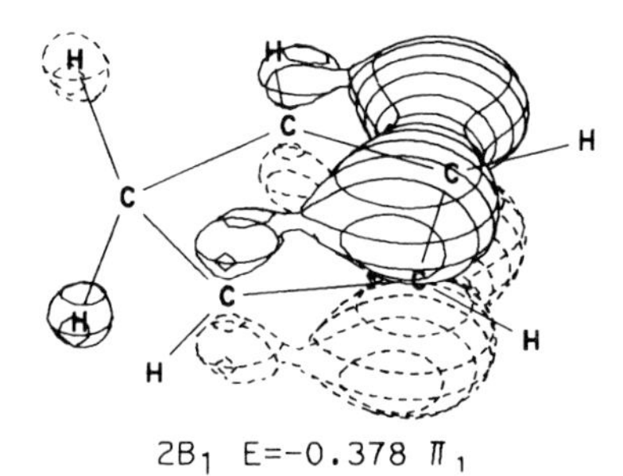

$2B_1$ E=-0.378 π_1

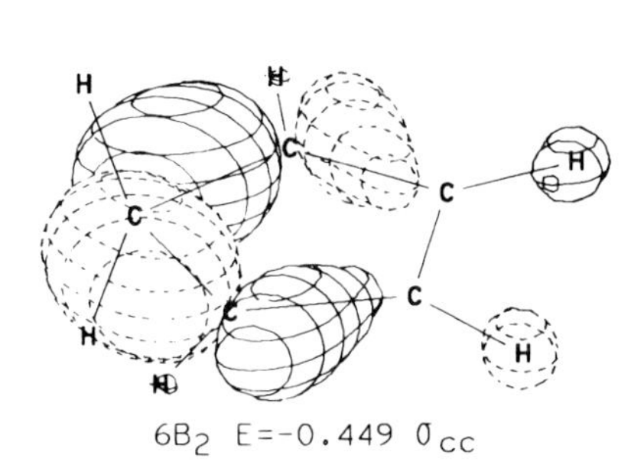

$6B_2$ E=-0.449 σ_{CC}

84. (2.1.0)-Bicyclopentene-2

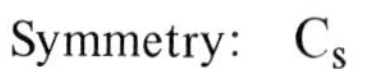
Symmetry: C_s

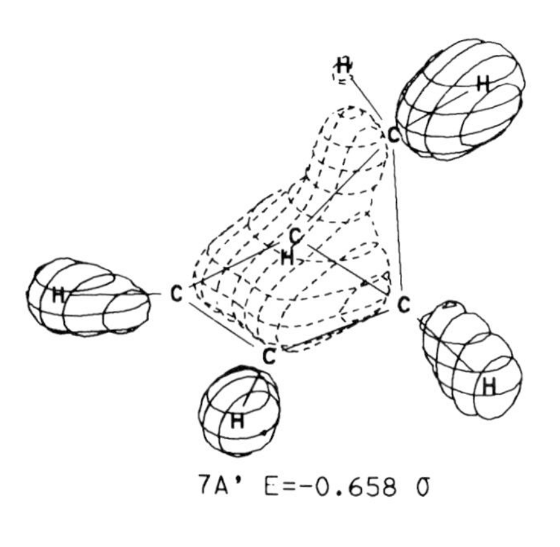

7A' E=-0.658 σ

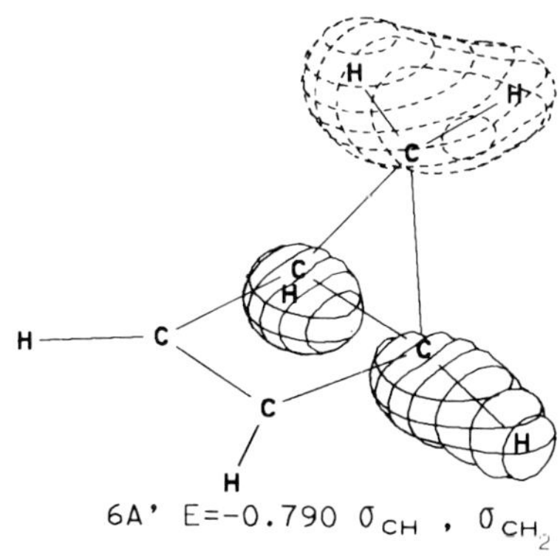

6A' E=-0.790 σ_{CH} , σ_{CH_2}

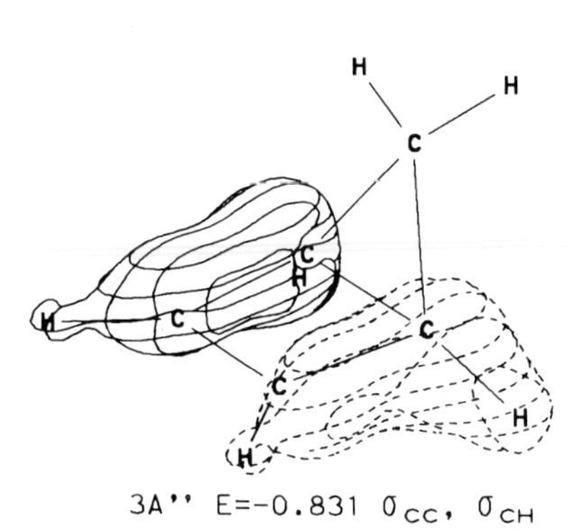

3A'' E=-0.831 σ_{CC}, σ_{CH}

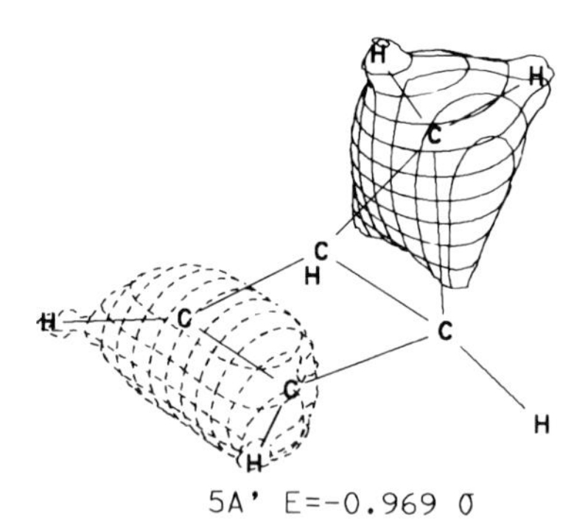

5A' E=-0.969 σ

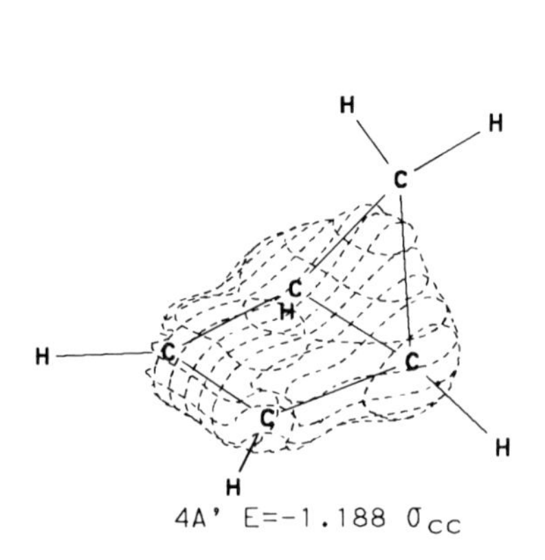

4A' E=-1.188 σ_{CC}

(2.1.0)-Bicyclopentene-2 (Continued)

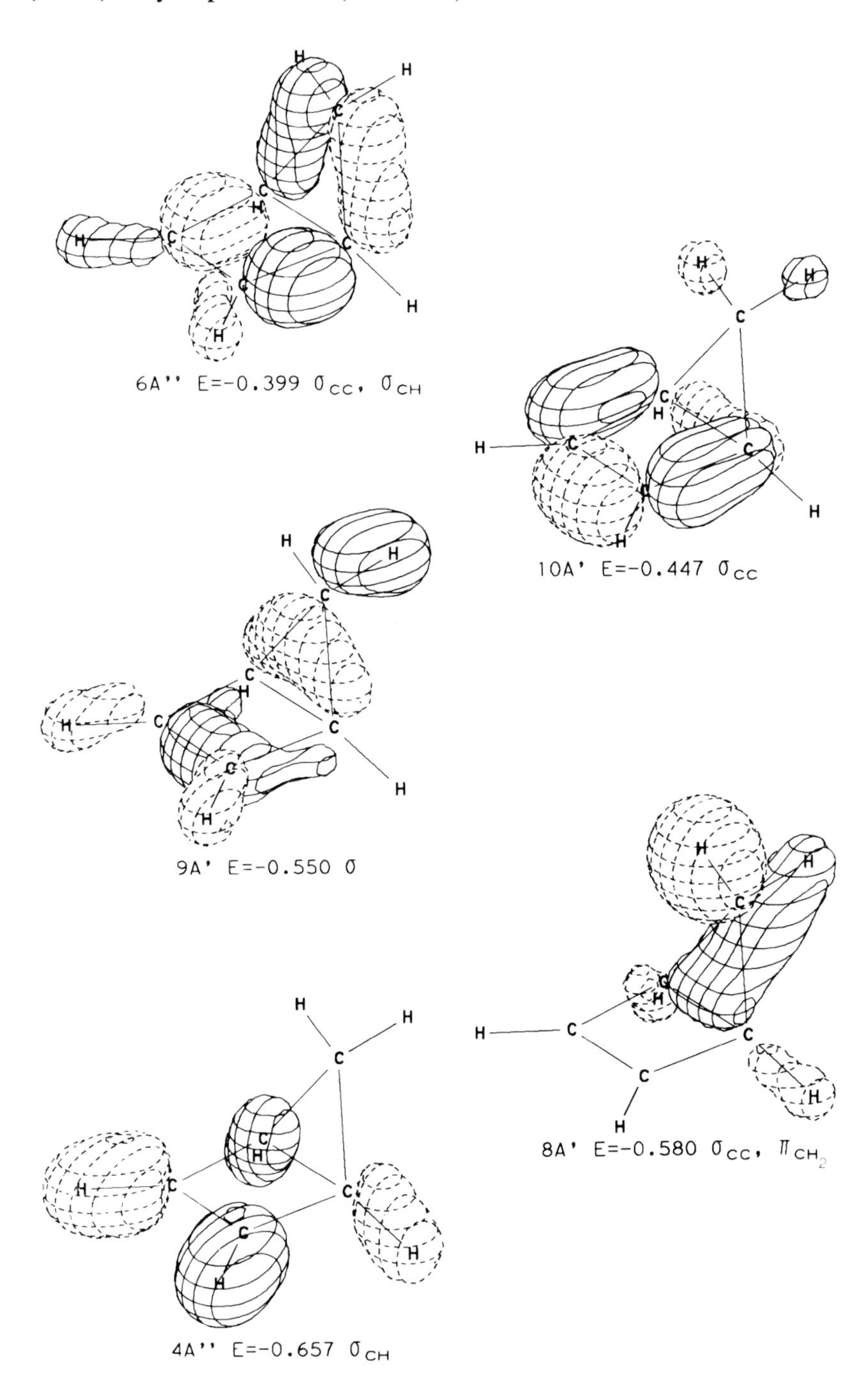

6A'' E=−0.399 σ_{CC}, σ_{CH}

10A' E=−0.447 σ_{CC}

9A' E=−0.550 σ

8A' E=−0.580 σ_{CC}, π_{CH_2}

4A'' E=−0.657 σ_{CH}

(2.1.0)-Bicyclopentene-2 (Continued)

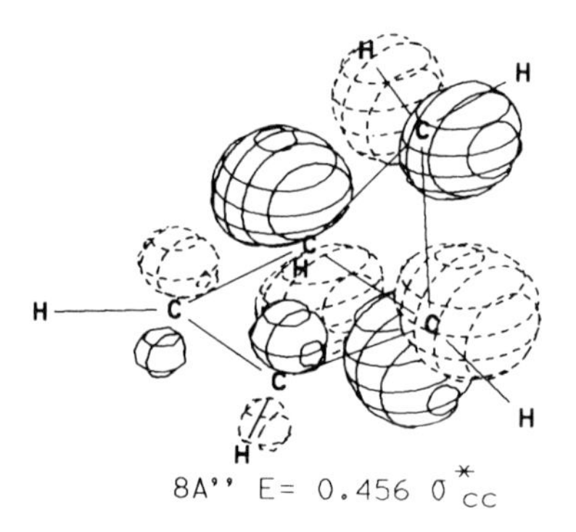

8A'' E= 0.456 σ^{*}_{CC}

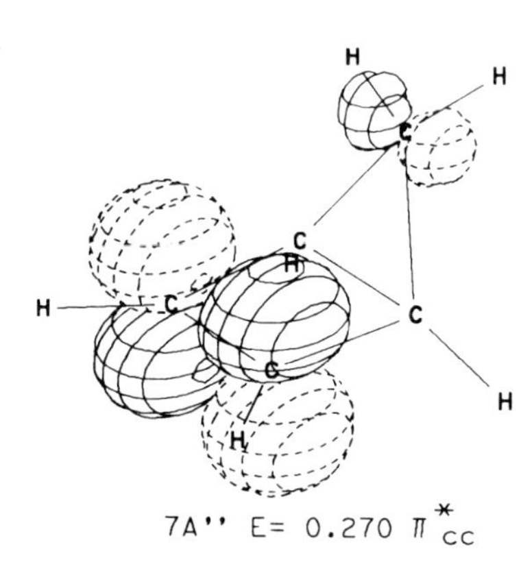

7A'' E= 0.270 π^{*}_{CC}

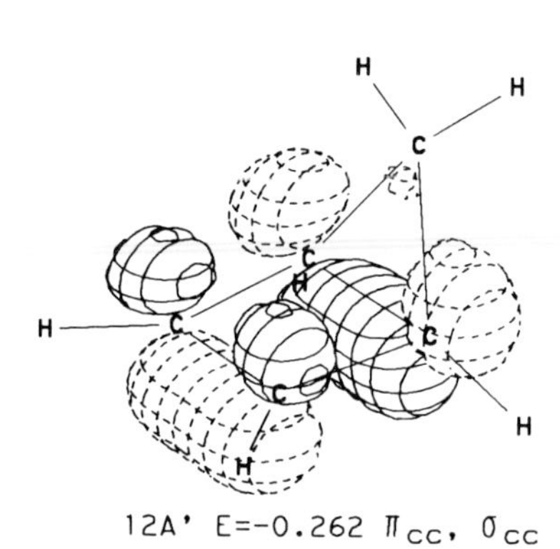

12A' E=-0.262 π_{CC}, σ_{CC}

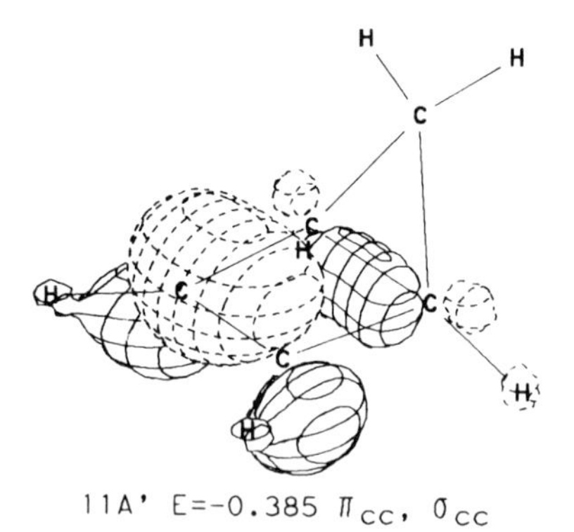

11A' E=-0.385 π_{CC}, σ_{CC}

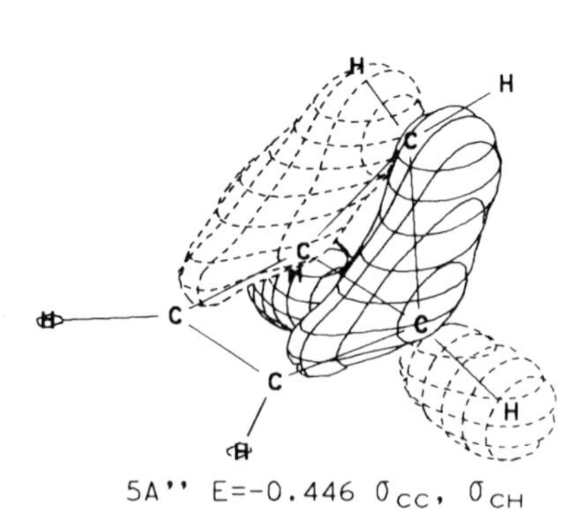

5A'' E=-0.446 σ_{CC}, σ_{CH}

85. Pyrrole

Symmetry: C_{2v}

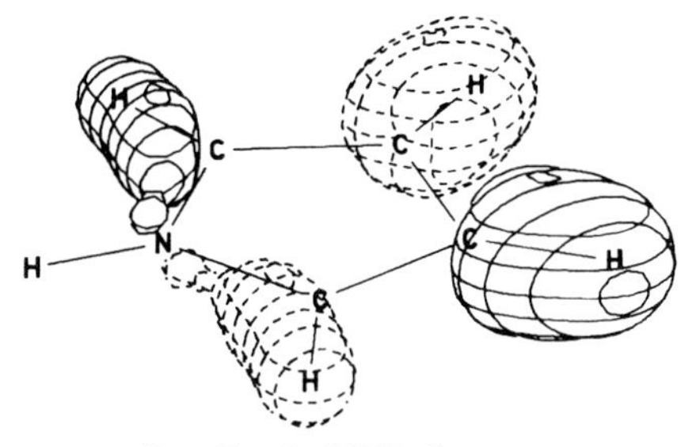

$4B_2$ E=-0.7970 σ_{CH}

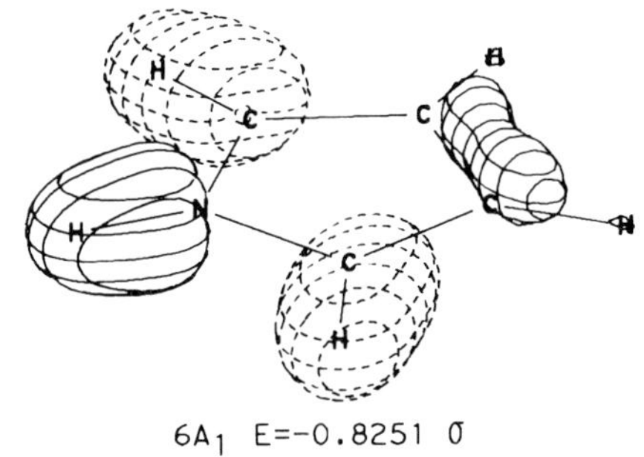

$6A_1$ E=-0.8251 σ

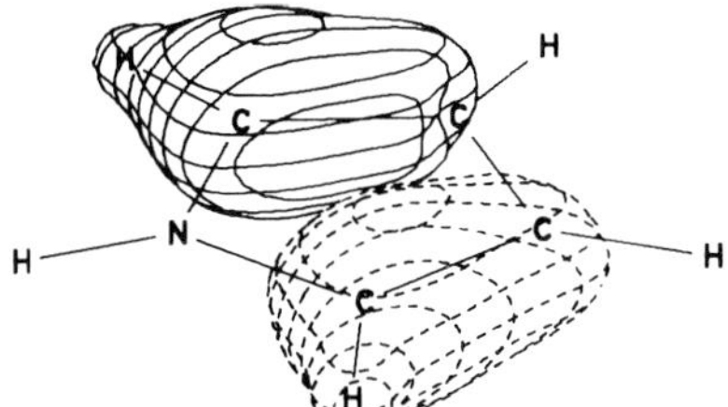

$3B_2$ E=-1.0345 σ_{CC}, σ_{CH}

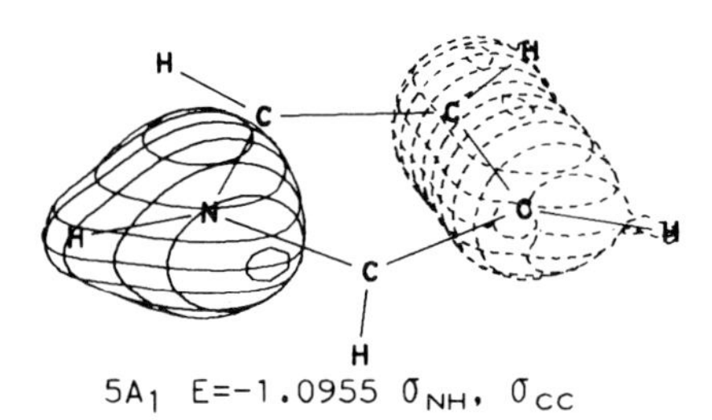

$5A_1$ E=-1.0955 σ_{NH}, σ_{CC}

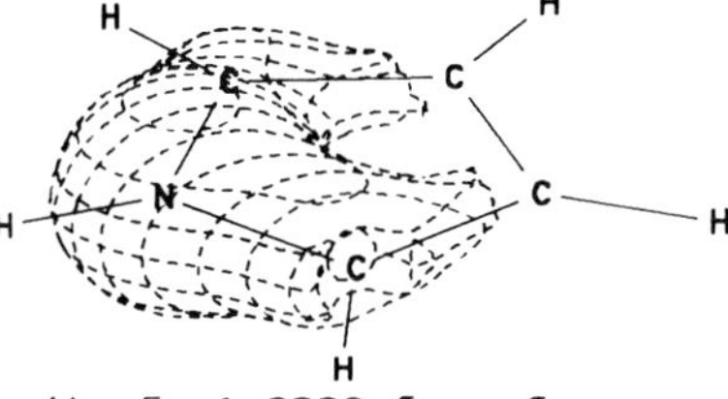

$4A_1$ E=-1.3239 σ_{NC}, σ_{CC}

Pyrrole (Continued)

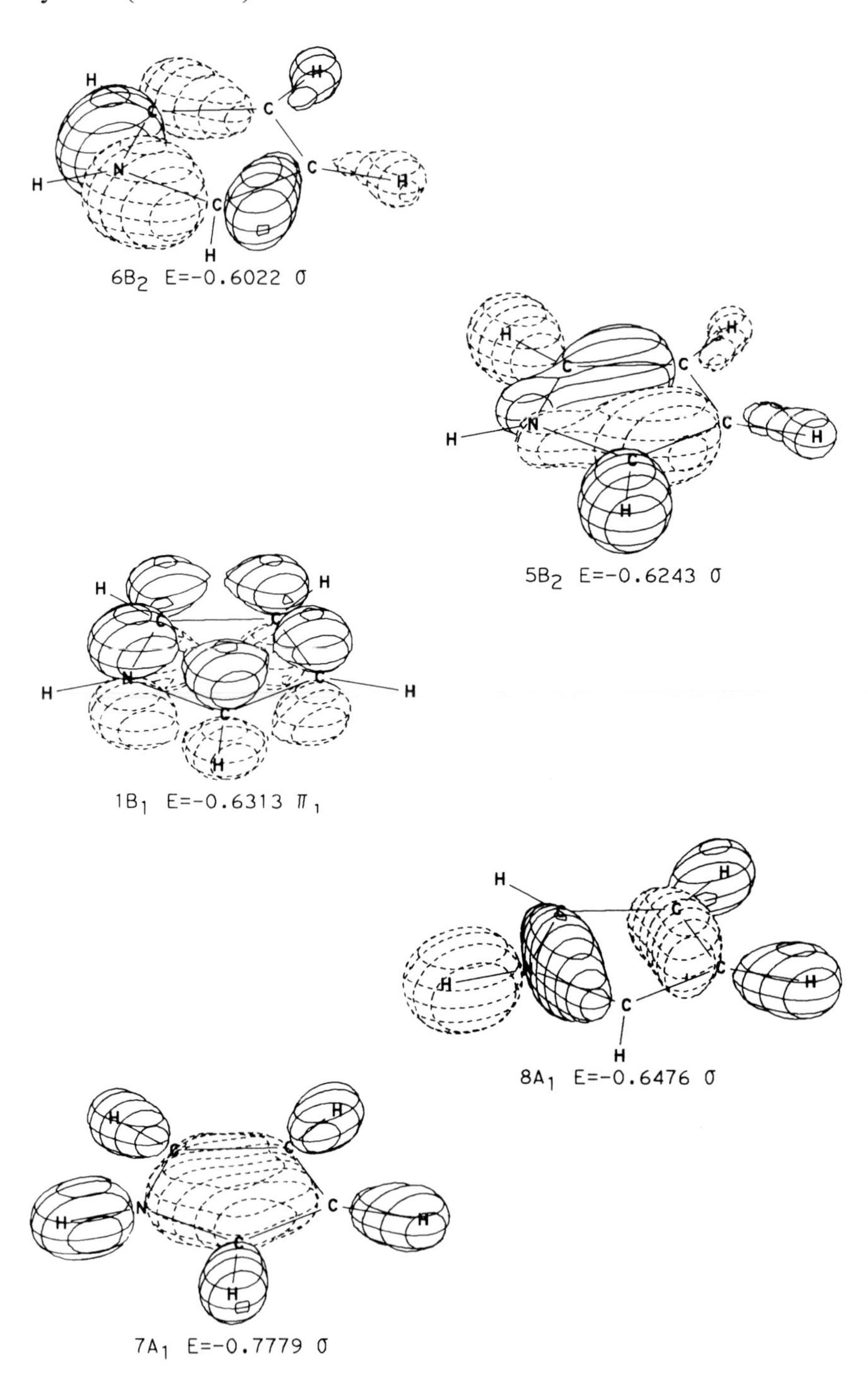

Pyrrole (Continued)

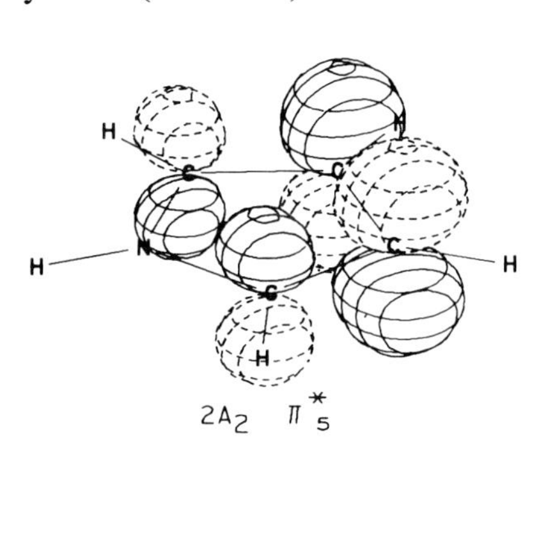

$2A_2$ π^*_5

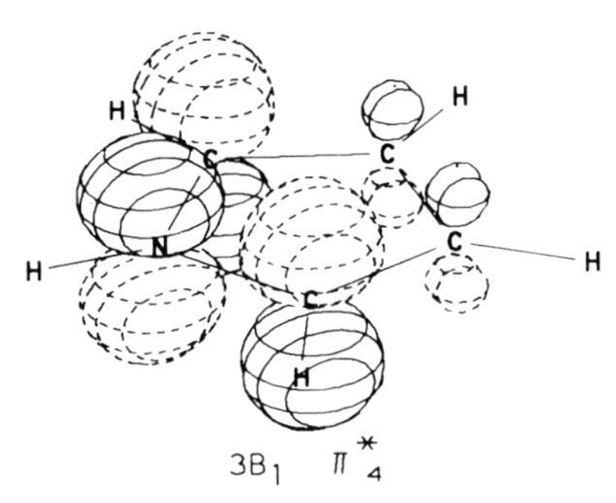

$3B_1$ π^*_4

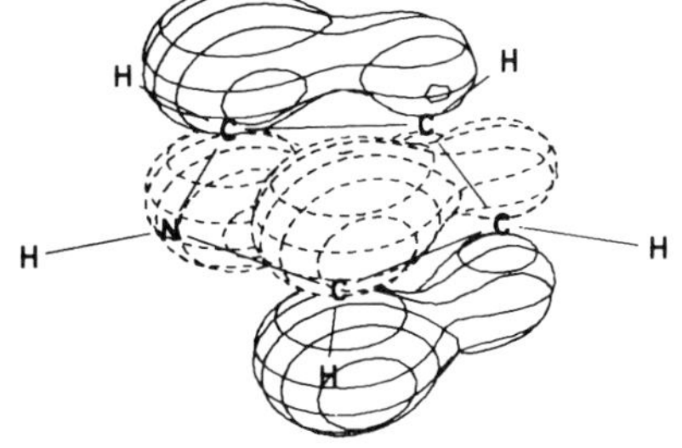

$1A_2$ E=-0.3879 π_3

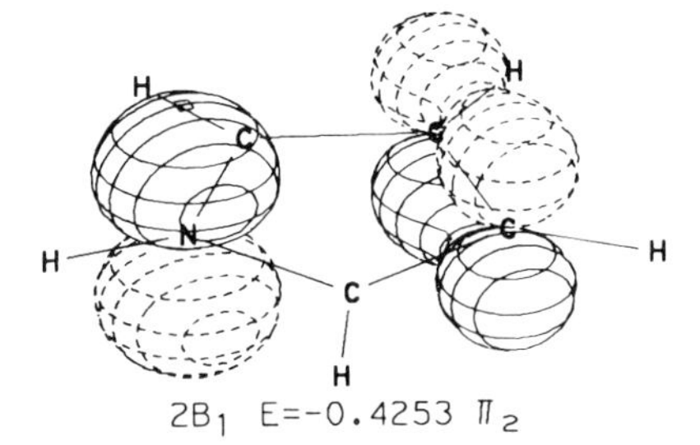

$2B_1$ E=-0.4253 π_2

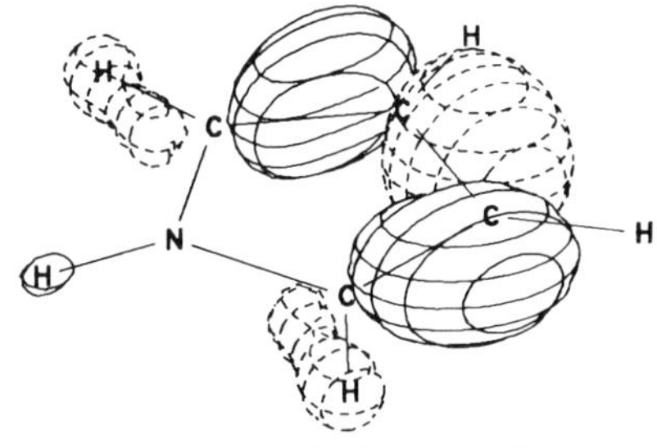

$9A_1$ E=-0.5766 σ_{CC}, σ_{CH}

86. Furan

Symmetry: C_{2v}

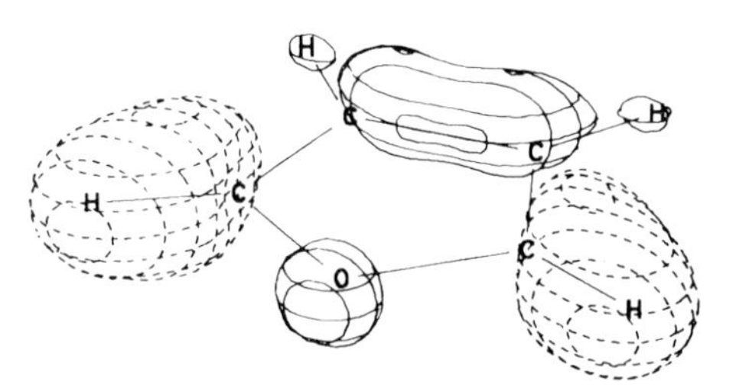

$6A_1$ E=-0.7846 σ_{CH}, σ_{CC}

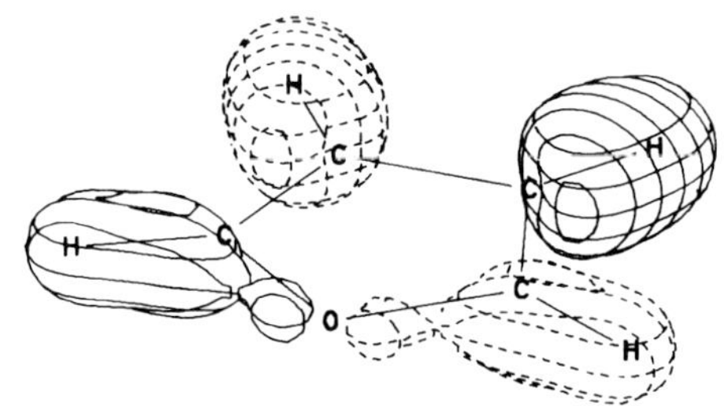

$4B_2$ E=-0.8115 σ_{CH}, σ_{CO}

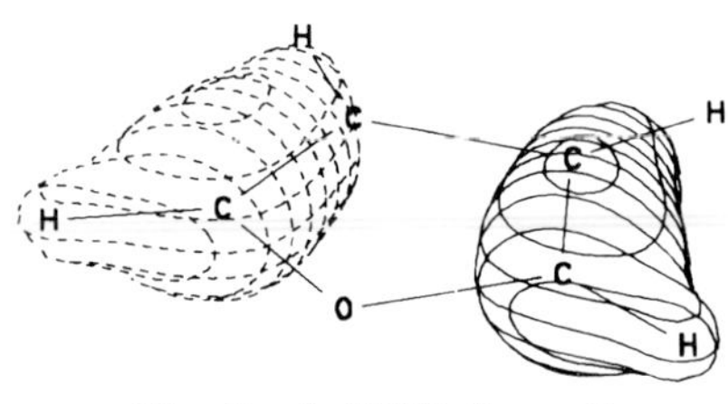

$3B_2$ E=-1.0198 σ_{CC}, σ_{CH}

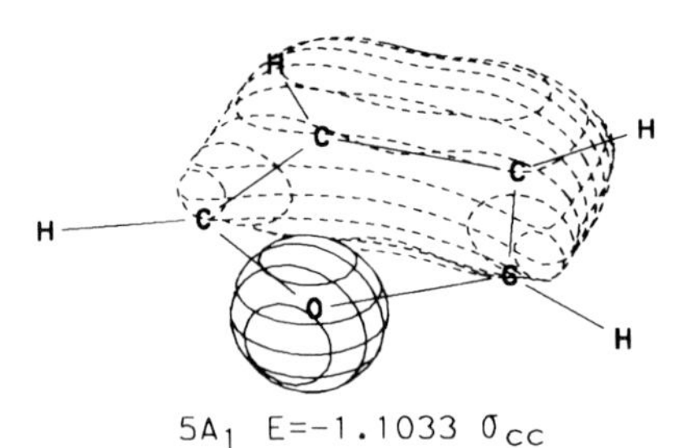

$5A_1$ E=-1.1033 σ_{CC}

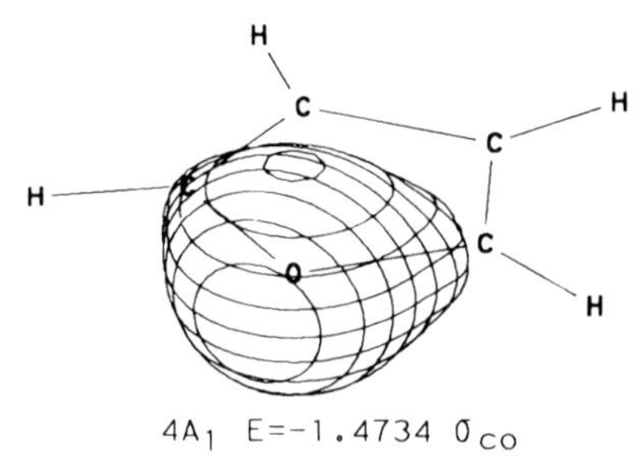

$4A_1$ E=-1.4734 σ_{CO}

Furan (Continued)

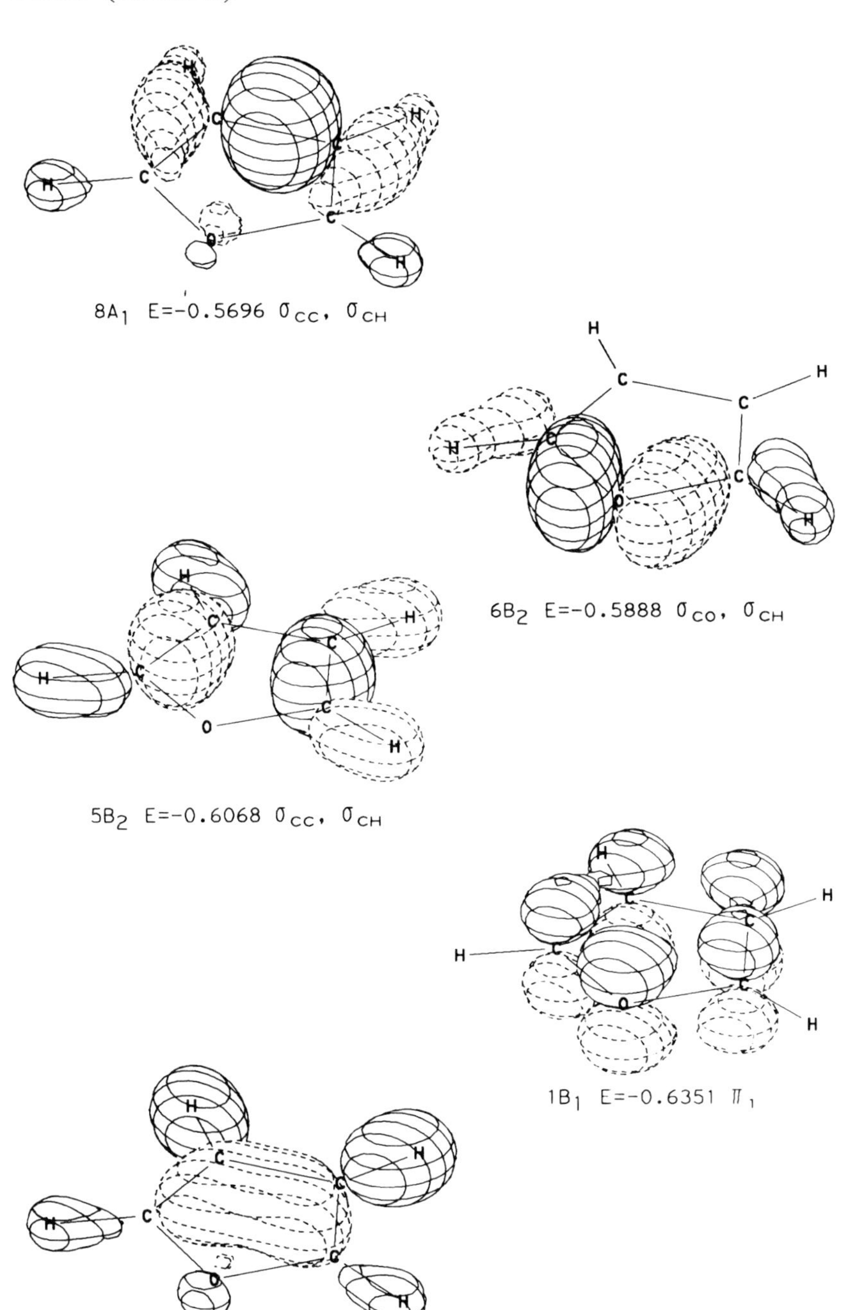

$8A_1$ E=−0.5696 σ_{CC}, σ_{CH}

$6B_2$ E=−0.5888 σ_{CO}, σ_{CH}

$5B_2$ E=−0.6068 σ_{CC}, σ_{CH}

$1B_1$ E=−0.6351 π_1

$7A_1$ E=−0.7446 σ_{CC}, σ_{CH}

Furan (Continued)

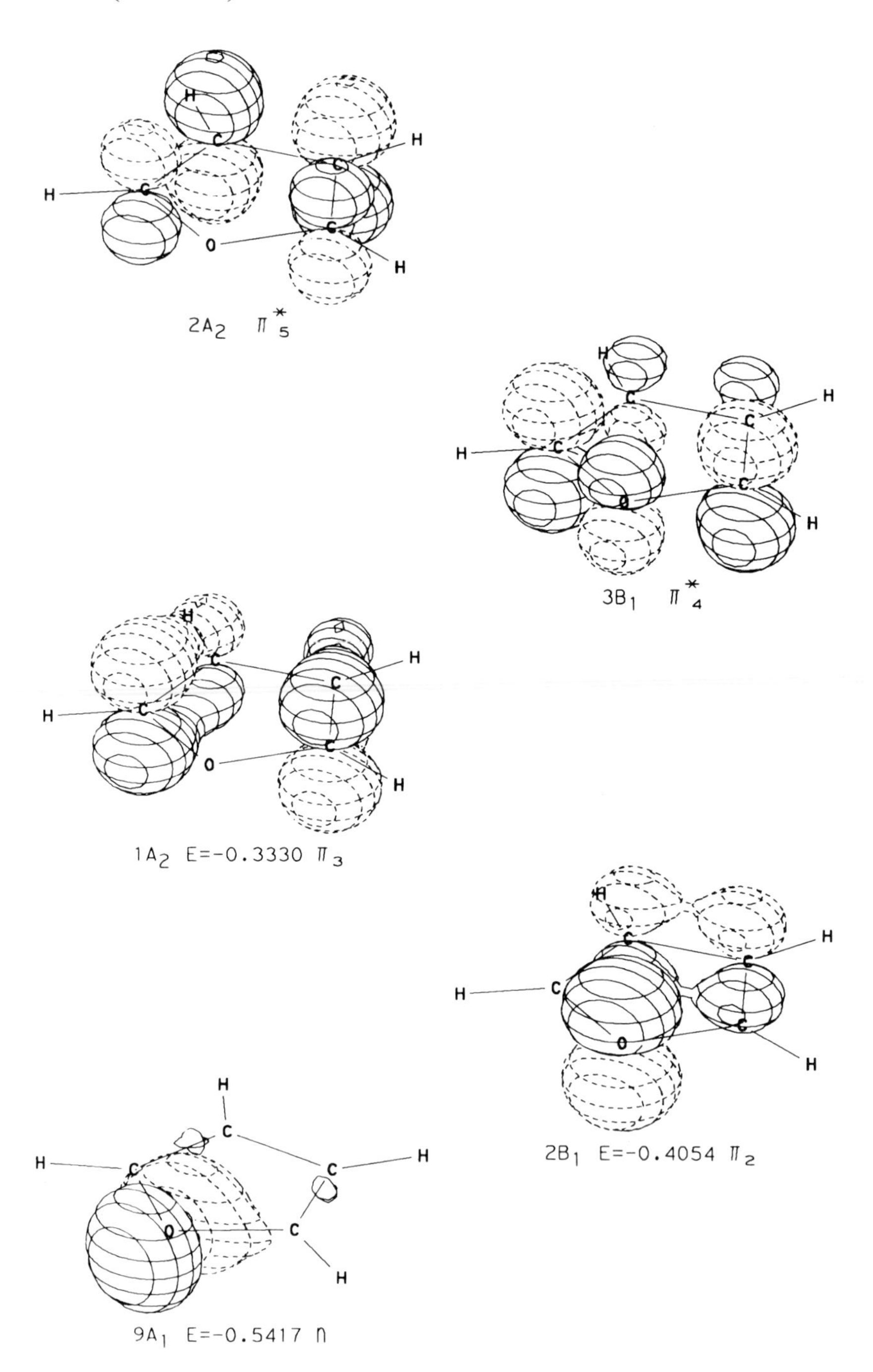

87. Cyclopentadienyl Anion

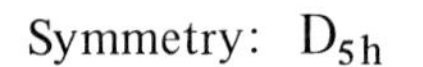
Symmetry: D_{5h}

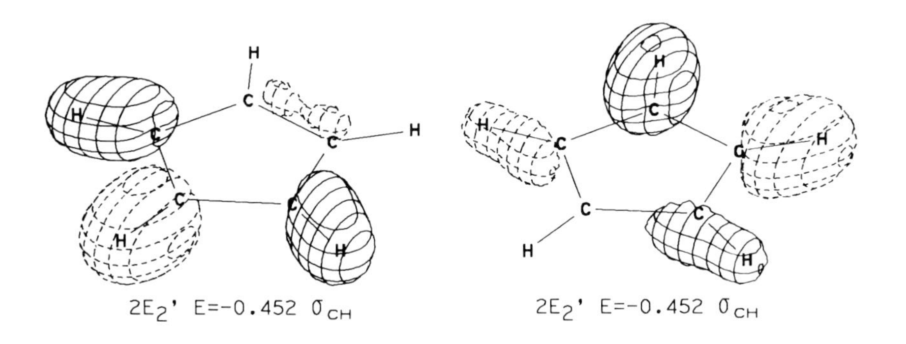

$2E_2'$ E=-0.452 σ_{CH} $2E_2'$ E=-0.452 σ_{CH}

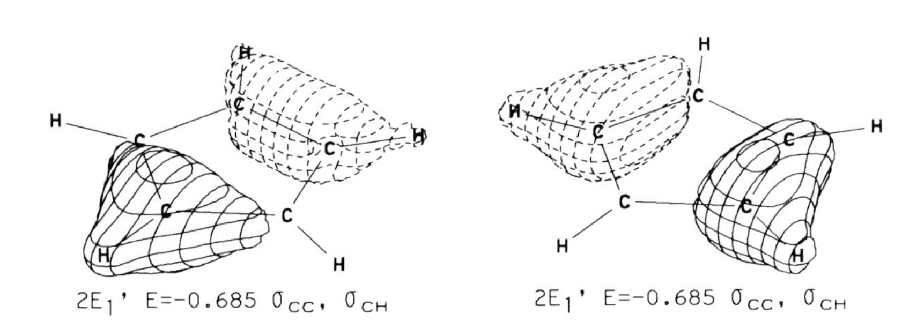

$2E_1'$ E=-0.685 σ_{CC}, σ_{CH} $2E_1'$ E=-0.685 σ_{CC}, σ_{CH}

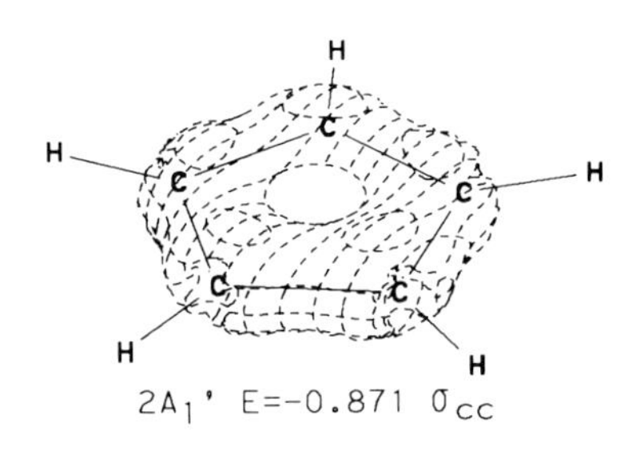

$2A_1'$ E=-0.871 σ_{CC}

Cyclopentadienyl Anion (Continued)

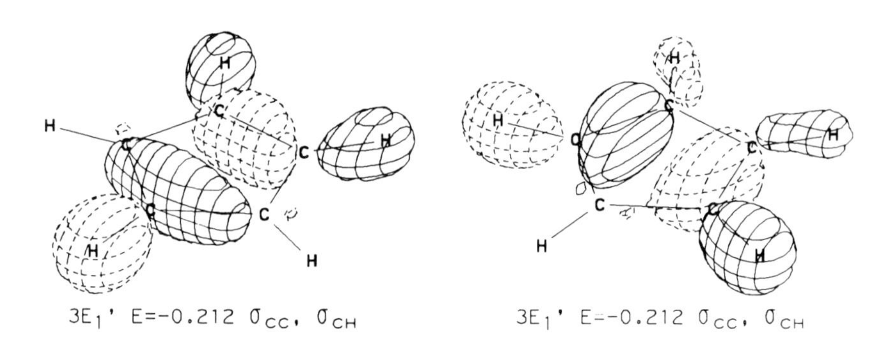

$3E_1'$ E=-0.212 σ_{CC}, σ_{CH} $3E_1'$ E=-0.212 σ_{CC}, σ_{CH}

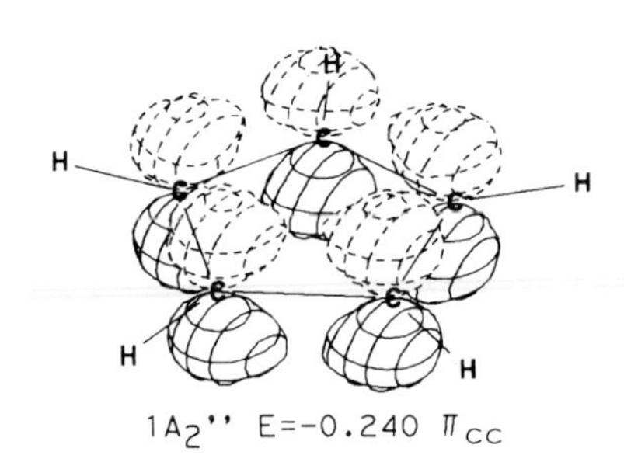

$1A_2''$ E=-0.240 π_{CC}

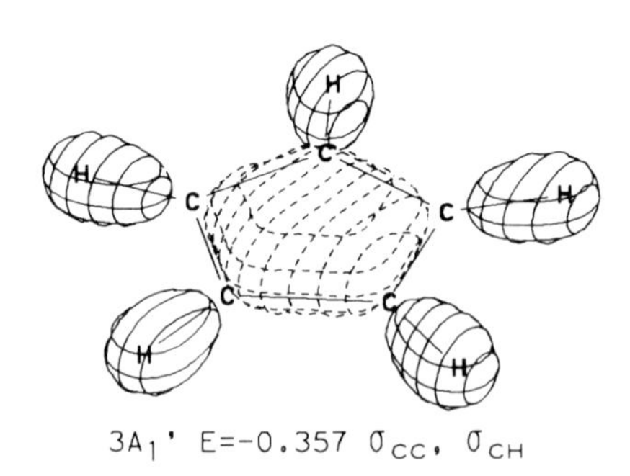

$3A_1'$ E=-0.357 σ_{CC}, σ_{CH}

Cyclopentadienyl Anion (Continued)

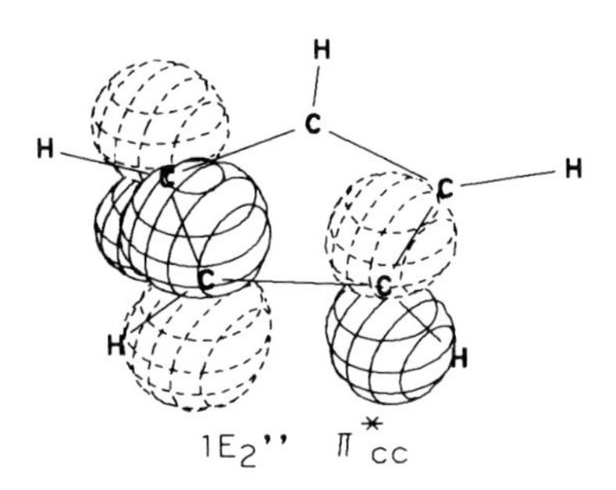

$1E_2''$ π^*_{CC}

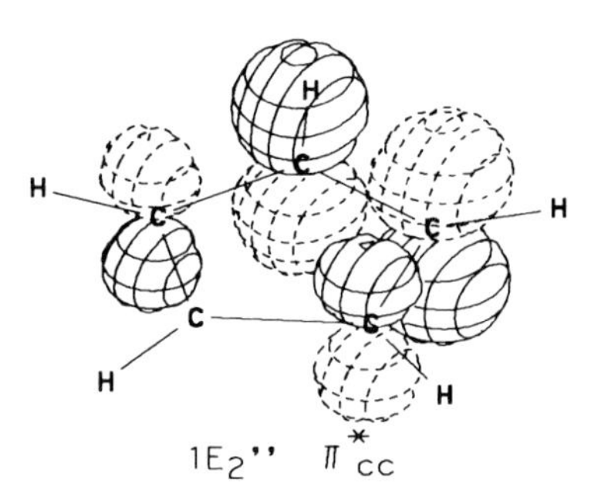

$1E_2''$ π^*_{CC}

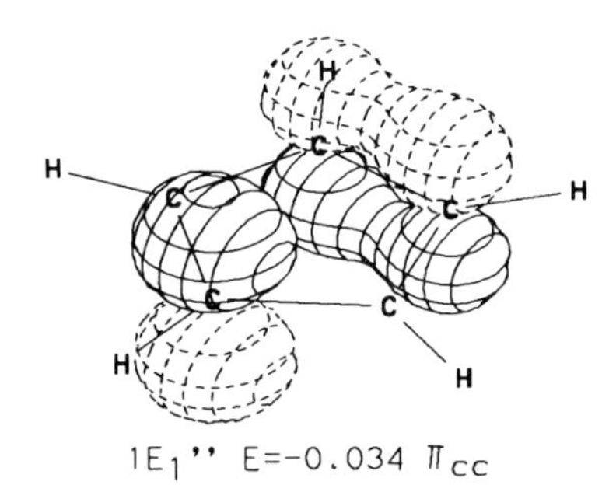

$1E_1''$ E=-0.034 π_{CC}

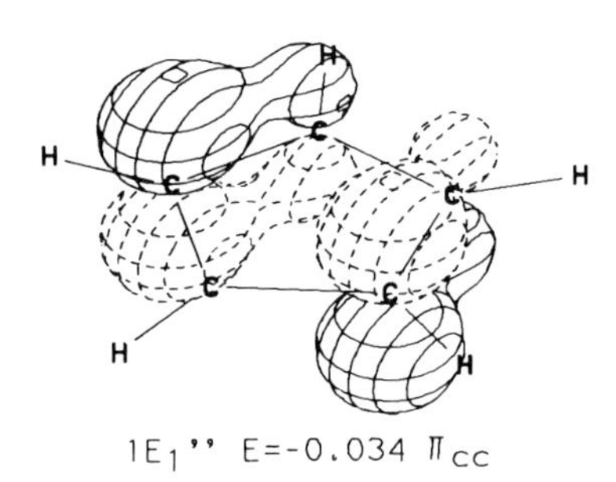

$1E_1''$ E=-0.034 π_{CC}

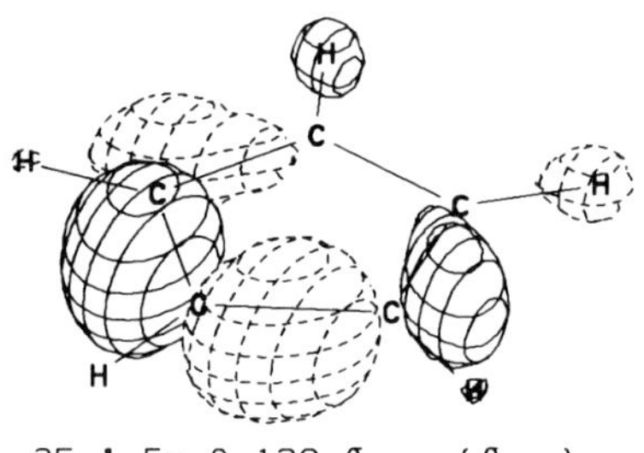

$3E_2'$ E=-0.129 σ_{CC}, (σ_{CH})

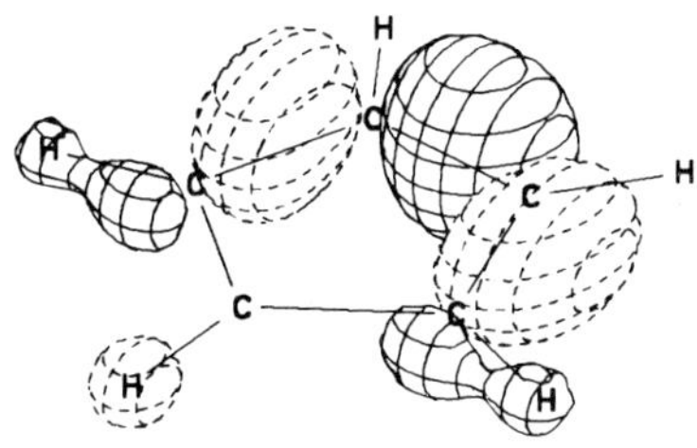

$3E_2'$ E=-0.129 σ_{CC}, (σ_{CH})

88. Pentadienyl Radical (π orbitals only) Symmetry: C_{2v}

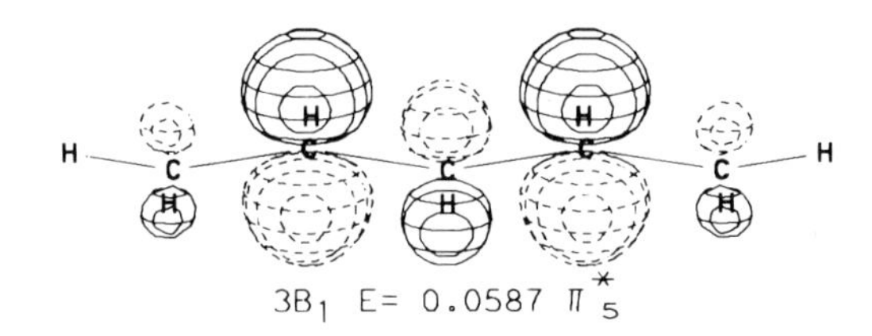

$3B_1$ E= 0.0587 π^*_5

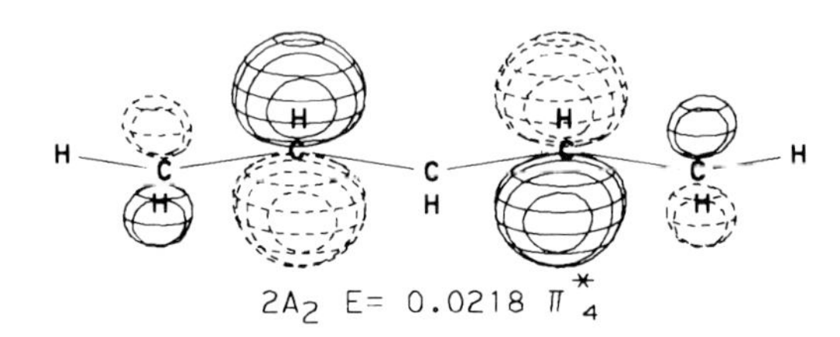

$2A_2$ E= 0.0218 π^*_4

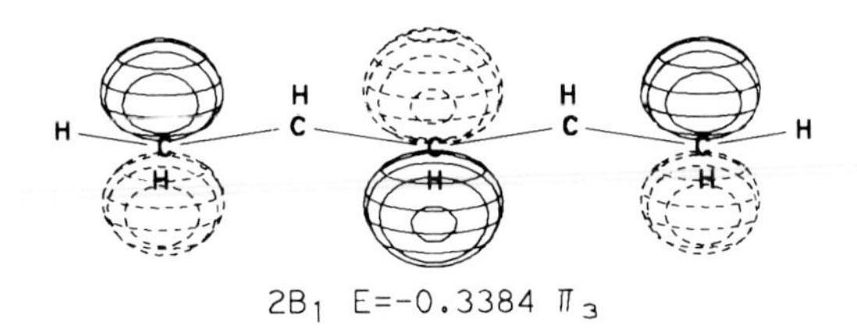

$2B_1$ E=-0.3384 π_3

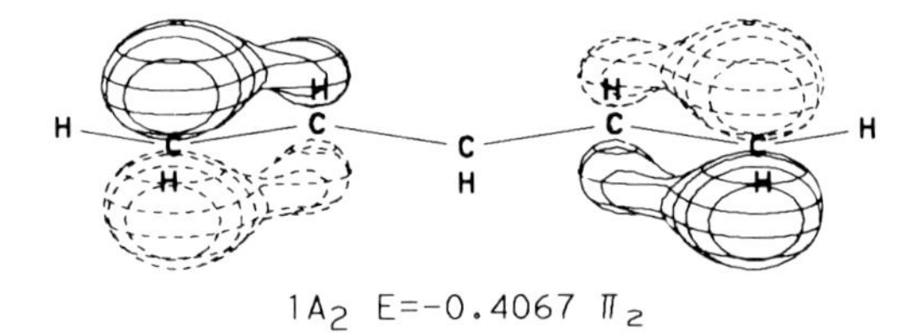

$1A_2$ E=-0.4067 π_2

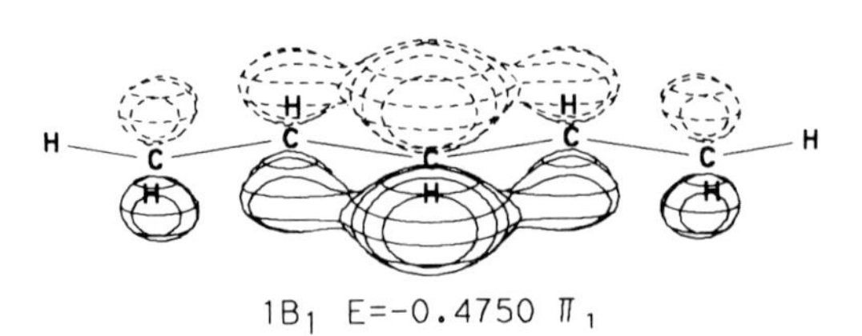

$1B_1$ E=-0.4750 π_1

89. Cyclopentene

Symmetry: C_s

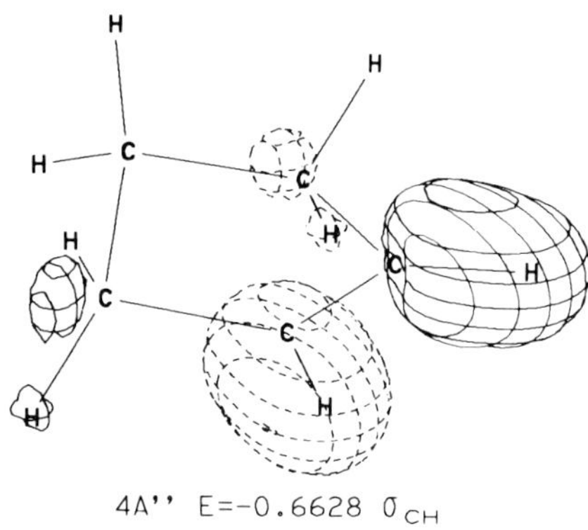

4A'' E=-0.6628 σ_{CH}

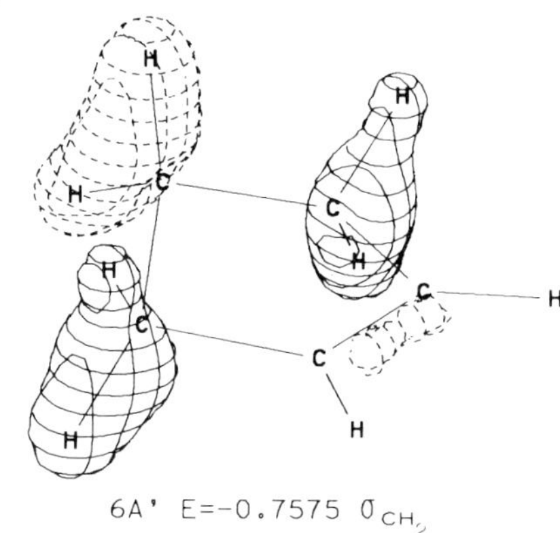

6A' E=-0.7575 σ_{CH_2}

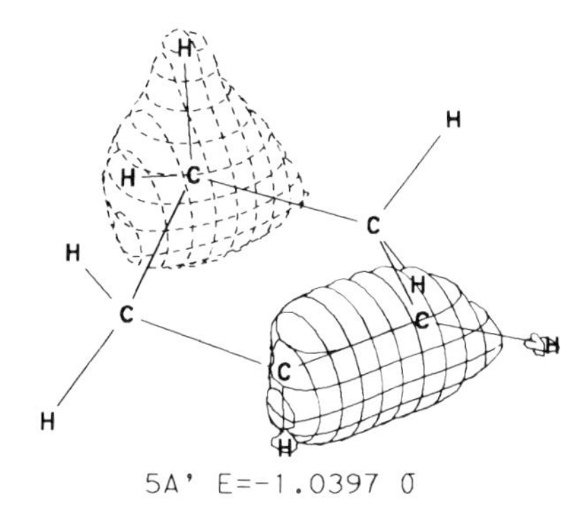

5A' E=-1.0397 σ

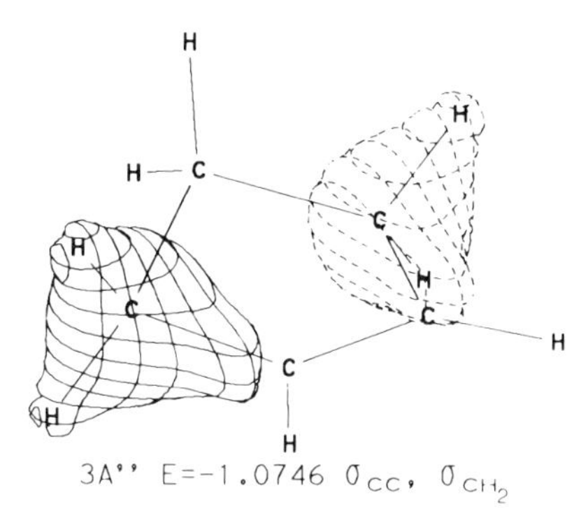

3A'' E=-1.0746 σ_{CC}, σ_{CH_2}

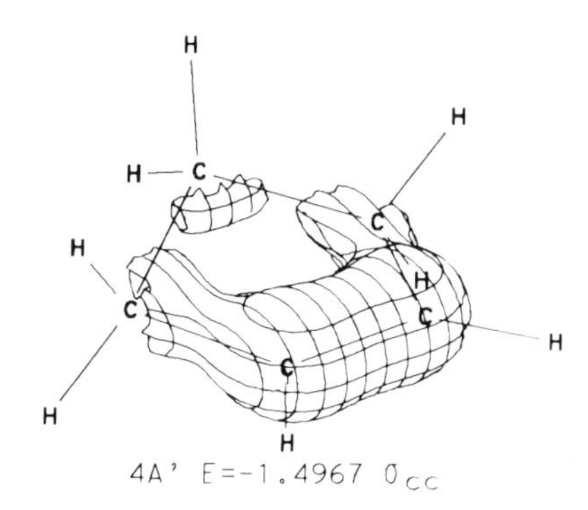

4A' E=-1.4967 σ_{CC}

Cyclopentene (Continued)

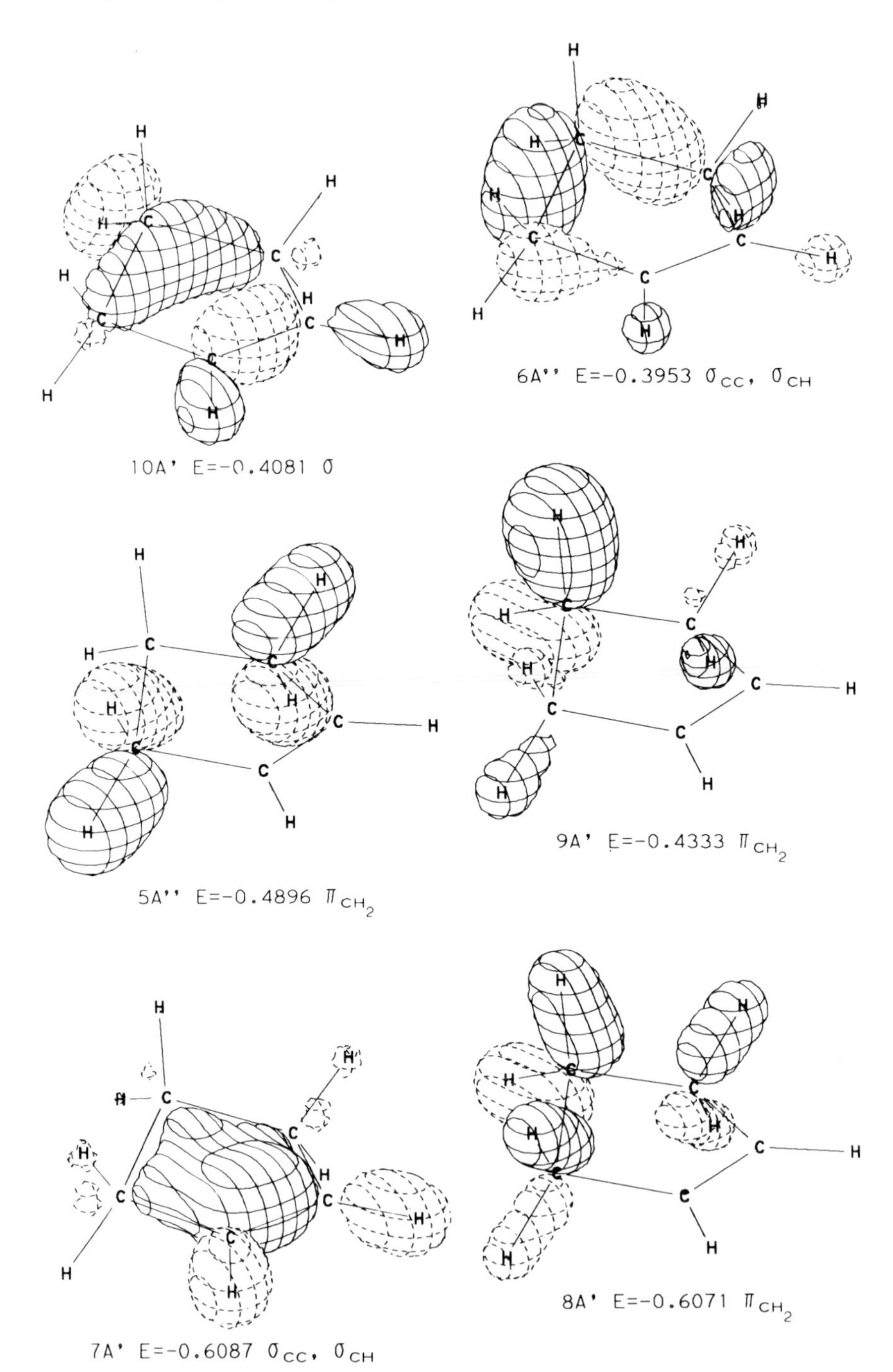

10A' E=−0.4081 σ

6A'' E=−0.3953 σ_{CC}, σ_{CH}

5A'' E=−0.4896 π_{CH_2}

9A' E=−0.4333 π_{CH_2}

7A' E=−0.6087 σ_{CC}, σ_{CH}

8A' E=−0.6071 π_{CH_2}

Cyclopentene (Continued)

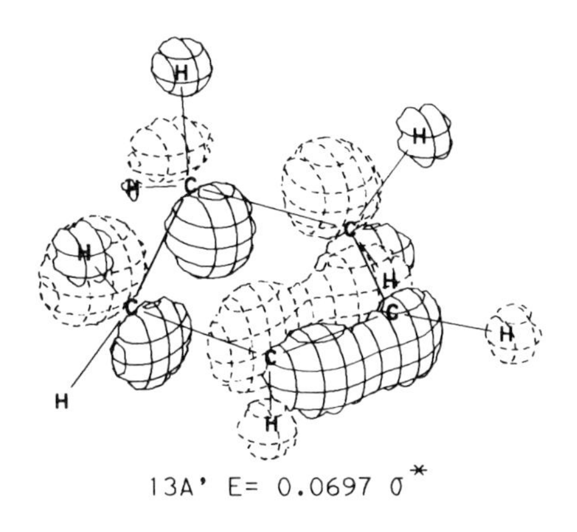

13A' E= 0.0697 σ^*

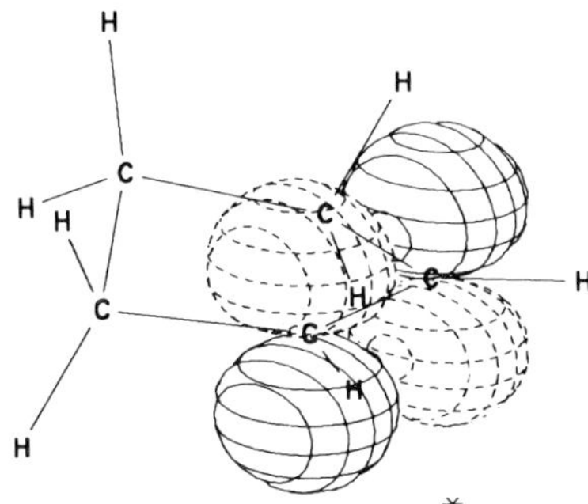

8A'' E= 0.0543 π^*_{cc}

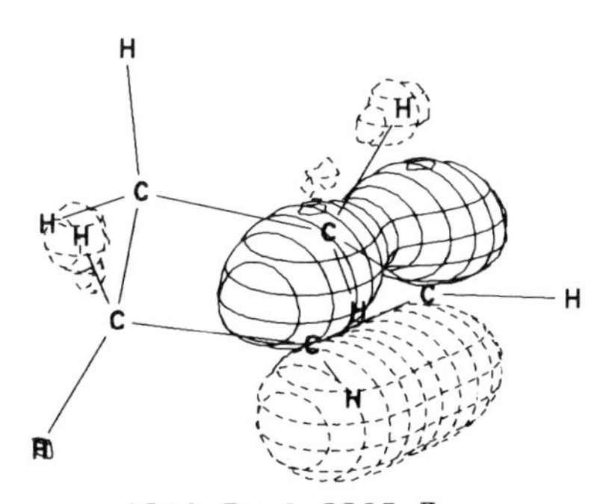

12A' E=-0.3285 π_{cc}

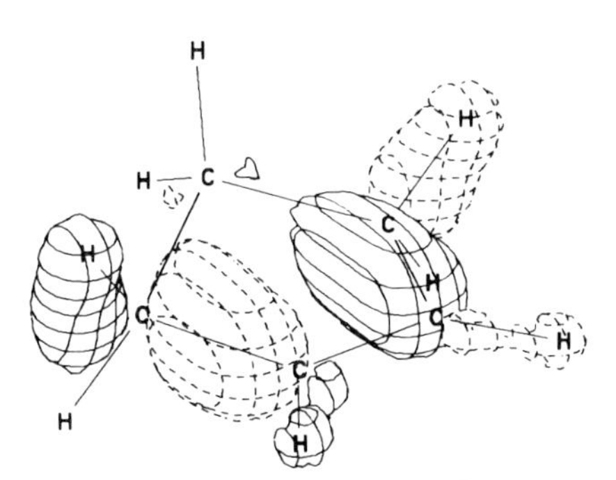

7A'' E=-0.3801 σ

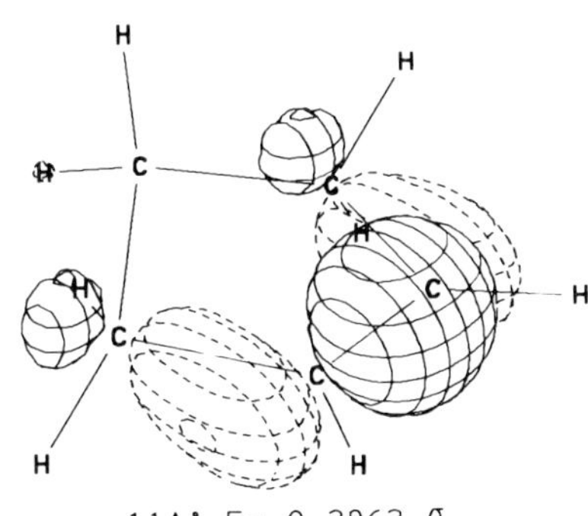

11A' E=-0.3862 σ_{cc}

90. (1.1.1)-Bicyclopentane

Symmetry: D_{3h}

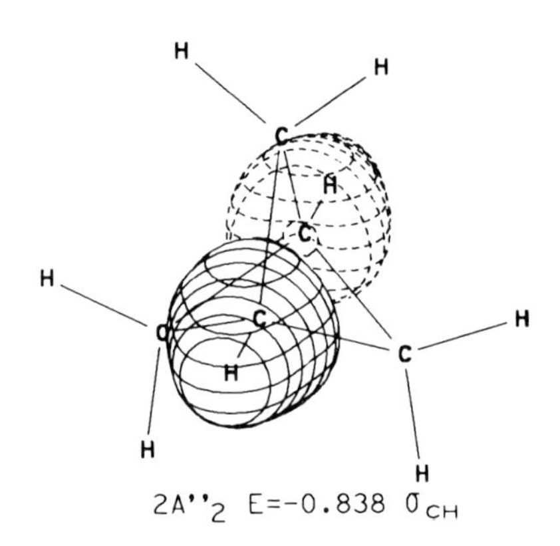

2A''$_2$ E=-0.838 σ_{CH}

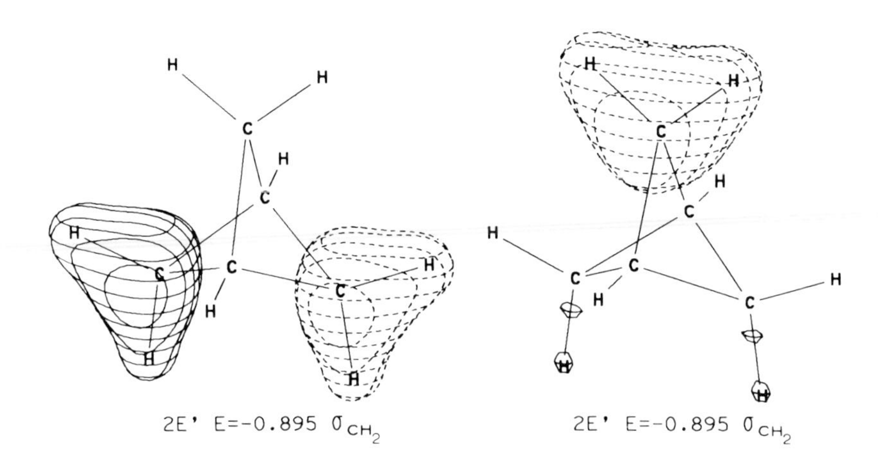

2E' E=-0.895 σ_{CH_2} 2E' E=-0.895 σ_{CH_2}

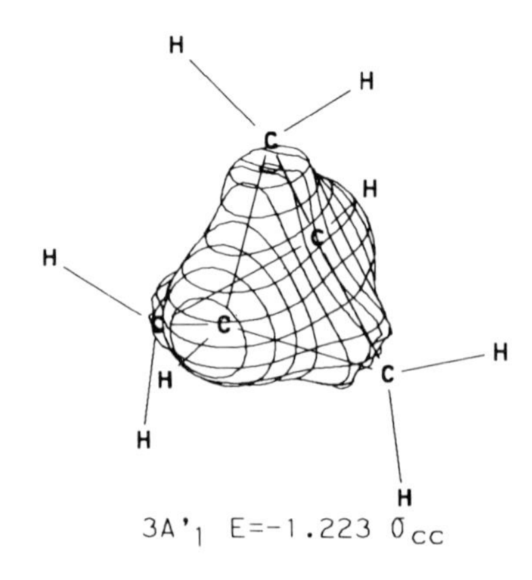

3A'$_1$ E=-1.223 σ_{CC}

(1.1.1)-Bicyclopentane (Continued)

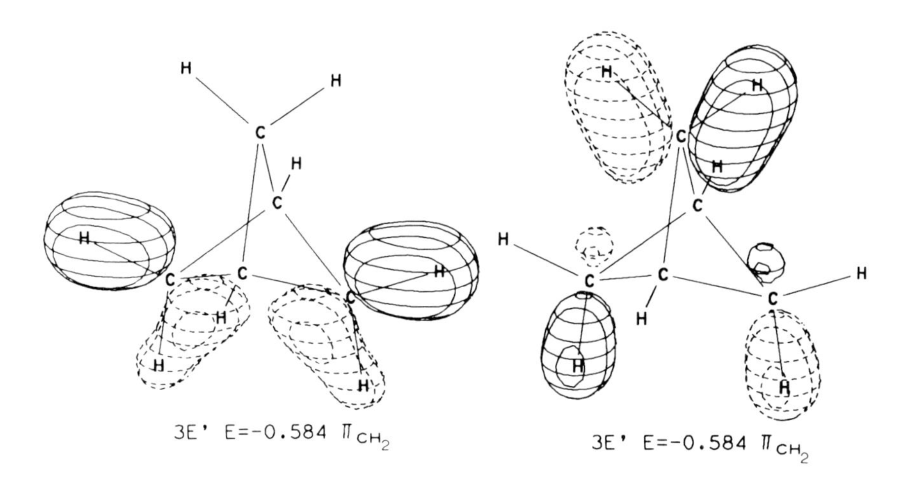

3E' E=-0.584 π_{CH_2} 3E' E=-0.584 π_{CH_2}

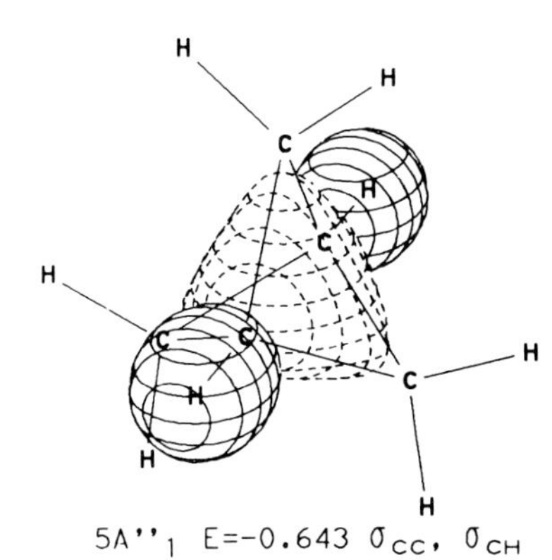

5A''$_1$ E=-0.643 σ_{CC}, σ_{CH}

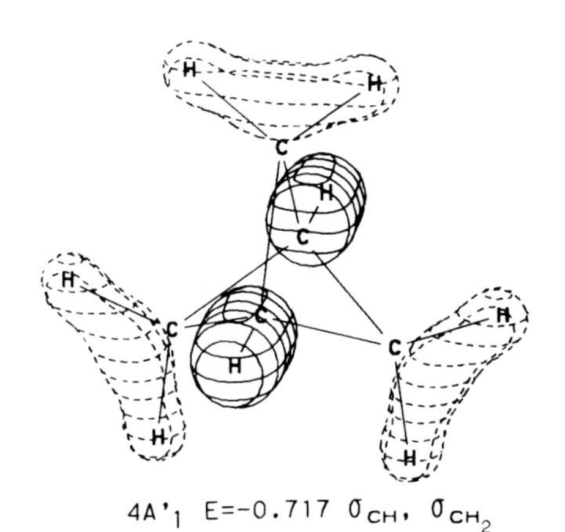

4A'$_1$ E=-0.717 σ_{CH}, σ_{CH_2}

(1.1.1)-Bicyclopentane (Continued)

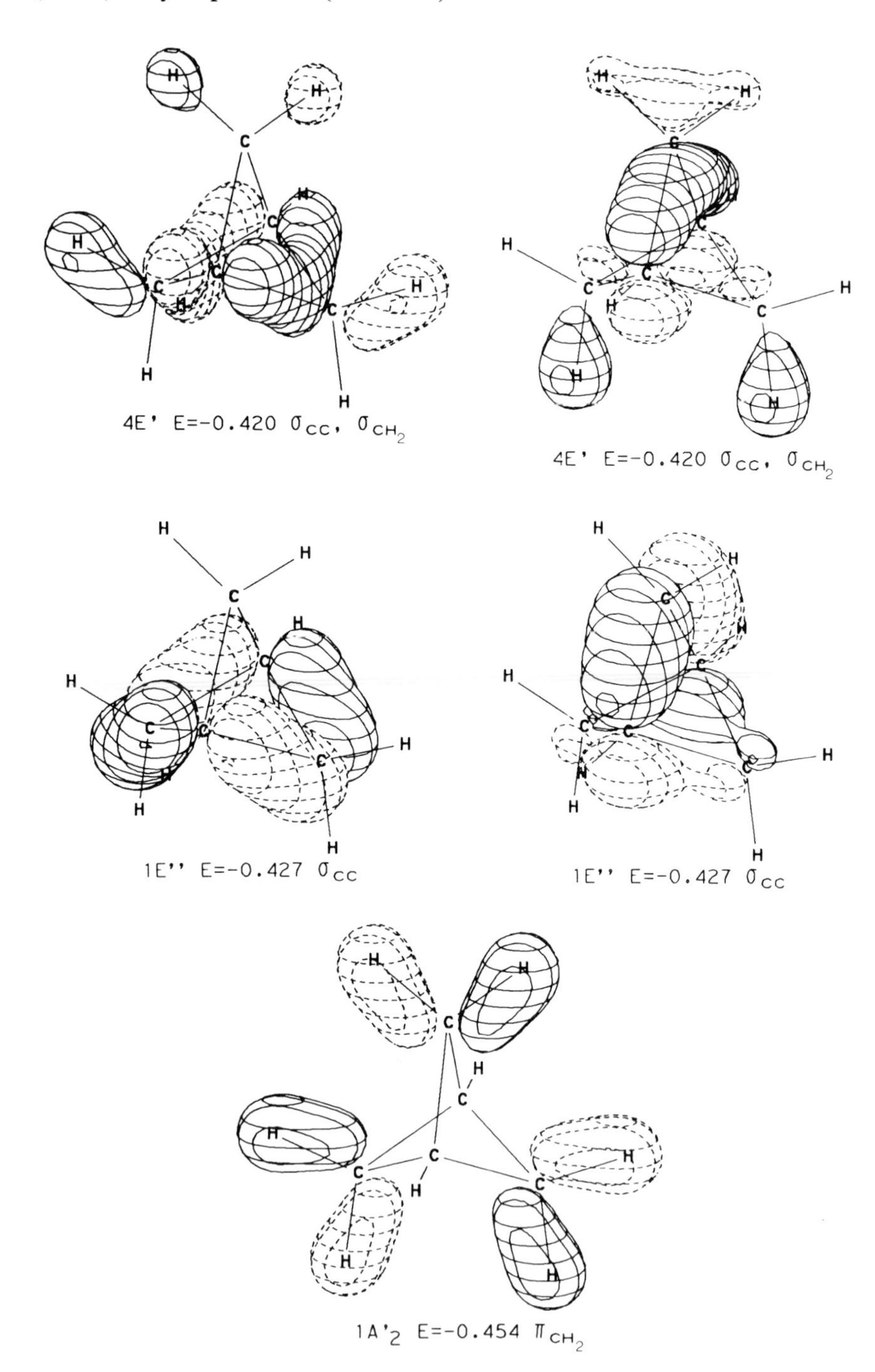

(1.1.1)-Bicyclopentane (Continued)

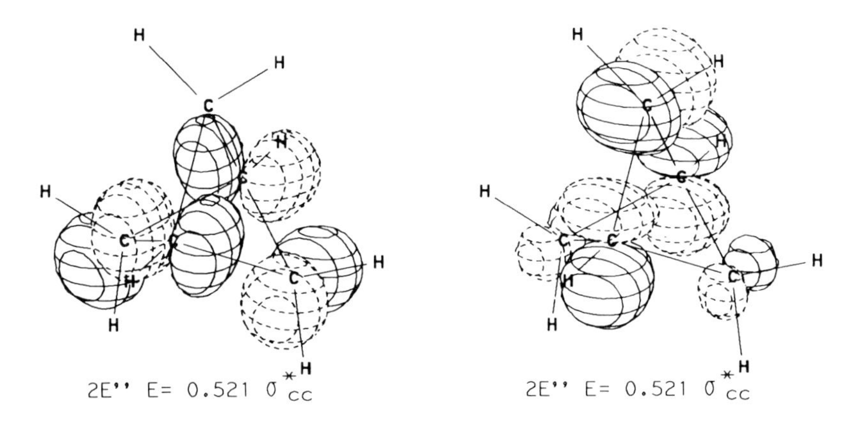

2E'' E= 0.521 σ^*_{CC} 2E'' E= 0.521 σ^*_{CC}

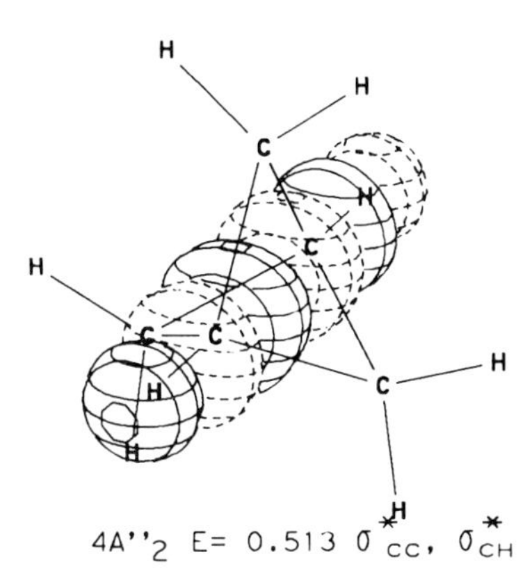

4A''$_2$ E= 0.513 σ^*_{CC}, σ^*_{CH}

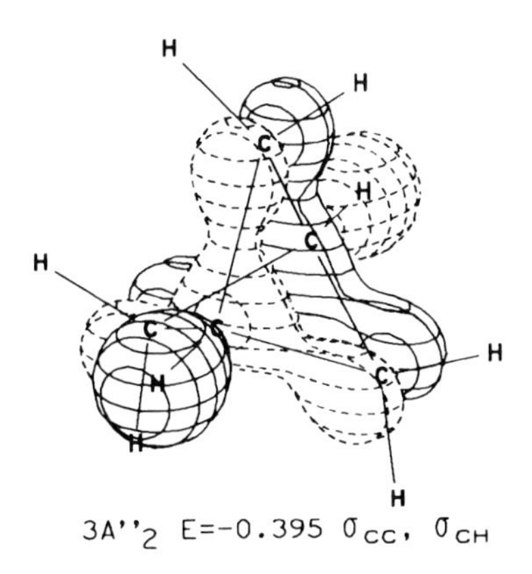

3A''$_2$ E=−0.395 σ_{CC}, σ_{CH}

91. Spiropentane

Symmetry: D_{2d}

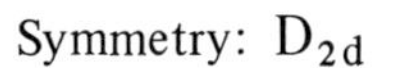

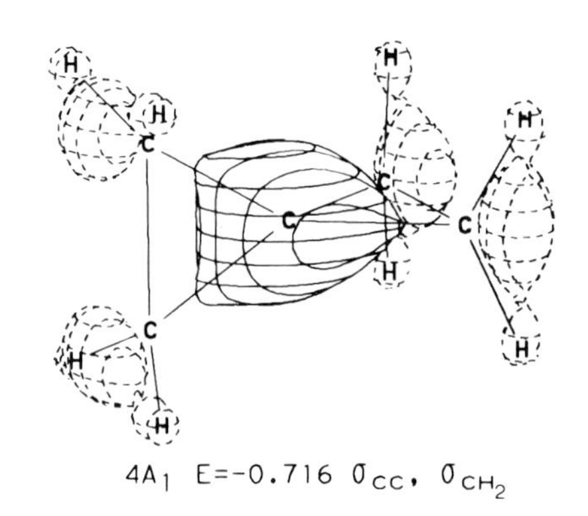

$4A_1$ E=−0.716 σ_{CC}, σ_{CH_2}

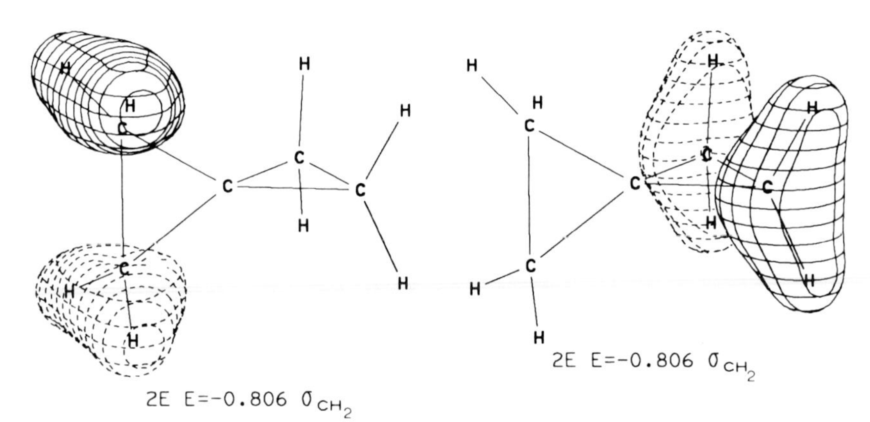

$2E$ E=−0.806 σ_{CH_2}

$2E$ E=−0.806 σ_{CH_2}

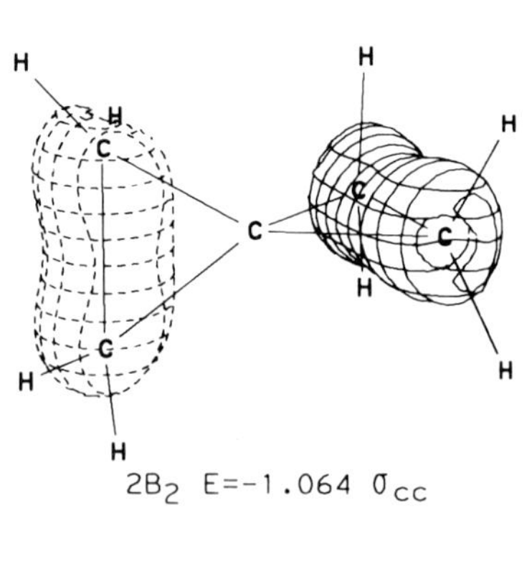

$2B_2$ E=−1.064 σ_{CC}

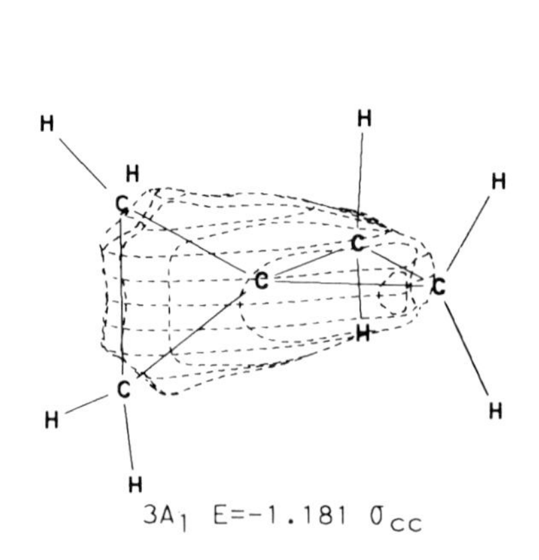

$3A_1$ E=−1.181 σ_{CC}

Spiropentane (Continued)

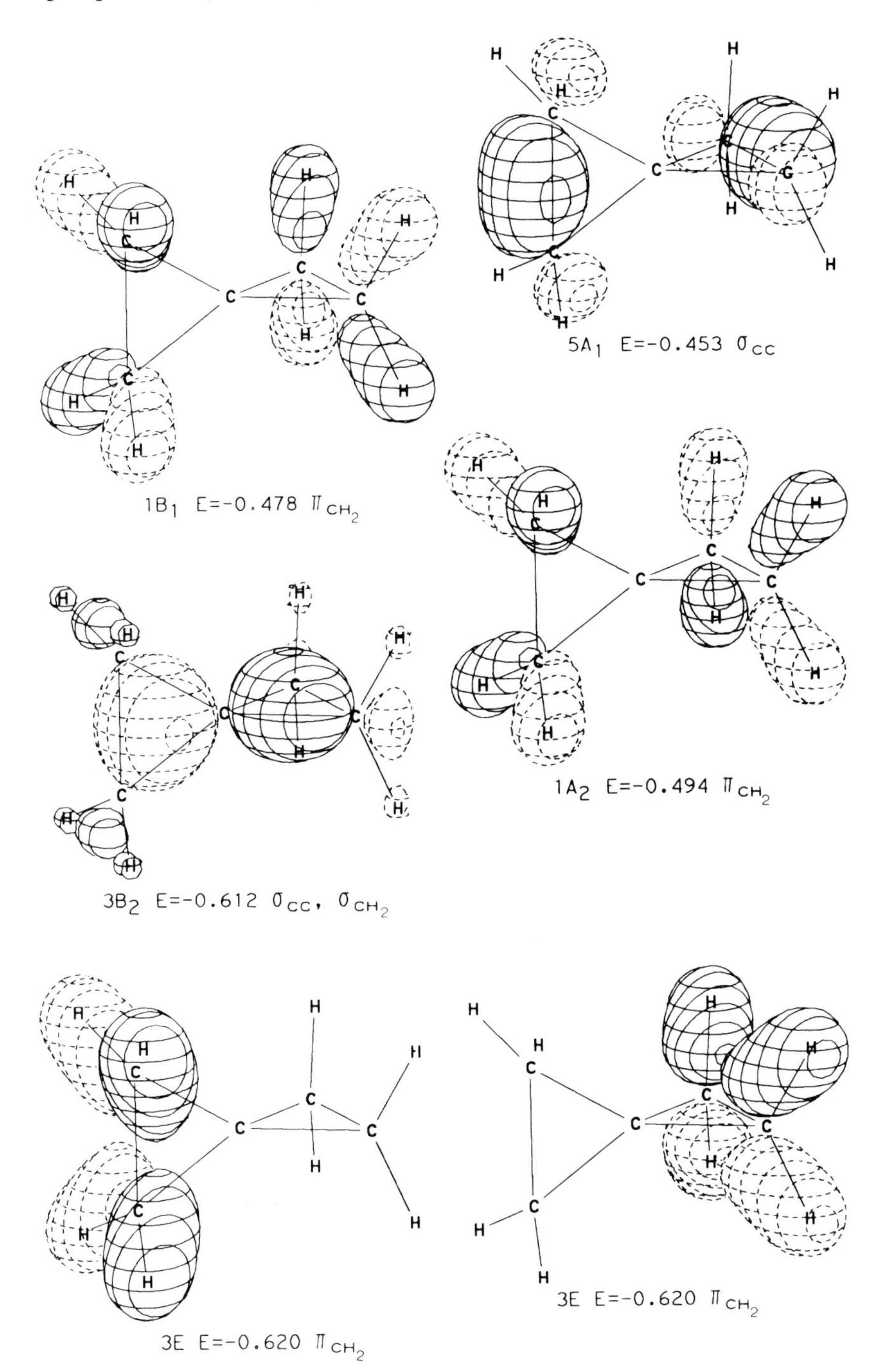

$5A_1$ E=−0.453 σ_{CC}

$1B_1$ E=−0.478 π_{CH_2}

$1A_2$ E=−0.494 π_{CH_2}

$3B_2$ E=−0.612 σ_{CC}, σ_{CH_2}

$3E$ E=−0.620 π_{CH_2}

$3E$ E=−0.620 π_{CH_2}

Spiropentane (Continued)

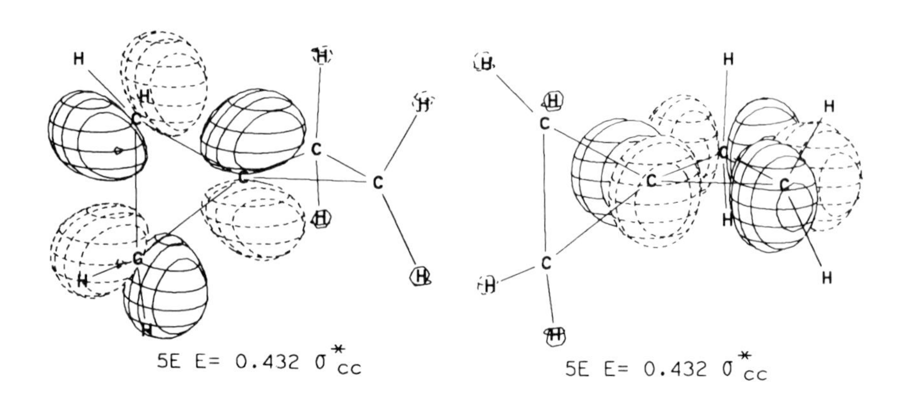

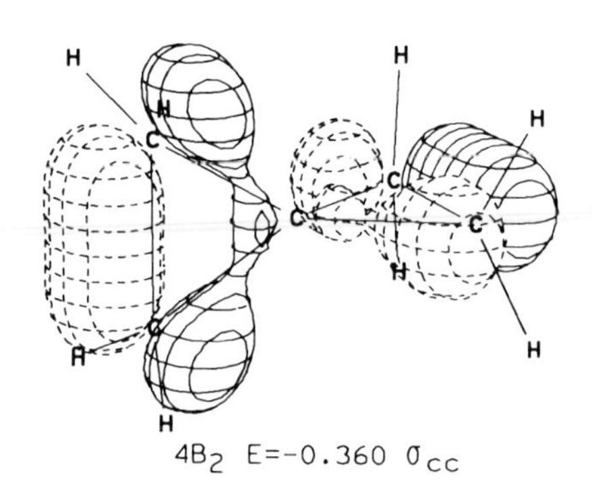

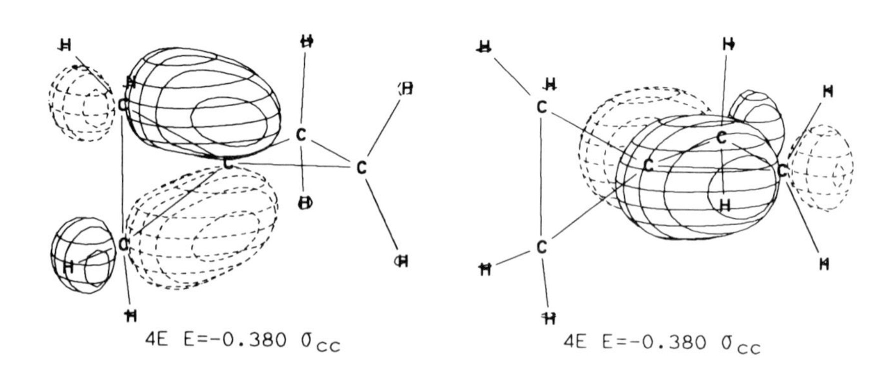

92. *Para*-Benzyne

Symmetry: D_{2h}

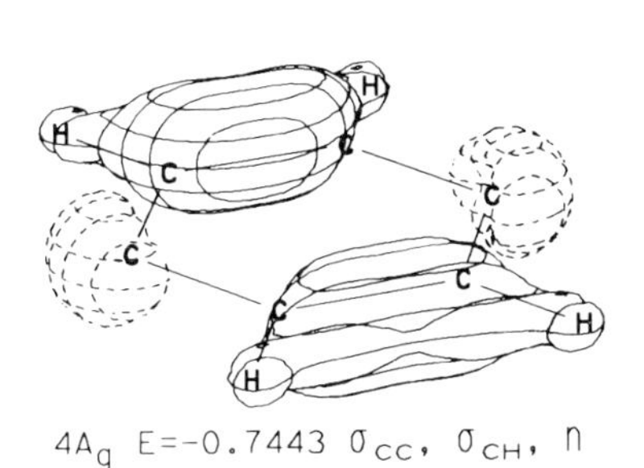

$4A_g$ E=-0.7443 σ_{CC}, σ_{CH}, n

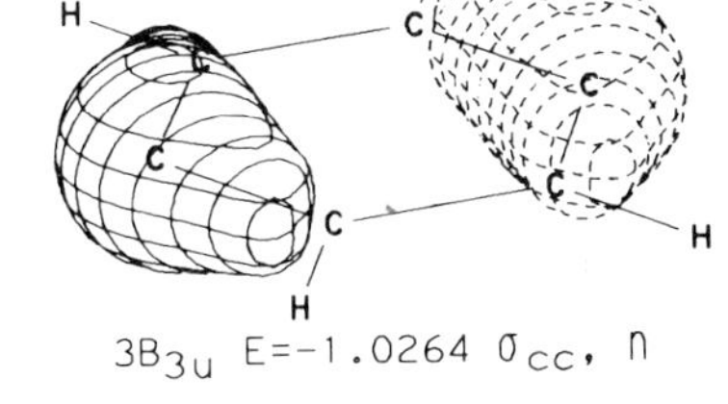

$3B_{3u}$ E=-1.0264 σ_{CC}, n

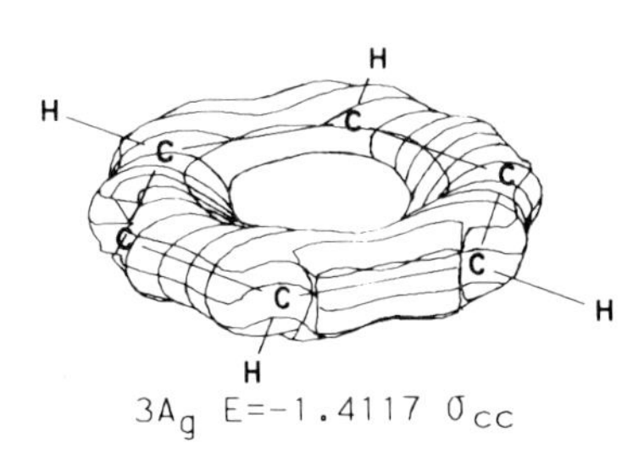

$3A_g$ E=-1.4117 σ_{CC}

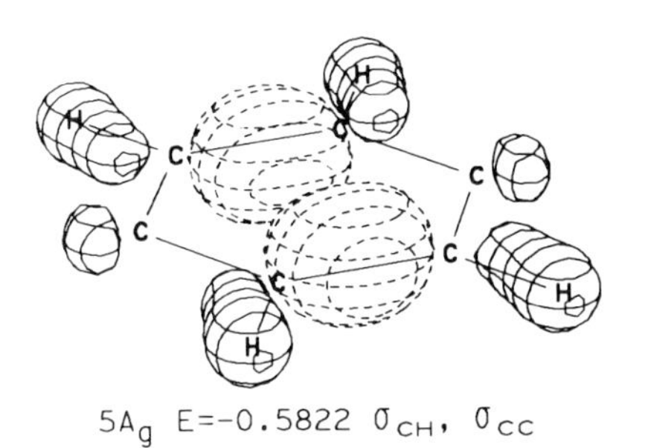

$5A_g$ E=-0.5822 σ_{CH}, σ_{CC}

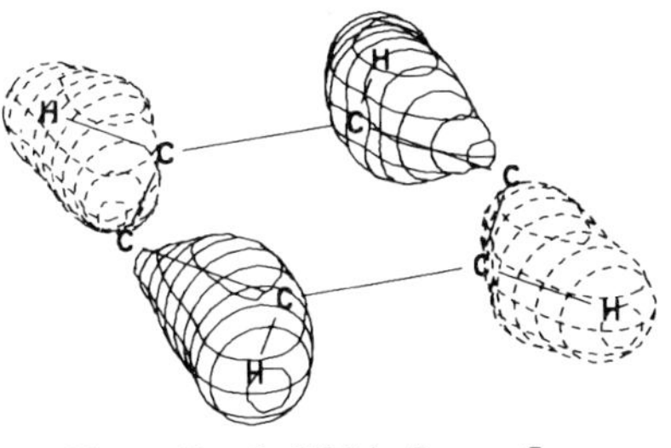

$2B_{1g}$ E=-0.7886 σ_{CH}, σ_{CC}

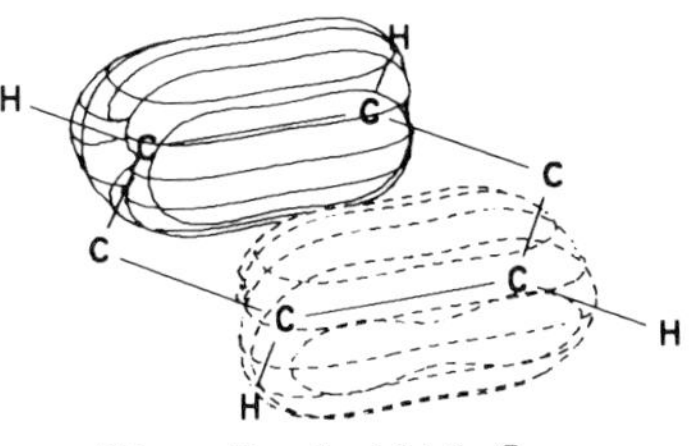

$2B_{2u}$ E=-1.1016 σ_{CC}

Para-Benzyne (Continued)

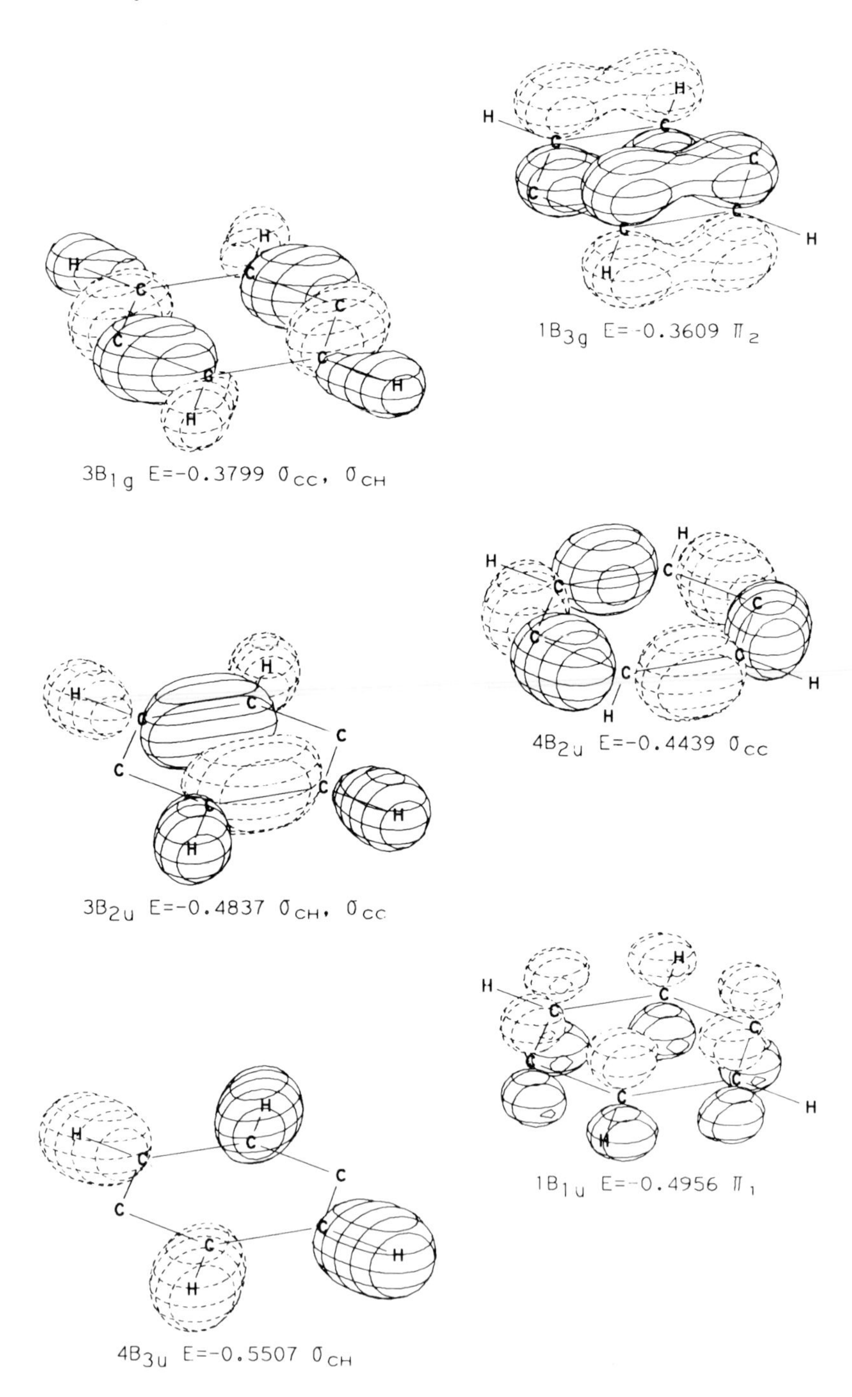

$1B_{3g}$ E=-0.3609 π_2

$3B_{1g}$ E=-0.3799 σ_{CC}, σ_{CH}

$4B_{2u}$ E=-0.4439 σ_{CC}

$3B_{2u}$ E=-0.4837 σ_{CH}, σ_{CC}

$1B_{1u}$ E=-0.4956 π_1

$4B_{3u}$ E=-0.5507 σ_{CH}

Para-Benzyne (Continued)

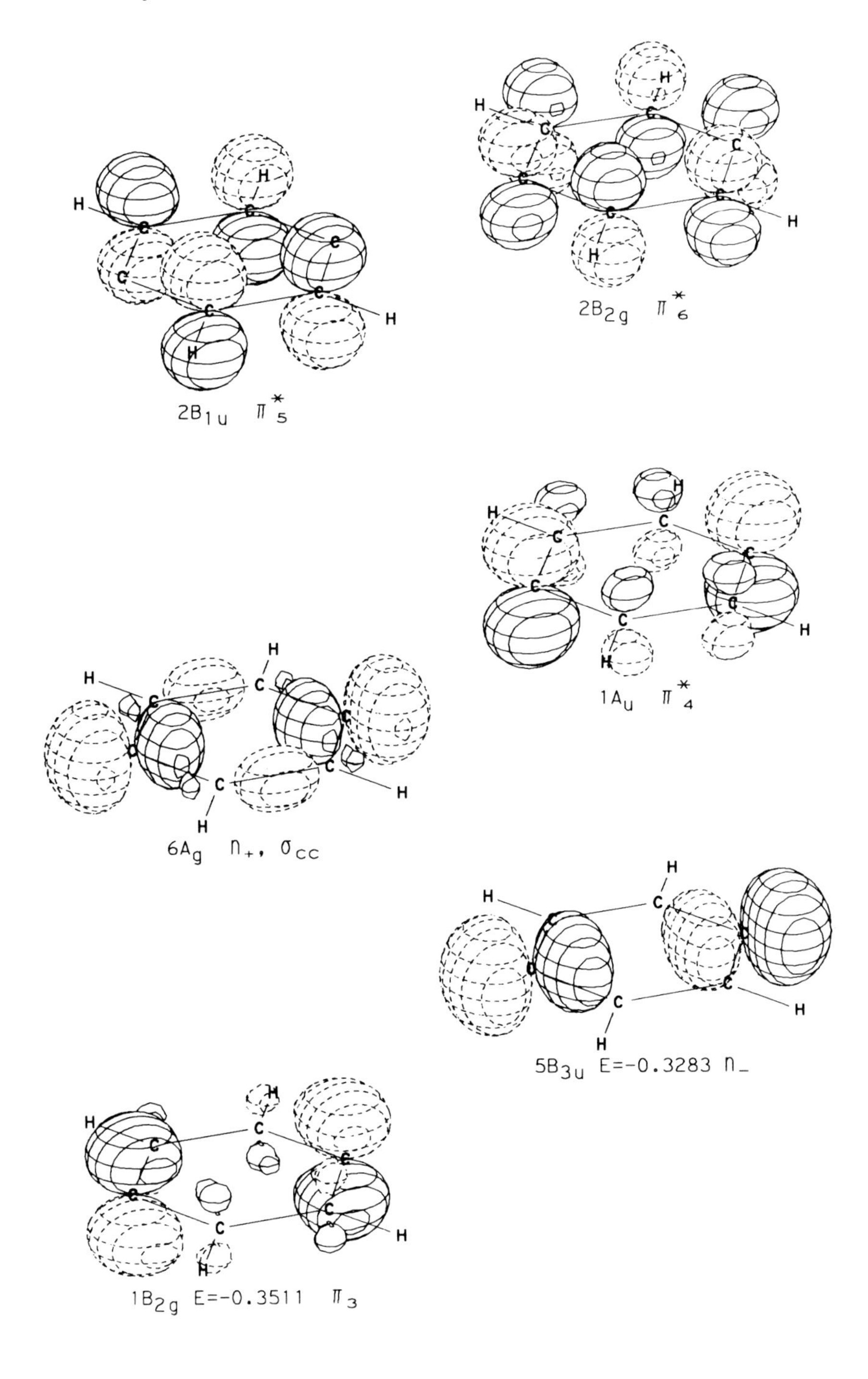

93. Cyclopentane

Symmetry: C_s

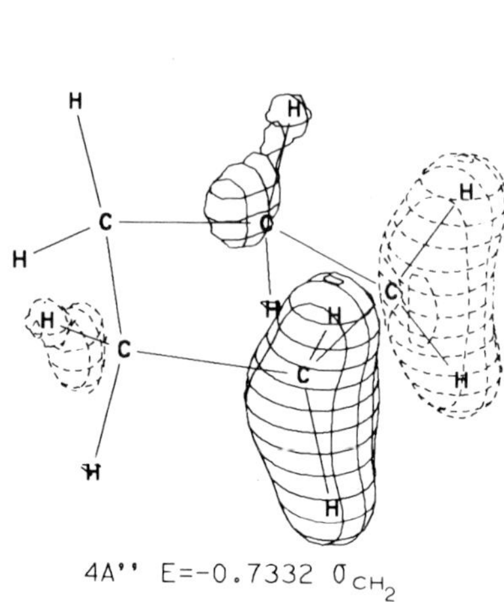

4A'' E=-0.7332 σ_{CH_2}

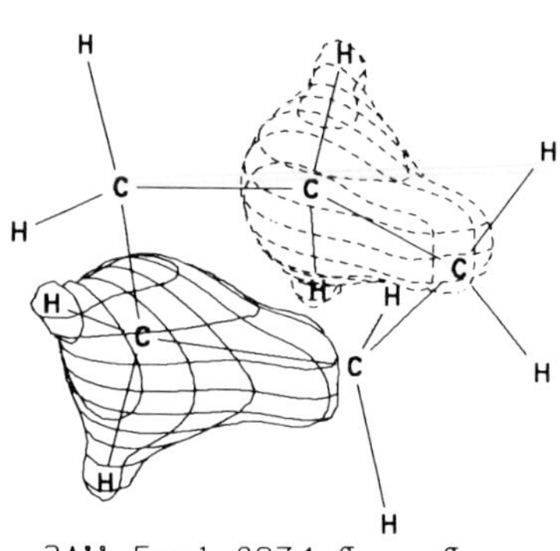

3A'' E=-1.0974 σ_{CC}, σ_{CH_2}

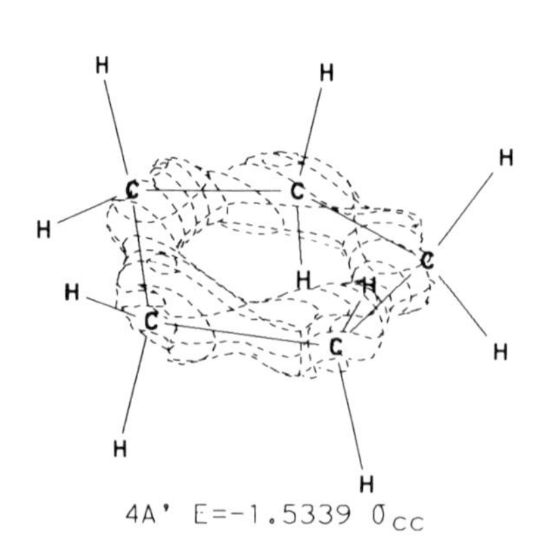

4A' E=-1.5339 σ_{CC}

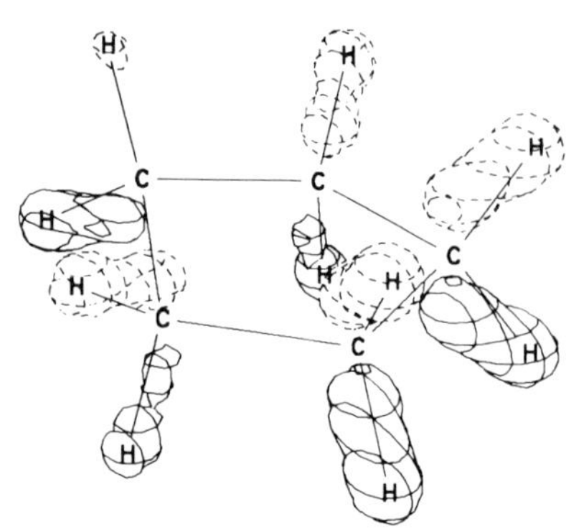

7A' E=-0.6426 π_{CH_2}

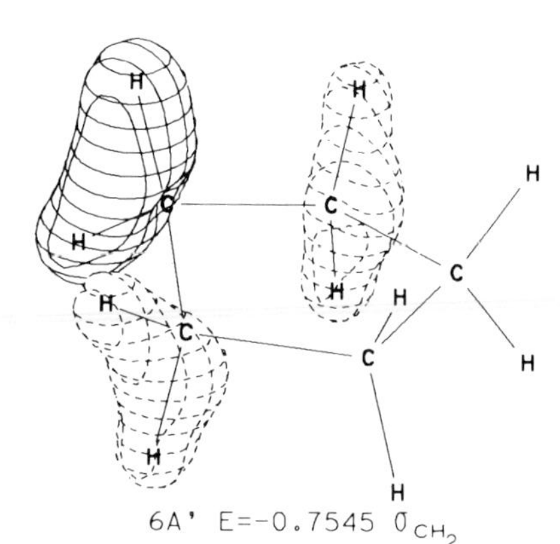

6A' E=-0.7545 σ_{CH_2}

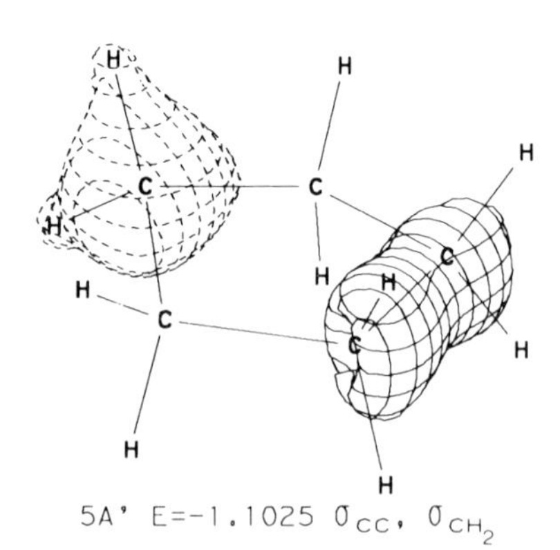

5A' E=-1.1025 σ_{CC}, σ_{CH_2}

Cyclopentane (Continued)

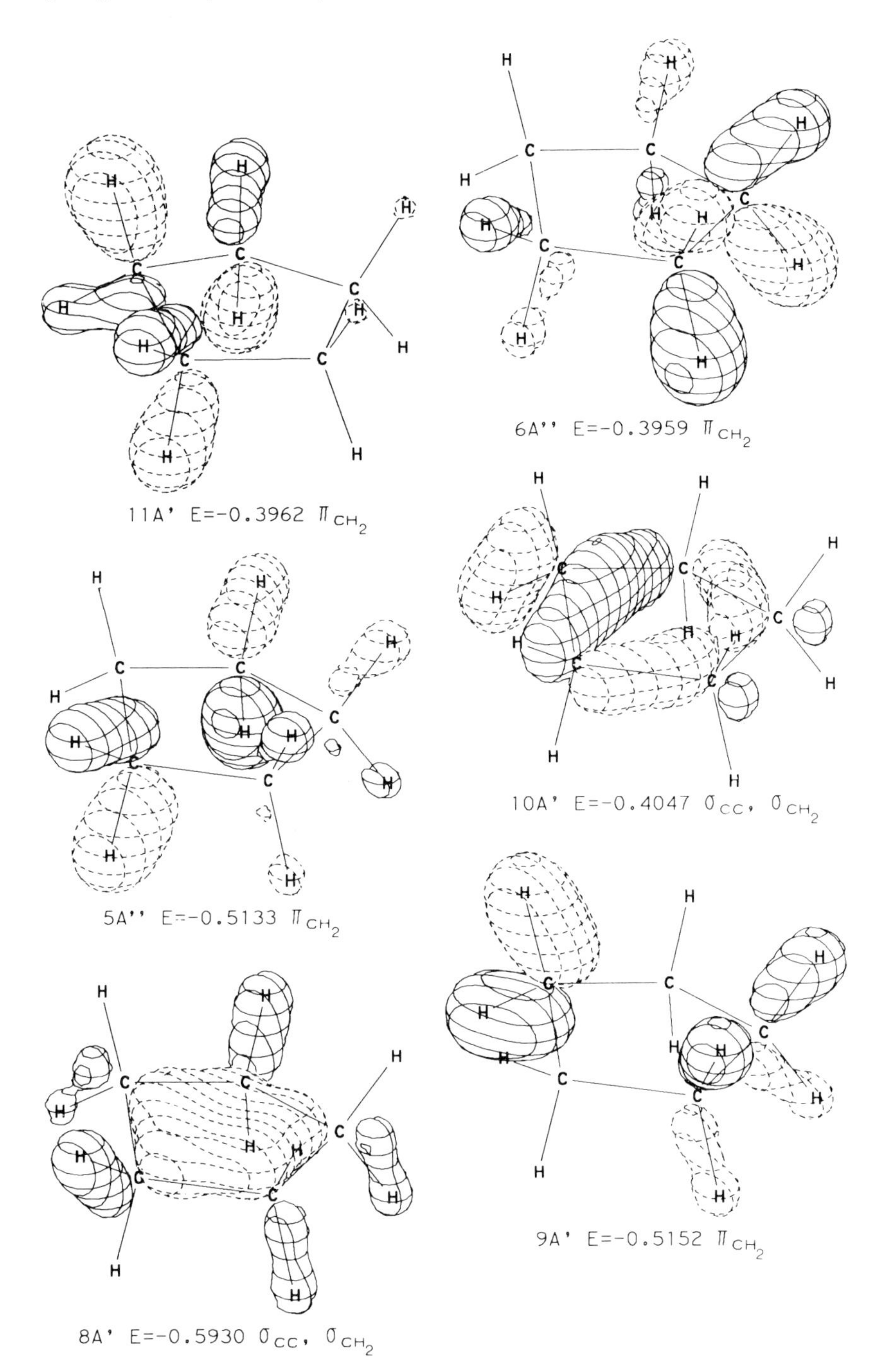

11A' E=-0.3962 π_{CH_2}

6A'' E=-0.3959 π_{CH_2}

5A'' E=-0.5133 π_{CH_2}

10A' E=-0.4047 σ_{CC}, σ_{CH_2}

8A' E=-0.5930 σ_{CC}, σ_{CH_2}

9A' E=-0.5152 π_{CH_2}

Cyclopentane (Continued)

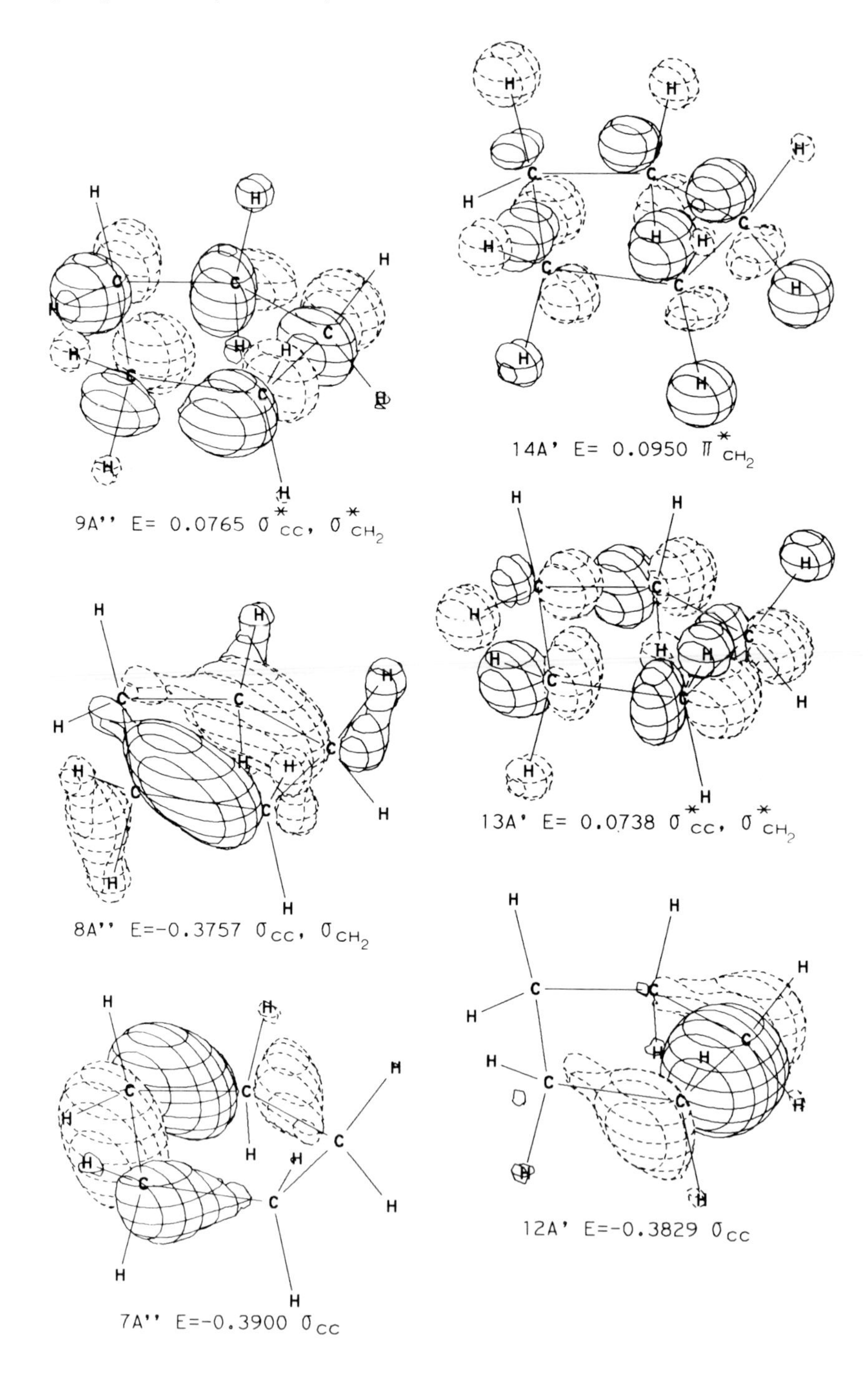

9A'' E= 0.0765 σ^*_{CC}, $\sigma^*_{CH_2}$

8A'' E=-0.3757 σ_{CC}, σ_{CH_2}

7A'' E=-0.3900 σ_{CC}

14A' E= 0.0950 $\pi^*_{CH_2}$

13A' E= 0.0738 σ^*_{CC}, $\sigma^*_{CH_2}$

12A' E=-0.3829 σ_{CC}

94. Benzene

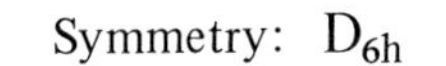

Symmetry: D_{6h}

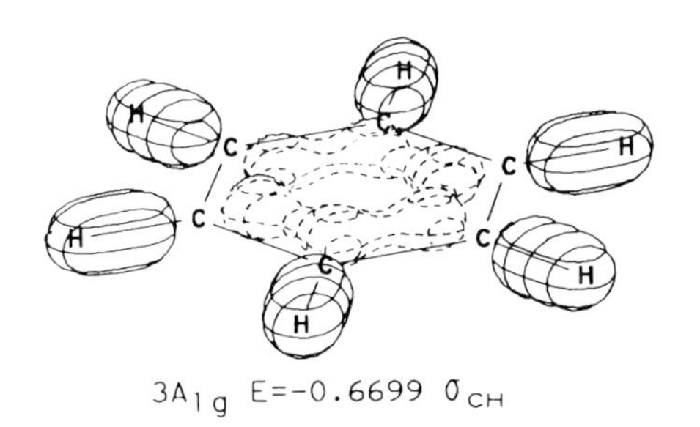

$3A_{1g}$ $E=-0.6699$ σ_{CH}

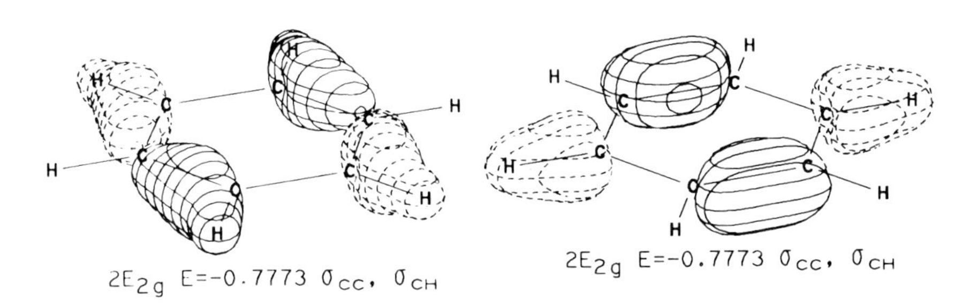

$2E_{2g}$ $E=-0.7773$ σ_{CC}, σ_{CH}

$2E_{2g}$ $E=-0.7773$ σ_{CC}, σ_{CH}

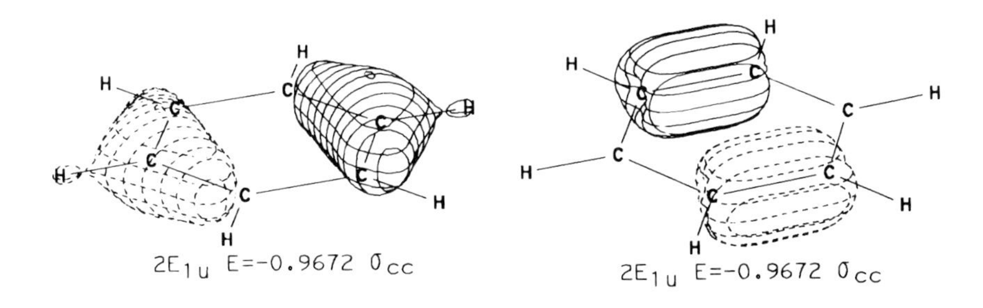

$2E_{1u}$ $E=-0.9672$ σ_{CC}

$2E_{1u}$ $E=-0.9672$ σ_{CC}

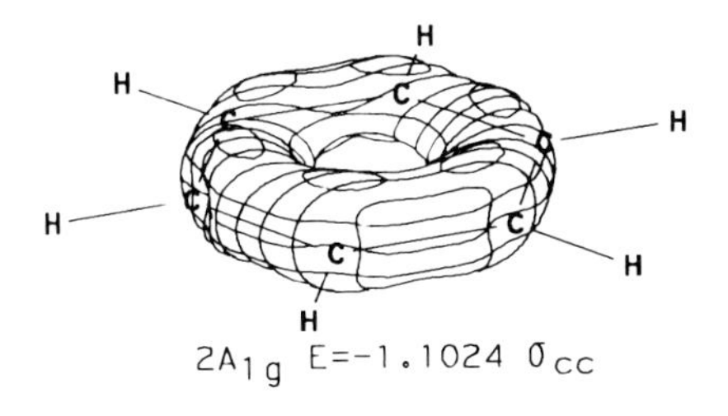

$2A_{1g}$ $E=-1.1024$ σ_{CC}

Benzene (Continued)

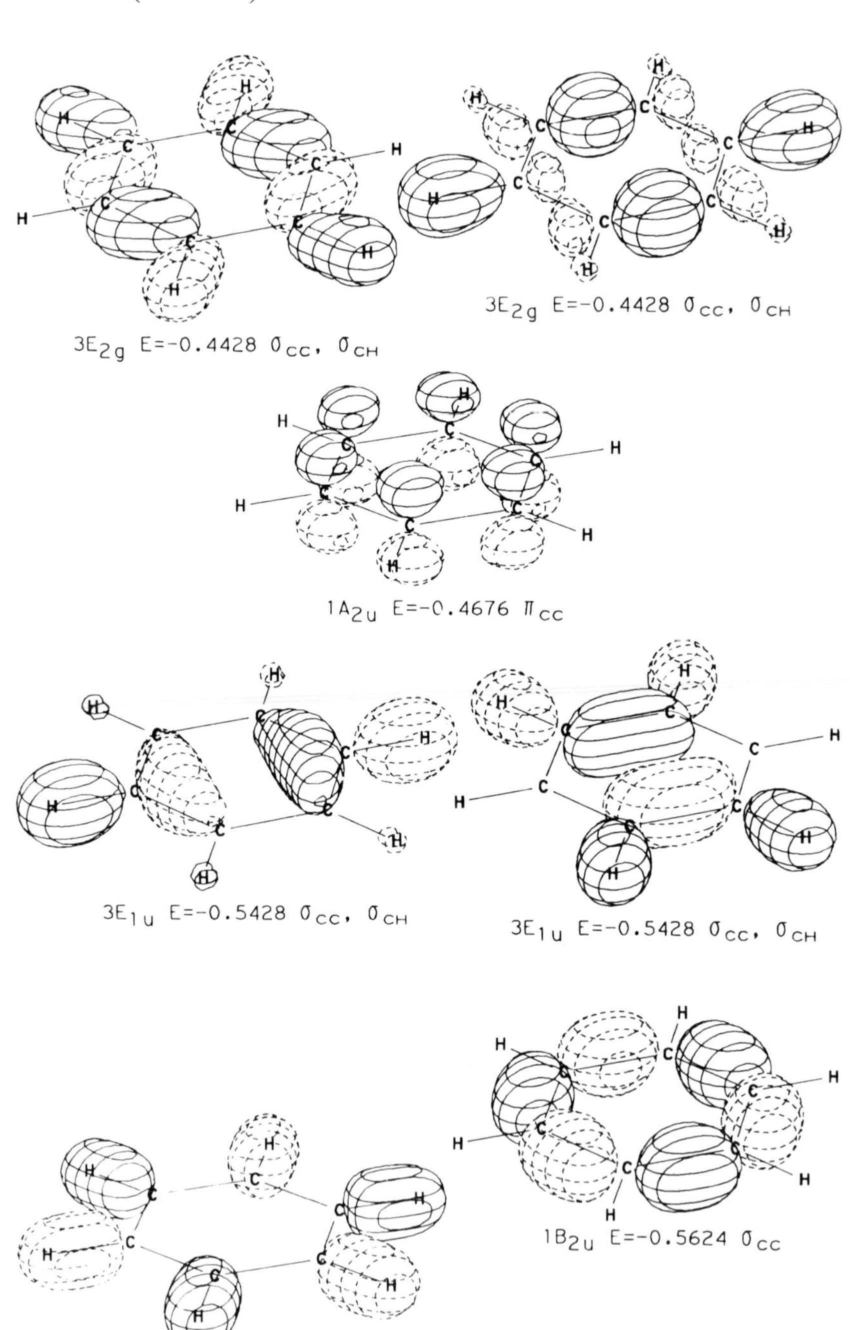

3E$_{2g}$ E=-0.4428 σ_{CC}, σ_{CH}

3E$_{2g}$ E=-0.4428 σ_{CC}, σ_{CH}

1A$_{2u}$ E=-0.4676 π_{CC}

3E$_{1u}$ E=-0.5428 σ_{CC}, σ_{CH}

3E$_{1u}$ E=-0.5428 σ_{CC}, σ_{CH}

1B$_{2u}$ E=-0.5624 σ_{CC}

2B$_{1u}$ E=-0.6040 σ_{CH}

Benzene (Continued)

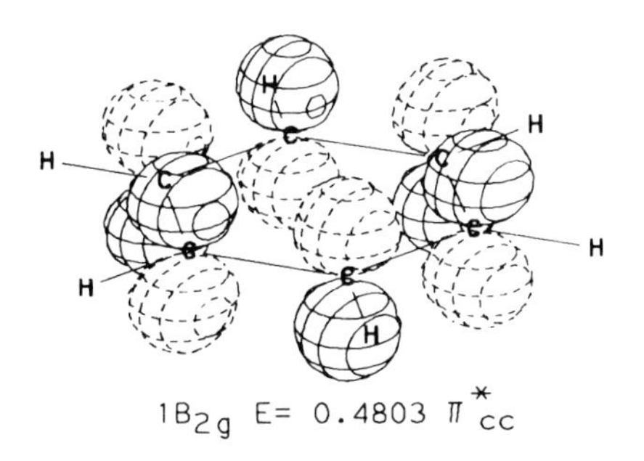

$1B_{2g}$ E= 0.4803 π^*_{cc}

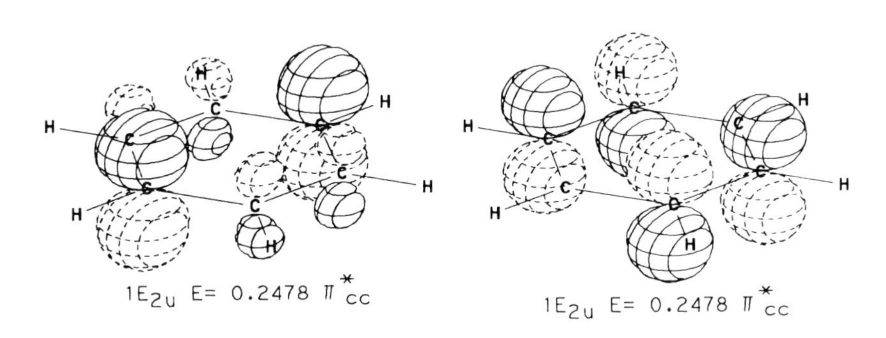

$1E_{2u}$ E= 0.2478 π^*_{cc} $1E_{2u}$ E= 0.2478 π^*_{cc}

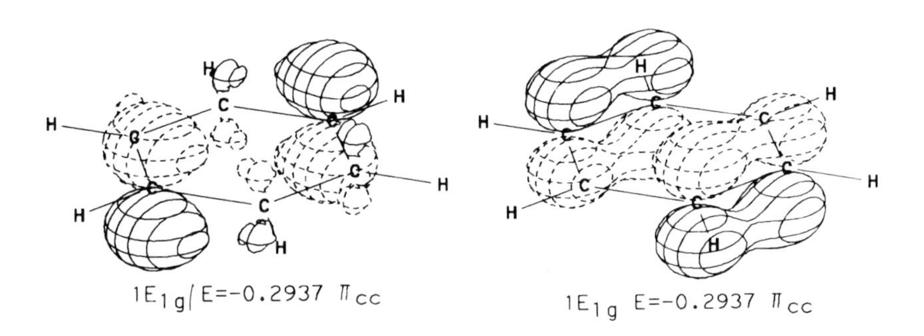

$1E_{1g}$ E=-0.2937 π_{cc} $1E_{1g}$ E=-0.2937 π_{cc}

95. Dewar Benzene

Symmetry: C_{2v}

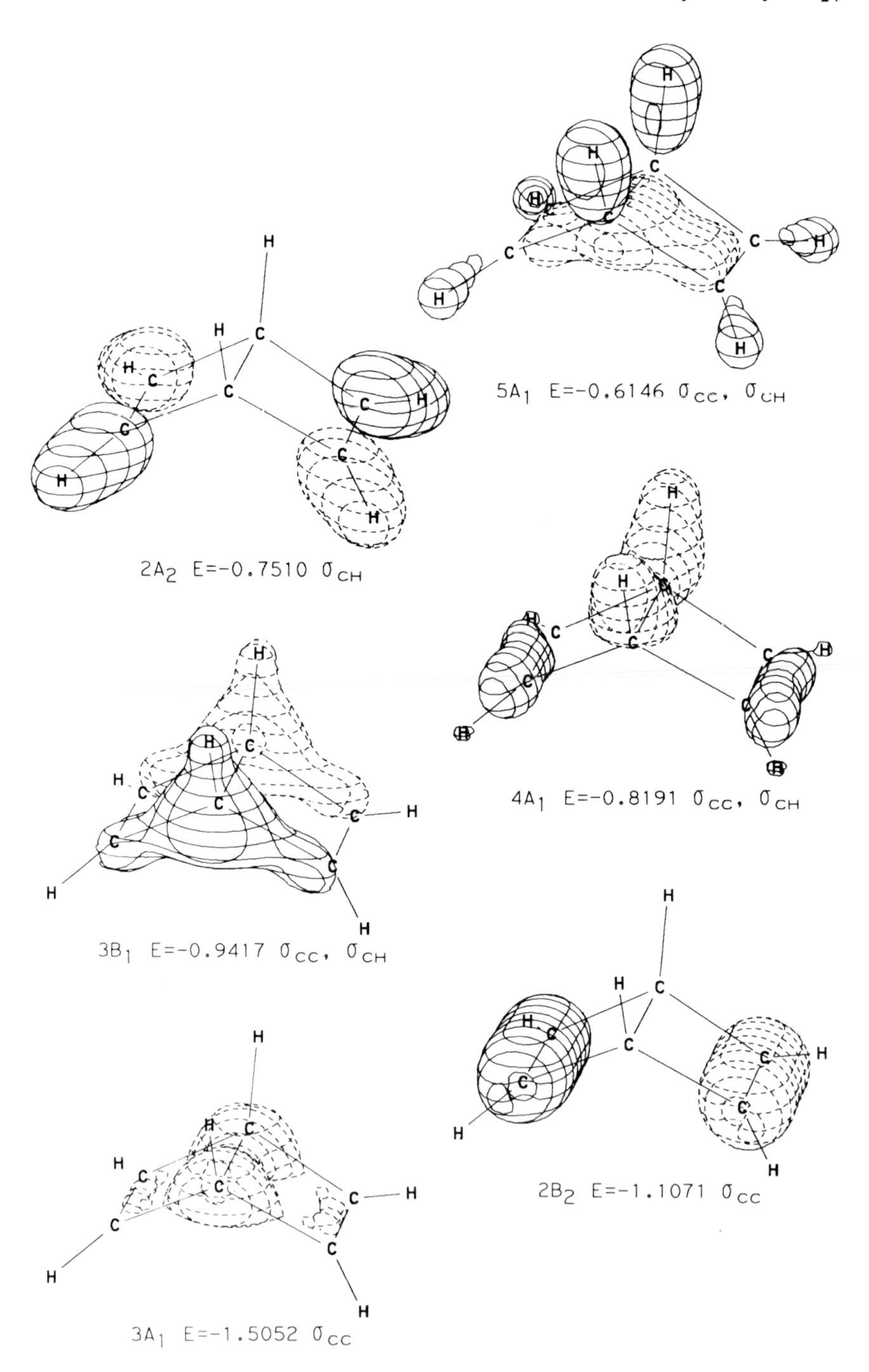

$5A_1$ E=-0.6146 σ_{CC}, σ_{CH}

$2A_2$ E=-0.7510 σ_{CH}

$4A_1$ E=-0.8191 σ_{CC}, σ_{CH}

$3B_1$ E=-0.9417 σ_{CC}, σ_{CH}

$2B_2$ E=-1.1071 σ_{CC}

$3A_1$ E=-1.5052 σ_{CC}

Dewar Benzene (Continued)

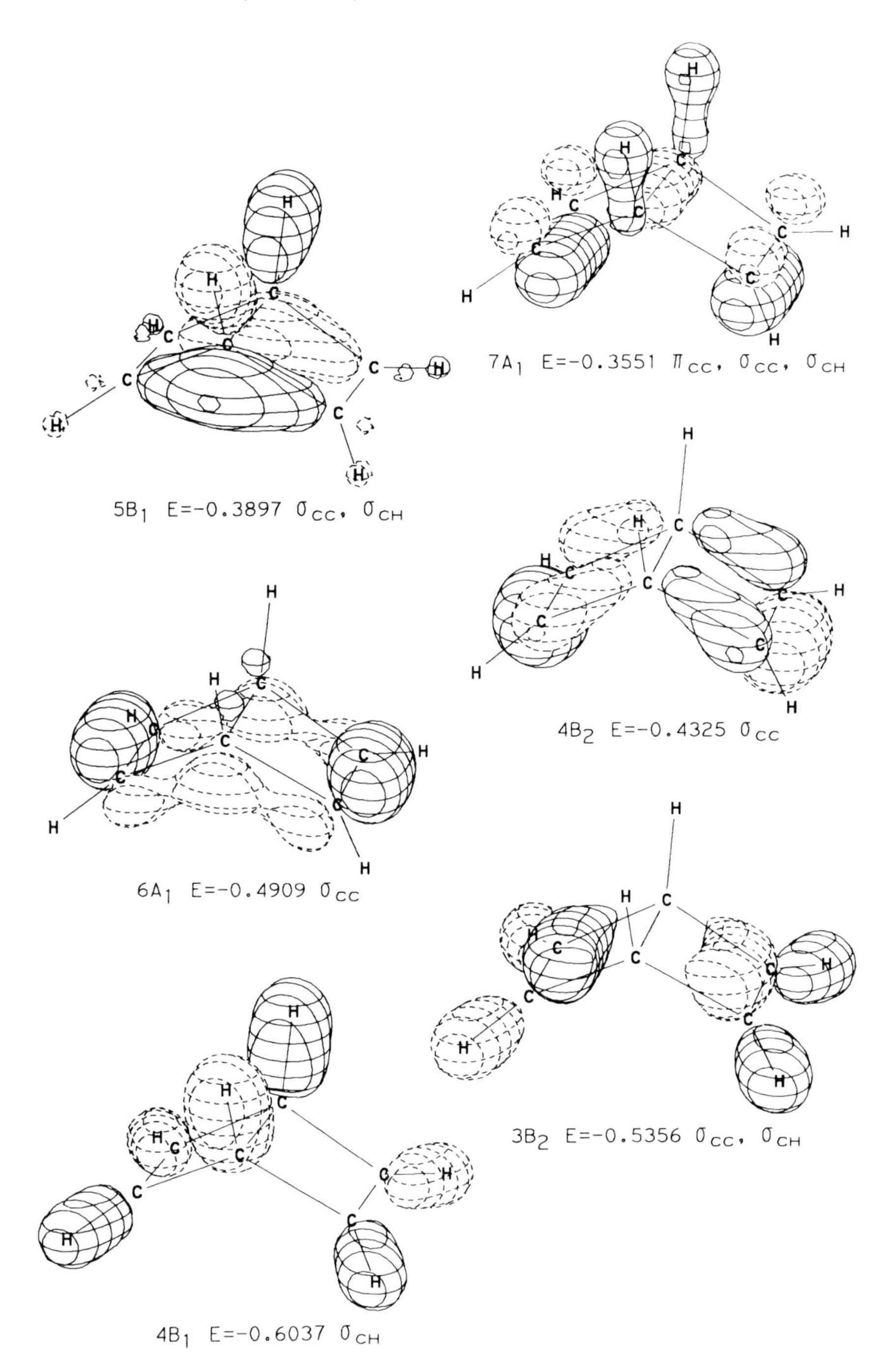

$5B_1$ E=-0.3897 σ_{CC}, σ_{CH}

$7A_1$ E=-0.3551 π_{CC}, σ_{CC}, σ_{CH}

$6A_1$ E=-0.4909 σ_{CC}

$4B_2$ E=-0.4325 σ_{CC}

$3B_2$ E=-0.5356 σ_{CC}, σ_{CH}

$4B_1$ E=-0.6037 σ_{CH}

Dewar Benzene (Continued)

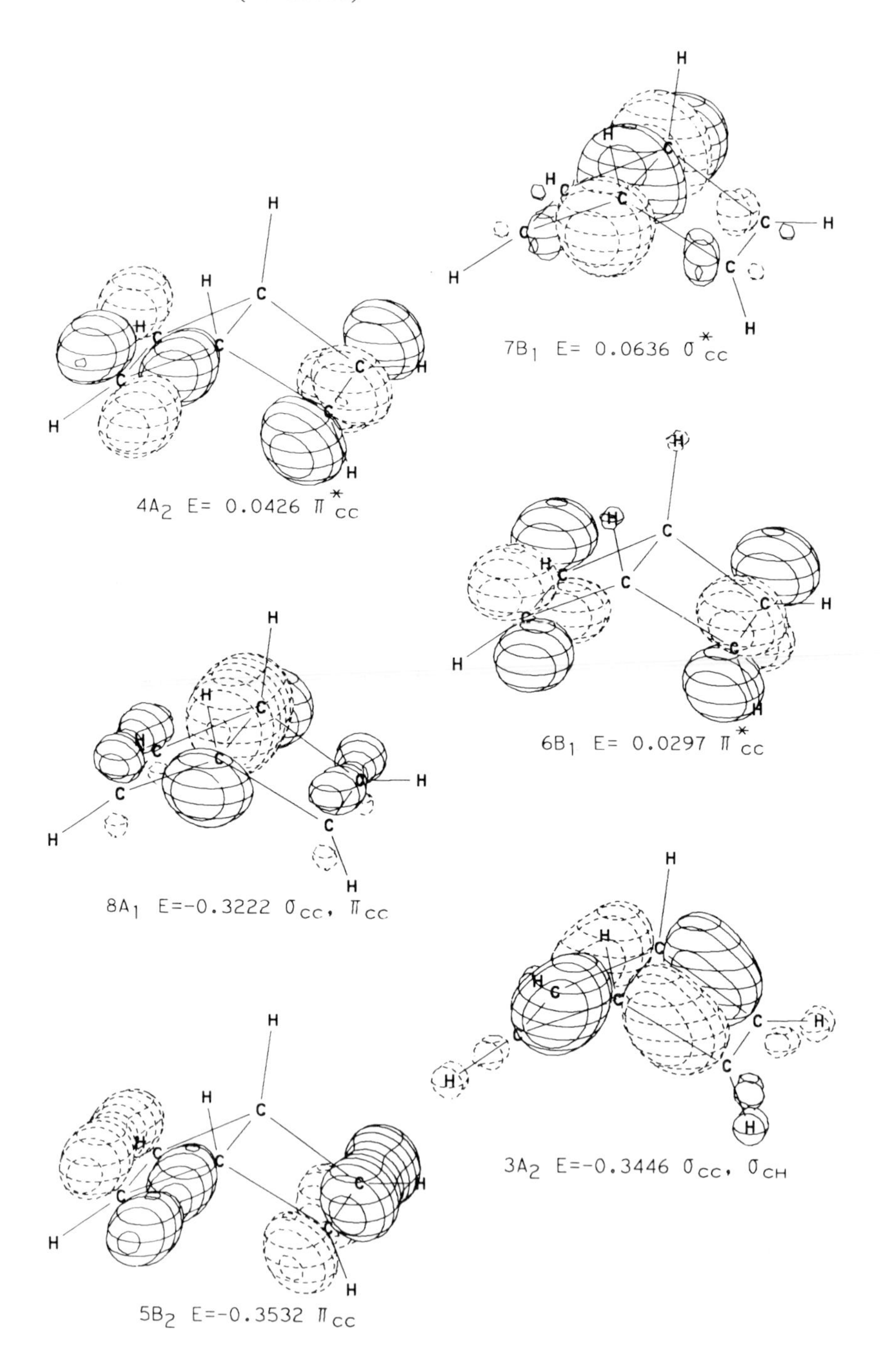

96. Pyridine

Symmetry: C_{2v}

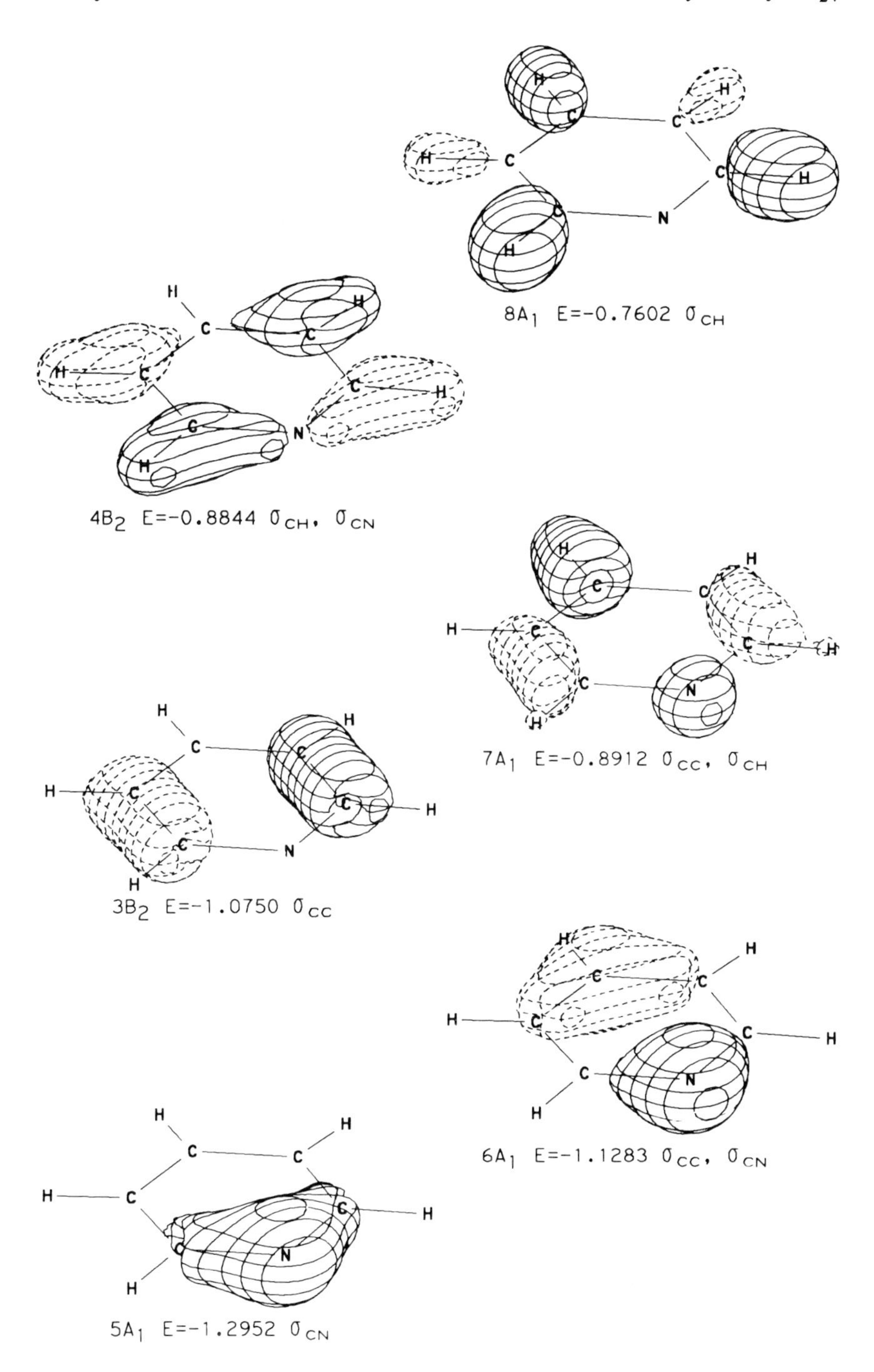

$8A_1$ E=-0.7602 σ_{CH}

$4B_2$ E=-0.8844 σ_{CH}, σ_{CN}

$7A_1$ E=-0.8912 σ_{CC}, σ_{CH}

$3B_2$ E=-1.0750 σ_{CC}

$6A_1$ E=-1.1283 σ_{CC}, σ_{CN}

$5A_1$ E=-1.2952 σ_{CN}

Pyridine (Continued)

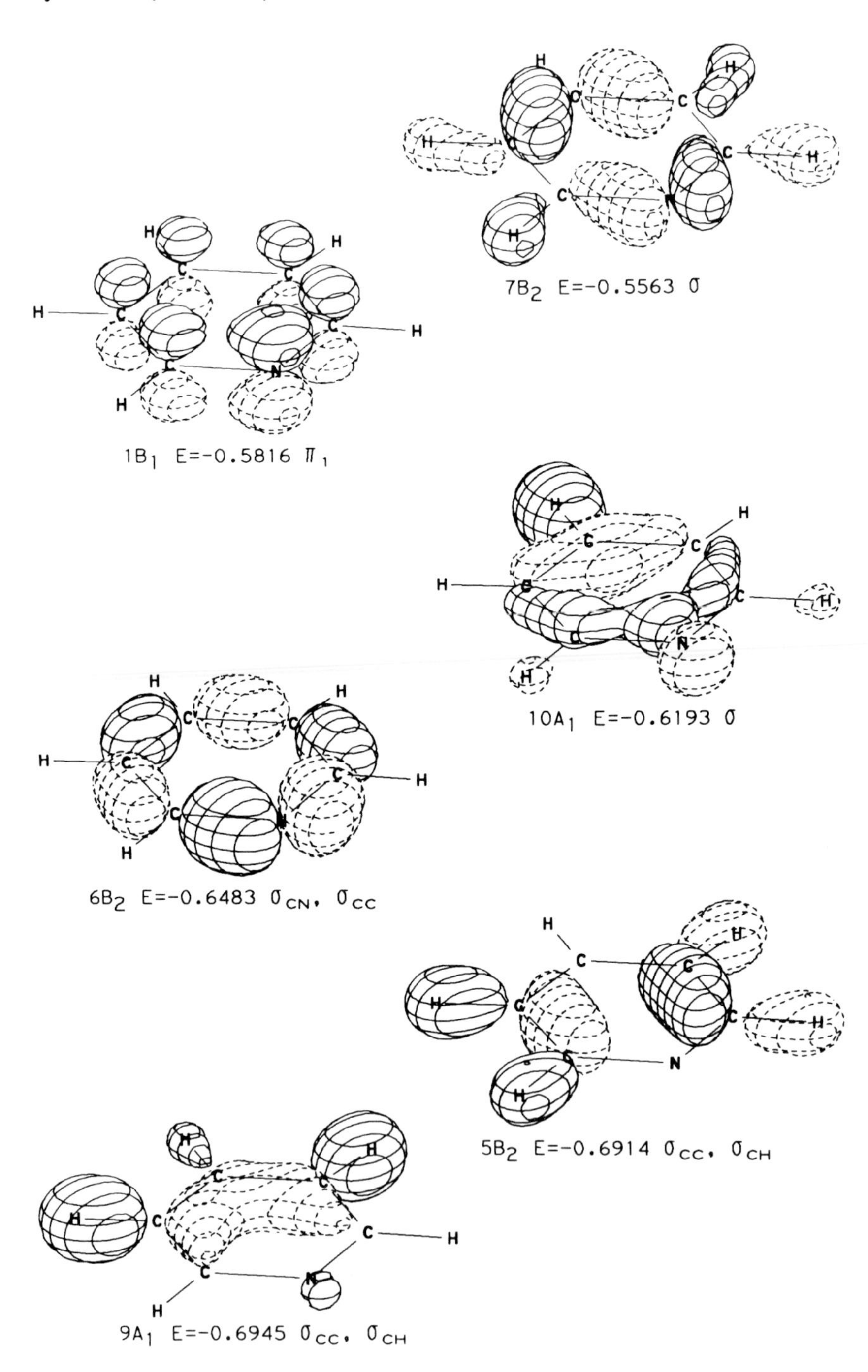

Pyridine (Continued)

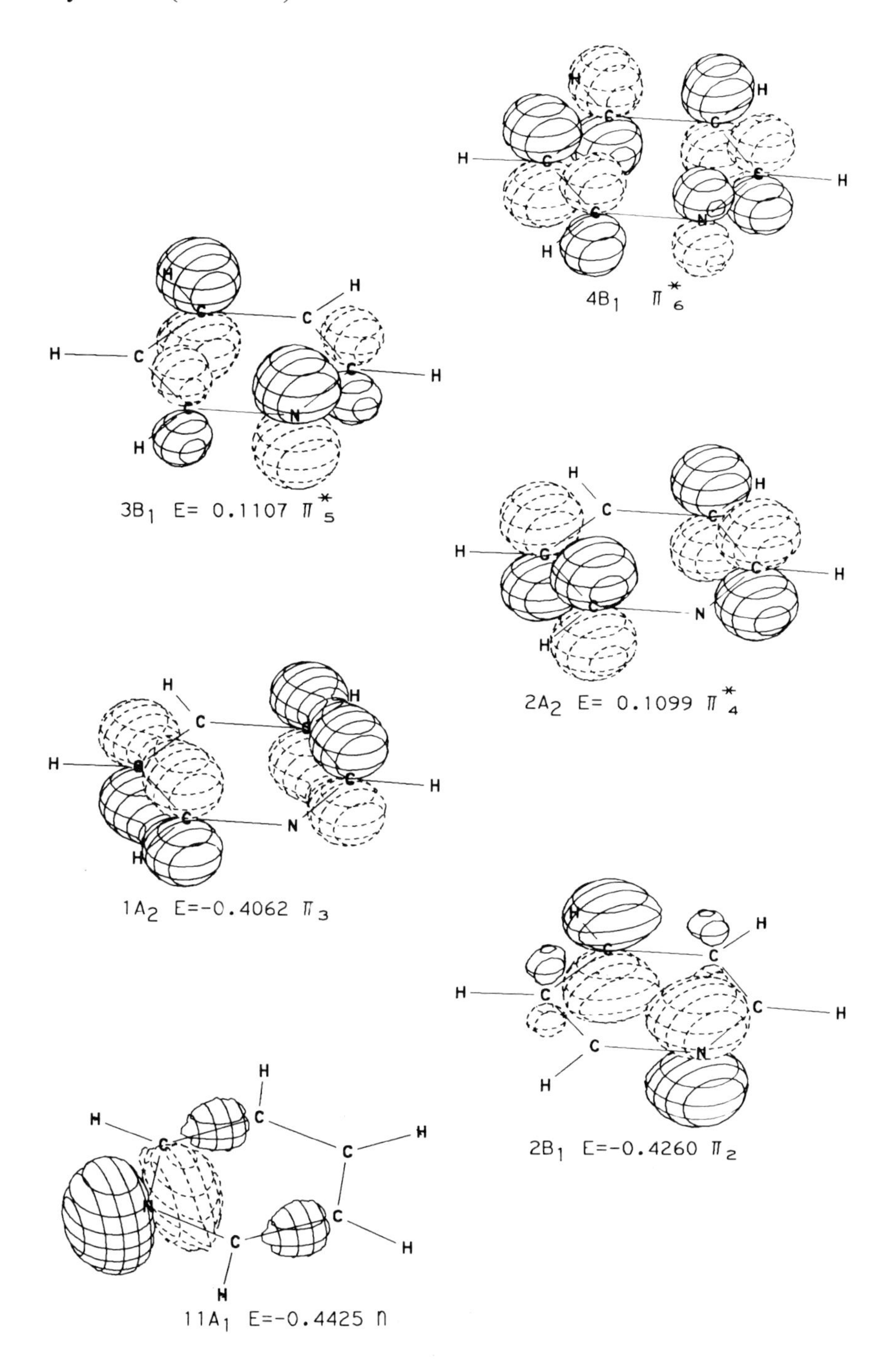

97. Pyrazine

Symmetry: D_{2h}

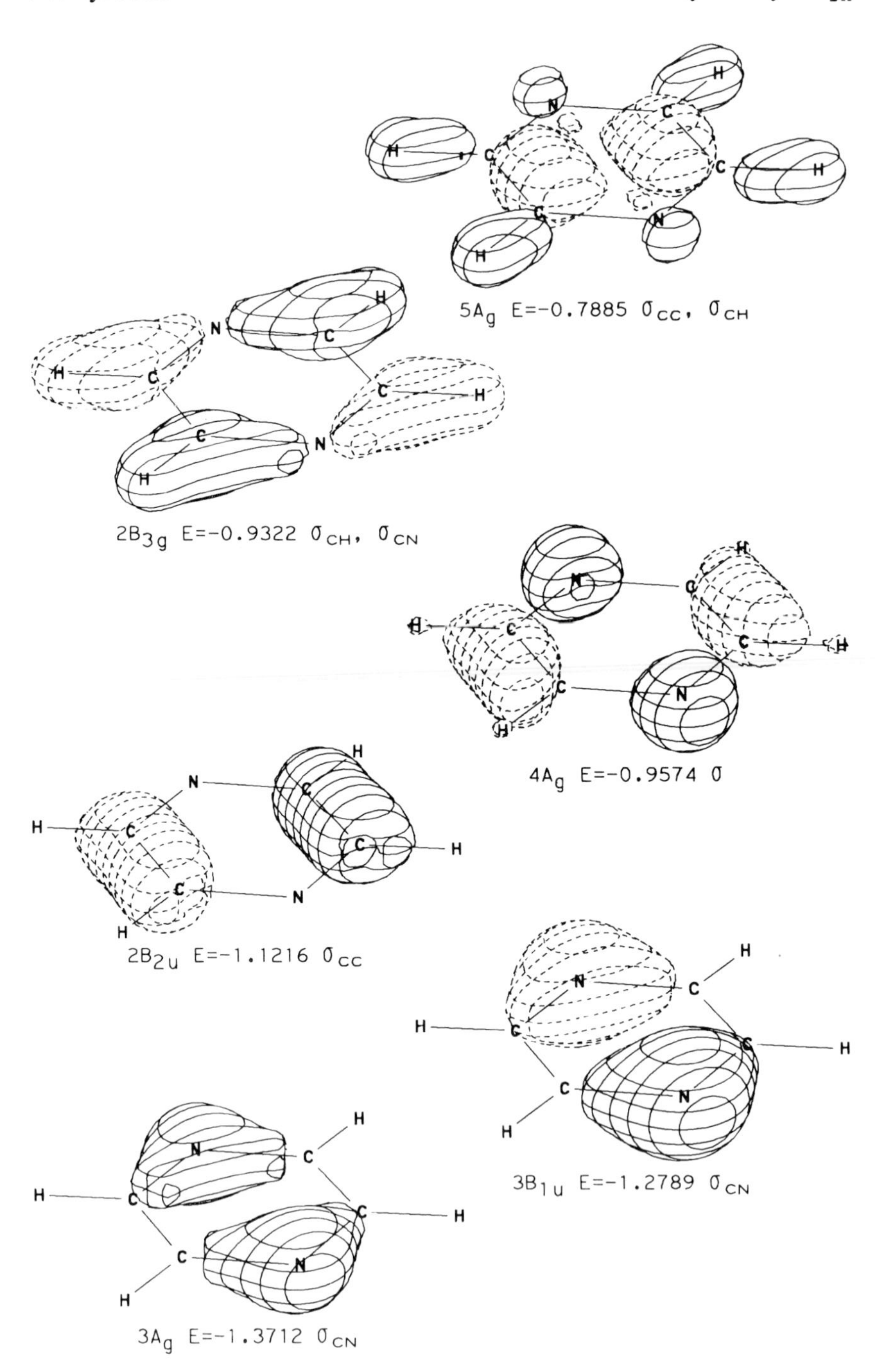

$5A_g$ E=-0.7885 σ_{CC}, σ_{CH}

$2B_{3g}$ E=-0.9322 σ_{CH}, σ_{CN}

$4A_g$ E=-0.9574 σ

$2B_{2u}$ E=-1.1216 σ_{CC}

$3B_{1u}$ E=-1.2789 σ_{CN}

$3A_g$ E=-1.3712 σ_{CN}

Pyrazine (Continued)

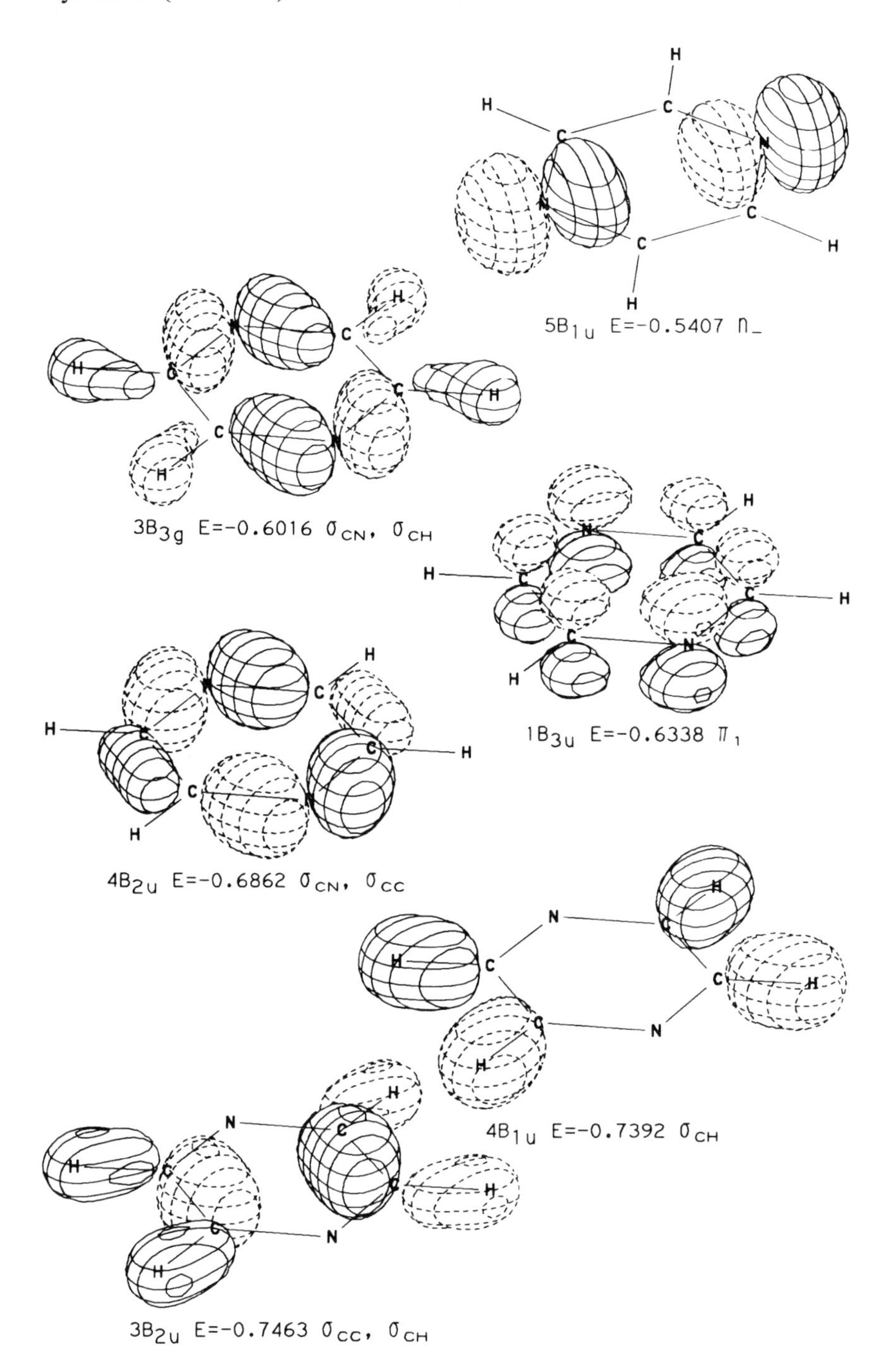

Pyrazine (Continued)

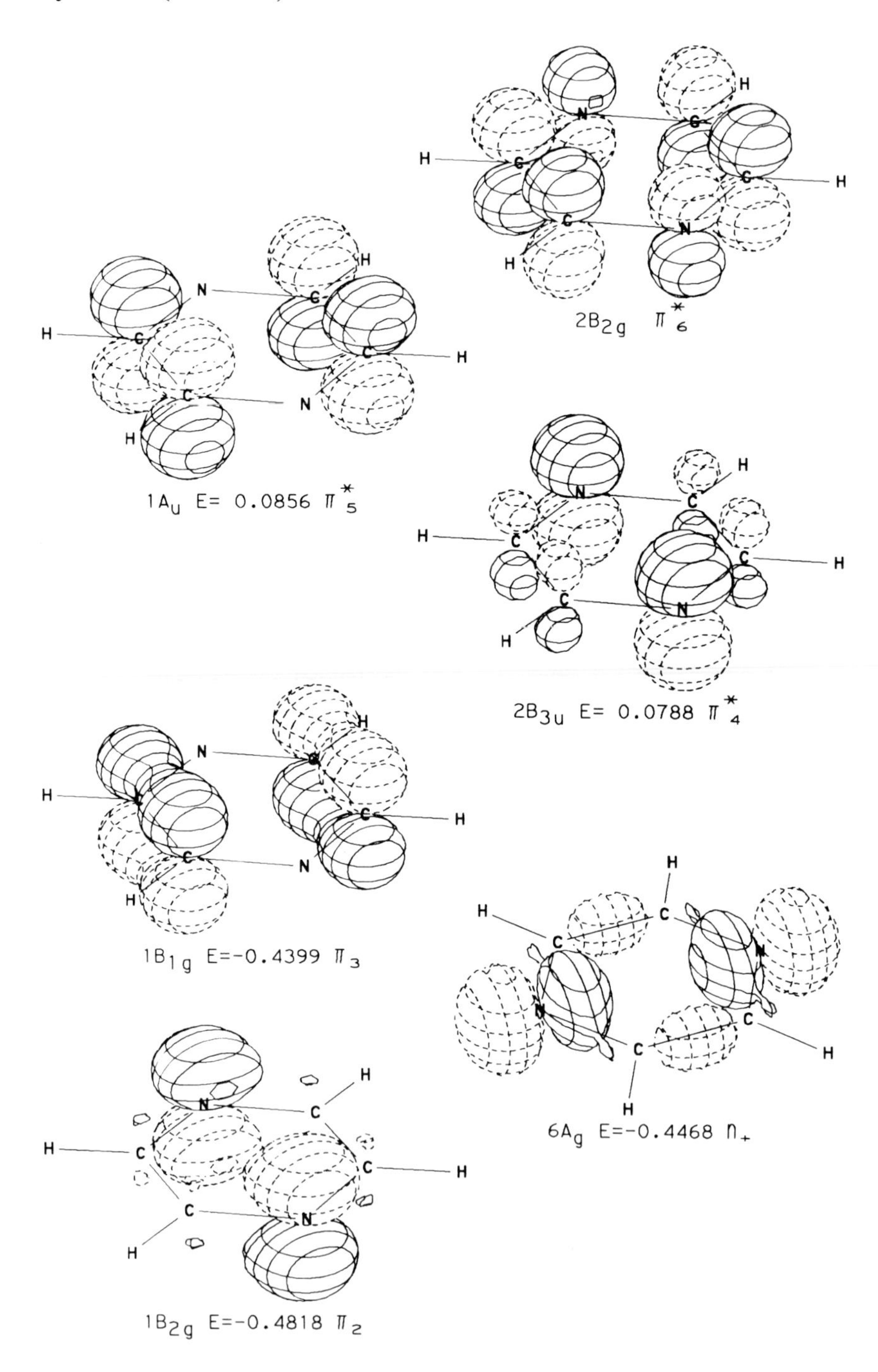

98. Cyclopentadienone (π orbitals only) Symmetry: C_{2v}

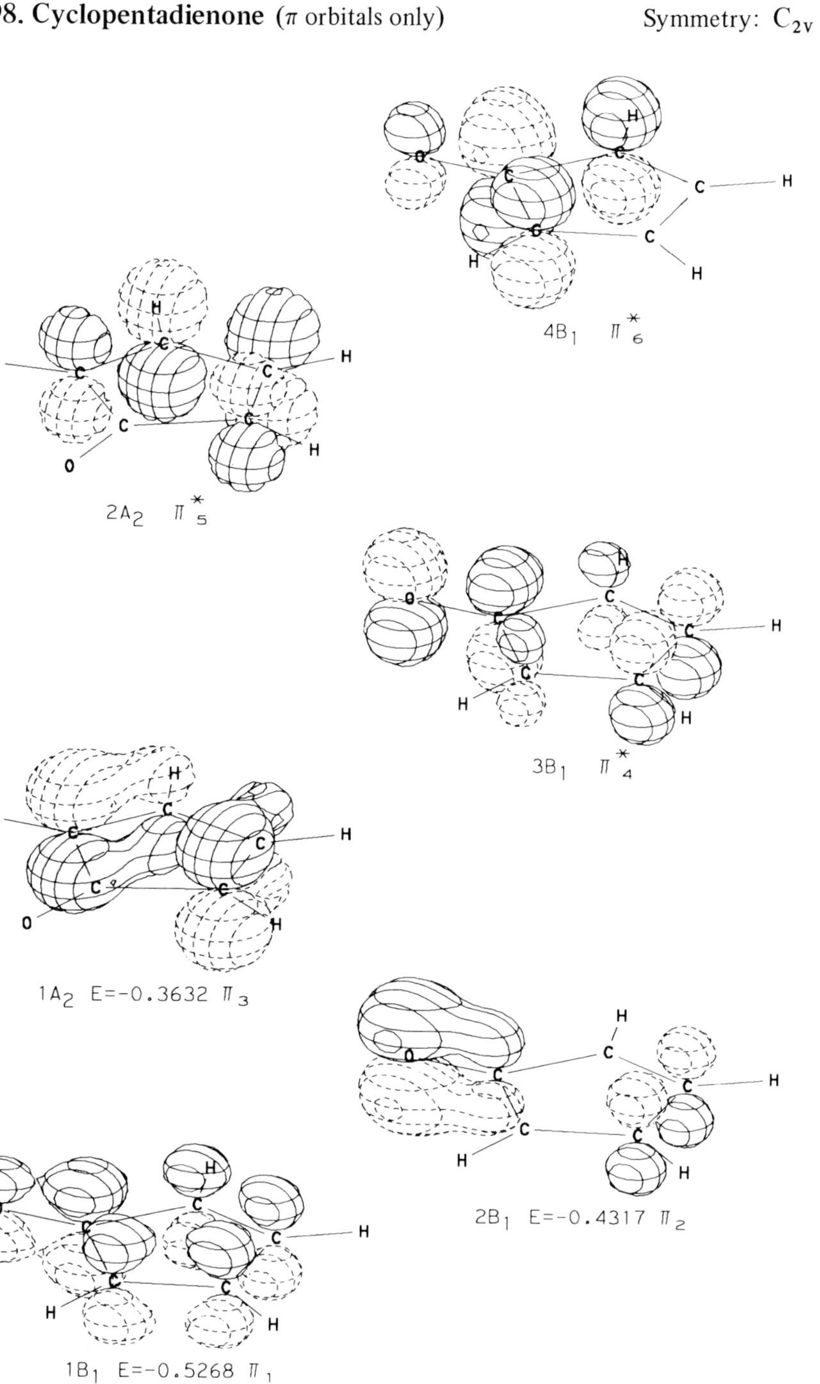

99. 1,3,5-Hexatriene (π orbitals only) Symmetry: C_{2h}

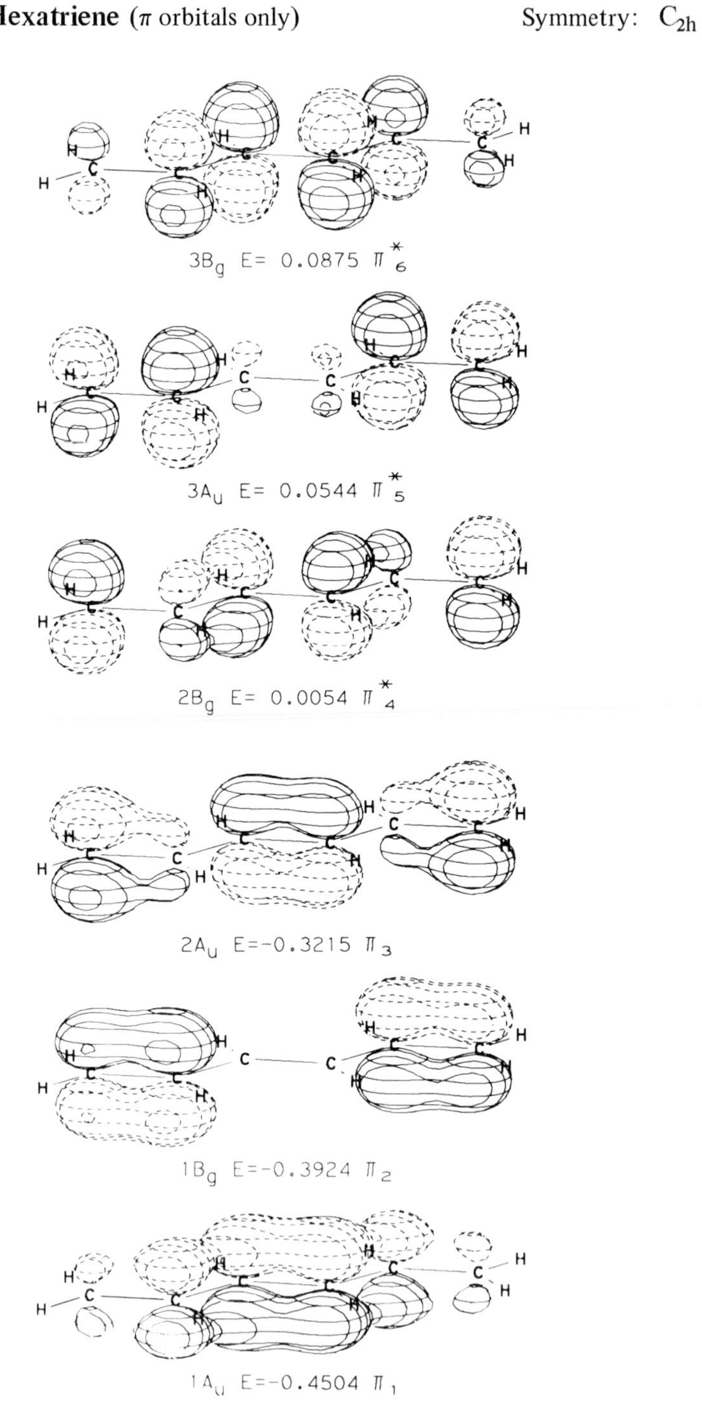

$3B_g$ E= 0.0875 π^*_6

$3A_u$ E= 0.0544 π^*_5

$2B_g$ E= 0.0054 π^*_4

$2A_u$ E=−0.3215 π_3

$1B_g$ E=−0.3924 π_2

$1A_u$ E=−0.4504 π_1

100. (2.1.1)-Bicyclohexene-2

Symmetry: C_{2v}

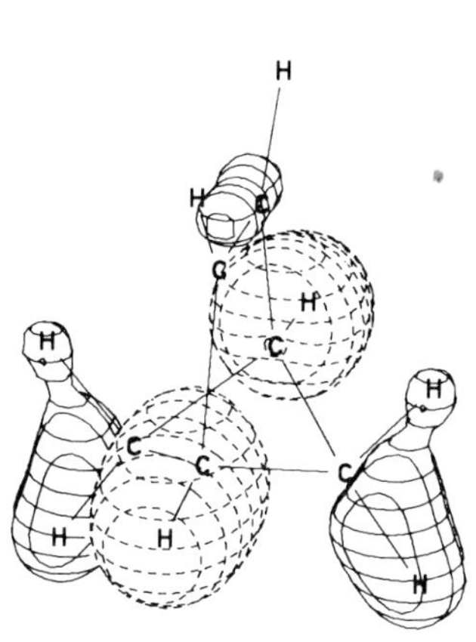

6A1 E=−0.727 σ_{CH}, σ_{CH_2}

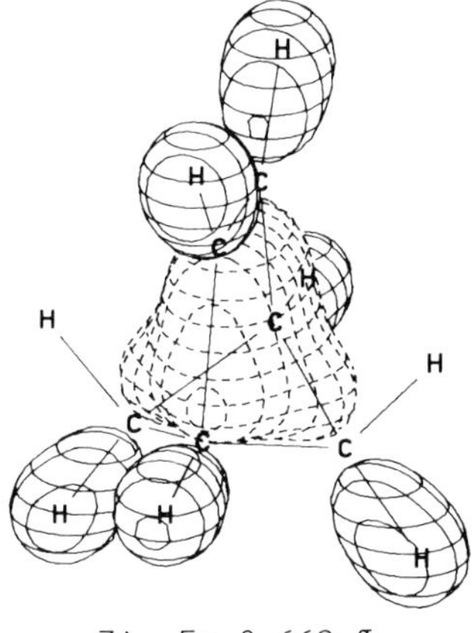

7A1 E=−0.669 σ

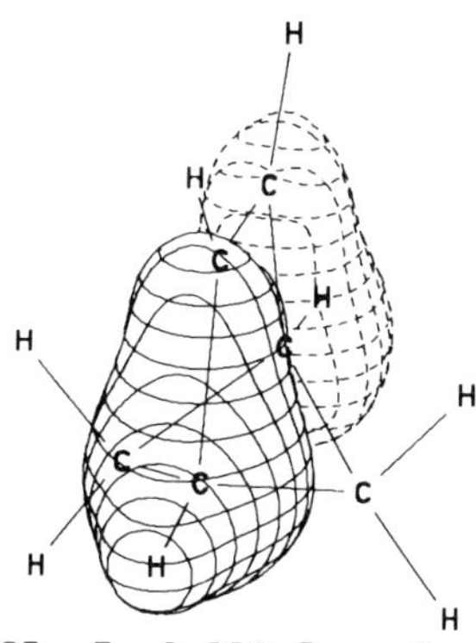

3B2 E=−0.920 σ_{CC}, σ_{CH}

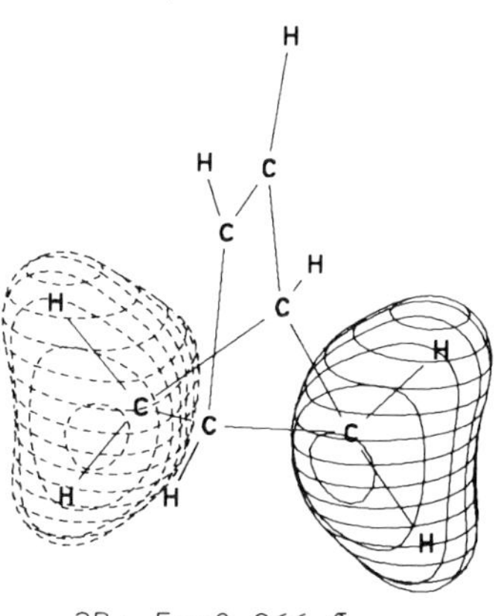

2B1 E=−0.866 σ_{CH_2}

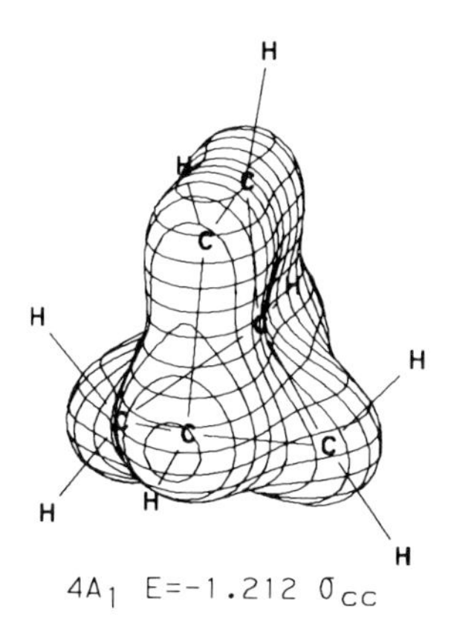

4A1 E=−1.212 σ_{CC}

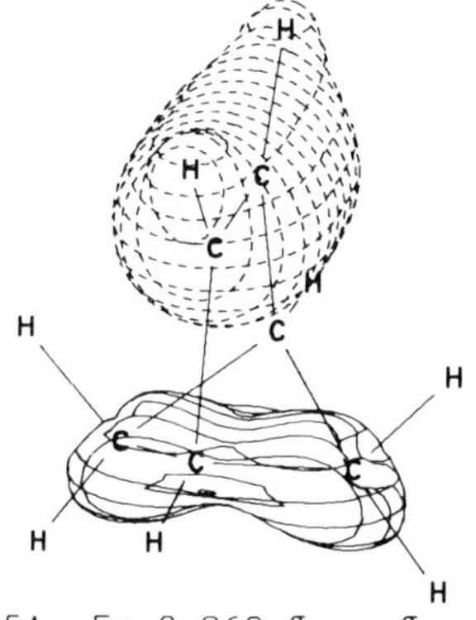

5A1 E=−0.968 σ_{CC}, σ_{CH}

(2.1.1)-Bicyclohexene-2 (Continued)

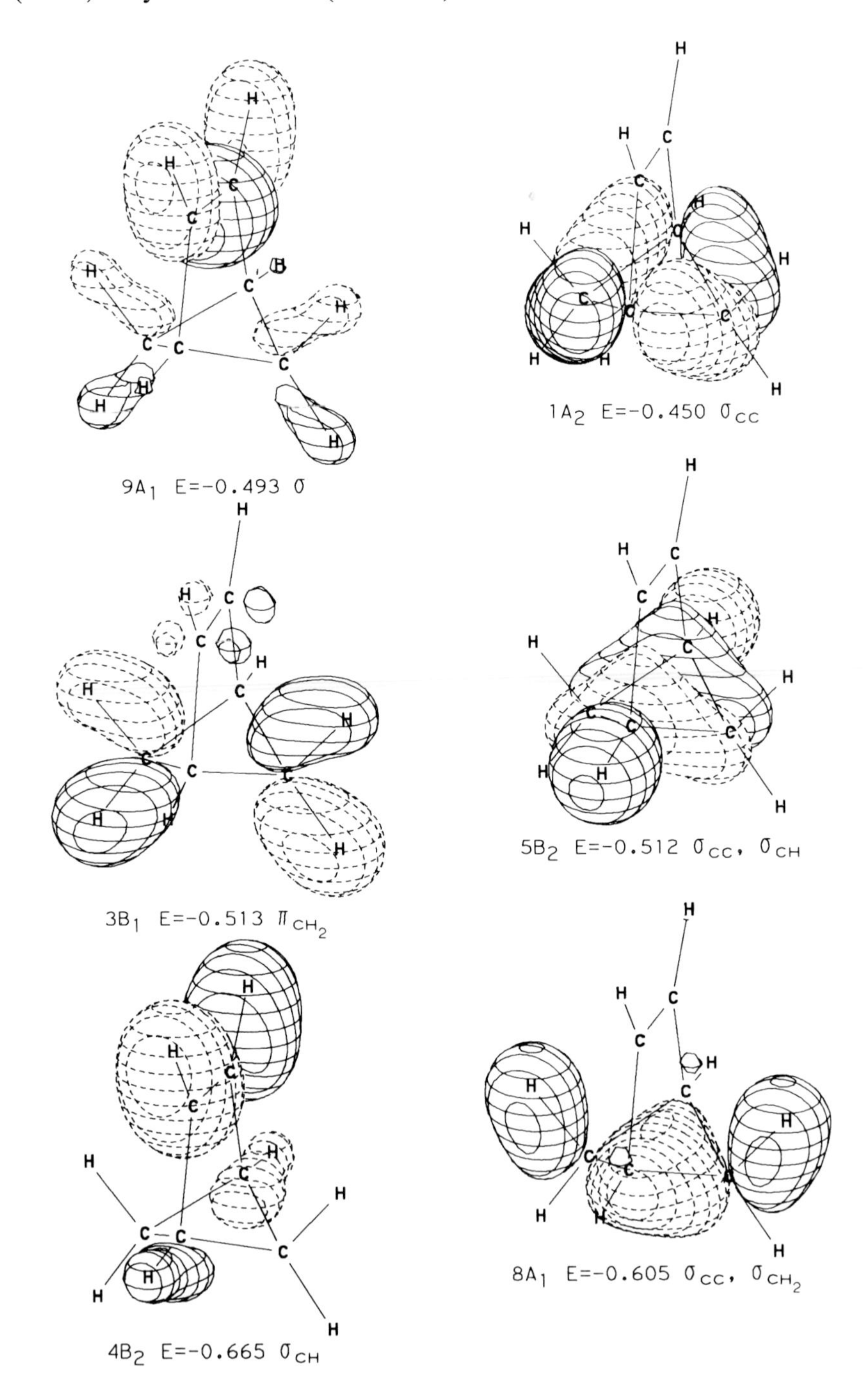

$9A_1$ E=−0.493 σ

$1A_2$ E=−0.450 σ_{CC}

$3B_1$ E=−0.513 π_{CH_2}

$5B_2$ E=−0.512 σ_{CC}, σ_{CH}

$4B_2$ E=−0.665 σ_{CH}

$8A_1$ E=−0.605 σ_{CC}, σ_{CH_2}

(2.1.1)-Bicyclohexene-2 (Continued)

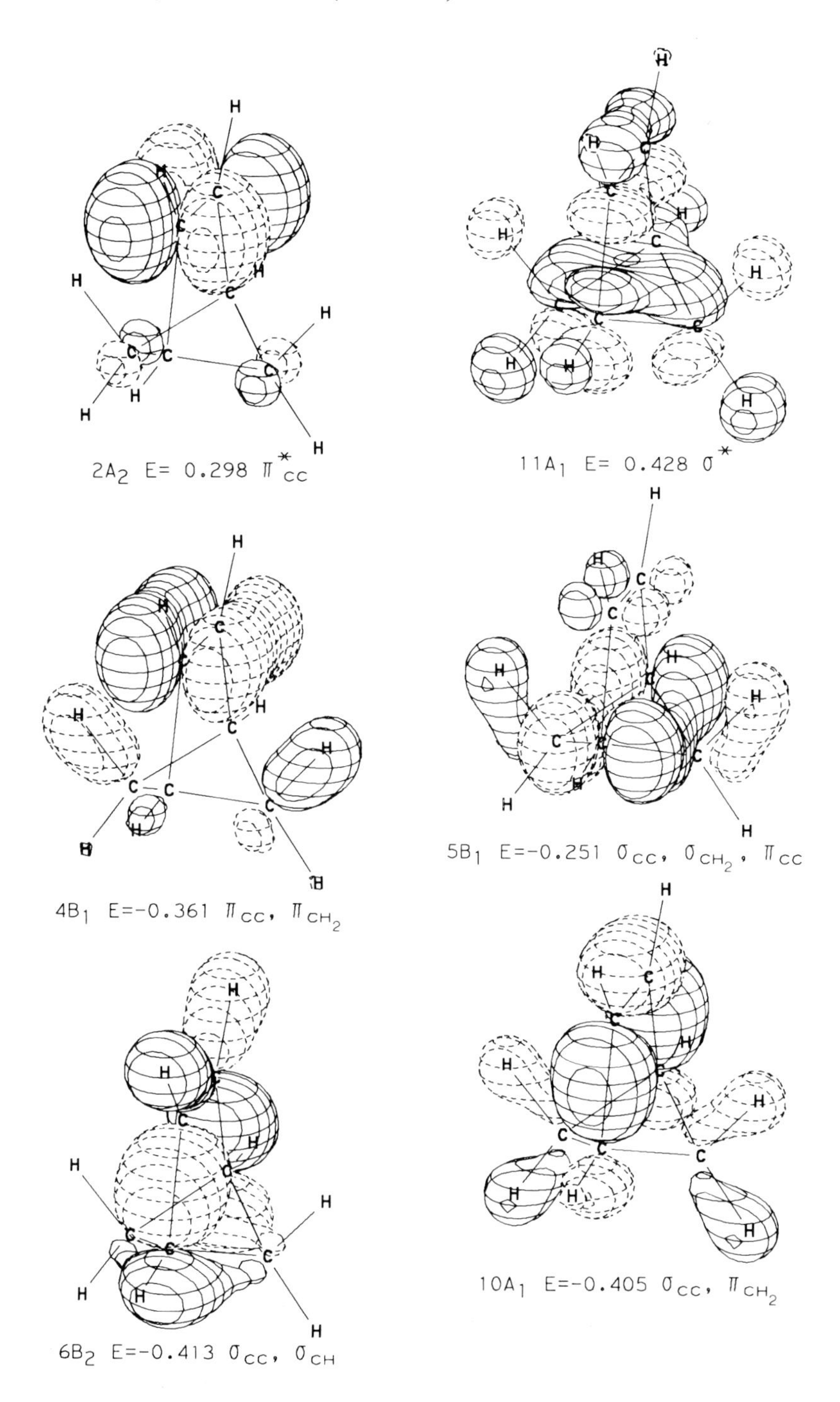

$2A_2$ E= 0.298 π^*_{CC}

$11A_1$ E= 0.428 σ^*

$4B_1$ E=-0.361 π_{CC}, π_{CH_2}

$5B_1$ E=-0.251 σ_{CC}, σ_{CH_2}, π_{CC}

$6B_2$ E=-0.413 σ_{CC}, σ_{CH}

$10A_1$ E=-0.405 σ_{CC}, π_{CH_2}

101. Cyclohexene, Half-Boat

Symmetry: C_s

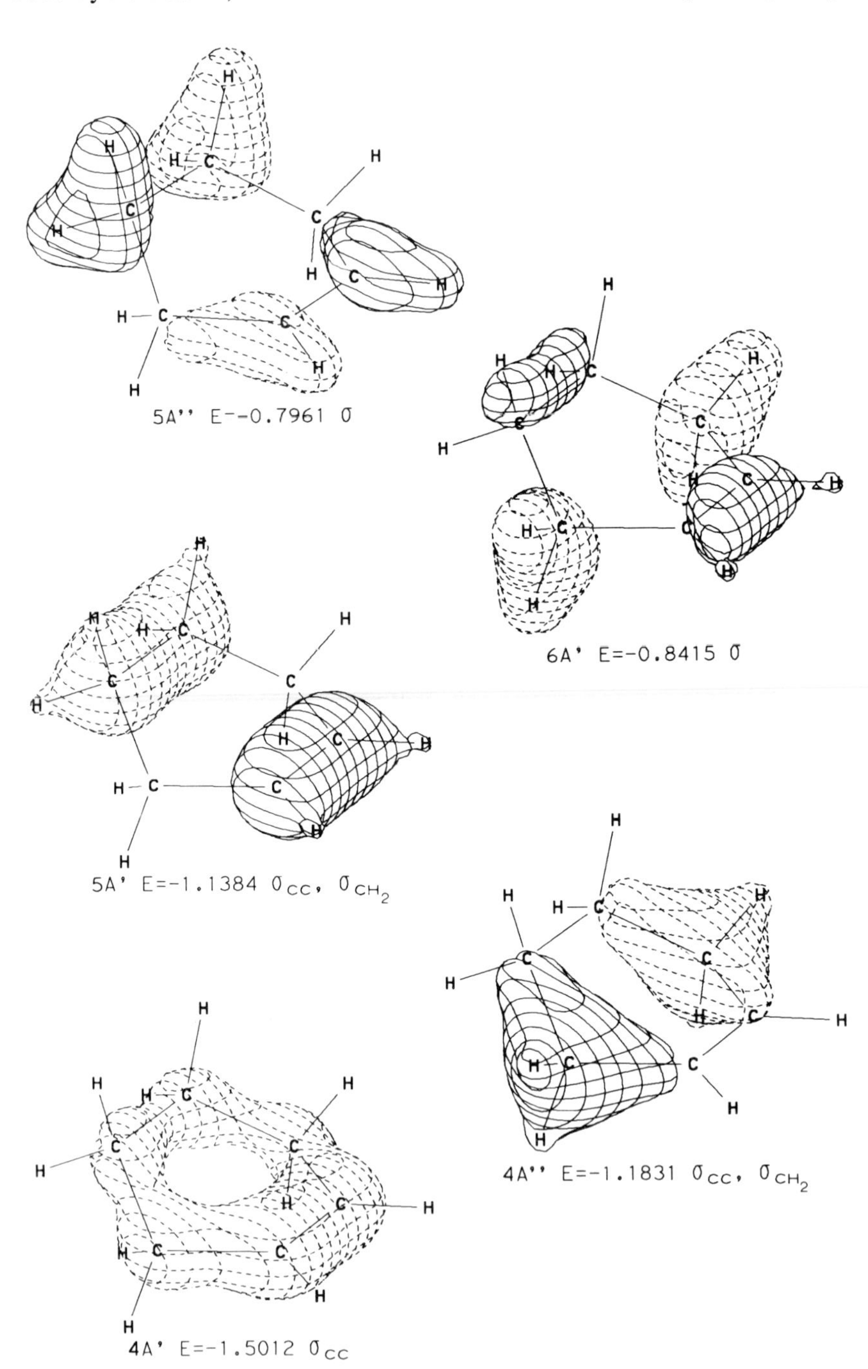

5A'' E--0.7961 σ

6A' E=-0.8415 σ

5A' E=-1.1384 σ_{CC}, σ_{CH_2}

4A'' E=-1.1831 σ_{CC}, σ_{CH_2}

4A' E=-1.5012 σ_{CC}

Cyclohexene, Half-Boat (Continued)

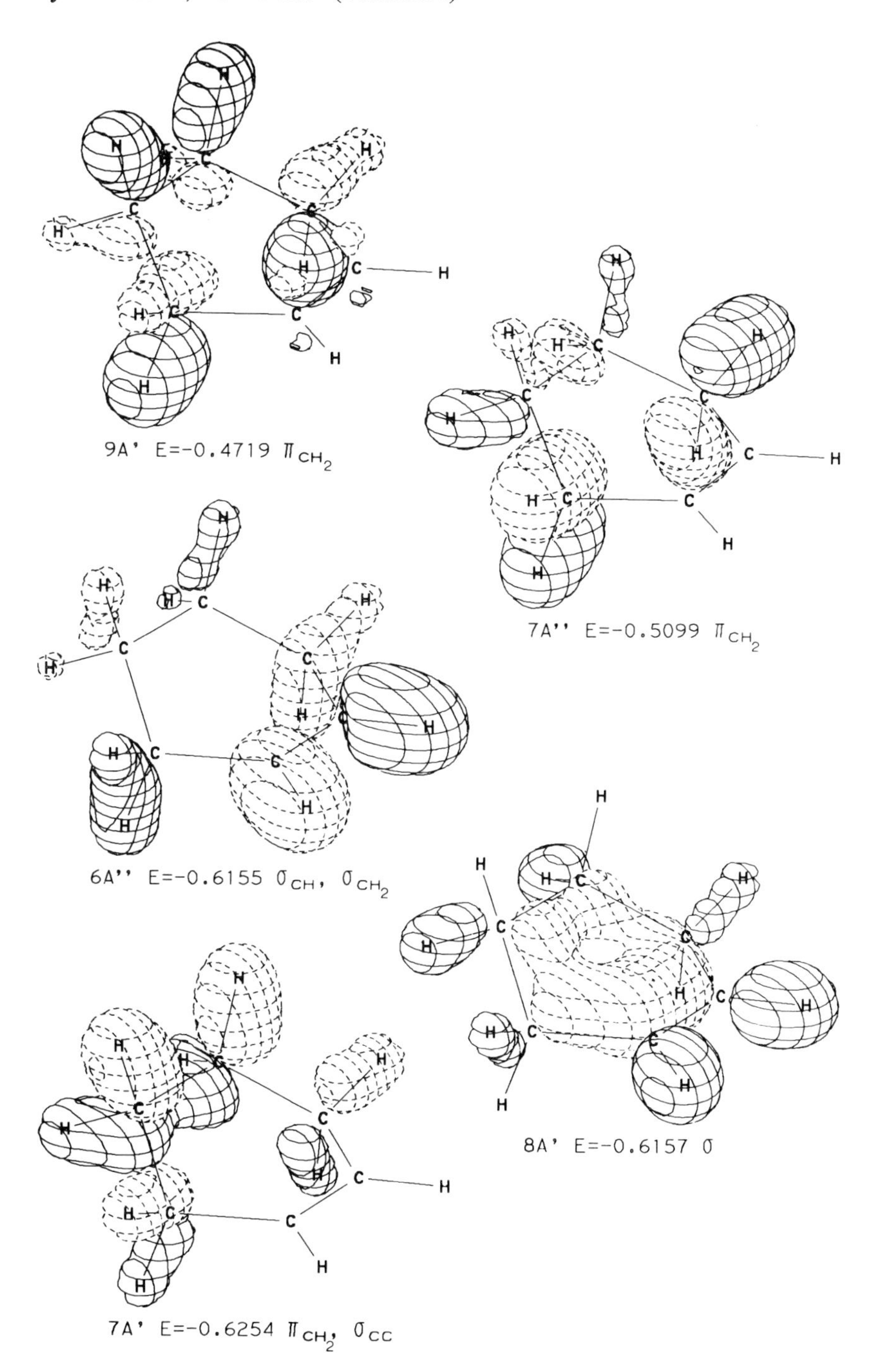

9A' $E=-0.4719$ π_{CH_2}

7A'' $E=-0.5099$ π_{CH_2}

6A'' $E=-0.6155$ σ_{CH}, σ_{CH_2}

8A' $E=-0.6157$ σ

7A' $E=-0.6254$ π_{CH_2}, σ_{CC}

Cyclohexene, Half-Boat (Continued)

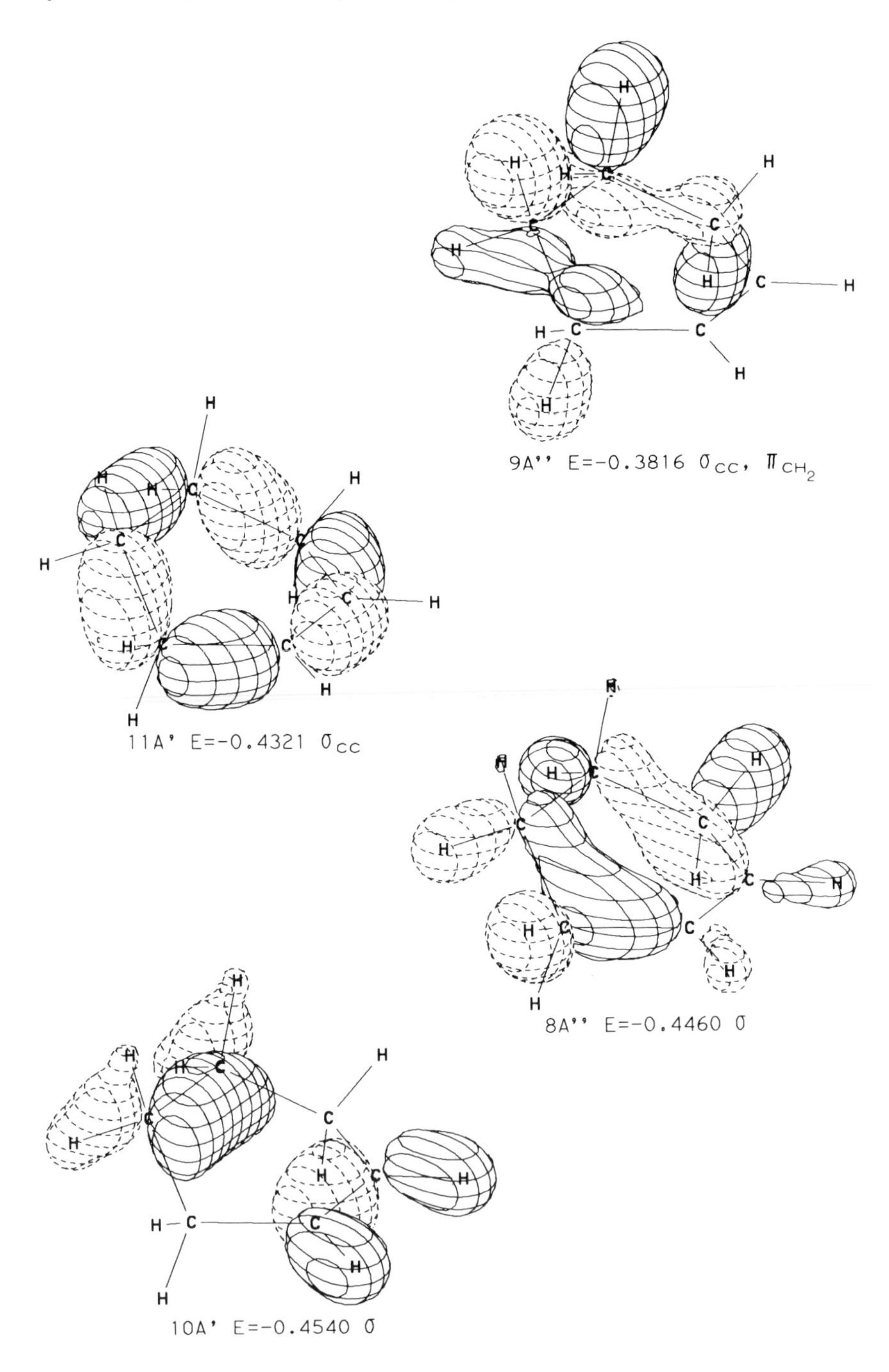

Cyclohexene, Half-Boat (Continued)

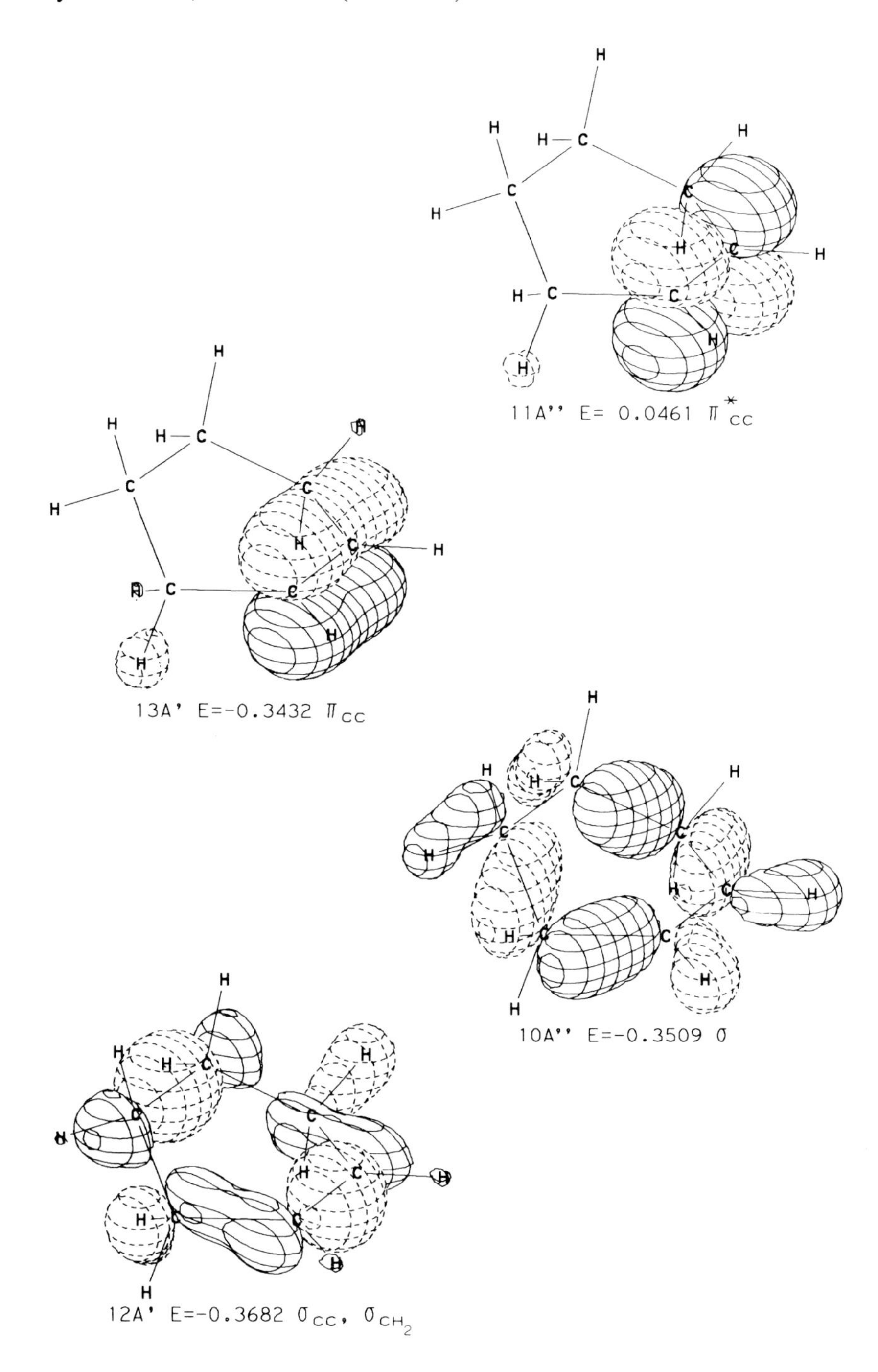

11A'' E= 0.0461 π^{*}_{CC}

13A' E=-0.3432 π_{CC}

10A'' E=-0.3509 σ

12A' E=-0.3682 σ_{CC}, σ_{CH_2}

102. Cyclohexane, Chair

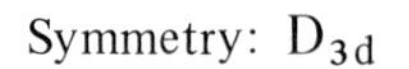

Symmetry: D_{3d}

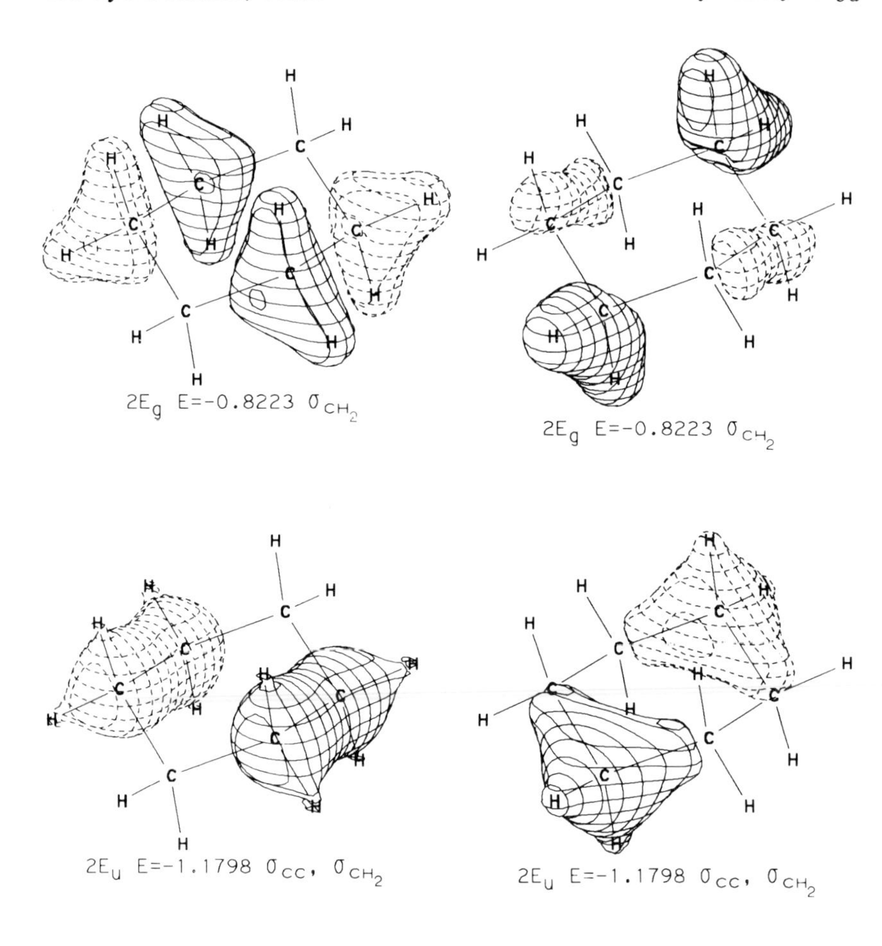

$2E_g$ E=-0.8223 σ_{CH_2}

$2E_g$ E=-0.8223 σ_{CH_2}

$2E_u$ E=-1.1798 σ_{CC}, σ_{CH_2}

$2E_u$ E=-1.1798 σ_{CC}, σ_{CH_2}

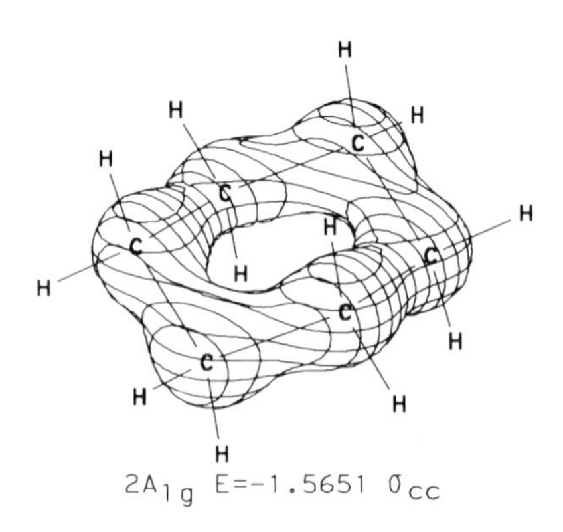

$2A_{1g}$ E=-1.5651 σ_{CC}

Cyclohexane, Chair (Continued)

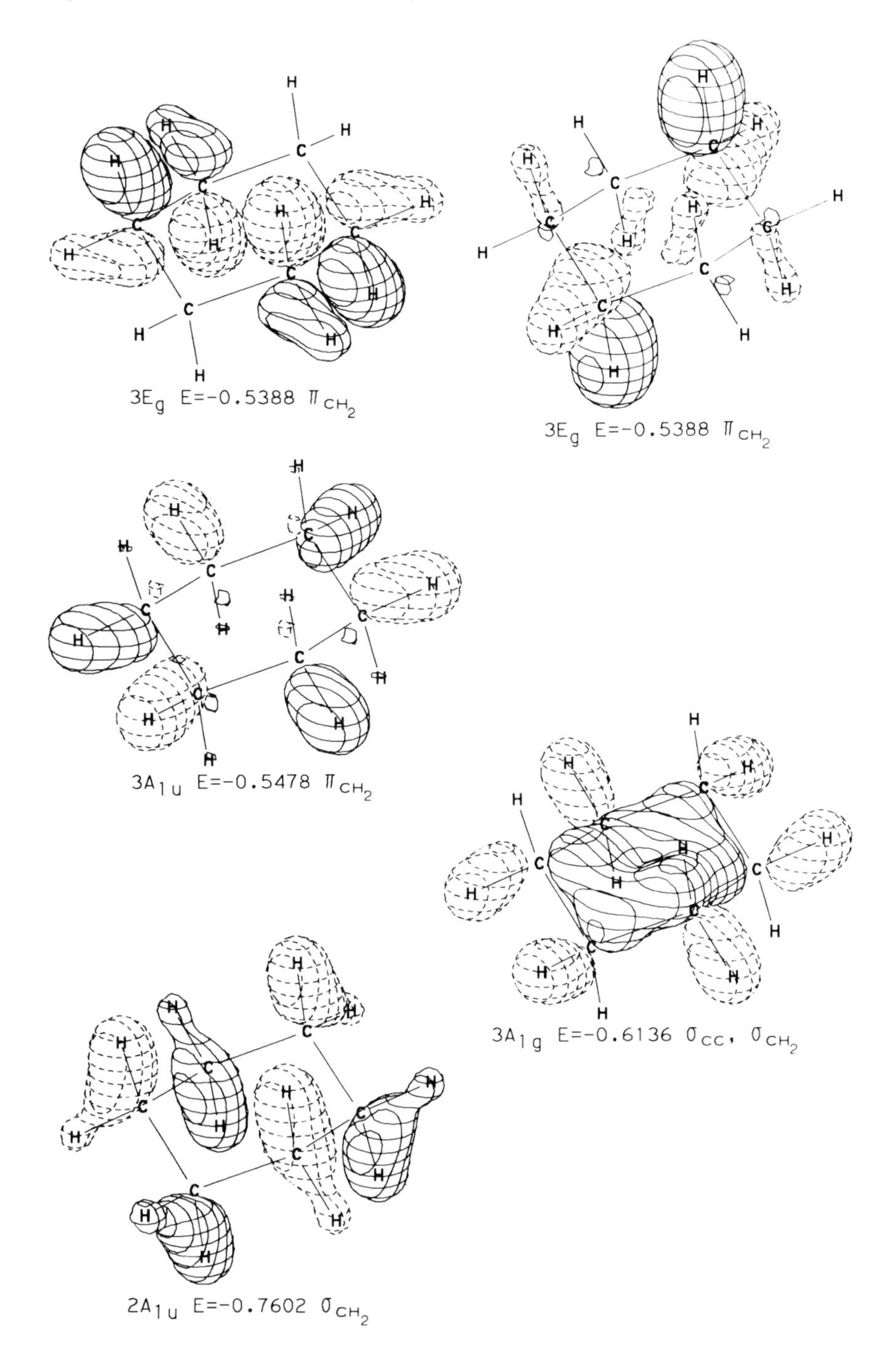

$3E_g$ E=−0.5388 π_{CH_2}

$3E_g$ E=−0.5388 π_{CH_2}

$3A_{1u}$ E=−0.5478 π_{CH_2}

$3A_{1g}$ E=−0.6136 σ_{CC}, σ_{CH_2}

$2A_{1u}$ E=−0.7602 σ_{CH_2}

Cyclohexane, Chair (Continued)

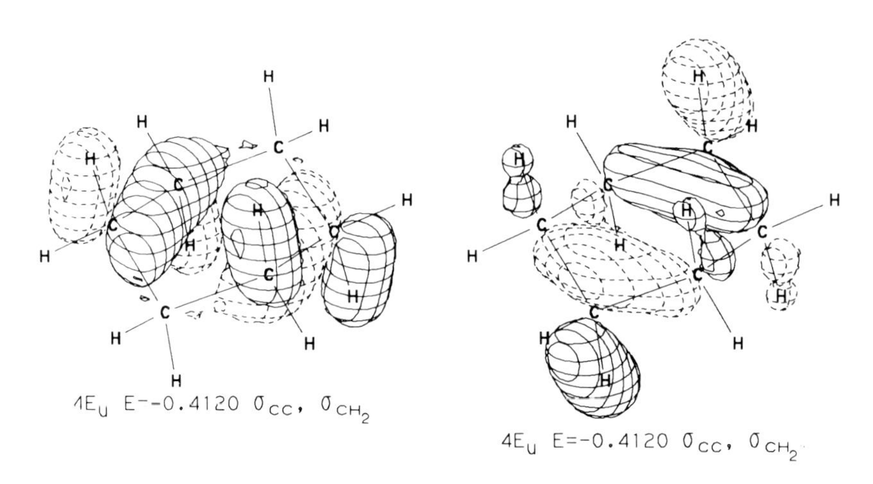

$4E_u$ E--0.4120 σ_{CC}, σ_{CH_2}

$4E_u$ E=-0.4120 σ_{CC}, σ_{CH_2}

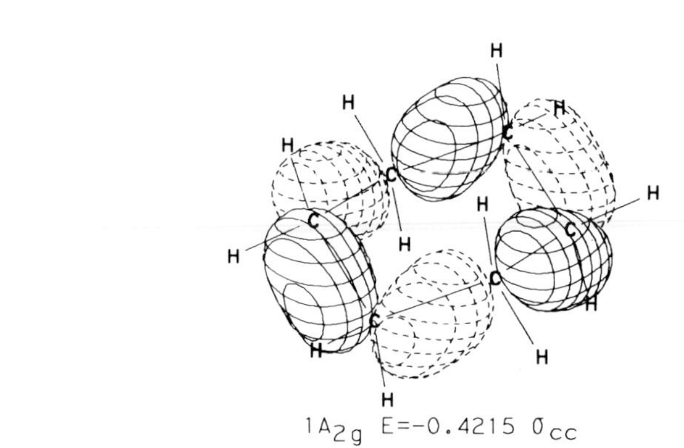

$1A_{2g}$ E=-0.4215 σ_{CC}

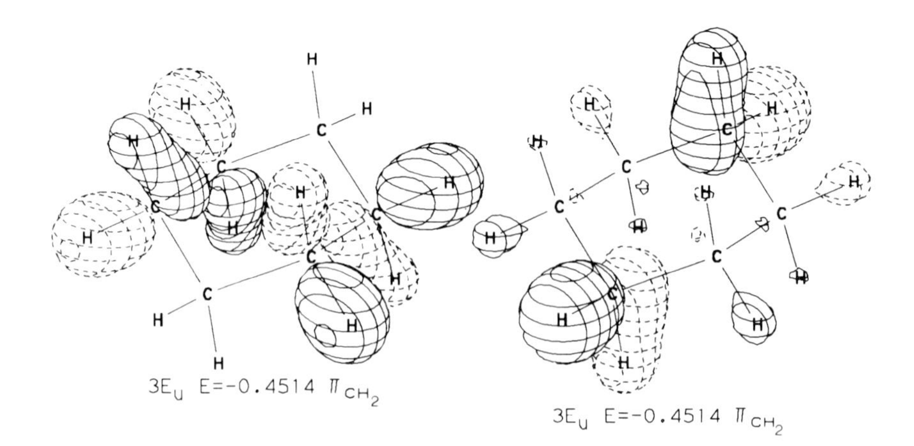

$3E_u$ E=-0.4514 π_{CH_2}

$3E_u$ E=-0.4514 π_{CH_2}

Cyclohexane, Chair (Continued)

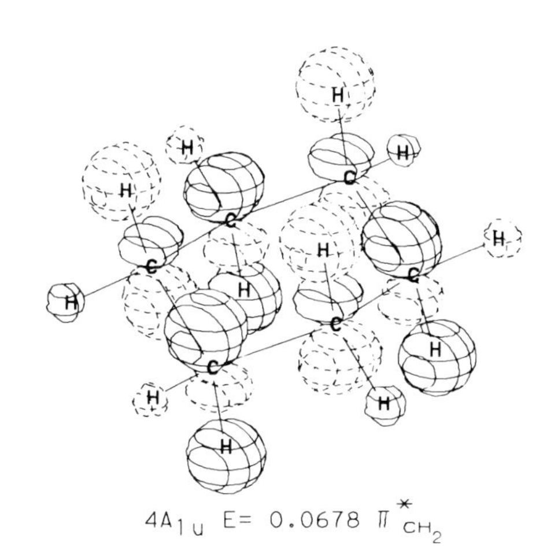

$4A_{1u}$ E= 0.0678 $\pi^{*}_{CH_2}$

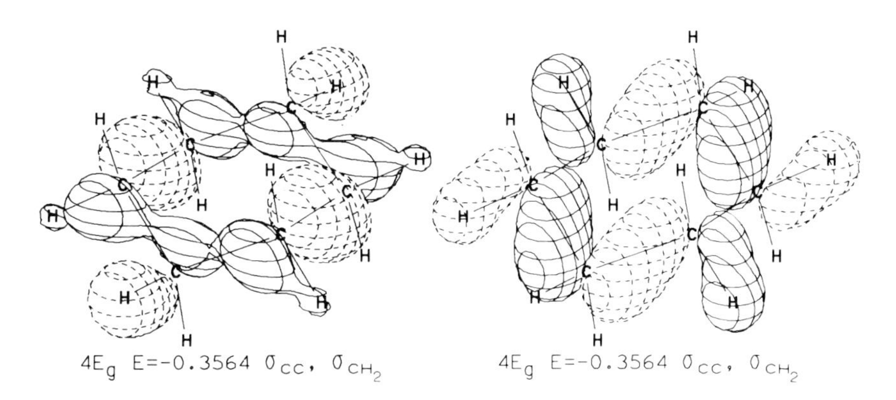

$4E_g$ E=-0.3564 σ_{CC}, σ_{CH_2}

$4E_g$ E=-0.3564 σ_{CC}, σ_{CH_2}

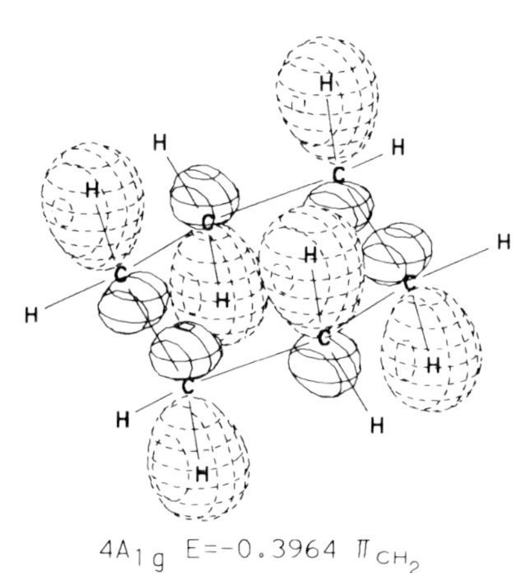

$4A_{1g}$ E=-0.3964 π_{CH_2}

103. Norbornadiene

Symmetry: C_{2v}

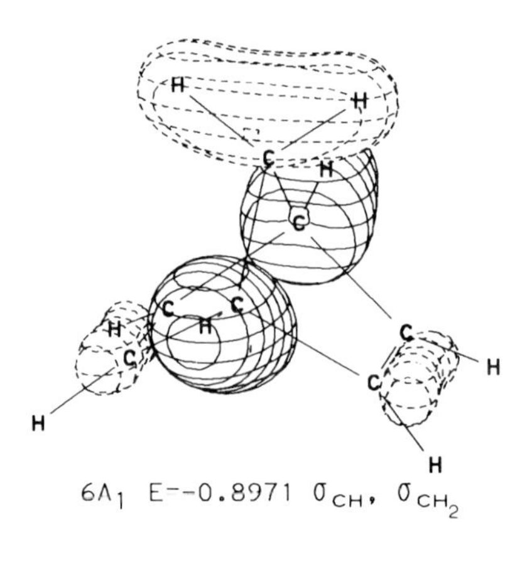

$6A_1$ E=−0.8971 σ_{CH}, σ_{CH_2}

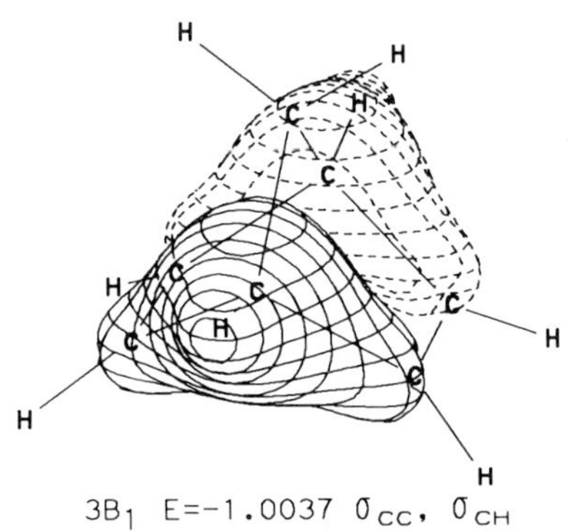

$3B_1$ E=−1.0037 σ_{CC}, σ_{CH}

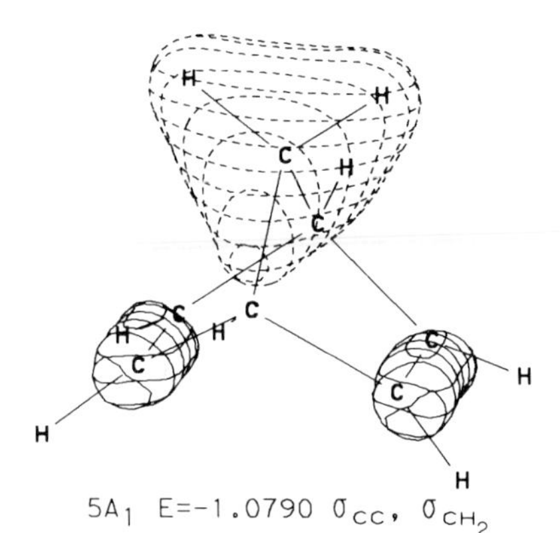

$5A_1$ E=−1.0790 σ_{CC}, σ_{CH_2}

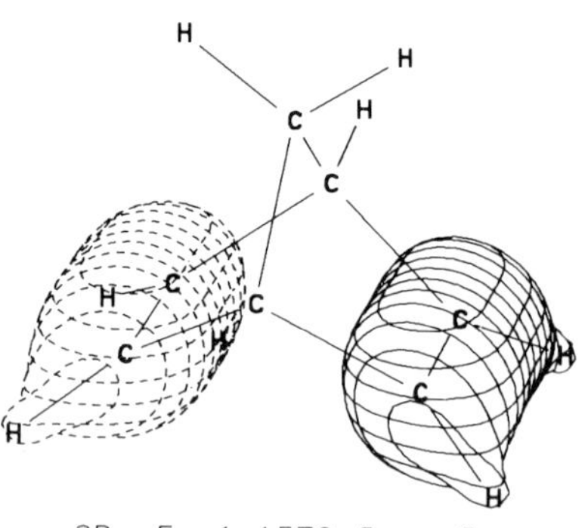

$2B_2$ E=−1.1573 σ_{CC}, σ_{CH}

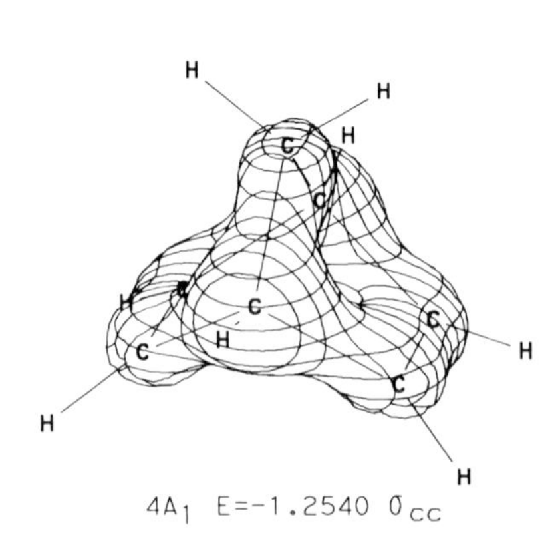

$4A_1$ E=−1.2540 σ_{CC}

Norbornadiene (Continued)

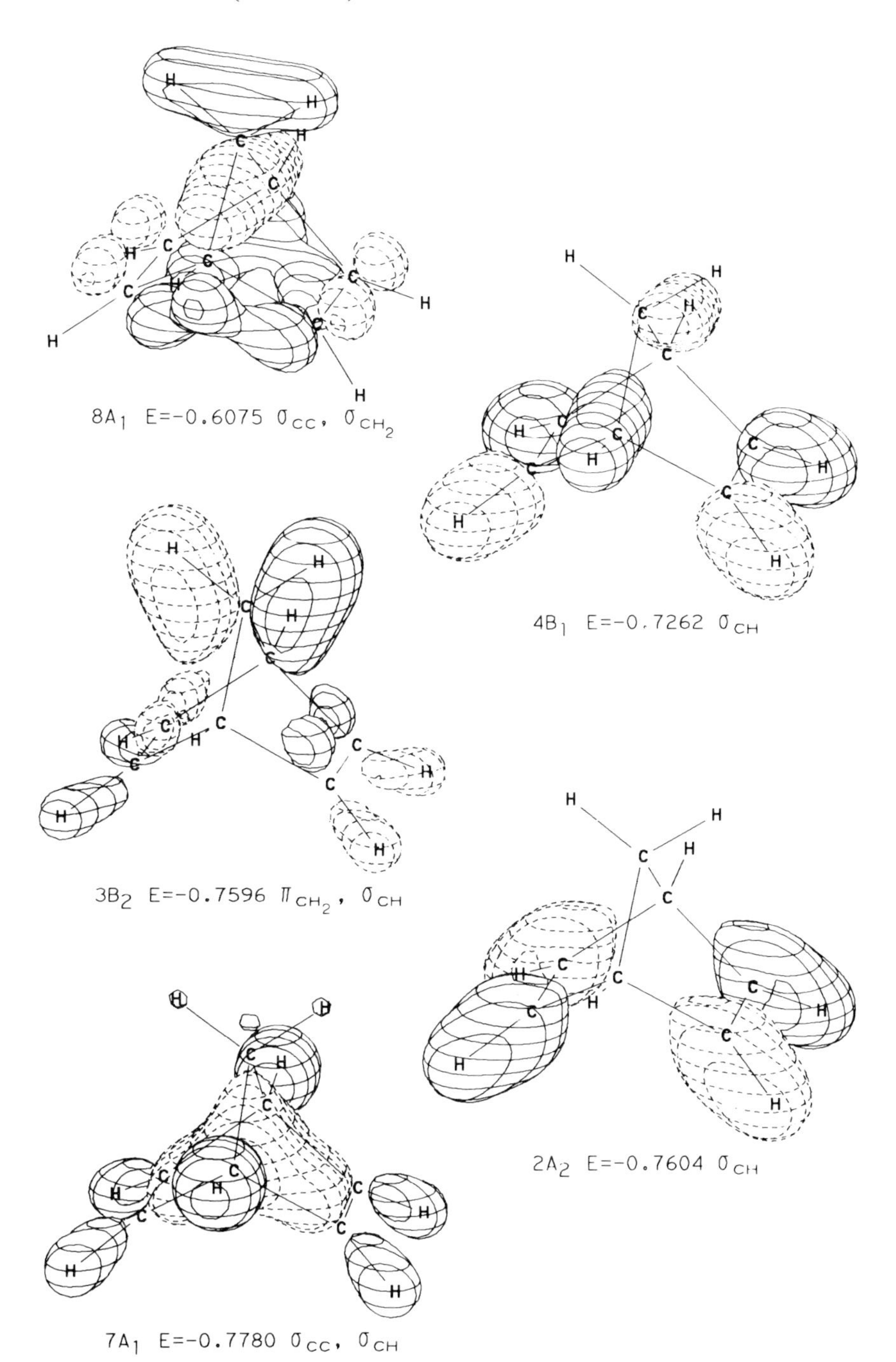

8A$_1$ E=-0.6075 σ_{CC}, σ_{CH_2}

4B$_1$ E=-0.7262 σ_{CH}

3B$_2$ E=-0.7596 π_{CH_2}, σ_{CH}

2A$_2$ E=-0.7604 σ_{CH}

7A$_1$ E=-0.7780 σ_{CC}, σ_{CH}

Norbornadiene (Continued)

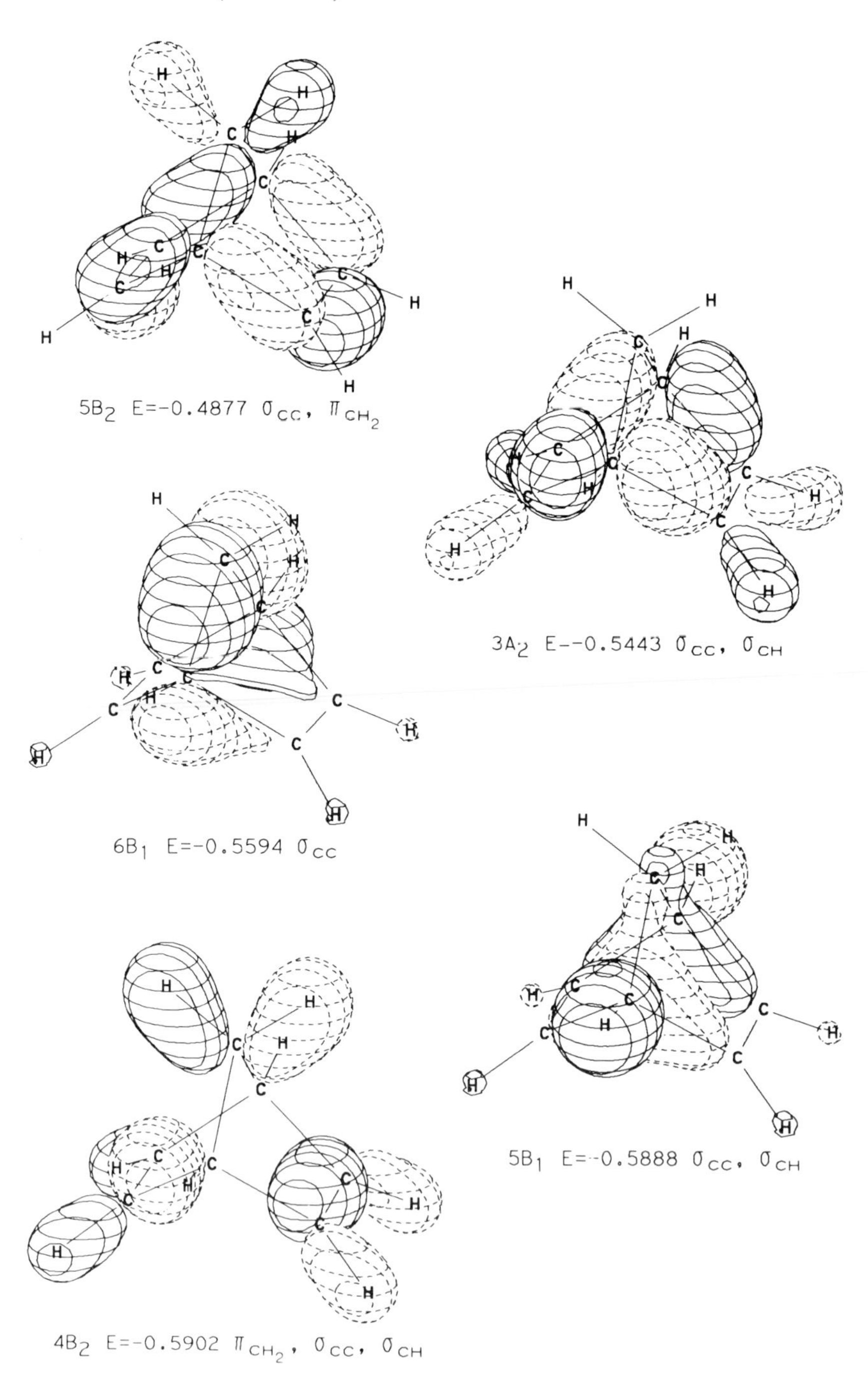

$5B_2$ E=-0.4877 σ_{CC}, π_{CH_2}

$3A_2$ E--0.5443 σ_{CC}, σ_{CH}

$6B_1$ E=-0.5594 σ_{CC}

$5B_1$ E=-0.5888 σ_{CC}, σ_{CH}

$4B_2$ E=-0.5902 π_{CH_2}, σ_{CC}, σ_{CH}

Norbornadiene (Continued)

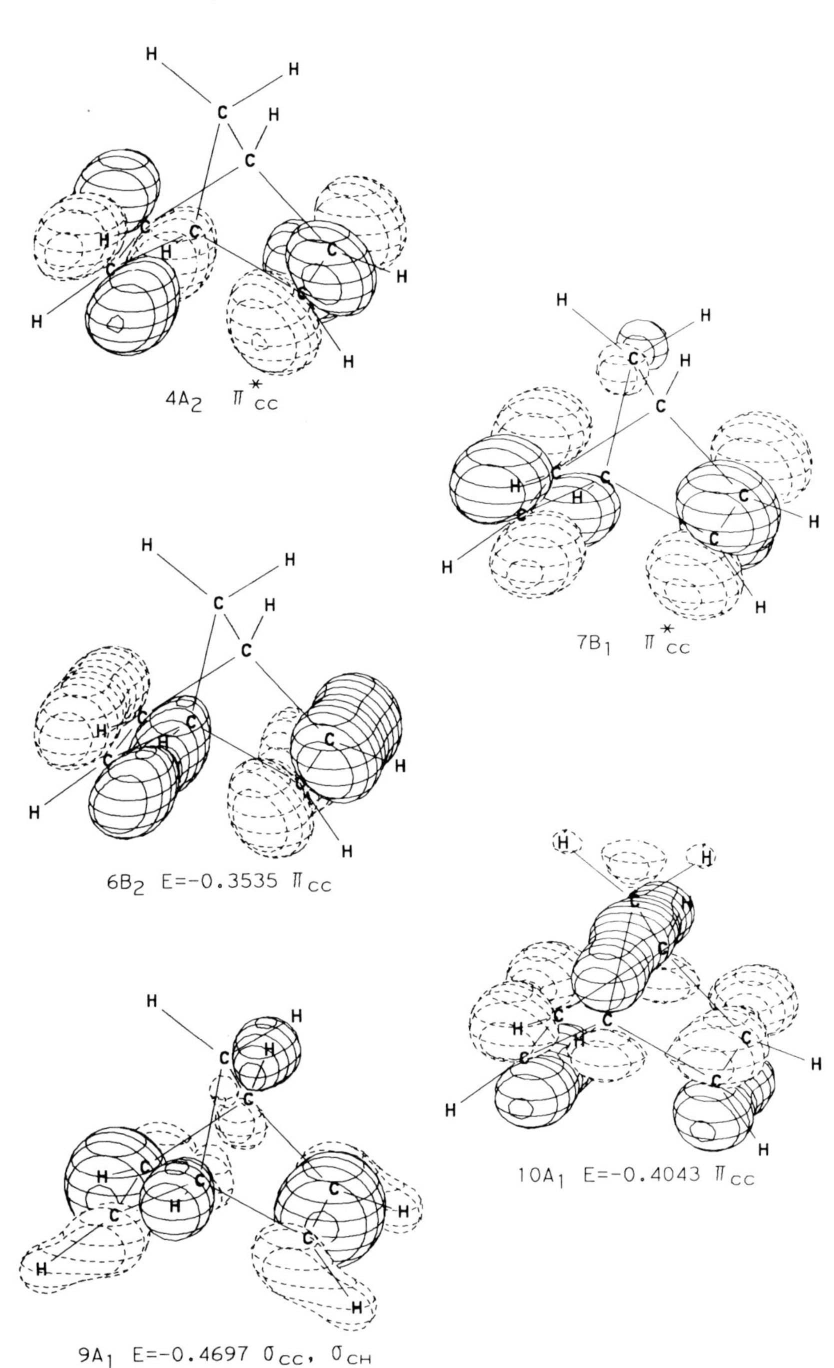

$4A_2$ π^*_{CC}

$7B_1$ π^*_{CC}

$6B_2$ E=-0.3535 π_{CC}

$10A_1$ E=-0.4043 π_{CC}

$9A_1$ E=-0.4697 σ_{CC}, σ_{CH}

104. Maleic Anhydride (π orbitals only) Symmetry: C_{2v}

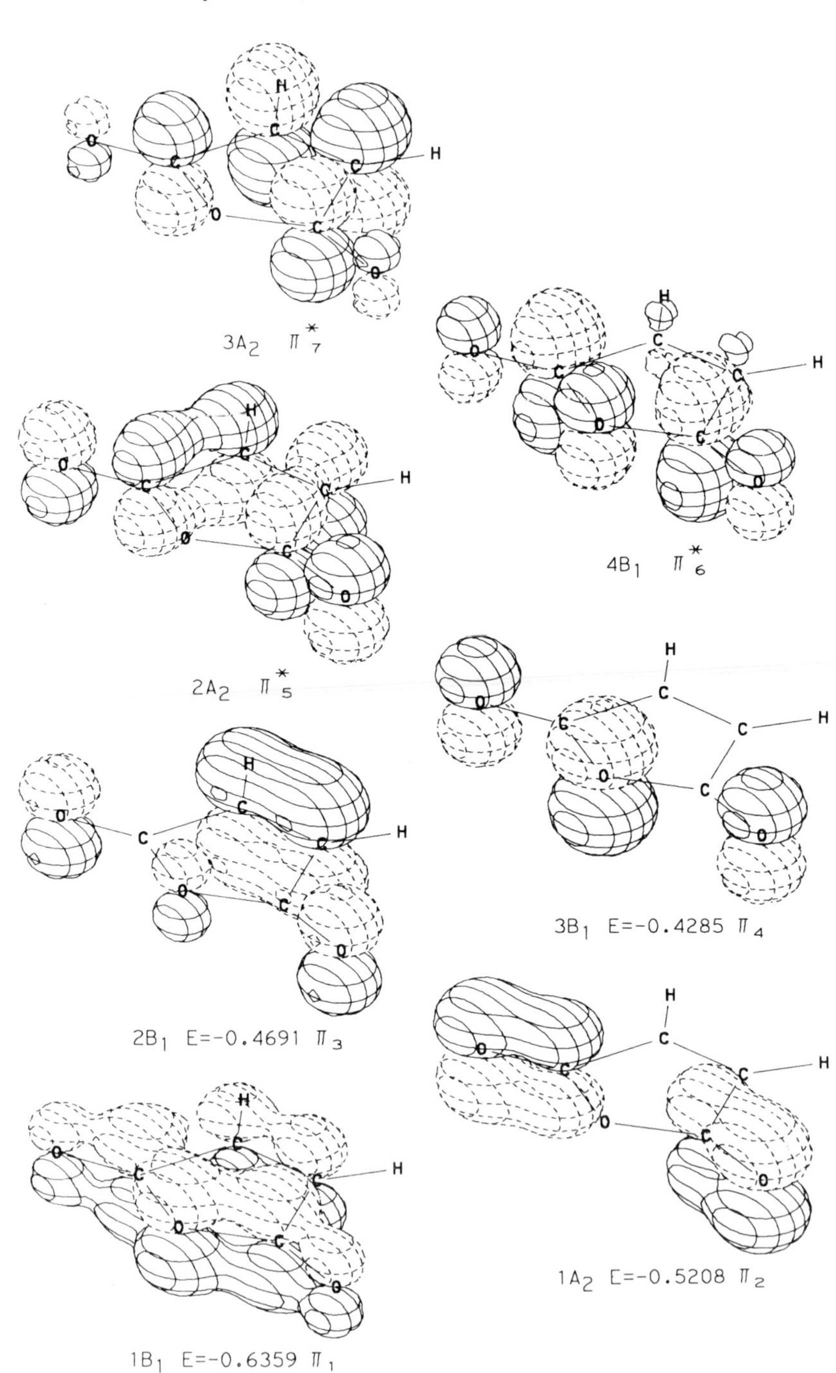

IV. Index of References

Geometries and Orbital Energies

The following is a list of references for the molecular geometries and orbital energies used in constructing the drawings of Chapter III. The ordering of the molecules is the same as in Chapter III.

1. Hydrogen
   - geometry — B. P. Stoicheff, *Can. J. Phys.*, **35,** 730 (1957).
   - energies — S. Fraga and B. J. Ransil, *J. Chem. Phys.*, **35,** 1967 (1961).
2. Triplet Methylene
   - geometry — G. Herzberg, "Electronic Spectra of Polyatomic Molecules." Van Nostrand, Princeton, New Jersey (1967).
   - energies — MINDO/2. See also M. Krauss, *J. Res. Natn. Bur. Stand.*, **68A,** 635 (1964).
3. Singlet Methylene
   - geometry — G. Herzberg, "Electronic Spectra of Polyatomic Molecules." Van Nostrand, Princeton, New Jersey (1967).
   - energies — MINDO/2.
4. Amino Cation (Triplet)
   - geometry — Calculated: W. A. Lathan, W. J. Hehre, L. A. Curtiss, and J. A. Pople, *J. Amer. Chem. Soc.*, **93,** 6377 (1971).
   - energies — MINDO/2.

5. PLANAR METHYL RADICAL

geometry — G. Herzberg, "Electronic Spectra of Polyatomic Molecules." Van Nostrand, Princeton, New Jersey (1967).

energies — P. Millie and G. Berthier, *Int. J. Quantum Chem.*, **2S,** 67 (1968).

6. PYRAMIDAL METHYL RADICAL

geometry — Adapted from planar methyl radical. Tetrahedral.

energies — MINDO/2.

7. METHANE

geometry — H. C. Allen, Jr., and E. K. Plyler, *J. Chem. Phys.*, **26,** 972 (1957).

energies — W. E. Palke and W. N. Lipscomb, *J. Amer. Chem. Soc.*, **88,** 2384 (1966).

8. AMMONIA

geometry — W. S. Benedict and E. K. Plyler, *Can. J. Phys.*, **35,** 1235 (1957).

energies — W. E. Palke and W. N. Lipscomb, *J. Amer. Chem. Soc.*, **88,** 2384 (1966).

9. WATER

geometry — W. S. Benedict, N. Gailar, and E. K. Plyler, *J. Chem. Phys.*, **24,** 1139 (1956).

energies — R. M. Pitzer and D. P. Merrifield, *J. Chem. Phys.*, **52,** 4782 (1970).

10. HYDROGEN FLUORIDE

geometry — G. Herzberg, "Diatomic Molecules." Van Nostrand, Princeton, New Jersey (1955).

energies — A. D. McLean and M. Yoshimine, Tables of Linear Molecule Wavefunctions. Supplement to *IBM J. Res. Develop.*, **3,** 206 (1968).

11. PROTONATED METHANE

geometry — Calculated: V. Dyczmons, V. Staemmler, and

W. Kutzelnigg, *Chem. Phys. Lett.*, **5**, 361 (1970).

energies — Same source as geometry.

12. Acetylene

geometry — J. H. Callomon and B. P. Stoicheff, *Can. J. Phys.*, **35**, 373 (1957).

energies — W. E. Palke and W. N. Lipscomb, *J. Amer. Chem. Soc.*, **88**, 2384 (1966).

13. Vinyl Cation

geometry — Calculated: P. C. Hariharan, W. A. Lathan, and J. A. Pople, *Chem. Phys. Lett.*, **14**, 385 (1972).

energies — MINDO/2.

14. Hydrogen Cyanide

geometry — H. Kaplan, *J. Chem. Phys.*, **26**, 1704 (1957).

energies — W. E. Palke and W. N. Lipscomb, *J. Amer. Chem. Soc.*, **88**, 2384 (1966).

15. Carbon Monoxide

geometry — G. Herzberg and K. N. Rao, *J. Chem. Phys.*, **17**, 1099 (1949).

energies — A. D. McLean and M. Yoshimine, Tables of Linear Molecule Wavefunctions. Supplement to *IBM J. Res. Develop.*, **3**, 206 (1968).

16. Nitrogen

geometry — P. G. Wilkinson, *Astrophys. J.*, **126**, 1 (1957).

energies — B. J. Ransil, *Rev. Mod. Phys.*, **32**, 245 (1960).

17. Nitric Oxide

geometry — G. Herzberg, "Diatomic Molecules." Van Nostrand, Princeton, New Jersey (1950).

energies — H. Brion and M. Yamazaki, *J. Chem. Phys.*, **30**, 673 (1959).

18. ETHYLENE

geometry — H. C. Allen and E. K. Plyler, *J. Amer. Chem. Soc.*, **80,** 2673 (1958).

energies — W. E. Palke and W. N. Lipscomb, *J. Amer. Chem. Soc.*, **88,** 2384 (1966).

19. METHYLENIMINE

geometry — K. Sastry and R. F. Curl, *J. Chem. Phys.*, **41,** 77 (1964).

energies — J. B. Moffat, *Can. J. Chem.*, **48,** 1820 (1970).

20. FORMALDEHYDE

geometry — K. Takagi and T. Oka, *J. Phys. Soc. Jap.*, **18,** 1174 (1963).

energies — S. Aung, R. M. Pitzer, and S. I. Chan, *J. Chem. Phys.*, **45,** 3457 (1966).

21. DIIMIDE

geometry — Calculated: B. Munsch, Dissertation, Strasbourg (1971).

energies — Same source as geometry.

22. DIBORANE

geometry — L. S. Bartell and B. L. Carroll, *J. Chem. Phys.*, **42,** 1135 (1965).

energies — E. A. Laws, R. M. Stevens, and W. N. Lipscomb, *J. Amer. Chem. Soc.*, **94,** 4461 (1972).

23. OXYGEN (TRIPLET)

geometry — I. L. Karle, *J. Chem. Phys.*, **23,** 1739 (1955).

energies — M. Krauss, NBS Tech. Note No. 438 (1967).

24. ETHYL CATION, BISECTED

geometry — Calculated: P. C. Hariharan, W. A. Lathan and J. A. Pople, *Chem. Phys. Lett.*, **14,** 385 (1972).

energies — MINDO/2. See also L. J. Massa, S. Ehrenson, and M. Wolfsberg, *Int. J. Quantum Chem.*, **4,** 625 (1970).

25. ETHYL CATION, ECLIPSED

geometry	Same reference as bisected (24).
energies	Same reference as bisected (24).

26. ETHYL CATION, BRIDGED

geometry	Same reference as bisected (24).
energies	MINDO/2.

27. ETHYL RADICAL, BISECTED

geometry	Calculated: W. A. Lathan, W. J. Hehre, and J. A. Pople, *J. Amer. Chem. Soc.*, **93,** 808 (1971).
energies	MINDO/2.

28. ETHYL RADICAL, ECLIPSED

geometry	Same reference as bisected (27).
energies	MINDO/2.

29. ETHANE, STAGGERED

geometry	H. C. Allen and E. K. Plyler, *J. Chem. Phys.*, **31,** 1062 (1959).
energies	W. H. Fink and L. C. Allen, *J. Chem. Phys.*, **46,** 2261 (1967).

30. ETHANE, ECLIPSED

geometry	Adapted from staggered (29) via rigid rotation.
energies	Same reference as staggered (29).

31. METHYLAMINE

geometry	D. R. Lide, Jr., *J. Chem. Phys.*, **27,** 343 (1957).
energies	W. H. Fink and L. C. Allen, *J. Chem. Phys.*, **46,** 2276 (1967).

32. METHANOL

geometry	K. Kimura and M. Kubo, *J. Chem. Phys.*, **30,** 151 (1959).
energies	W. H. Fink and L. C. Allen, *J. Chem. Phys.*, **46,** 2261 (1967).

33. METHYL FLUORIDE

geometry C. C. Costain, *J. Chem. Phys.*, **29,** 864 (1958).

energies G. Berthier, D.-J. David, and A. Veillard, *Theor. Chim. Acta,* **14,** 329 (1969).

34. HYDRAZINE

geometry Y. Morino, T. Iijima, and Y. Murata, *Bull. Chem. Soc. Jap.*, **33,** 46 (1960); T. Kasuya and T. Kohima, *J. Phys. Soc. Jap.*, **18,** 364 (1963).

energies W. H. Fink, D. C. Pan, and L. C. Allen, *J. Chem. Phys.*, **47,** 895 (1967).

35. HYDROGEN PEROXIDE

geometry R. L. Redington, W. B. Olson, and P. C. Cross, *J. Chem. Phys.*, **36,** 1311 (1962).

energies W. H. Fink and L. C. Allen, *J. Chem. Phys.*, **46,** 2261 (1967).

36. FLUORINE

geometry D. Andrychuck, *Can. J. Phys.*, **29,** 151 (1951).

energies B. J. Ransil, *Rev. Mod. Phys.*, **32,** 245 (1960).

37. CYCLOPROPENIUM CATION

geometry Adapted from benzene.

energies MINDO/2.

38. METHYL ACETYLENE

geometry C. C. Costain, *J. Chem. Phys.*, **29,** 864 (1958).

energies M. D. Newton and W. N. Lipscomb, *J. Amer. Chem. Soc.*, **89, 4261** (1967).

39. ACETONITRILE

geometry C. C. Costain, *J. Chem. Phys.*, **29,** 864 (1958).

energies W. J. Hehre, Private communication (STO-3G).

40. METHYL ISOCYANIDE

geometry C. C. Costain, *J. Chem. Phys.*, **29,** 864 (1958).

energies W. J. Hehre, Private communication (STO-

3G). See also E. Clementi and D. Klint, *J. Chem. Phys.*, **50,** 4899 (1969).

41. ALLENE
 geometry — A. G. Maki and R. A. Toth, *J. Mol. Spectry.*, **17,** 136 (1965).
 energies — L. J. Schaad, *Tet.*, **26,** 4115 (1970).

42. KETENE
 geometry — H. R. Johnson and M. W. P. Strandberg, *J. Chem. Phys.*, **20,** 687 (1952).
 energies — J. H. Letcher, M. L. Unland, and J. R. Van Wazer, *J. Chem. Phys.*, **50,** 2185 (1969).

43. DIAZOMETHANE
 geometry — A. P. Cox, L. F. Thomas, and J. Sheridan, *Nature*, **181,** 1000 (1958).
 energies — J. M. Andre, M. C. Andre, G. Leroy, and J. Weiler, *Int. J. Quantum Chem.*, **3,** 1013 (1969).

44. CARBODIIMIDE
 geometry — Calculated: B. Munsch, Dissertation, Strasbourg (1971).
 energies — Same source as geometry.

45. CARBON DIOXIDE
 geometry — C. P. Courtoy, *Ann. Soc. Sci. Brux.*, **73,** 5 (1959).
 energies — A. D. McLean and M. Yoshimine, Tables of Linear Molecule Wavefunctions. Supplement to *IBM J. Res. Develop.*, **3,** 206 (1968).

46. CYCLOPROPENE
 geometry — P. H. Kasai, R. J. Meyers, D. F. Eggers, Jr., and K. B. Wiberg, *J. Chem. Phys.*, **30,** 512 (1959).
 energies — M. B. Robin, H. Basch, N. A. Kuebler, K. B. Wiberg, and G. B. Ellison, *J. Chem. Phys.*, **51,** 45 (1969).

47. Diazirine

geometry	L. Pierce and V. Dobyns, *J. Amer. Chem. Soc.*, **84,** 2651 (1962).
energies	M. B. Robin, H. Basch, N. A. Kuebler, K. B. Wiberg, and G. B. Ellison, *J. Chem. Phys.*, **51,** 45 (1969).

48. Allyl Cation

geometry	Estimated: D. T. Clark and D. R. Armstrong, *Theor. Chim. Acta*, **13,** 365 (1969).
energies	S. D. Peyerimhoff and R. J. Buenker, *J. Chem. Phys.*, **51,** 2528 (1969).

49. Propylene

geometry	D. R. Lide, Jr., and D. Christensen, *J. Chem. Phys.*, **35,** 1374 (1961).
energies	M. L. Unland, J. R. Van Wazer, and J. M. Letcher, *J. Amer. Chem. Soc.*, **91,** 1045 (1969).

50. Acetaldehyde

geometry	R. W. Kilb, C. C. Lin, and E. B. Wilson, Jr., *J. Chem. Phys.*, **26,** 1695 (1957).
energies	W. J. Hehre, Private communication (STO-3G).

51. Formamide

geometry	R. J. Kurland and E. B. Wilson, Jr., *J. Chem. Phys.*, **27,** 585 (1957).
energies	H. Basch, M. B. Robin, and N. A. Kuebler, *J. Chem. Phys.*, **49,** 5007 (1968).

52. Formic Acid

geometry	Tables of Interatomic Distances, The Chemical Society, London (1958) and Supplement (1965).
energies	H. Basch, M. B. Robin, and N. A. Kuebler, *J. Chem. Phys.*, **49,** 5007 (1968).

53. FORMYL FLUORIDE

geometry E. Ferronato, L. Grifons, A. Guarniere, and G. Zuliani, *Adv. Mol. Spec.*, **3,** 1153 (1962).

energies I. G. Csizmadia, M. C. Harrison, and B. T. Sutcliffe, *Theor. Chim. Acta*, **6,** 217 (1966).

54. NITROSOMETHANE

geometry D. Coffey, Jr., C. O. Britt, and J. E. Boggs, *J. Chem. Phys.*, **49,** 591 (1968).

energies P. A. Kollmann and L. C. Allen, *Chem. Phys. Lett.*, **5,** 75 (1970).

55. OZONE

geometry R. H. Hughes, *J. Chem. Phys.*, **24,** 131 (1956).

energies A. Devaquet, Private communication.

56. CYCLOPROPANE

geometry O. Bastiansen, F. N. Fritsch, and K. Hedberg, *Acta Cryst.*, **17,** 538 (1964); W. J. Jones and B. P. Stoicheff, *Can. J. Phys.*, **42,** 2259 (1964).

energies H. Basch, M. B. Robin, N. A. Kuebler, C. Baker, and D. W. Turner, *J. Chem. Phys.*, **51,** 52 (1969).

57. AZIRIDINE

geometry T. E. Turner, V. C. Fiora, and W. M. Kendrick, *J. Chem. Phys.*, **23,** 1966 (1955).

energies H. Basch, M. B. Robin, N. A. Kuebler, C. Baker, and D. W. Turner, *J. Chem. Phys.*, **51,** 52 (1969).

58. ETHYLENE OXIDE

geometry G. L. Cunningham, Jr., A. W. Boyd, R. J. Meyers, W. D. Gwinn, and W. I. Le Van, *J. Chem. Phys.*, **19,** 676 (1951).

energies H. Basch, M. B. Robin, N. A. Kuebler, C. Baker, and D. W. Turner, *J. Chem. Phys.*, **51,** 52 (1969).

59. TRIMETHYLENE, EDGE-TO-EDGE

geometry — Calculated: Y. Jean, L. Salem, J. S. Wright, J. A. Horsley, C. Moser, and R. M. Stevens, *Pure Appl. Chem., Suppl.* (23rd Congr.), **1,** 197 (1971); *J. Amer. Chem. Soc.*, **94,** 279 (1972).

energies — Y. Jean, L. Salem, J. S. Wright, J. A. Horsley, C. Moser, and R. M. Stevens, unpublished results.

60. *n*-PROPYL CATION, BISECTED

geometry — Calculated: L. Radom, J. A. Pople, V. Buss, and P. v.R. Schleyer, *J. Amer. Chem. Soc.*, **93,** 1813 (1971).

energies — MINDO/2.

61. PROPANE

geometry — D. R. Lide, Jr., *J. Chem. Phys.*, **33,** 1514 (1960).

energies — W. J. Hehre, Private communication (STO-3G).

62. DIMETHYLETHER

geometry — K. Kimura and M. Kubo, *J. Chem. Phys.*, **30,** 151 (1959).

energies — W. J. Hehre, Private communication (STO-3G).

63. ETHYL FLUORIDE

geometry — J. Kraitchman and B. P. Dailey, *J. Chem. Phys.*, **23,** 184 (1955).

energies — L. C. Allen and H. Basch, *J. Amer. Chem. Soc.*, **93,** 6373 (1971).

64. CYCLOBUTADIENE (RECTANGULAR SINGLET)

geometry — Calculated: R. J. Buenker and S. D. Peyerimhoff, *J. Chem. Phys.*, **48,** 354 (1968).

energies — Same source as geometry.

65. 1,3-Butadiene, Transoid

geometry A. Almennigen, O. Bastiansen and M. Traetteberg, *Acta Chem. Scand.*, **12,** 1221 (1958).

energies R. J. Buenker and J. L. Whitten, *J. Chem. Phys.*, **49,** 5381 (1968).

66. 1,3-Butadiene, Cisoid

geometry Adapted from transoid (65) via rigid rotation.

energies Same reference as transoid (65).

67. Acrolein, Transoid

geometry M. Traetteberg, *Acta Chem. Scand.*, **24,** 373 (1970).

energies Extended Hückel.

68. Acrolein, Cisoid

geometry Adapted from transoid (67) via rigid rotation.

energies Extended Hückel.

69. Glyoxal, Transoid

geometry K. Kuchitsu, T. Fukuyama, and Y. Morino, *J. Mol. Struct.*, **1,** 463 (1967).

energies U. Pincelli, B. Cadioli, and D. J. David, *J. Mol. Struct.*, **9,** 173 (1971).

70. Glyoxal, Cisoid

geometry Adapted from transoid (69) via rigid rotation.

energies Same reference as transoid (69).

71. Methylazide

geometry R. L. Livingston and C. N. R. Rao, *J. Phys. Chem.*, **64,** 756 (1960).

energies W. J. Hehre, Private communication (STO-3G).

72. Methylenecyclopropane

geometry V. W. Laurie and W. M. Stigliani, *J. Amer. Chem. Soc.*, **92,** 1485 (1970).

energies Extended Hückel.

73. Cyclopropanone

geometry J. M. Pochan, J. E. Baldwin, and W. H. Flygare, *J. Amer. Chem. Soc.*, **91,** 1896 (1969).

energies Extended Hückel.

74. Cyclobutene

geometry B. Bak, J. J. Led, L. Nygaard, J. Rastrup-Andersen, and G. O. Sorensen, *J. Mol. Struct.*, **3,** 369 (1969).

energies J. M. André, M. C. André, and G. Leroy, *Bull. Soc. Chim. Belges*, **78,** 539 (1969).

75. Bicyclobutane

geometry K. W. Cox, M. D. Harmony, G. Nelson, and K. B. Wiberg, *J. Chem. Phys.*, **50,** 1976 (1968).

energies G. Wipff and J. M. Lehn, Private communication (GTO).

76. Cyclopropylcarbinyl Cation, Bisected

geometry Calculated: W. J. Hehre and P. C. Hiberty, *J. Amer. Chem. Soc.*, **94,** 5917 (1972).

energies MINDO/2.

77. Cyclopropylcarbinyl Cation, Perpendicular

geometry Adapted from bisected (76) via rigid rotation.

energies MINDO/2.

78. Trans-2-Butene

geometry A. Almenningen, I. M. Anfinsen, and A. Haaland, *Acta Chem. Scand.*, **24,** 43 (1970).

energies MINDO/2.

79. Acetone

geometry R. Nelson and L. Pierce, *J. Mol. Spectry.*, **18,** 344 (1965).

energies W. J. Hehre, Private communication (STO-3G).

80. Isopropenol (2-Hydroxy-Propylene)
geometry Adapted from isobutene and formic acid.

energies W. J. Hehre, Private communication (STO-3G).

81. Nitromethane
geometry Tables of Interatomic Distances, The Chemical Society, London (1958) and Supplement (1965).

energies W. J. Hehre, Private communication (STO-3G).

82. Cyclobutane, Planar
geometry A. Almenningen and O. Bastiansen, *Acta. Chem. Scand.*, **15,** 711 (1961).

energies J. S. Wright and L. Salem, *J. Amer. Chem. Soc.*, **94,** 322 (1972).

83. Cyclopentadiene
geometry L. H. Scharpen and V. W. Laurie, *J. Chem. Phys.*, **43,** 2765 (1965).

energies G. Wipff and J. M. Lehn, Private communication (GTO).

84. (2.1.0)-Bicyclopentene-2
geometry S. L. Hsu, A. H. Andrist, T. D. Gierke, R. C. Benson, W. H. Flygare, and J. E. Baldwin, *J. Amer. Chem. Soc.*, **92,** 5250 (1970).

energies G. Wipff and J. M. Lehn, Private communication (GTO).

85. Pyrrole
geometry B. Bak, D. Christensen, L. Hansen, and J. Rastrup-Andersen, *J. Chem. Phys.*, **24,** 720 (1956).

energies E. Clementi, H. Clementi, and D. R. Davis, *J. Chem. Phys.*, **46,** 4725 (1967).

86. Furan

geometry B. Bak, L. Hansen, and J. Rastrup-Andersen, *Disc. Farad. Soc.*, **19,** 30 (1955).

energies P. Siegbahn, *Chem. Phys. Lett.*, **8,** 245 (1971).

87. Cyclopentadienyl Anion

geometry Adapted: H. Preuss and G. Diercksen, Int. J. Quantum Chem., **1,** 349 (1967).

energies Same source as geometry.

88. Pentadienyl Radical

geometry Adapted (C—C = 1.366 Å, 1.424 Å; C—H = 1.085 Å; Angles trigonal).

energies MINDO/2.

89. Cyclopentene

geometry M. I. Davis and T. W. Muecke, *J. Phys. Chem.*, **74,** 1104 (1970).

energies MINDO/2.

90. (1.1.1)-Bicyclopentane

geometry J. F. Chiang and S. H. Bauer. *J. Amer. Chem. Soc.*, **92,** 1614 (1970).

energies G. Wipff and J. M. Lehn, Private communication (GTO).

91. Spiropentane

geometry G. Dallinga, R. K. Vander Draai and C. H. Toneman, *Rec. Trav. Chim.*, **87,** 897 (1968).

energies G. Wipff and J. M. Lehn, Private communication (GTO).

92. Para-Benzyne

geometry Adapted from benzene.

energies MINDO/2.

93. Cyclopentane

geometry W. J. Adams, H. J. Geise, and L. S. Bartell, *J. Amer. Chem. Soc.*, **93,** 5013 (1970).

energies MINDO/2.

94. BENZENE

geometry A. Langseth and B. P. Stoicheff, *Can. J. Phys.*, **34**, 350 (1956).

energies R. M. Stevens, E. Switkes, E. A. Laws, and W. N. Lipscomb, *J. Amer. Chem. Soc.*, **93**, 2603 (1971).

95. DEWAR BENZENE

geometry Adapted from hexamethyl Dewar benzene. M. J. Cardillo and S. H. Bauer, *J. Amer. Chem. Soc.*, **92**, 2399 (1970).

energies MINDO/2.

96. PYRIDINE

geometry B. Bak, L. Hansen-Nygaard, and J. Rastrup-Andersen, *J. Mol. Spectry.*, **2**, 361 (1958).

energies J. D. Petke, J. L. Whitten, and J. A. Ryan, *J. Chem. Phys.*, **48**, 953 (1968).

97. PYRAZINE

geometry P. J. Wheatley, *Acta Cryst.*, **10**, 182 (1957); V. Schomaker and L. Pauling, *J. Amer. Chem. Soc.*, **61**, 1769 (1939).

energies J. D. Petke, J. L. Whitten, and J. A. Ryan, *J. Chem. Phys.*, **48**, 953 (1968). See also M. Hackmeyer and J. L. Whitten, *J. Chem. Phys.*, **54**, 3739 (1971).

98. CYCLOPENTADIENONE

geometry Adapted from cyclopentadiene and acetone.

energies MINDO/2.

99. 1,3,5-HEXATRIENE

geometry Adapted from 1,3-butadiene.

energies MINDO/2.

100. (2.1.1)-BICYCLOHEXENE-2

geometry D. L. Zebelmann, S. H. Bauer, and J. F. Chiang, *Tet.*, **28**, 2727 (1972).

energies G. Wipff and J. M. Lehn, Private communication (GTO).

101. CYCLOHEXENE, HALF-BOAT

geometry Adapted from half-chair: J. F. Chiang and S. H. Bauer, *J. Amer. Chem. Soc.*, **91,** 1898 (1969); L. H. Scharpen, J. E. Wollrab, and D. P. Ames, *J. Chem. Phys.*, **49,** 2368 (1968).

energies MINDO/2.

102. CYCLOHEXANE, CHAIR

geometry M. Davis and O. Hassel, *Acta Chem. Scand.*, **17,** 1181 (1963).

energies MINDO/2.

103. NORBORNADIENE

geometry G. Dallinga and L. H. Toneman, *Rec. Trav. Chim.*, **87,** 805 (1968).

energies M. H. Palmer and R. H. Findlay, *Chem. Phys. Lett.*, **15,** 416 (1972).

104. MALEIC ANHYDRIDE

geometry Adapted from acrolein and formic acid.

energies MINDO/2.

Subject Index